Errata /Corrections

D1240136

p 31 — Col. 1, first paragraph, change to: Data set 3 included ADG predicted, using model level 2, from independent trials with 96 different diets fed to a total of 943 Bos indicus (Nellore breed) steers and bulls.... (Lanna et al., 1996). FSBW...content. For the NRC 1984 and the present systems, the r^2 was 0.58 and 0.72, and the bias was –20 percent and –2 percent respectively.

p 36 — Col. 1, line 3 change 427 to 407

p 36 — Col. 1, line 6 change 585 to 497

p 38 — Col. 1, Reference Lanna et al., 1994, change to: Lanna, D. P. D., D. G. Fox, C. Boin, M. J. Traxler, and M. C. Barry. 1996. Validation of the CNCPS estimates of nutrient requirements of growing and lactating zebu germplasm in tropical conditions. J. Anim. Sci. 76 (suppl 1): 287.

p 43 — Equation 4-7, change to:
$$NEm = 0.576 \text{ birth weight } (0.4504 – 0.000766t)e^{(0.03233 – 0.0000275t)t}$$

p 55 — Col. 1, under Calcium Requirements, first paragraph, last sentence, change to: This requirement was distributed over the last 3 months of pregnancy.

p 58 — Col. 1, last sentence of paragraph at top of page, change to: This requirement was distributed over the last 3 months of pregnancy.

p 59 — Col. 1, last paragraph, delete heading Factors Affecting Potassium Requirements

p 61 — Col. 1, heading Sulfur Sources should be Factors Affecting Sulfur Requirements, Second paragraph should be headed Sulfur Sources.

p 81 — Col. 2, water intake equation should read: Water intake (L/day) =
$$–18.67 + (0.3937 \cdot MT) + (2.432 \cdot DMI) – (3.870 \cdot PP) – (4.437 \cdot DS)$$

p 99 — Table 8-2, second column, change ppm to mg/kg, under Daily Nutrient Intake heading, change the following:

	Suggested Range	Daily Nutrient Intake for 250-kg Calf	
		0–7 days	0–14 days
Dry Matter		3.88	4.75
Copper		39–58	47–71
Iron		388–775	475–950
Manganese		155–271	190–332
Zinc		290–387	356–475
Cobalt		0.4–0.8	0.5–1.0
Selenium		0.4–0.8	0.5–1.0
Iodine		1.2–2.3	1.4–2.9
Vitamin E	75–100	291–388	356–475

p 114 — Col. 2, under ENERGY heading, change NE_m equation to:
$$NE_m = (a_1 + a_2)SBW^{0.75} (BE)(L)(SEX)(COMP)$$

p 115 — Col. 2, Table 10-1, under heading NE_m (BE), Gelbvieh, change 1.00 to: 1.10

p 115 — Col. 2, below Table 10-1, definition of terms, change k_m = diet NE_m/diet ME to:
$$k_m = \text{diet } NE_m/\text{diet ME; (assumed 0.576 in derivation)}$$

p 117 Col. 2, under PREGNANCY heading, change NE_m equation to:

$$NE_{m, preg}, kcal/day = CBW \circ (k_m/0.13) \circ (0.05855 - 0.0000996t) \circ e^{((0.03233 - 0.0000275t)t)}$$

OR

$$= k_m \circ CBW \circ (0.4504 - 0.000766) \circ e^{((0.03233 - 0.0000275t)t)}$$

p 119 Col. 2, under equation: If FA > (DMI $\circ$ 4)..., DMI = ..., change to DMI = predicted dry matter intake using previous equations, expressed as g/kg SBW, and FA = daily forage allowance in g/kg SBW/day

p 120 Col. 1, line 16 change equation to: $= 1.0 - ((20 - eNDF) \circ 0.025)$

p 133 Col. 1, first paragraph, second sentence should read: Data were also extracted from the 1989 Nutrient Requirements of Horses (National Research Council, 1989)

p 133 Col. 1, third paragraph, first sentence should read: Data from this table is intended to help producers evaluate whether data they receive <u>on</u> their own feedstuffs....

p 136 Table 11-1, Entry No. 27 (Canola meal, sun-cured) insert row with standard deviations (SD) as follows:

	Dry Matter %	Crude Protein %	Rumen Unde-graded %	Ether Extract %	Fiber %	NDF %	ADF %	Ash %	Calcium %
Value	82	40.9	28	3.47	13.3	27.2	17.0	7.10	0.70
N	154	129	10	105	120	24	19	31	102
SD	1.63	4.32	17	1.13	1.95	4.81	3.36	0.38	0,10

	Phos-phorus %	Mag-nesium %	Potas-sium %	Sodium %	Sulfur %	Copper mg/kg	Iron mg/kg	Manga-nese mg/kg	Zinc mg/kg	Molyb-denum mg/kg
Value	1.20	0.57	1.37	0.03	1.17	7.95	211	55.8	71.5	1.79
N	133	27	38	25	14	14	25	27	27	22
SD	0.11	0.11	0.20	0.07	0.04	0.94	88	12.6	6.0	0.35

p 138 Table 11-1, Entry No. 43, Fat, animal, hydrolyzed, add the following standard deviations (SD) for Dry matter %, SD = 0.28, for ether extract, %, SD = 1.04

First row of SDs under Entry No. 44 should be deleted.

Computer disk: Warning

Two potential problems have been noted in the software.

1) For level 1, ME, NE_m and NE_g values are calculated directly from TDN using the California net energy equations. UIP is calculated as 100 – DIP. In the software (model and feed library), the cells where these calculations are performed were left unprotected. This could lead to the possibility of overwriting the formulas, rendering the model unresponsive to changes to TDN with respect to energy values. Cells where this occurred will appear in yellow on the screen. To reset the formulas in these cells, press{/}, {W}, {G}, {P}, {E}, and save the file.

2) An error was found in the calculation of potential grazing intake. DMI is specified in the software as kg/DM, when in fact it should be expressed as g/kg SBW. The result of this error is that in almost all cases, the graze factor evaluates to 1.0, indicating that full voluntary intake can be met with pasture. Consequently, potential pasture intake and the grazing activity adjustment for the NE_m requirement could be overestimated in cases where forage availability is limited.

Nutrient Requirements of Beef Cattle

Seventh Revised Edition, 1996

Subcommittee on Beef Cattle Nutrition
Committee on Animal Nutrition
Board on Agriculture
National Research Council

NATIONAL ACADEMY PRESS
Washington, D.C.   1996

National Academy Press • 2101 Constitution Avenue • Washington, D.C. 20418

NOTICE: The project that is the subject of this report was approved by the Governing Board of the National Research Council, whose members are drawn from the councils of the National Academy of Sciences, the National Academy of Engineering, and the Institute of Medicine. The members of the committee responsible for the report were chosen for their special competencies and with regard for appropriate balance.

This report has been reviewed by a group other than the authors according to procedures approved by a Report Review Committee consisting of members of the National Academy of Sciences, the National Academy of Engineering, and the Institute of Medicine.

This study was supported by the Agricultural Research Service of the U.S. Department of Agriculture, under Agreement No. 59-32U4-5-6, and by the Center for Veterinary Medicine, Food and Drug Administration of the U.S. Department of Health and Human Services, under Cooperative Agreement No. FD-U-000006-10. Additional support was provided by the American Feed Industry Association.

The National Academy of Sciences is a private, nonprofit, self-perpetuating society of distinguished scholars engaged in scientific and engineering research, dedicated to the furtherance of science and technology and to their use for the general welfare. Upon the authority of the charter granted to it by the Congress in 1863, the Academy has a mandate that requires it to advise the federal government on scientific and technical matters. Dr. Bruce M. Alberts is president of the National Academy of Sciences.

The National Academy of Engineering was established in 1964, under the charter of the National Academy of Sciences, as a parallel organization of outstanding engineers. It is autonomous in its administration and in the selection of its members, sharing with the National Academy of Sciences the responsibility for advising the federal government. The National Academy of Engineering also sponsors engineering programs aimed at meeting national needs, encourages education and research, and recognizes the superior achievements of engineers. Dr. Harold Liebowitz is president of the National Academy of Engineering.

The Institute of Medicine was established in 1970 by the National Academy of Sciences to secure the services of eminent members of appropriate professions in the examination of policy matters pertaining to the health of the public. The Institute acts under the responsibility given to the National Academy of Sciences by its congressional charter to be an adviser to the federal government and, upon its own initiative, to identify issues of medical care, research, and education. Dr. Kenneth I. Shine is president of the Institute of Medicine.

The National Research Council was organized by the National Academy of Sciences in 1916 to associate the broad community of science and technology with the Academy's purposes of furthering knowledge and advising the federal government. Functioning in accordance with general policies determined by the Academy, the Council has become the principal operating agency of both the National Academy of Sciences and the National Academy of Engineering in providing services to the government, the public, and the scientific and engineering communities. The Council is administered jointly by both Academies and the Institute of Medicine. Dr. Bruce M. Alberts and Dr. Harold Liebowitz are chairman and vice chairman, respectively, of the National Research Council.

Library of Congress Cataloging-in-Publication Data

Nutrient requirements of beef cattle / Subcommittee on Beef Cattle Nutrition, Committee on Animal Nutrition, Board on Agriculture, National Research Council.—7th rev. ed.
    p.   cm.—(Nutrient requirements of domestic animals)
Includes bibliographical references and index.
ISBN 0-309-05426-5 (alk. paper)
1. Beef cattle—Feeding and feeds.   I. National Research Council (U.S.). Subcommittee on Beef Cattle Nutrition.   II. Series: Nutrient requirements of domestic animals (Unnumbered)
SF203.N88   1996
636.2′13—dc20
                                                96-13244
                                                  CIP

Any opinions, findings, conclusions, or recommendations expressed in this publication are those of the author(s) and do not necessarily reflect the view of the organizations or agencies that provided support for this project.

Printed in the United States of America

Preface

Among domestic livestock in North America, the range in type and condition of beef cattle as well as production environment is relatively extreme. Since the 1984 publication of *Nutrient Requirements of Beef Cattle, Sixth Revised Edition*, a great amount of effort in our universities and research stations has been put into defining the impact of cattle's biological, production, and environmental diversities and variations on nutrient utilization and requirements. Variations in cattle type are reflected in the large number of cattle breeds. However, within breed groupings, for example, beef vs dairy, many breed effects on nutrient requirements can be accounted for by reference to animal mass, and this principle is adopted in the recommendations presented in this volume, with qualifications noted where necessary. Other animal variations relate to body condition, and this impacts on the ability of growing animals to make compensatory growth. A substantial amount of information has been published concerning the effects condition or finish in cows has on energy requirements. For this edition, effects of body condition and compensatory growth have been described more completely than has been done previously. Environmental variation can also have a significant effect on nutrient requirements, and our knowledge of this has grown substantially as well. In particular, the important effects of the environment and stress on food intake are documented.

Calculating the effects of variation has benefitted from the development of mathematical models to predict and understand relationships between nutrient inputs and animal outputs and from experience in using them. For this edition, the subcommittee has evaluated modeling in concert with experimental data. This publication is not a revolution but only one step on the ladder to a more complete understanding of beef cattle and their nutrient requirements.

To this end, and very early in the development of this edition, the subcommittee decided to include with the publication a computer model on diskette to formulate requirements at two modeling levels. The first level is more empirical and presented in a format similar to that provided with the previous edition. By defining some fundamental relationships concerning nutrient digestion, the second level is more mechanistic and provided for students of beef cattle nutrition, young and old, to provoke discussion and continuing evolution of knowledge in the subject. The goal at the first level is to obtain the greatest predictive accuracy; whereas the goal for the second level is less concerned with predictive accuracy than with developing an understanding of the process.

The subcommittee expects that the two levels will often be compared. Where the second level can predict animal performance successfully, it could then be used for purposes other than those traditionally envisioned for nutrient requirement standards. For example, the model level 2 could be helpful in diagnosing why animal performance on a given diet is less than expected. For both levels, requirements are documented and summarized in equations. To facilitate adoption of requirements, examples are illustrated in the publication in tabular format. In addition, a table generator is provided with the computer models and all of these accompany this publication on a diskette, together with a user's guide.

There are specific aspects about feedstuffs and nutrient requirements that have been emphasized in the revision of the previous edition. There was a need to develop a data base on feed composition that is current and widely applicable. This was done by obtaining information from analytical laboratories throughout North America. Compositional data on feedstuffs have been summarized to show not only the average values but also their variances. Detailed information collected recently on development of the fetus and conceptus during pregnancy has been used to prescribe nutrient requirements of gestating cattle more precisely. The same was possible for lactation because of a better understanding of factors affecting the level of milk production and the lactation curve in beef cows. Protein requirements have been established to reflect our current knowledge of nitrogen usage in ruminants; and the systems prescribed in this publication have been advanced to incorporate more recent developments, for example, host requirements for amino acids and microbial requirements in the rumen for nitrogen constituents in addition to ammonia. Revision of mineral and vitamin requirements has incorporated a significant body of new information on phosphorus, magnesium, and the B vitamins and vitamin E. The fact that trace mineral requirements for young cattle are not always identical to mature animals has been documented as well. A more complete documentation of water requirements is given in this publication and a table on water intake by beef cattle is included. As in previous editions, nutrient requirements are expressed on a per animal per day basis.

To formulate diets and predict performance of cattle on any given feeding program, it is necessary to predict intake. The general relationship of liveweight and diet quality to intake, as presented in the sixth edition, has been retained in this publication; but the effect of diet quality and animal state has been defined more precisely.

In undertaking its work, the subcommittee considered current issues in beef cattle production inasmuch as they affect nutrient requirements. One of these involves product quality and the trend toward carcasses containing more lean meat and less fat. The subcommittee has not attempted to define what level of fat is appropriate; however, users of this publication should be able to define nutrient requirements for feeding cattle to different levels of fatness more precisely than has been possible with previous editions. Another issue involves environmental awareness. Nutrient requirements bear only a peripheral role in this problem; however, the models presented can be used indirectly to predict the loss of nutrients—for example, nitrogen and phosphorus—in animal manure and promote responsible feeding to prevent pollution from beef cattle operations.

The subcommittee would like to remind readers, when using the recommendations presented in this report, that animal observation can be as useful as direct adoption of what is recommended in this book. In the preparation of this report, the subcommittee acknowledges the assistance of many colleagues who have provided data on which recommendations are based or provided commentary. Without their input, this publication would not have been possible.

Jock Buchanan-Smith, *Chair*
Subcommittee on Beef Cattle Nutrition

Acknowledgments

The Subcommittee on Beef Cattle Nutrition is grateful to the many individuals and organizations who provided specific input, data, and critiques of this revision during its development. In particular, appreciation is noted to the beef cattle specialists who provided input on changes needed from the previous edition, to the analytical laboratories that provided valuable information on nutrient composition of commonly used feeds, and to the many animal science students and graduate students of subcommittee members who served as test pilots for the model and provided feedback to increase its user friendliness.

The subcommittee reserves a special acknowledgment and thanks to Michael C. Barry, Cornell University, for the many hours spent on software development, testing, and refinement and for his unfailing helpfulness in guiding the members through program installation and use. Without his able technological assistance, this major step forward in the evolution of predicting nutrient requirements of beef cattle would not have been possible.

Contents

Tables and Figures

FIGURES

Overview

This seventh revised edition of *Nutrient Requirements of Beef Cattle* is a significant revision of the sixth edition. One major improvement involves expansion in describing different cattle. Requirements are also defined in terms of a greater variety of management and environmental conditions than was possible in previous editions. One result of this innovation is that there is now a greater responsibility for the user to define the animals and their conditions before proceeding to determine nutrient requirements.

A second major improvement involves presentation of requirements using computer models. Computer models are the only effective way to take animal variation into account. Also important is the fact that computer models can describe the dynamic state of the animal, which is not possible with the presentation of discrete tabular values of nutrient requirements alone. The dynamic state of describing nutrient requirements of ruminants refers here to the fact that feed ingredients can affect absorbable nutrients, hence potential performance, which has a feedback on requirements. This situation is best illustrated with protein. Diet can have a major effect on protein, which is degraded in the rumen or undegraded and bypassed to the lower portion of the gastrointestinal tract. Energy in the diet affects the amount of microbial protein that can be synthesized in the rumen. Hence, the amount of total true protein that the animal absorbs from the gut, equivalent to metabolizable protein, depends on the energy and degradable protein level of the diet. Net energy and metabolizable protein set the potential growth, reproductive, or lactational performance, which then dictates the need for other nutrients, such as calcium and phosphorus. For these reasons, the subcommittee chose to present nutrient requirements in terms of evaluating rations or diets, rather than as discrete recommendations for nutrients to fulfill a given level of performance. Net energy is used to evaluate ration and diet energy, which is the same format used in the sixth edition. To evaluate protein requirements in this edition, it is necessary to know both the crude protein concentration in the feedstuffs being used as well as the rumen degradability of that protein.

Modeling the dynamic state of nutrient requirements in cattle is a major departure for those familiar with seeing nutrient requirements in tabular form only. To satisfy those who might wish to use the information in a format similar to the previous edition, a table generator is provided. However, because of the dynamic state of protein digestion, protein requirements in the table generator are expressed as metabolizable rather than crude protein.

The model prepared in this publication is at two levels. For the first level, equations are very similar to equations used in the sixth edition. Revision of requirements at this level have, for the most part, only updated equations when there was sufficient new information to justify this. The subcommittee chose to add a second modeling level, which is more mechanistic than level 1 and was included to describe the dynamic state of digestion in and passage of digesta through the reticulo-rumen. Level 1 is recommended for users who were comfortable with using nutrient requirements recommended in the previous edition of this publication and who want the greatest accuracy in evaluating requirements. Level 2 is offered as a model to give a greater interpretation of the results, for example to diagnose underperformance of animals on a given diet. The subcommittee anticipates that as users become comfortable with level 2 in accuracy of prediction, level 2 will become the modeling level used to evaluate rations as well.

Chapter 1 contains a discussion of energy as a nutrient by providing basic definitions and terms used to describe energy content of feedstuffs. There is also an extensive discussion of maintenance energy and factors such as cattle breed, sex, physiological state, and environment that can alter maintenance requirements. This chapter concludes with a discussion of use of energy from body weight loss.

1

Chapter 2 is a review of protein digestion and metabolism and presents the basis for considering metabolizable protein (MP), or amino acids absorbed from the gastrointestinal tract, and the utilization of MP in setting protein requirements for beef cattle. There is a discussion of factors affecting microbial protein synthesis, which includes consideration of needs for energy and degradable protein. A value for the maintenance requirement for protein, based on metabolic and endogenous losses of nitrogen, is proposed as well as an equation to estimate conversion of MP to net protein. Chapter 2 concludes with a section on validation of recommendations for protein requirements of both growing-finishing and breeding cattle.

A discussion of cattle size and body composition with reference to energy and protein begins Chapter 3. This discussion provides the basis for using mature size as a reference point to unify description of nutrient requirements across animals of different mature size and as affected by liveweight, age, and physiological state. To use a system of nutrient requirements based on a constant body fat composition, it is necessary to understand factors that affect rate of growth such as use of anabolic implants or ionophores, and a discussion of these is included. Chapter 3 also includes a discussion of compensatory growth and validation of the energy and protein requirement system. In addition, Chapter 3 considers predicting target weight gains for replacement heifers and discusses variables that affect nutrient requirements of breeding females. Chapter 3 concludes by describing a mechanism to predict energy reserves of beef cows through the use of body condition score, body weight, and body composition. It provides a relationship between condition score and percent fat in the body.

Unique considerations in setting nutrient requirements for breeding animals are considered in Chapter 4. It includes a discussion of factors affecting calf birth weight, energy, and protein requirements for gestation, and nutrient metabolism by the gravid uterus and placenta. In the discussion of lactation requirements, Chapter 4 reviews the literature on determining milk yield in beef cows and energy requirements for milk production. This chapter also describes factors affecting heifer development and breeding performance of mature cows and bulls.

Macromineral and micromineral requirements are presented in Chapter 5. Where possible, discussion of each mineral includes the role of the mineral in physiological processes of cattle, the bases for setting requirements of these nutrients, and relevant aspects about digestion, absorption, and metabolism. Signs of deficiency, factors affecting requirements, and toxicity and maximum tolerable concentrations in diets are discussed also. A table summarizing recommended dietary concentrations and maximal tolerable concentrations of some minerals is included in this chapter and differentiates recommendations according to physiological function. Sufficient information exists to specify higher levels of magnesium, potassium, sodium (salt), and manganese in diets for breeding cattle, particularly lactating animals, compared to growing and finishing cattle. Calcium and phosphorus requirements, as in the previous edition, are presented in equation format (in Chapter 7) to calculate recommended daily intakes for a comprehensive description of cattle types and management circumstances. Calcium requirements are similar to those established in the previous edition of this publication, but phosphorus requirements have been modified slightly from the previous edition and are discussed in the context of some recent studies on these minerals.

Maximum tolerable concentrations of other minerals have been listed in Chapter 5. In the case of chromium, molybdenum, and nickel, evidence that these minerals are essential to cattle has been presented, but there are insufficient data on which to base dietary requirements.

Requirements for vitamins and water have been considered in Chapter 6. Besides the fat-soluble vitamins, for which the evidence to support a required concentration in the diet is very strong, the literature on the water-soluble B vitamins is reviewed to document where supplementation of diets for beef cattle may be beneficial. A discussion of water requirements of beef cattle includes a table detailing these requirements as affected by ambient temperature and physiological function and liveweight.

Factors affecting feed intake of beef cattle are reviewed in Chapter 7. This chapter includes a review of how physiological factors affect feed intake. There is a section on prediction of feed intake by beef cattle and this includes validation of equations used for the models of requirements. There is a special section in this chapter to consider intake of all-forage diets.

Chapter 8 provides an overview of the effects of stress on nutrient requirements. Effects on energy, protein, mineral, and vitamin requirements are addressed.

Chapter 9 presents the application of new information to formulate equations and models for nutrient requirements. Tables of requirements generated by the model are provided for growing-finishing steers or heifers, for pregnant replacement heifers, for lactating cows, and for bulls. A step-by-step example of how to predict average daily gain and crude protein requirements is also presented.

Chapter 10 provides all of the equations used in the model plus a thorough description of the data contained in the feed library on the model disk.

Chapter 11 provides tables of nutrient composition of feedstuffs commonly used in beef cattle diets, including estimates of variation of nutrient content and discussion of processing effects.

1 Energy

ENERGY UNITS

Energy is defined as the potential to do work and can be measured only in reference to defined, standard conditions; thus, all defined units are equally absolute. The joule is the preferred unit of expressing electrical, mechanical, and chemical energy. The joule can be converted to ergs, watt-seconds, and calories; the converse is also true. Nutritionists now standardize their combustion calorimeters using specifically purified benzoic acid, the energy content of which has been determined in electrical units and computed in terms of joules/g mole. The calorie has been standardized to equal 4.184 joules and is approximately equal to the heat required to raise the temperature of 1 g of water from 16.5° to 17.5° C. In practice the calorie is a small amount of energy; thus, the kilocalorie (1 kcal = 1,000 calories) and megacalorie (1 Mcal = 1,000 kcal) are more convenient for use in conjunction with animal feeding standards.

A number of abbreviations have been used to describe energy fractions in the animal system. Many of the abbreviations used throughout this text are those recommended in *Nutritional Energetics of Domestic Animals and Glossary of Energy Terms* (National Research Council, 1981a). Gross energy (E) or heat of combustion is the energy released as heat when an organic substance is completely oxidized to carbon dioxide and water. E is related to chemical composition, but it does not provide any information regarding availability of that energy to the animal. Thus, E is of limited use for assessing the value of a particular diet or dietary ingredient as an energy source for the animal.

Expressing Energy Values of Feeds

E of the food minus the energy lost in the feces is termed digestible energy (DE). DE as a proportion of E may vary from 0.3 for a very mature, weathered forage to nearly 0.9 for processed, high-quality cereal grains. DE has some value for feed evaluation because it reflects diet digestibility and can be measured with relative ease; however, DE fails to consider several major losses of energy associated with digestion and metabolism of food. As a result, DE overestimates the value of high-fiber feedstuffs such as hays or straws relative to low-fiber, highly digestible feedstuffs such as grains. Total digestible nutrients (TDN) is similar to DE but includes a correction for digestible protein. TDN has no particular advantages or disadvantages over DE as the unit to describe feed values or to express the energy requirements of the animal. TDN can be converted to DE by the equation

$$1 \text{ kg TDN} = 4.4 \text{ Mcal DE.}$$

Metabolizable energy (ME) is defined as E minus fecal energy (FE), urinary energy (UE), and gaseous energy (GE) losses, or ME = DE − (UE + GE). ME is an estimate of the energy available to the animal and represents an accounting progression to assess food energy values and animal requirements. ME, however, has many of the same weaknesses as DE; and because UE and GE are highly predictable from DE, ME and DE are strongly correlated. Also, the main source of GE (the primary gas being methane) is microbial fermentation, which also results in heat production. This heat is useful in helping to maintain body temperature in cold-stressed animals but is otherwise an energy loss not accounted for by ME. For most forages and mixtures of forages and cereal grains, the ratio of ME to DE is about 0.8 but can vary considerably (Agricultural Research Council, 1980; Commonwealth Scientific and Industrial Research Organization, 1990) depending on intake, age of animal, and feed source. The definition of ME and the energy balance identity indicate ME can appear only as heat production (HE) or retained energy (RE), that is, ME = HE + RE. As indicated by this relationship, a major value of ME is used as a reference

3

unit and as a starting point for most systems based on the net energy (NE) concept.

The value of feed energy for the promotion of energy retention is measured by determining the RE at two or more amounts of intake energy (IE). The NE of a feed or diet has classically been illustrated by the equation:

$$NE = \Delta RE/\Delta IE.$$

Determination of NE by this method assumes the relationship between RE and feed intake is linear. Actually the relationship is curvilinear and shows a diminishing return effect (Garrett and Johnson, 1983). The relationship is conventionally approximated by two straight lines. The intersection of the two lines is the point at which RE = 0 and is defined as maintenance (M). Conversely, when RE = 0, ME = HE. The relationship between feed intake and body tissue loss (negative RE) comprises one portion of the curve and the relationship between body tissue gain (positive RE) comprises a second portion of the curve. The heat production at zero feed intake (H_eE) is equivalent to the animal's NE requirement for maintenance. The ability of the food consumed to meet the NE required for maintenance is expressed as NE_m and is represented by the following expression:

$$NE_m = H_eE/I_m,$$

where I_m is the amount of feed consumed at RE = 0. Similarly, the value of feed consumed to promote energy retention is represented by the expression NE_r and is determined as

$$NE_r = RE/(I - I_m),$$

where $(I - I_m)$ represents the amount of feed consumed in excess of maintenance requirements.

The relationship ME = RE + HE can be rewritten in terms of NE. Thus, HE can be partitioned into H_eE, H_jE and H_iE (heat increment of intake energy) as

$$ME = RE + H_eE + H_jE + H_iE.$$

Because in practical situations the heat of activity associated with obtaining feed (H_jE) is often included with H_eE, the expression becomes

$$ME = NE_r + NE_m + H_iE.$$

The NE_r used in this expression does not distinguish among different forms in which energy may be retained, such as body tissue (TE), milk, (LE) or tissues of the conceptus (YE). Thus, the former expression might be expanded such that in a pregnant lactating heifer it becomes:

$$RE = LE + YE + TE, \text{ or}$$
$$NE_r = NE_l + NE_y + NE_g,$$

where NE_r, NE_l, and NE_g are equivalent to RE, LE, YE, and TE, respectively.

Thus,

$$ME = NE_g + NE_l + NE_y + NE_m + H_iE.$$

In this expression, a portion of the heat increment (H_iE) is associated with the feed consumed for maintenance and each of the productive functions.

The primary advantages of an NE system are that animal requirements stated as net energy are independent of the diet, and the energy value of feeds for different physiological functions are estimated separately—for example, NE_m, NE_g, NE_l, NE_y. This requires, however, that each feed must be assigned multiple NE values because the value varies with the function for which energy is used by the animal. Alternatively, the animal's energy requirement for various physiological functions may be expressed in terms of a single NE value, provided the relationships among efficiencies of utilization of ME for different functions are known.

Relationships for converting ME values to NE_m and NE_g (Mcal/kg DM) have been reported by Garrett (1980) and are

$$NE_m = 1.37\ ME - 0.138\ ME^2 + 0.0105\ ME^3 - 1.12$$
$$NE_g = 1.42\ ME - 0.174\ ME^2 + 0.0122\ ME^3 - 1.65$$

The NE_m and NE_g values used in the derivation of these equations were based on comparative slaughter studies involving 2,766 animals fed complete, mixed diets at or near ad libitum intake for 100 to 200 days. Digestion trials were conducted on most diets fed at about 1.1 times the maintenance amount. The ME values were estimated as DE * 0.82. Data were not uniformly distributed across the range of ME concentrations encountered in practical situations (1 percent, <1.9 Mcal/kg; 22 percent, 1.9 − 2.6 Mcal/kg; 65 percent, 2.6 − 2.9 Mcal/kg; 12 percent, >2.9 Mcal/kg). Caution should be exercised in use of these equations for predicting NE_m or NE_g values for individual feed ingredients or for feeds outside the ranges indicated above. The relationship between DE and ME can vary considerably among feed ingredients or diets as a result of differences in intake, rate of digestion and passage, and composition (for example, fiber vs starch vs fat). In addition, conversion of ME to NE_m or NE_g may vary beyond that associated with variation in dietary ME in part because of differences in composition of absorbed nutrients.

Available data, as discussed in subsequent sections, indicate efficiencies of ME use for lactation and maintenance are similar in beef cattle; thus, energy requirements for lactation have been expressed in NE_m units. Efficiency of utilization of ME for accretion of energy in gravid uterine tissues is, likewise, discussed in a subsequent section. Some evidence is available to indicate that the efficiency of utilization of ME for maintenance (k_m) and pregnancy (k_y) vary similarly with changes in ME concentration in the diet (Robinson et al., 1980). For convenience, estimates of requirements for beef cows were converted to NE_m equivalents. Conversion of requirements for lactation and pregnancy to NE_m equivalents allow the energy value of feedstuffs to be adequately described by only two NE values (NE_m and NE_g).

REQUIREMENTS FOR ENERGY

Measurement of Maintenance Requirements

The maintenance requirement for energy has been defined as the amount of feed energy intake that will result in no net loss or gain of energy from the tissues of the animal body. Processes or functions comprising maintenance energy requirements include body temperature regulation, essential metabolic processes, and physical activity. Energy maintenance does not necessarily equate to maintenance of body fat, body protein, or body weight. Although for many practical situations maintenance may be considered a theoretical condition, it is useful and appropriate to consider maintenance energy requirements separate from energy requirements for "production." ME required for maintenance functions represents approximately 70 percent of the total ME required by mature, producing beef cows (Ferrell and Jenkins, 1987) and more than 90 percent of the energy required by breeding bulls. The fraction of total ME intake that growing cattle use for maintenance functions is rarely less than 0.40, even at maximum intake. Successful management of beef cattle, whether for survival and production in poor nutritive environments or for maximal production, depends on knowledge of and understanding their maintenance requirements.

Basically, three methods have been used to measure maintenance energy requirements. These include the use of

- long-term feeding trials to determine the quantity of feed required to maintain body weight or, conversely, determine body weight maintained after feeding a predetermined amount of feed for an extended period of time (Taylor et al., 1981, 1986);
- calorimetric methods (Agricultural Research Council, 1965, 1980); or
- comparative slaughter (Lofgreen, 1965; Lofgreen and Garrett, 1968).

Each approach has advantages as well as limitations.

Estimates of feed required for maintenance of body weight, usually measured in long-term feeding trials, are obtainable with relative ease and can be determined with large numbers of cattle. Values obtained generally correlate well with energy maintenance in mature, nonpregnant, nonlactating cattle (Jenkins and Ferrell, 1983; Ferrell and Jenkins, 1985a; Laurenz et al., 1991; Solis et al., 1988).

Changes in body composition and composition of weight change in growing, pregnant, or lactating cattle are problematic with this approach. Expression of the results in terms of ME or NE requirements depends on use of information from other approaches.

The energy feeding systems of the Agricultural Research Council (ARC) (1965, 1980), Ministry of Agriculture, Fisheries, and Food (MAFF) (1976, 1984), Commonwealth Scientific and Industrial Research Organization (CSIRO) (1990), and Agricultural and Food Research Council (AFRC) (1993), and the energy requirements of dairy cows (National Research Council, 1989) are primarily based on calorimetric methods. Fasting heat production (FHP) measured by calorimetry plus urinary energy lost during the same period provide measures of fasting metabolism (FM), which by definition, equates to net energy required for maintenance (NE_m). Measurement conditions are standardized such that animals are fed a specified diet at approximately maintenance for 3 weeks prior to measurement. Animals are trained to the calorimeter and kept in a thermoneutral environment. Measurements are usually made during the third and fourth day after withdrawal of feed. For practical use, FM values are adjusted for the difference between fasted weight of an animal and its liveweight when fed. In addition, recognizing that fasted animals are less physically active than fed animals, ARC (1980) adjusts FM by adding an activity allowance of 1 kcal/kg liveweight for cattle. CSIRO (1990) has incorporated additional corrections for breed, sex, proportional contribution of milk to the diet, energy intake, grazing activity, and cold stress.

Because of the complexity and cost of measurements, numbers of animals that can be used is limited. With this approach, measurements are basically acute in that they are made over one or at most a few days. Practical limitations of these systems stem largely from difficulties in adjusting data obtained in well-controlled laboratory environments to the practical feeding situation.

The California Net Energy System, proposed by Lofgreen and Garrett (1968) and adopted in the two preceding editions of this volume (National Research Council, 1976, 1984), is based on comparative slaughter methods. In contrast to calorimetry, in which ME intake and HE are measured and RE is determined by difference, comparative slaughter procedures measure ME and RE directly and HE by difference. RE is measured as the change in body energy content of animals fed at two or more levels of intake (one of which approximates maintenance) during a feeding period. RE equates, by definition, to NE_g in a growing animal. The slope of the linear regression of RE on ME intake provides an estimate of efficiency of utilization of ME for RE and in growing animals equates to k_g. The ME intake at which RE = 0 provides an estimate of ME required for maintenance (ME_m). By convention, the intercept of the regression of log HE on ME intake is used to calculate an estimate of FHP, which equates to NE_m. The efficiency of utilization of ME for maintenance (k_m) is calculated as the ratio of NE_m to ME_m. These approaches have an advantage over calorimetric methods because they allow experiments to be conducted under situations more similar to those found in the beef cattle industry. They must be conducted over extended time periods, however, to allow accurate assessment of body energy changes. Accurate assessment of body composition at the beginning and end of the feeding period is required.

The NE_m requirements of beef cattle have been estimated as

$$NE_m = 0.077 \text{ Mcal/EBW}^{0.75};$$

EBW is the average empty body weight in kilograms (Lofgreen and Garrett, 1968; Garrett, 1980). This expression was derived using data from, primarily, growing steers and heifers of British ancestry that were penned in generally nonstressful environments. Effects of activity and environment are implicitly incorporated into NE_m in this system. Similarly, influences of increased feed during the feeding period, altered activity, or environmental effects differing from those at maintenance are implicitly incorporated into estimates of NE_g. Application to differing situations requires appropriate adjustments.

Variation in Energy Requirements for Maintenance

Maintenance energy expenditures vary with body weight, breed or genotype, sex, age, season, temperature, physiological state, and previous nutrition. FHP or NE_m is more closely related to a fractional power of EBW than to $EBW^{1.0}$ (Brody, 1945; Kleiber, 1961); the most proper power has been the subject of much debate. $EBW^{0.75}$, often referred to as metabolic body weight, was originally used to confer proportionality on measurements of H_eE made in species differing considerably in mature weight (for example, mice to elephants). The convention generally adopted is to use $EBW^{0.75}$ to scale energy requirements for body weight, even though other functions may be more appropriate for specific applications.

BREED DIFFERENCES IN MAINTENANCE

Armsby and Fries (1911) reported that "scrub" steers utilized energy less efficiently than "good" beef animals. Subsequently, numerous researchers noted differences in energy requirements or efficiencies of energy utilization among breeds of cattle. However, because of differences in procedures and approaches as well as diversity of breeds compared, direct comparison among available data is difficult. Blaxter and Wainman (1966), using calorimetry, noted that Ayrshire steers had 20 percent higher FHP (kcal/

$BW^{0.75}$) than black (Angus type) steers and 6 percent higher than crosses of those breeds. Results of Garrett (1971), using comparative slaughter, indicated that Holstein steers required 23 percent more feed to maintain body energy than Hereford steers. Similarly, Jenkins and Ferrell (1984b) and Ferrell and Jenkins (1985a) indicated feed required for weight or energy stasis in young bulls and heifers was greater in the Simmental breed than in those of the Hereford breed. Those data indicated ME_m was, averaged across sexes, 19 percent (126 vs 106 kcal/$BW^{0.75}$) greater for Simmental than Hereford cattle. Estimates reported for Simmental bulls were equal to those reported by Stetter et al. (1989). Values reported by Andersen (1980) and Byers (1982) indicated Simmental had 6 and 3 percent higher requirements than Herefords, respectively. Conversely, Old and Garrett (1987) and Andersen (1980) found maintenance requirements of Charolais and Hereford steers to be similar. Estimates for growing Friesian cattle average approximately 13 percent higher (5 to 20 percent) than for Charolais (Robelin and Geay, 1976; Vermorel et al., 1976; Geay et al., 1980; Vermorel et al., 1982). Webster et al. (1976, 1982) reported predicted basal metabolism rates of Friesian cattle to be greater than Angus (10 percent), Hereford (31 percent), or Friesian × Hereford (8 percent). Chestnutt et al. (1975) estimated maintenance requirements of Friesian to be 20 percent higher than Friesian × Hereford and 14 percent greater than Angus steers, whereas estimates of Truscott et al. (1983) were 7 percent higher for Friesian than for Hereford steers. Wurgler and Bickel (1985) found no consistent difference in estimates of maintenance requirements among Angus × Braunvieh, Braunvieh, or Friesian steers. Estimates of maintenance requirements of Limousin have been similar to those of Angus (Byers, 1982), Hereford, and Charolais (Andersen, 1980). Results of Webster et al. (1982) and Andersen (1980) indicated Chianina had about 30 percent higher energy expenditures than Angus and Hereford. Several other reports (Vercoe, 1970; Vercoe and Frisch, 1974; Patle and Mudgal, 1975; Frisch and Vercoe 1976, 1977, 1982; van der Merwe and van Rooyen, 1980; Carstens et al., 1989a) indicate that maintenance energy requirements of *Bos indicus* breeds of cattle, including Africander, Barzona, Brahman, and Sahiwal, are about 10 percent lower, and British crosses with those breeds about 5 percent lower than British breeds. In contrast, data of Ledger (1977) and Ledger and Sayers (1977) suggest maintenance requirements of the Boran may be about 5 percent higher than for Herefords. However, those results appear to conflict with those in the report of Rogerson et al. (1968).

Results of Jenkins and Ferrell (1983) and Ferrell and Jenkins (1984a,b,c) indicated maintenance requirements differed among genotypes of mature crossbred cows. ME required for energy stasis (kcal/$BW^{0.75}$) of nonpregnant, nonlactating Jersey, Simmental, and Charolais sired cows (from Angus or Hereford dams) was 112, 123, and 99 percent that of Angus-Hereford (130 kcal/BW$^{0.75}$) cross cows. Similarly, the results of Lemenager et al. (1980) suggested that energy needs of Simmental × Hereford cows was about 25 percent higher than Hereford cows during gestation, whereas Angus × Hereford and Charolais × Hereford required about 5 and 7 percent more than Herefords. Laurenz et al. (1991) reported that Simmental cows required 21 percent more ME (kcal/BW$^{0.75}$) than Angus cows. Klosterman et al. (1968) observed no difference in estimated energy requirements to maintain weight of mature nonpregnant nonlactating Hereford and Charolais cows when adjusted for body condition. Similarly, when adjusted for body condition, Hereford × Friesian and White Shorthorn × Galloway cows required similar amounts of energy to maintain liveweight (Russel and Wright, 1983). Estimates of ME (kcal/BW$^{0.75}$) for energy stasis of nonpregnant, nonlactating Red Poll, Brown Swiss, Gelbvieh, Maine Anjou, and Chianina sired cows (C. L. Ferrell and T. G. Jenkins, unpublished data) were 112, 122, 117, 113, and 108 percent of values for Angus-Hereford (126 kcal/BW$^{0.75}$) cross cows. Similar values were reported for weight stasis of those cows, with the exception of Gelbvieh and Chianina, which were higher (Ferrell and Jenkins, 1987). In that study, ME (kcal/BW$^{0.75}$) required for weight stasis of purebred Angus, Hereford, and Brown Swiss were 116, 115 and 155 percent of that estimated for Angus-Hereford crossbreds (119 kcal/BW$^{0.75}$). Results of Taylor and Young (1968) and Taylor et al. (1986) indicated energy required (recalculated as kcal/BW$^{0.75}$) for long-term weight equilibrium of British Friesian, Jersey, and Ayrshire cows to be 20 percent higher than that of Angus and Hereford cows. Energy required by Dexter cows was 9 percent higher than the average of Angus and Hereford cows. Thompson et al. (1983) reported estimates indicating ME required for energy stasis was 9 percent higher in Angus × Holstein than in Angus × Hereford cows. Ritzman and Benedict (1938) observed no difference between energy required by Jersey and Holstein cows, whereas Brody (1945) observed slightly higher requirements by Holstein cows than Jersey cows. Solis et al. (1988) reported estimates of ME required for weight and energy stasis for 15 breed or breed crosses from a 5-breed diallel. Simple correlation between the two estimates was 0.84 and the slope of the linear regression was 0.99, indicating good agreement between the two estimates. When pooled, estimates of ME required for energy stasis were 104, 96, 96, 112, and 106 kcal/BW$^{0.75}$/day for 1/2 Angus, 1/2 Brahman, 1/2 Hereford, 1/2 Holstein, and 1/2 Jersey cows, respectively.

Most of these reports observed differences between or among breeds compared and serve to document that considerable variation exists in maintenance requirements among cattle germ plasm resources. However, because of the diversity of breeds, methodologies, conditions, etc.,

direct comparisons between studies are often tenuous. As a result, the subcommittee selected studies in which British breeds or British breed crosses were compared with other breeds or breed crosses and expressed the results as relative values. It is believed the following generalizations can be made with some confidence, based on the data reviewed in the preceding paragraphs. In growing cattle, *Bos indicus* breeds of cattle (for example, Africander, Barzona, Brahman, Sahiwal) require about 10 percent less energy than beef breeds of *Bos taurus* cattle (for example, Angus, Hereford, Shorthorn, Charolais, Limousin) for maintenance, with crossbreds being intermediate. Conversely, dairy or dual-purpose breeds of *Bos taurus* cattle (for example, Ayrshire, Brown Swiss, Braunvieh, Friesian, Holstein, Simmental) apparently require about 20 percent more energy than beef breeds, with crosses being intermediate. Data involving straightbred, mature cows are more limited. However, available data with straightbreds combined with those of crossbreds, indicate that relative differences between breeds in mature cows is similar to that observed in growing animals. This may be generalized further to indicate, in both adult and growing cattle, that a positive relationship exists between maintenance requirement and genetic potential for measures of productivity (for example, rate of growth or milk production; Webster et al., 1977; Taylor et al., 1986; Ferrell and Jenkins, 1987; Montano-Bermudez et al., 1990).

Consistent with this concept, available data also suggest that animals having genetic potential for high-productivity may have less advantage or be at a disadvantage in nutritionally or environmentally restrictive environments (Kennedy and Chirchir, 1971; Baker et al., 1973; Frisch, 1973; Moran, 1976; O'Donovan et al., 1978; Jenkins and Ferrell, 1984b; Ferrell and Jenkins, 1985a,b; Jenkins et al., 1986). This concept is further supported by the reports of Peacock et al. (1976), Ledger and Sayers (1977), and Frisch and Vercoe (1977). Frisch and Vercoe (1980, 1982) have subsequently shown that selection for increased growth in a high-stress environment results in decreased FHP. Results from these and other studies show that correlated responses to selection may result in a genotype/environment interaction. Selection may result in a population of animals highly adapted to a specific environment but less adapted to different environments and with decreased adaptability to environmental changes (Frisch and Vercoe, 1977; Taylor et al., 1986; Jenkins et al., 1991).

SEX DIFFERENCES IN MAINTENANCE

Garrett (1970) found little difference in estimated fasting HE or ME required for maintenance between steers and heifers. Subsequently, Garrett (1980), in a study based on comparative slaughter experiments involving 341 heifers and 708 steers, concluded that FHP (net energy required for maintenance) of steers and heifers is similar. ARC (1980) and CSIRO (1990) similarly concluded fasting metabolism of castrate males and heifers was similar.

Ferrell and Jenkins (1985a) estimated similar FHP (kcal/ $BW^{0.75}$/day) for Hereford bulls (70.4) and heifers (69.3), but estimates for Simmental bulls (80.8) were 9 percent higher than for Simmental heifers (74.1). When expressed as ME required for maintenance, Hereford bulls and heifers differed by only 2 percent, but estimates for Simmental bulls were 16.5 percent higher than for Simmental heifers. Pooled across breeds, estimated ME required for energy stasis was 12 percent higher for intact males than for females (123 vs 110 kcal ME/$BW^{0.75}$/day). Webster et al. (1977) reported that Hereford × Friesian bulls had predicted basal metabolism values about 20 percent higher than steers of the same breed cross. In a subsequent report (Webster et al., 1982), values presented indicated bulls had 13 to 15 percent higher predicted basal metabolism than steers. Geay et al. (1980) also suggested higher maintenance requirements of bulls than heifers. ARC (1980) and CSIRO (1990), cited the report of Graham (1968) as indicating rams had 18 percent higher fasting metabolism than wethers and ewes. However, Bull et al. (1976) and Ferrell et al. (1979) estimated the ME required for maintenance of rams to be only 2 to 3 percent higher than for ewe lambs. The average of available data, if the sheep data of Bull et al. (1976) and Ferrell et al. (1979) are excluded, support the conclusion of ARC (1980) and CSIRO (1990) that maintenance requirements of bulls are 15 percent higher than that of steers or heifers of the same genotype.

AGE EFFECTS ON MAINTENANCE

The concept that maintenance per unit of size declines with age in cattle and sheep (Blaxter, 1962; Graham et al., 1974) has been generally accepted. Data from sheep, predominately castrate males, generally support this view (Graham and Searle, 1972a,b; Graham, 1980). The equation of Graham et al. (1974) indicated maintenance decreased exponentially and was related to age by the relationship $e^{-0.08age}$, which indicates the decrease was 8 percent per year. The generalized equation reported by Corbett et al. (1985) for sheep and cattle, which was later adopted by CSIRO (1990), indicates maintenance decreases 3 percent per year. CSIRO (1990) indicated a minimum of 84 percent of initial values to be attained at about 6 years. Young et al. (1989) noted metabolic rate deviated substantially from allometric relationships; deviations were greatest during times of highest relative growth rate. They further suggested that significant deviations may also occur in association with other productive functions. Data reported from cattle are less consistent. Blaxter et al. (1966) found little influence of age (15 to 81 weeks), other than that associated with weight, on maintenance of

steers. Results of Blaxter and Wainman (1966), Taylor et al. (1981) and Birkelo et al. (1989) were consistent with those findings. Vermorel et al. (1980) indicated maintenance requirements of cattle changed little between 5 and 34 weeks of age, but data of Carstens et al. (1989a) indicate a 6 percent decrease in FHP and an 8 percent decrease in ME required for maintenance between 9 and 20 months. Conversely, data reported by Tyrrell and Reynolds (1988) indicated ME required for maintenance (kcal/$SBW^{0.75}$) increased 14 percent in beef heifers as weight increased from 275 to 475 kg. To our knowledge, direct comparisons of mature, productive females to younger or nonreproducing animals are not available. Indirect evidence (see above) suggests that maintenance of mature, productive cows is not less than that of younger, growing animals postweaning.

SEASONAL EFFECTS ON MAINTENANCE

Although, typically, effects of season have been associated with effects of temperature, it has become increasingly evident that season per se may have significant effects on maintenance requirements of cattle and sheep. Christopherson et al. (1979), Blaxter and Boyne (1982), and Webster et al. (1982) noted lower maintenance requirements of sheep, cattle, and bison during the fall of the year. Predicted basal metabolism of cattle was 90.3, 92.0, 78.9, and 86.3 kcal/$BW^{0.75}$ during weeks 0 to 16, 17 to 32, 33 to 48 and 49 to 52, respectively, in Scotland (Webster et al., 1982). Data reported from Colorado by Birkelo et al. (1989) indicate FHP during fall, winter, and spring measurements were 90.7, 95.6, and 96.2 percent of FHP measured during the summer, but ME_m did not consistently follow this pattern. Estimates of energy required for weight stasis of mature cows by Byers et al. (1985) for fall, winter, and spring were 86, 86, and 92 percent and those for energy stasis were 94, 102, and 100 percent of estimates made during the summer. Laurenz et al. (1991) reported similar effects of season on energy required for weight stasis of Angus and Simmental cows and for energy stasis of Angus cows but a dissimilar pattern for energy stasis of Simmental cows. Byers and Carstens (1991) reported further observations and indicated that as cow fatness increased, maintenance requirements increased during the spring and summer but decreased during the fall and winter. Walker et al. (1991) clearly demonstrated that seasonal effects in ewes are related to photoperiod. Possible season/genotype or latitude effects have not been quantified.

TEMPERATURE EFFECTS ON MAINTENANCE

For a detailed review, the reader is referred to the report, *Effect of Environment on Nutrient Requirements of Domestic Animals* (National Research Council, 1981b). Heat production in cattle arises from tissue metabolism and from fermentation in the digestive tract. Animals dissipate heat by evaporation, radiation, convection, and conduction. Both heat production and dissipation are regulated to maintain a nearly constant body temperature. Within the zone of thermoneutrality, HE is essentially independent of temperature and is determined by feed intake and the efficiency of use; body temperature control is primarily via regulation of heat dissipation. When effective ambient temperature increases above the zone of thermoneutrality—that is, higher than the upper critical temperature (UCT)—productivity decreases, primarily as a result of reduced feed intake. In addition, elevated body temperature results in increased tissue metabolic rate and increased "work" of dissipating heat (for example, increased respiration and heart rates); consequently, energy requirements for maintenance increase. Conversely, when effective ambient temperature decreases below the zone of thermoneutrality—that is, below the lower critical temperature (LCT)—HE produced from "normal" tissue metabolism and fermentation is inadequate to maintain body temperature. As a result, animal metabolism must increase to provide adequate heat to maintain body temperature. Consequently energy requirements for maintenance increase. Both UCT and LCT vary with the rate of heat production in thermoneutral conditions and the animals ability to dissipate or conserve heat. As noted in other sections of this report, heat production of animals in thermoneutral conditions may differ substantially as functions of feed intake, physiological state, genotype, sex, and activity.

The word *acclimatization* is used to describe adaptive changes in response to changes in the climatic conditions and include behavioral as well as physiological changes. Behavioral modification includes using variation in terrain or other topographical features such as windbreaks, huddling in groups, or changing posture to minimize heat loss in cold and during decreased activity, seeking shade to decrease exposure to radiant heat, seeking a hill to increase exposure to wind, or wading in water to increase heat dissipation in high temperatures. Physiological adaptations include changes in basal metabolism, respiration rate, distribution of blood flow to skin and lungs, feed and water consumption, rate of passage of feed through the digestive tract, hair coat, and body composition. Physiological changes usually associated with acute temperature changes include shivering and sweating as well as acute changes in feed and water consumption, respiration rate, heart rate, and activity. It should also be noted that animals differ greatly in their behavioral responses and in their ability to physiologically adapt to the thermal environment. Genotype differences are particularly evident in this regard.

Recognizing the importance of adaptation, the National Research Council committee (1981b), relying primarily on the results of Young (1975a,b), concluded that required NE_m of cattle adapted to the thermal environment is related to the previous ambient (air) temperature (T_p, °C) in the following manner:

$$NE_m = (0.0007 * (20 - T_p)) + 0.077 \text{ Mcal/BW}^{0.75}.$$

This equation indicates that the NE_m requirement of cattle changes by 0.0007 Mcal/BW$^{0.75}$ for each degree that previous ambient temperature differed from 20° C. It should be noted that these corrections for previous temperature are largely opposite the photoperiod effect discussed previously.

Heat or cold stress occur when effective ambient temperature is higher than UCT or less than LCT. UCT and LCT are functions of how much heat the animal produces and how much heat is lost to the environment. HE of the animal may be calculated as shown previously:

$$HE = ME - RE, \text{ or}$$
$$HE = NE_m/k_m + (RE(1 - k_g))$$

where ME is ME intake and RE is retained energy, which may include NE_g, NE_l, NE_y, etc. (all expressed relative to BW$^{0.75}$).

Cold Stress   Both environmental and animal factors contribute to differences in heat loss from the animal. Environmental factors include air movement, precipitation, humidity, contact surfaces, and thermal radiation. Although results are not totally satisfactory, numerous efforts have been made to integrate these effects with animal responses.

Factors contributing to differences in animal heat loss from conduction, convection, and radiation are surface area (SA), which includes surface or external insulation (EI), and internal or tissue insulation (TI). Evaporative losses are affected by respiration volume as well as SA, EI, and TI. Respiratory losses, although not quantified by National Research Council (1981b), represent 5 to 25 percent and total evaporative heat losses represent 20 to 80 percent of total heat losses (Ehrlemark, 1991).

Surface area is related to body weight by the equation

$$SA, \text{ m}^2 = 0.09 \text{ BW}^{0.67},$$

thus,

$$HE/SA = HE/BW^{0.75} * BW^{0.75}/SA.$$

TI (°C/Mcal/m²/day) is primarily a function of subcutaneous fat and skin thicknesses. Typical values are 2.5 for a newborn calf, 6.5 for a 1-month old calf, 5.5 to 8.0 for yearling cattle and 6.0 to 12 for adult cattle. EI is provided by hair coat plus the layer of air surrounding the body. Thus, external insulation is related to hair depth. However, the effectiveness of hair as external insulation is influenced by wind, precipitation, mud, and hide thickness. These effects have been described as follows:

$$EI = (7.36 - 0.296 \text{ WIND} + 2.55 \text{ HAIR}) * \text{MUD} * \text{HIDE}$$

where EI is expressed as °C/Mcal/m², WIND is wind speed (kph), and HAIR is effective hair depth (cm). MUD and HIDE are adjustments for mud and hide thickness. Total insulation (IN) is

$$IN = TI + EI,$$

and LCT may be calculated as (National Research Council, 1981b):

$$LCT = 39 - IN * (HE/SA - H_e),$$

where LCT, IN, and HE/SA are as described previously. The term H_e represents the minimal total evaporative heat loss and is estimated (Ehrlemark, 1991) as:

$$H_e = HE/SA * 0.15.$$

The animal can receive or lose heat by solar or long-wave radiation. The net impact of thermal radiation on the animal depends on the difference between the combined solar and long-wave radiation received by the animal and the long-wave radiation emitted by the animal. For animals in bright sunlight, a net gain of heat by thermal radiation usually exists, resulting in an increased effective ambient temperature (EAT) of 3° to 5° C (National Research Council, 1981b). In bright sunlight, this effect lowers LCT by 3° to 5° C. Conversely, CSIRO (1990) have indicated that the rate of heat loss by long-wave radiation increases on cold clear nights resulting in an increase in the LCT. Within the temperature range of −10° to 10° C this effect is about 5° C.

The increase in energy required to maintain productivity in an environment colder than the animal's LCT may be estimated as

$$ME_c = SA(LCT - EAT)/IN,$$

where ME_c is the increase in maintenance energy requirement (Mcal/day), SA is surface area (m²), LCT is lower critical temperature (°C), EAT is effective ambient temperature (°C) adjusted for thermal radiation, and IN is total insulation (°C/Mcal/m²/day).

$$NE_c = k_m * ME_c/EBW^{0.75}.$$

Total net energy for maintenance under conditions of cold stress (NE_{mc}) becomes

$$NE_{mc} = NE_m + NE_c.$$

Heat Stress If ambient temperature and thermal radiation exceed the temperature of the skin surface, the animal cannot lose heat by sensible means (conduction, convection, and radiation) and will gain heat by these routes. Evaporative heat loss occurs from the skin (cutaneous) or through respiration. The effectiveness of both cutaneous and respiratory evaporative heat loss diminishes as relative humidity (RH) of the air increases and is totally ineffective

when RH = 100. Animals can store some heat in their bodies during the day and dissipate the stored heat during cooler daytime periods or at night, if the animal's heat production exceeds its ability to dissipate heat; but if hyperthermia persists, animals cannot survive.

There has been much study of the various aspects of heat stress on animal performance, but there are no established bases for quantitative description of effects. Ehrlemark (1991), for example, developed a regression of respiratory heat loss on the ratio of ambient temperature minus LCT to body temperature minus LCT but did not include cutaneous evaporative heat loss or the influence of RH. It is generally agreed that adjustments to maintenance energy requirement for heat stress should be based on the severity of heat stress; however, severity can vary considerably among animals, depending on animal behavior, acclimatization, diet, level of productivity, radiant heat load, or genotype. The type and intensity of panting by an animal can provide an index for appropriate adjustment in maintenance requirement—an increase of 7 percent when there is rapid shallow breathing and 11 to 25 percent when there is deep, open-mouth panting (National Research Council, 1981b). With severe heat, feed consumption is reduced and consequently metabolic heat production and productivity are reduced.

EFFECTS OF PHYSIOLOGICAL STATE ON MAINTENANCE

Total heat production increases during gestation (Brody, 1945). Although indirect evidence is available to suggest maintenance requirements of cows increase during gestation (Brody, 1945; Kleiber, 1961; Ferrell and Reynolds, 1985), an increase has not been directly measurable by comparative slaughter evaluations (Ferrell et al., 1976). Increased heat production associated with pregnancy, for the purpose of estimating energy requirements, may be assumed to be attributable to the productive process of pregnancy.

In contrast, Moe et al. (1970) estimated ME requirements for maintenance to be 22 percent higher in lactating than in nonlactating cows (primarily Holstein). A similar difference (23 percent) was reported by Flatt et al. (1969), whereas Ritzman and Benedict (1938) reported a larger (49 percent) difference. Neville and McCullough (1969) and Neville (1974) using Hereford cows and different approaches, estimated the maintenance requirement of lactating cows to be more than 30 percent higher than nonlactating cows. The reports of Patle and Mudgal (1975, 1977) agree with those observations, whereas data of Ferrell and Jenkins (1985b, 1987; and unpublished data) suggest a difference of 10 to 20 percent. Taken in total, available data indicate maintenance requirements of lactating cows to be about 20 percent higher than those of nonlactating cows.

EFFECTS OF ACTIVITY ON MAINTENANCE

Few data are available regarding efficiency of ME use for muscular work. In addition, it may be debated whether activity is a maintenance or productive function. It is highly probable that grazing cattle walk considerably further than penned animals and, therefore, expend more energy for work; however, the extent to which grazing animals expend more energy standing, changing positions, eating, or ruminating than penned cattle is not well documented. It is recognized that energy expenditure for work by grazing cattle is influenced by numerous factors including herbage quality and availability, topography, weather, distribution of water, genotype, or interactions among these factors. Variation among individuals may be substantial. In a review of available literature, CSIRO (1990) estimated the increase in maintenance energy requirements of grazing as compared to penned cattle to be 10 to 20 percent in best grazing conditions and about 50 percent for cattle on extensive, hilly pastures where animals walk considerable distances to preferred grazing areas and water. An alternative approach to estimate the NE required for activity (NE_{ma}; CSIRO, 1990) was devised as follows:

$$NE_{ma}, Mcal/day = [(0.006 * DMI(0.9-D)) + (0.05T/(GF + 3))] * W/4.184.$$

where DMI is dry matter intake from pasture (kg/day); D is digestibility of dry matter (as a decimal); T is terrain (level, 1.0l; undulating, 1.5; or hilly, 2.0), and GF is green forage availability (ton/ha). If no green forage is available, replacement of GF with total forage available (TF) was suggested on the premise that selectivity, hence distance walked, decreases when no green forage is available.

Effects of Previous Nutrition/Compensatory Gain

The phenomenon of compensatory gain is described as a period of faster or more efficient rate of growth following a period of nutritional or environmental stress. Numerous reports are available to document this phenomena in cattle and other species (Wilson and Osbourn, 1960; Carrol et al., 1963; Lawrence and Pierce, 1964; Hironaka and Kozub, 1973; Lopez-Sanbidet and Verde, 1976; O'Donovan, 1984; Hovell et al., 1987; Abdalla et al., 1988; Drouillard et al., 1991). The response to previous nutritional deprivation is highly variable, however. Data are available, for example, that show that at similar body weights, body fat is decreased (Smith et al., 1977; Mader et al., 1989; Carstens et al., 1991), not changed (Fox et al., 1972; Burton et al., 1974; Rompala et al., 1985) or increased (Searle and Graham, 1975; Tudor et al., 1980; Abdalla et al., 1988) after a period of realimentation. Differences among animal genotypes; severity, nature, and duration of restriction; and nutritional regime and interval of measurement of the response during realimentation are among the many variables contributing to differences.

A major component of compensatory growth by animals given abundant feed after a period of restriction is increased feed intake. This component is discussed in more detail in a later section. This response will cause increased gut fill and liveweight, but there is also evidence for higher efficiency of energy use. Several reports (Graham and Searle, 1979; Thompsen et al., 1980; Carstens et al., 1991) have provided evidence to suggest higher net efficiency of ME use for body energy gain. The duration of these effects is subject to debate, however (Butler-Hogg, 1984; Ryan et al., 1993a,b).

Results of studies reported by Marston (1948) have contributed to an understanding of the other possible mechanisms involved in compensatory growth. Those results showed that level of feed intake may affect the metabolic rate of sheep and cattle. These and other reports (Graham and Searle, 1972a,b; Graham et al., 1974; Graham and Searle, 1975; Thomson et al., 1980; Ferrell and Koong, 1987; Ferrell et al., 1986) have shown that fasting heat production decreases in response to decreased feed intake. Similarly, several reports (Wilson and Osbourn, 1960; Walker and Garrett, 1970; Foot and Tulloh, 1977; Ledger, 1977; Ledger and Sayers, 1977; Gray and McCracken, 1980; Andersen, 1980; Corbett et al., 1982) have shown that maintenance in rats, swine, cattle, and sheep is decreased after periods of decreased nutritional intake. Some of the possible explanations for altered metabolism associated with different planes of nutrition have been discussed by Milligan and Summers (1986), Ferrell (1988), and Johnson et al. (1990). Briefly, metabolic bases for changes include altered rates of ion pumping and metabolite cycling (Milligan and Summers, 1986; Harris et al., 1989; Summers et al., 1988; McBride and Kelly, 1990; Lobley et al., 1992) and altered size and metabolic rate of visceral organs (Canas et al., 1982; Koong et al., 1982, 1985; Burrin et al., 1989).

There is much, although not total, support for the general conclusion that maintenance is reduced during and for some time after a period of feed restriction (Graham and Searle, 1972a; Thorbek and Henckel, 1976; Andersen, 1980; Ledger and Sayers, 1977; Schnyder et al., 1982; Stetter et al., 1989); however, reports on the extent of reduction have been variable, and range from about 10 percent to more than 50 percent. Little definitive information is available regarding the duration of the reduced maintenance or, stated another way, the length of time that an animal exhibits compensatory gain after it has access to abundant feed is not well defined. Further, critical description of animals such that expected degree of compensation can be predicted with confidence, without knowing their genotype and history (the nature and severity of restriction, etc.) is lacking. Because of these types of problems, generalizations are difficult, although several mathematical descriptions have been proposed (Baldwin et al., 1980; Corbett et al., 1985; Koong et al., 1985; Frisch and Vercoe, 1977). A reduction in maintenance of 20 percent for a compensating animal seems a reasonable generalization (Thorbek and Henckel, 1976; Crabtree et al., 1976; Frisch and Vercoe, 1977; Andersen, 1980; Baldwin et al., 1980; Thomson et al., 1980; Vermorel et al., 1982; Schnyder et al., 1982; Koong et al., 1982, 1985; Koong and Nienaber, 1987; Webster et al., 1982; Wurgler and Bickel, 1985; Ferrell et al., 1986; Birkelo et al., 1989; Burrin et al., 1989; Carstens et al., 1989b). The duration of reduced maintenance is subject to the extent and duration of restricted growth and to nutritional regimen during the recovery periods; typically, 60 to 90 days of compensation is expected.

Use of Energy from Weight Loss

Animals, particularly in a pasture or range situation, intermittently lose body weight when feed quantity or quality is inadequate to meet the animal's nutrient requirements. Available data indicate composition of liveweight loss is approximately equal to the composition of liveweight gain in animals (Agricultural Research Council, 1980; Commonwealth Scientific and Industrial Research Organization, 1990). Thus, the energy content of liveweight loss and gain are similar. Energy content and composition of weight gain are discussed in subsequent sections.

Buskirk et al. (1992) argued that the energy content of empty body weight gain in mature cows varies, depending on cow body condition. They estimated energy content of empty body weight change in cows with body condition scores (1 to 5 scale) of 1, 2, 3, 4, and 5 to be 2.57, 3.82, 5.06, 6.32, and 7.57 Mcal/kg, respectively. Similarly, CSIRO (1990) adopted relationships established by Hulme et al. (1986) that indicate energy content of liveweight change in dairy cattle increases linearly from 3.0 to 7.1 Mcal/kg as condition score increases from 1 to 8 (on a scale of 1 to 8). Composition of weight change in mature cows is discussed in greater detail in Chapter 3.

Although limited data are available, data from sheep (Marston, 1948), dairy cows (Flatt et al., 1965; Moe et al., 1970) and beef cows (Russel and Wright, 1983) indicate the efficiency of use of energy from body tissue loss for maintenance or milk production to be 77 to 84 percent with the mean being approximately 80 percent.

REFERENCES

Abdalla, H. O., D. G. Fox, and M. L. Thonney. 1988. Compensatory gain by Holstein calves after underfeeding protein. J. Anim. Sci. 66:2687–2695.

Agricultural and Food Research Council. 1993. Energy and Protein Requirements of Ruminants. Wallingford, U.K.: CAB International.

Agricultural Research Council. 1965. The Nutrient Requirements of Farm Livestock. No. 2. Ruminants. London, U.K.: Agricultural Research Council.

Agricultural Research Council. 1980. The Nutrient Requirements of Ruminant Livestock: Technical Review. Farnham Royal, U.K.: Commonwealth Agricultural Bureaux.

Andersen, B. B. 1980. Feeding trials describing net requirements for maintenance as dependent on weight, feeding level, sex and genotype. Ann. Zootech. 29:85–92.

Armsby, H. P., and J. A. Fries. 1911. The Influence of Type and of Age upon the Utilization of Feed by Cattle. Bull. No. 128. Washington, D.C.: U.S. Department of Agriculture, Bureau of Animal Industry.

Baker, F. S., Jr., A. Z. Palmer, and J. W. Carpenter. 1973. Brahman × European crosses vs British breeds. Pp. 227–284 in Crossbreeding Beef Cattle, Series 2, M. Koger, T. J. Cunha, and P. C. Wainick, eds. Gainesville, Fla.: University of Florida Press.

Baldwin, R. L., N. E. Smith, J. Taylor, and M. Sharp. 1980. Manipulating metabolic parameters to improve growth rate and milk secretion. J. Anim. Sci. 51:1416–1428.

Birkelo, C. P., D. E. Johnson, and H. W. Phetteplace. 1989. Plane of nutrition and season effects on energy maintenance requirements of beef cattle. Energy Metab. Proc. Symp. 43:263–266.

Blaxter, K. L. 1962. The Energy Metabolism of Ruminants. Springfield, Ill.: Charles C Thomas.

Blaxter, K. L., and A. W. Boyne. 1982. Fasting and maintenance metabolism of sheep. J. Agric. Sci. Camb. 99:611–620.

Blaxter, K. L., and F. W. Wainman. 1966. The fasting metabolism of cattle. Br. J. Nutr. 20:103–111.

Blaxter, K. L., J. L. Clapperton, and F. W. Wainman. 1966. Utilization of the energy and protein of the same diet by cattle of different ages. J. Agric. Sci. Camb. 67:67–75.

Brody, S. 1945. Bioenergetics and Growth. New York: Hafner.

Bull, L. S., H. F. Tyrrell, and J. T. Reid. 1976. Energy utilization by growing male and female sheep and rats, by comparative slaughter and respiration techniques. Energy Metab. Proc. Symp. 19:137–140.

Burrin, D. G., C. L. Ferrell, and R. A. Britton. 1989. Effect of feed intake of lambs on visceral organ growth and metabolism. Energy Metab. Proc. Symp. 43:103–106.

Burton, J. H., M. Anderson, and J. T. Reid. 1974. Some biological aspects of partial starvation: The effect of weight loss and regrowth on body composition in sheep. Br. J. Nutr. 32:515–527.

Buskirk, D. D., R. P. Lemenager, and L. A. Horstman. 1992. Estimation of net energy requirements (NE$_m$ and NE change) of lactating beef cows. J. Anim. Sci. 70:3867–3876.

Butler-Hogg, B. W. 1984. Growth patterns in sheep: Changes in the chemical composition of the empty body and its constituent parts during weight loss and compensatory growth. J. Agric. Sci. Camb. 103:17–24.

Byers, F. M. 1982. Patterns of energetic efficiency of tissue growth in beef cattle of four breeds. Energy Metab. Proc. Symp. 29:92–95.

Byers, F. M., and G. E. Carstens. 1991. Seasonality of maintenance requirements in beef cows. Energy Metab. Proc. Symp. 58:450–453.

Byers, F. M., G. T. Schelling, and R. D. Goodrich. 1985. Maintenance requirements of beef cows with respect to genotype and environment. Energy Metab. Proc. Symp. 32:230–233.

Canas, R., J. J. Romero, and R. L. Baldwin. 1982. Maintenance energy requirements during lactation in rats. J. Nutr. 112:1876–1880.

Carrol, F. D., J. E. Ellsworth, and D. Kroger. 1963. Compensatory carcass growth in steers following protein and energy restriction. J. Anim. Sci. 22:197–201.

Carstens, G. E., D. E. Johnson, K. A. Johnson, S. K. Hotovy, and T.J. Szymanski. 1989a. Genetic variation in energy expenditures of monozygous twin beef cattle at 9 and 20 months of age. Energy Metab. Proc. Symp. 43:312–315.

Carstens, G. E., D. E. Johnson, and M. A. Ellenberger. 1989b. Energy metabolism and composition of gain in beef steers exhibiting normal and compensatory growth. Energy Metab. Proc. Symp. 43:131–134.

Carstens, G. E., D. E. Johnson, M. A. Ellenberger, and J. D. Tatam. 1991. Physical and chemical components of the empty body during compensatory growth in beef steers. J. Anim. Sci. 69:3251–3264.

Chestnutt, D. M. B., R. Marsh, J. G. Wilson, T. A. Steqart, T. A. McCillough, and T. McCallion. 1975. Effects of breed of cattle on energy requirements for growth. Anim. Prod. 21:109–119.

Christopherson, R. J., R. J. Hudson, and M. K. Christopherson. 1979. Seasonal energy expenditures and thermoregulatory response of bison and cattle. Can J. Anim. Sci. 59:611–617.

Commonwealth Scientific and Industrial Research Organization. 1990. Feeding Standards for Australian Livestock: Ruminants. East Melbourne, Victoria, Australia: CSIRO Publications.

Corbett, J. L., E. P. Furnival, and R. S. Pickering. 1982. Energy expenditure at pasture of shorn and unshorn border Leichester ewes during late pregnancy and lactation. Energy Metab. Proc. Symp. 29:34–37.

Corbett, J. L., M. Freer, and N. M. Graham. 1985. A generalized equation to predict the varying maintenance metabolism of sheep and cattle. Energy Metab. Proc. Symp. 32:62–65.

Crabtree, R. M., M. Kay, and A. J. F. Webster. 1976. The net availabilities of ME for body gain of two pelleted diets offered to Hereford × Friesian castrate males over different live-weight ranges. Anim. Prod. 22:156–157.

Drouillard, J. S., C. L. Ferrell, T. J. Klopfenstein, and R. A. Britton. 1991. Compensatory growth following metabolizable protein or energy restrictions in beef steers. J. Anim. Sci. 69:811–818.

Ehrlemark, A. 1991. Heat and Moisture Dissipation from Cattle: Measurements and Simulation Model. Ph.D. dissertation. Swedish University of Agricultural Sciences, Uppsala, Sweden.

Ferrell, C. L. 1988. Contribution of visceral organs to animal energy expenditures. J. Anim. Sci. 66(Suppl. 3):23–34.

Ferrell, C. L., and T. G. Jenkins. 1984a. A note on energy requirements for maintenance of lean and fat Angus, Hereford and Simmental cows. Anim. Prod. 35:305–309.

Ferrell, C. L., and T. G. Jenkins. 1984b. Relationships among various body components of mature cows. J. Anim. Sci. 58:222–233.

Ferrell, C. L., and T. G. Jenkins. 1984c. Energy utilization by mature, nonpregnant, nonlactating cows of different breeds. J. Anim. Sci. 58:234–243.

Ferrell, C. L., and T. G. Jenkins. 1985a. Energy utilization by Hereford and Simmental males and females. Anim. Prod. 41:53–61.

Ferrell, C. L., and T. G. Jenkins. 1985b. Cow type and the nutritional environment: Nutritional aspects. J. Anim. Sci. 61:725–741.

Ferrell, C. L., and L. P. Reynolds. 1985. Oxidative metabolism of gravid uterine tissues of the cow. Energy Metab. Proc. Symp. 32:298–301.

Ferrell, C. L., and T. G. Jenkins. 1987. Influence of biological type on energy requirements. Pp. 1–7 in Proceedings of the Grazing Livestock Nutrition Conference. Misc. Publ. Stillwater, Okla.: Agricultural Experiment. Station, Oklahoma State University.

Ferrell, C. L., and L. J. Koong. 1987. Response of body organs of lambs to differing nutritional treatments. Energy Metab. Proc. Symp. 32:26–29.

Ferrell, C. L., W. N. Garrett, N. Hinman, and G. Gritchting. 1976. Energy utilization by pregnant and nonpregnant heifers. J. Anim. Sci. 42:937–950.

Ferrell, C. L., J. D. Crouse, R. A. Field, and J. L. Chant. 1979. Effects of sex, diet and stage of growth upon energy utilization by lambs. J. Anim. Sci. 49:790–801.

Ferrell, C. L., L. J. Koong, and J. A. Nienaber. 1986. Effect of previous nutrition on body composition and maintenance energy costs of growing lambs. Br. J. Nutr. 56:595–605.

Flatt, W. P., L. A. Moore, N. W. Hooven, and R. D. Plowman. 1965. Energy metabolism studies with a high-producing lactating dairy cow. J. Dairy Sci. 48:797–798.

Flatt, W. P., P. W. Moe, A. W. Munson, and T. Cooper. 1969. Energy utilization by high-producing dairy cows. II. Summary of energy balance experiments with lactating Holstein cows. Energy Metab. Proc. Symp. 12:235–239.

Foot, J. Z., and N. M. Tulloh. 1977. Effects of two paths of liveweight change on efficiency of feed use and on body composition of Angus steers. J. Agric. Sci. Camb. 88:135–142.

Fox, D. G., R. R. Johnson, R. L. Preston, T. R. Dockerty, and E. W. Klosterman. 1972. Protein and energy utilization during compensatory growth in beef cattle. J. Anim. Sci. 34:310–318.

Frisch, J. E. 1973. Comparative drought resistance of Bos indicus and Bos taurus crossbred herds in Central Queensland. 2. Relative mortality rates, calf birth weights and weights and weight changes of breeding cows. Aust. J. Exp. Agric. Anim. Husb. 13:117–126.

Frisch, J. E., and J. E. Vercoe. 1976. Maintenance requirements, fasting metabolism and body composition in different cattle breeds. Energy Metab. Proc. Symp. 19:209–212.

Frisch, J. E., and J. E. Vercoe. 1977. Feed intake, eating rate, weight gains, metabolic rate and efficiency of feed utilization in Bos taurus and Bos indicus crossbred cattle. Anim. Prod. 25:343–358.

Frisch, J. E., and J. E. Vercoe. 1980. Changes in fasting metabolism of cattle as a consequence of selection for growth rate. Energy Metab. Proc. Symp. 26:431–434.

Frisch, J. E., and J. E. Vercoe. 1982. The effect of previous exposure to parasites on the fasting metabolism and feed intake of three cattle breeds. Energy Metab. Proc. Symp. 29:100–103.

Garrett, W. N. 1970. The influence of sex on the energy requirements of cattle for maintenance and growth. Energy Metab. Proc. Symp. 13:101–104.

Garrett, W. N. 1971. Energetic efficiency of beef and dairy steers. J. Anim. Sci. 32:451–456.

Garrett, W. N. 1980. Energy utilization by growing cattle as determined in 72 comparative slaughter experiments. Energy Metab. Proc. Symp. 26:3–7.

Garrett, W. N., and D. E. Johnson. 1983. Nutritional energetics of ruminants. J. Anim. Sci. 57(Suppl. 2):478–497.

Geay, Y., J. Robelin, and M. Vermorel. 1980. Influence of the ME content of the diet on energy utilization for growth in bulls and heifers. Energy Metab. Proc. Symp. 26:9–12.

Graham, N. M. 1968. Effects of undernutrition in late pregnancy on the nitrogen and energy metabolism of ewes. Aust. J. Agric. Res. 19:555–565.

Graham, N. M. 1980. Variation in energy and nitrogen utilization by sheep between weaning and maturity. Aust. J. Agric. Res. 31:335–345.

Graham, N. M., and T. W. Searle. 1972a. Balance of energy and matter in growing sheep at several ages, body weights and planes of nutrition. Aust. J. Agric. Res. 23:97–108.

Graham, N. M., and T. W. Searle. 1972b. Growth in sheep. II. Efficiency of energy and nitrogen utilization from birth to 2 years. J. Agric. Sci. Camb. 79:383–389.

Graham, N. M., and T. W. Searle. 1975. Studies on weaner sheep during and after a period of weight stasis. I. Energy and nitrogen utilization. Aust. J. Agric. Res. 26:343–353.

Graham, N. M., and T. W. Searle. 1979. Studies of weaned lambs before, during and after weight loss. I. Energy and nitrogen utilization. Aust. J. Agric. Res. 30:513–523.

Graham, N. M., T. W. Searle, and D. A. Griffiths. 1974. Basal metabolic rate in lambs and young sheep. Aust. J. Agric. Res. 25:957–971.

Gray, R., and K. J. McCracken. 1980. Plane of nutrition and the maintenance requirement. Energy Metab. Proc. Symp. 26:163–167.

Harris, P. M., P. J. Garlick, and G. E. Lobley. 1989. Interactions between energy and protein metabolism in the whole body and hind limb of sheep in response to intake. Energy Metab. Proc. Symp. 43:167–170.

Hironaka, R., and G. C. Kozub. 1973. Compensatory growth of beef cattle restricted at two energy levels for two periods. Can. J. Anim. Sci. 53:709–715.

Hovell, F. D. DeB., E. R. Orskov, D. J. Kyle, and N. A. MacLeod. 1987. Undernutrition in sheep: Nitrogen repletion by N-depleted sheep. Br. J. Nutr. 57:77–88.

Hulme, D. J., R. C. Kellaway, and P. J. Booth. 1986. The CAMDAIRY model for formulating and analyzing dairy cow rations. Agric. Systems 22:81–108.

Jenkins, T. G., and C. L. Ferrell. 1983. Nutrient requirements to maintain weight of mature, nonlactating, nonpregnant cows of four diverse breed types. J. Anim. Sci. 56:761–770.

Jenkins, T. G., and C. L. Ferrell. 1984a. Output/input differences among biological types. Pp. 15–37 in Proceedings of the Beef Cow Efficiency Symposium. East Lansing: Michigan State University.

Jenkins, T. G., and C. L. Ferrell. 1984b. Characterization of postweaning traits of Simmental and Hereford bulls and heifers. Anim. Prod. 39:355–364.

Jenkins, T. G., C. L Ferrell, and L. V. Cundiff. 1986. Relationship of components of the body among mature cows as related to size, lactation potential and possible effects on productivity. Anim. Prod. 43:245–254.

Jenkins, T. G., J. A. Nienaber, and C. L. Ferrell. 1991. Heat production of mature Hereford and Simmental cows. Energy Metab. Proc. Symp. 58:296–299.

Johnson, D. E., K. A. Johnson, and R. L. Baldwin. 1990. Changes in liver and gastrointestinal tract energy demands in response to physiological workload in ruminants. J. Nutr. 120:649–655.

Kennedy, J. F., and G. I. K. Chirchir. 1971. A study of the growth rates of F_1 and F_2 Africander cross, Brahman cross and British cross cattle from birth to 18 months old in a tropical environment. Aust. J. Exp. Agric. Anim. Husb. 11:593–598.

Kleiber, M. 1961. The Fire of Life. New York: Wiley & Sons.

Klosterman, E. W., L. G. Sanford, and C. F. Parker. 1968. Effect of cow size and condition and ration protein content upon maintenance requirements of mature beef cows. J. Anim. Sci. 27:242–246.

Koong, L. J., and J. A. Nienaber. 1987. Changes in fasting heat production and organ size of pigs during prolonged weight maintenance. Energy Metab. Proc. Symp. 32:46–49.

Koong, L. J., C. L. Ferrell, and J. A. Nienaber. 1982. Effects of plane of nutrition on organ size and fasting heat production in swine and sheep. Energy Metab. Proc. Symp. 29:245–248.

Koong, L. J., C. L. Ferrell, and J. A. Nienaber. 1985. Assessment of interrelationships among level of intake and production, organ size and fasting heat production in growing animals. J. Nutr. 115:1383–1390.

Laurenz, J. C., F. M. Byers, G. T. Schelling, and L. W. Green. 1991. Effects of seasonal environment on the maintenance requirement of mature beef cows. J. Anim. Sci. 69:2168–2176.

Lawrence, T. L. J., and J. Pierce. 1964. Some effects of wintering yearling cattle on different planes of nutrition. II. Slaughter data and carcass evaluation. J. Agric. Sci. Camb. 63:23–34.

Ledger, H. P. 1977. The utilization of dietary energy by steers during periods of restricted feed intake and subsequent realimentation. 2. The comparative energy requirements of penned and exercised steers for long term maintenance at constant liveweight. J. Agric. Sci. Camb. 88:27–33.

Ledger, H. P., and A. R. Sayers. 1977. The utilization of dietary energy by steers during periods of restricted feed intake and subsequent realimentation. 1. The effect of time on the maintenance requirements of steers held at constant liveweight. J. Agric. Sci. Camb. 88:11–26.

Lemenager, R. P., L. A. Nelson, and K. S. Hendrix. 1980. Influence of cow size and breed type on energy requirements. J. Anim. Sci. 51:566–576.

Lobley, G. E., P. M. Harris, P. A. Skene, D. Brown, E. Milne, A. G. Calder, S. E. Anderson, P. J. Garlick, I. Nevison, and A. Connell. 1992. Responses in tissue protein synthesis to sub- and supra-maintenance intake in young, growing sheep: Comparison of large-dose and continuous-infusion techniques. Br. J. Nutr. 68:373–388.

Lofgreen, G. P. 1965. A comparative slaughter technique for determining net energy values with beef cattle. Energy Metab. Proc. Symp. 11:309–317.

Lofgreen, G. P., and W. N. Garrett. 1968. A system for expressing net energy requirements and feed values for growing and finishing cattle. J. Anim. Sci. 27:793–806.

Lopez-Sanbidet, C., and L. S. Verde. 1976. Relationship between liveweight, age and dry matter intake for beef cattle after different levels of feed restriction. Anim. Prod. 22:61–69.

Mader, T. L., O. A. Turgeon, Jr., T. J. Klopfenstein, D. R. Brink, and R. R. Oltjen. 1989. Effects of previous nutrition, feedlot regimen and protein level on feedlot performance of beef cattle. J. Anim. Sci. 67:318–328.

Marston, H. R. 1948. Energy transactions in sheep. I. The basal heat production and heat increment. Aust. J. Sci. Res. B1. 93–129.

McBride, B. W., and J. M. Kelly. 1990. Energy cost of absorption and metabolism in the ruminant gastrointestinal tract and liver: A review. J. Anim. Sci. 68:2997–3010.

Milligan, L. P., and M. Summers. 1986. The biological basis for maintenance and its relevance to assessing responses to nutrients. Proc. Nutr. Soc. 45:185–193.

Ministry of Agriculture, Fisheries and Food. 1975. Energy Allowances and Feeding Systems for Ruminants. Technical Bull. No. 33. London: Her Majesty's Stationery Office.

Ministry of Agriculture, Fisheries and Food. 1976. Energy Allowances and Feeding Systems for Ruminants. Technical Bull. No. 33. London: Her Majesty's Stationery Office.

Ministry of Agriculture, Fisheries and Food. 1984. Energy Allowances and Feeding Systems for Ruminants. ADAS Reference Book 433. London: Her Majesty's Stationery Office.

Moe, P. W., H. F. Tyrrell, and W. P. Flatt. 1970. Partial efficiency of energy use for maintenance, lactation, body gain and gestation in the dairy cow. Energy Metab. Proc. Symp. 13:65–68.

Montano-Bermudez, M., M. K. Nielsen, and G. Deutscher. 1990. Energy requirements for maintenance of crossbred beef cattle with different genetic potential for milk. J. Anim. Sci. 68:2279–2288.

Moran, J. B. 1976. The grazing feed intake of Hereford and Brahman cross cattle in a cool temperate environment. J. Agric. Sci. Camb. 86:131–134.

National Research Council. 1976. Nutrient Requirements of Beef Cattle, Fifth Rev. Ed. Washington, D.C.: National Academy of Sciences.

National Research Council. 1981a. Nutritional Energetics of Domestic Animals and Glossary of Energy Terms. Washington, D.C.: National Academy Press.

National Research Council. 1981b. Effect of Environment on Nutrient Requirements of Domestic Animals. Washington, D.C.: National Academy Press.

National Research Council. 1984. Nutrient Requirements of Beef Cattle, Sixth Rev. Ed. Washington, D.C.: National Academy Press.

National Research Council. 1989. Nutrient Requirements of Dairy Cattle, Sixth Rev. Ed. Update. Washington, D.C.: National Academy Press.

Neville, W. E., Jr. 1974. Comparison of energy requirements of nonlactating and lactating Hereford cows and estimates of energetic efficiency of milk production. J. Anim. Sci. 38:681–686.

Neville, W. E., Jr., and M. E. McCullough. 1969. Calculated net energy requirements of lactating and nonlactating Hereford cows. J. Anim. Sci. 29:823–829.

O'Donovan, P. B. 1984. Compensatory gain in cattle and sheep. Nutr. Abstr. Rev. 54:389–410.

O'Donovan, P. B., A. Gebrewolde, B. Kebede, and E. S. E. Galal. 1978. Fattening studies with crossbred (European × Zebu) bulls. 1. Performance on diets of native hay and concentrate. J. Agric. Camb. 90:425–429.

Old, C. A., and W. N. Garrett. 1987. Effects of energy intake on energetic efficiency and body composition of beef steers differing in size at maturity. J. Anim. Sci. 65:1371–1380.

Patle, B. R., and V. D. Mudgal. 1975. Maintenance requirements for energy in crossbred cattle. Br. J. Nutr. 33:127–139.

Patle, B. R., and V. D. Mudgal. 1977. Utilization of dietary energy requirements for maintenance, milk production and lipogenesis by lactating crossbred cows during their midstage of lactation. Br. J. Nutr. 37:23–33.

Peacock, F. M., M. Koger, W. G. Kirk, E. M. Hodges, and J. R. Crockett. 1976. Beef production of Brahman, Shorthorn, and their crosses on different pasture programs. Tech. Bull. 780. Gainesville: University of Florida Agricultural Experiment Station.

Ritzman, E. G., and F. G. Benedict. 1938. Nutritional Physiology of the Adult Ruminant. Washington, D.C.: Carnegie Institute.

Robelin, J., and Y. Geay. 1976. Changes with age (9, 13, 16, 19 months) of protein and energy retention, and energy utilization by growing Limousin bulls. Energy Metab. Proc. Symp. 19:213–216.

Robinson, J. J., I. McDonald, C. Frazer, and J. G. Gordon. 1980. Studies on reproduction in prolific ewes. 6. The efficiency of energy utilization for conceptus growth. J. Agric. Sci. Camb. 94:331–338.

Rogerson, A., H. P. Ledger, and G. H. Freeman. 1968. Feed intake and liveweight gain comparisons of Bos indicus and Bos taurus steers on a high plane of nutrition. Anim. Prod. 10:373–380.

Rompala, R. E., S. D. M. Jones, J. G. Buchanan-Smith, and H. S. Bayley. 1985. Feedlot performance and composition of gain in late maturing steers exhibiting normal and compensatory growth. J. Anim. Sci. 61:637–646.

Russel, A. J. F., and I. A. Wright. 1983. Factors affecting maintenance requirements of beef cows. Anim. Prod. 37:329–334.

Ryan, W. J., I. H. Williams, and R. J. Moir. 1993a. Compensatory growth in sheep and cattle. I. Growth pattern and feed intake. Aust. J. Agric. Res. 44:1609–1621.

Ryan, W. J., I. H. Williams, and R. J. Moir. 1993b. Compensatory growth in sheep and cattle. II. Changes in body composition and tissue weights. Aust. J. Agric. Res. 44:1623–1633.

Schnyder, W., H. Bickel, and A. Schurch. 1982. Energy metabolism during retarded and compensatory growth of Braunvieh steers. Energy Metab. Proc. Symp. 29:96–99.

Searle, T. W., and N. M. Graham. 1975. Studies of weaner sheep during and after a period of weight stasis. II. Body composition. Aust. J. Agric. Res. 26:355–361.

Smith, G. M., J. D. Crouse, R. W. Mandigo, and K. L. Neer. 1977. Influence of feeding regime and biological type on growth, composition and palatability of steers. J. Anim. Sci. 45:236–253.

Solis, J. C., F. M. Byers, G. T. Schelling, C. R. Long, and L. W. Green. 1988. Maintenance requirements and energetic efficiency of cows of different breeds. J. Anim. Sci. 66:764–773.

Stetter, R., A. Susenbeth, and K. H. Menke. 1989. Energy metabolism and protein retention of Simmental bulls fed four different feeding levels from maintenance to ad libitum at 250, 450 and 550 kg body mass. Energy Metab. Proc. Symp. 43:21–24.

Summers, M., B. W. McBride, and L. P. Milligan. 1988. Components of basal energy expenditure. Pp. 257–286 in Aspects of Digestive Physiology in Ruminants, A. Dobson and M. J. Dobson, eds. Ithaca, N.Y.: Cornell University Press.

Taylor, C. S., and G. B. Young. 1968. Equilibrium weight in relation to feed intake and genotype in twin cattle. Anim. Prod. 10:393–412.

Taylor, C. S., H. G. Turner, and G. B. Young. 1981. Genetic control of equilibrium maintenance efficiency in cattle. Anim. Prod. 33:179–194.

Taylor, C. S., R. B. Theissen, and J. Murray. 1986. Inter-breed relationship of maintenance efficiency to milk yield in cattle. Anim. Prod. 43:37–61.

Thompson, W. R., D. H. Theuninck, J. C. Meiske, R. D. Goodrich, J. R. Rust, and F. M. Byers. 1983. Linear measurements and visual appraisal as estimators of percentage empty body fat of beef cows. J. Anim. Sci. 56:755–760.

Thomson, E. F., M. Gingins, J. W. Blum, H. Bickel, and A. Schurch. 1980. Energy metabolism of sheep during nutritional limitation and realimentation. Energy Metab. Proc. Symp. 26:427–430.

Thomson, E. F., M. Gingins, J. W. Blum, H. Bickel, and A. Schurch. 1980. Energy metabolism of sheep during nutritional limitation and realimentation. Energy Metab. Proc. Symp. 26:427–430.

Thorbek, G., and S. Henckel. 1976. Studies on energy requirements for maintenance in farm animals. Energy Metab. Proc. Symp. 19:117–120.

Truscott, T. G., J. D. Wood, N. G. Gregory, and I. C. Hart. 1983. Fat deposition in Hereford and Friesian steers. 3. Growth efficiency and fat mobilization. J. Agric. Sci. Camb. 100:277–284.

Tudor, G. D., D. W. Utling, and P. K. O'Rourke. 1980. The effect of pre- and post-natal nutrition on the growth of beef cattle. III. The effect of severe restriction in early postnatal life on the development of body components and chemical composition. Aust. J. Agric. Res. 31:191–204.

Tyrrell, H. F., and C. K. Reynolds. 1988. Effect of stage of growth on utilization of energy by beef heifers. Energy Metab. Proc. Symp. 43:17–20.

van der Merwe, F. J., and P. van Rooyen. 1980. Estimates of ME needs for maintenance and gain in beef steers of four genotypes. Energy Metab. Proc. Symp. 26:135–136.

Vercoe, J. E. 1970. Fasting metabolism and heat increment of feeding in Brahman × British and British cross cattle. Energy Metab. Proc. Symp. 13:85–88.

Vercoe, J. E., and J. E. Frisch. 1974. Fasting metabolism, liveweight and voluntary intake of different breeds of cattle. Energy Metab. Proc. Symp. 14:131–134.

Vermorel, M., J. C. Bouvier, and Y. Geay. 1976. The effect of genotype (normal and double muscled Charolais and Friesian) on energy utilization at 2 and 16 months of age. Energy Metab. Proc. Symp. 19:217–220.

Vermorel, M., J. C. Bouvier, and Y. Geay. 1980. Energy utilization by growing calves: Effects of age, milk intake and feed level. Energy Metab. Proc. Symp. 26:9–53.

Vermorel, M., Y. Geay, and J. Robelin. 1982. Energy utilization by growing bulls, variations with genotype, liveweight, feeding level and between animals. Energy Metab. Proc. Symp. 29:88–91.

Walker, J. J., and W. N. Garrett. 1970. Shifts in energy metabolism of male rats during adaptation to prolonged undernutrition and during their subsequent realimentation. Energy Metab. Proc. Symp. 13:193–196.

Walker, V. A., B. A. Young, and B. Walker. 1991. Does seasonal photoperiod directly influence energy metabolism. Energy Metab. Proc. Symp. 58:372–375.

Webster, A. J. F., J. S. Smith, and G. Mollison. 1976. On the prediction of heat production in growing cattle. Energy Metab. Proc. Symp. 29:221–224.

Webster, A. J. F., J. S. Smith, and G. S. Mollison. 1977. Prediction of the energy requirements for growth in beef cattle. 3. Body weight and heat production in Hereford × British Friesian bulls and steers. Anim. Prod. 24:237–244.

Webster, A. J. F., J. S. Smith, and G. S. Mollison. 1982. Energy requirements of growing cattle: Effects of sire breed, plane of nutrition, sex and season on predicted basal metabolism. Energy Metab. Proc. Symp. 29:84–87.

Wilson, P. N., and D. F. Osbourn. 1960. Compensatory growth after undernutrition in mammals and birds. Biol. Rev. 35:324–363.

Wurgler, F., and H. Bickel. 1985. The partial efficiency of energy utilization in steers of different breeds. Energy Metab. Proc. Symp. 32:90–93.

Young, B. A. 1975a. Effects of winter acclimatization on resting metabolism of beef cows. Can. J. Anim. Sci. 55:619–625.

Young, B. A. 1975b. Temperature-induced changes in metabolism and body weight of cattle (Bos taurus). Can. J. Physiol. Pharmacol. 53:947–953.

Young, B. A., A. W. Bell, and R. T. Hardin. 1989. Mass specific metabolic rate of sheep from fetal life to maturity. Energy Metab. Proc. Symp. 43:155–158.

2 Protein

The previous edition of *Nutrient Requirements of Beef Cattle* (National Research Council, 1984) expressed protein requirements in terms of crude protein (CP). In 1985, the Subcommittee on Nitrogen Usage in Ruminants (National Research Council, 1985) presented an excellent rationale for expressing protein requirements in terms of absorbed protein, a rationale adopted in 1989 by the Subcommittee on Dairy Cattle Nutrition (National Research Council, 1989). Since then absorbed protein (AP) has become synonymous with metabolizable protein (MP), a system that accounts for rumen degradation of protein and separates requirements into the needs of microorganisms and the needs of the animal. MP is defined as the true protein absorbed by the intestine, supplied by microbial protein and undegraded intake protein (UIP).

There are basically two reasons for using the MP system rather than the CP system. The first is that there is more useable information about the two components of the MP system—bacterial (microbial) crude protein (BCP) synthesis and UIP, which allows more accurate prediction of BCP and UIP than was possible in 1984. The second reason is that the CP system is based on an invalid assumption—that all feedstuffs have an equal extent of protein degradation in the rumen, with CP being converted to MP with equal efficiency in all diets. The change from the CP system to the MP system was adopted in the *Nutrient Requirements of Dairy Cattle* (National Research Council, 1989) and by the Agricultural and Food Research Council (1992). Crude protein can be calculated from the sum of UIP and degraded intake protein (DIP), both of which are determined in both levels of the model. The table generator presents MP requirements in amounts required per day and checks diet adequacy when crude and degradable protein levels are entered. In addition to this, estimates of daily crude protein requirements can be obtained by dividing MP amounts by a value between 0.64 and 0.80, depending on degradability of protein in the feed. The coefficients of 0.64 and 0.80 apply when all of the protein is degradable and undegradable, respectively.

Protein requirements are best determined using model levels 1 or 2. Model level 1 uses UIP and DIP values of feeds from the feed library. Level 2 is mechanistic and uses rates of protein degradation of various protein fractions to estimate DIP and UIP. BCP synthesis is estimated from rates of digestion of various carbohydrate fractions. In both cases, rates of passage are also used. Level 2 also includes supply and requirements for amino acids.

MICROBIAL PROTEIN SYNTHESIS

Bacterial crude protein (BCP) can supply from 50 percent (National Research Council, 1985; Spicer et al., 1986) to essentially all the MP required by beef cattle, depending on the UIP content of the diet. Clearly, efficiency of synthesis of BCP is critical to meeting the protein requirements of beef cattle economically; therefore, prediction of BCP synthesis is an important component of the MP system. Burroughs et al. (1974) proposed that BCP synthesis averaged 13.05 percent of total digestible nutrients (TDN). In *Ruminant Nitrogen Usage* (National Research Council, 1985), two equations were developed to predict BCP synthesis—one for diets containing more than 40 percent forage and one for diets containing less than 40 percent forage. Both equations are more complex than that of Burroughs et al. (1974). Both forage and concentrate intakes (percent of body weight) are needed to calculate the less than 40 percent forage equation

$$\text{BCP (g/day)} = 6.25 \text{ TDN (kg intake/day)} \qquad \text{Eq. 2-1}$$
$$(8.63 + 14.6 * \text{forage intake} -$$
$$5.18 \text{ forage intake}^2 + 0.59 \text{ concentrate intake}).$$

The more than 40 percent forage equation was developed primarily for dairy cattle:

BCP (g/day) = 6.25 (−31.86 + 26.12 TDN (kg intake/day)).　　　Eq. 2-2

Its negative intercept is not biologically logical. The main fallacy is that it assumes a constant efficiency at all TDN concentrations. This is misleading because it suggests that both intake of TDN and concentration of TDN yield change in a similar direction. The equation underpredicts BCP production with low-TDN intakes commonly fed to beef cows and stocker calves. TDN intakes can be low either because body weight is low (young cattle) or because TDN concentration in the diet is low. Low-TDN diets might reduce passage rate and microbial efficiency; conversely, a lower intake of a higher TDN diet might give maximum microbial efficiency. The average BCP value for the data set from which Eq. 2-2 (>40 percent forage) was developed (National Research Council, 1985) is BCP = 12.8 of TDN intake. This should not be interpreted, however, as a constant.

The value 13 g BCP/100 g TDN for BCP synthesis is a good generalization but it does not fit all situations. At both high- and low-ration digestibilities, efficiency may be lower but for different reasons. Logically, the higher digestibility diets are based primarily on grain. High-grain finishing diets have lower rumen pH values and slower microbial turnover, which leads to lower efficiency for converting fermented protein and energy to BCP.

Eq. 2-1 (<40 percent forage; National Research Council, 1985) predicts about 8 percent BCP as a percentage of TDN on a 10 percent roughage diet. Spicer et al. (1986) found a somewhat higher value (10.8 percent of digestible organic matter). These researchers used the lysine to leucine ratio as the bacterial marker; purines were used as the marker by the Subcommittee on Nitrogen Usage in Ruminants (National Research Council, 1985). Russell et al. (1992) proposed that microbial yield is reduced 2.2 percent for every 1 percent decrease in forage effective neutral detergent fiber (eNDF) below 20 percent NDF. This gives values similar to those proposed in *Ruminant Nitrogen Usage* (National Research Council, 1985).

The synthesis of BCP is also likely to be lower on low-quality forage diets. With slow rates of passage, more digested energy is used for microbial maintenance—including cell lysis (Russell and Wallace, 1988; Russell et al., 1992). Therefore, the efficiency of synthesis of BCP from digestible energy is reduced. To summarize previous reports (Stokes et al., 1988; Krysl et al., 1989; Hannah et al., 1991; Lintzenick et al., 1993; Villalobos, 1993), BCP averaged 7.82 percent of total tract digestible organic matter; the range was 5 to 11.4 percent. The range of total tract organic matter digestibilities was 49.8 to 64.7 percent, and BCP synthesis efficiency was not related to digestibility differences. Intake levels may have been sufficiently low to influence rate of passage and microbial efficiency. The

difficulty in obtaining absolute results (Agricultural and Food and Research Council, 1992) makes it difficult to estimate BCP synthesis efficiency in low-quality diets. Most of the beef cows in the world are fed such diets during mid-gestation, so it is important to have more accurate estimates. Russell et al. (1992) predicted an efficiency of 11 percent of TDN for diets containing 50 percent TDN.

A review of the international literature (Agricultural and Food and Research Council, 1992) reveals that BCP synthesis was 12.6 to 17 g/100 g TDN. Some of the differences are compensated for by predicted differences in bacterial true protein (BTP) content and in intestinal digestibility of BTP. Because developers of many of the systems have based their systems on the summarized literature, many of the systems have a similar data base; consequently, values do not vary much from Burroughs et al. (1974) value of 13.05 percent of TDN. Therefore to simplify the NRC (1985) system, 13 percent of TDN was used here for diets containing more than 40 percent forage. For diets containing less than 40 percent forage, the equation of Russell et al. (1992) is used—2.2 percent reduction in BCP synthesis for every 1 percent decrease in forage eNDF less than 20 percent NDF. This provides consistency between model levels 1 and 2.

Currently there are no generalized empirical equations to predict BCP synthesis efficiency at low passage rates. Level 1 of the model with this publication assumes 0.13 efficiency on all forage diets; however, the user is able to reduce that efficiency value in the model. The data reviewed suggests that this value is as low as 0.08 with intakes of low TDN (50 to 60 percent) diets at 1.9 to 2.1 percent of BW. Low values may also be expected with low (limited) intakes of higher energy diets. Level 2 of the model estimates lower synthesis of BCP because of the low predicted rates of passage.

The consequence of using 0.13 BCP synthesis efficiency in level 1 and in the tables is that the BCP supply may be overestimated. Subsequently, DIP requirement would be overestimated and the UIP requirement would be underestimated. This would have little impact on the CP requirement.

Many factors affect efficiency of BCP synthesis (National Research Council, 1985; Russell et al., 1992). Compared to ammonia, ruminal amino acids and peptides may increase the rate and amount of BCP synthesized. In most cases, natural diets contain sufficient DIP to meet microbial needs for amino acids, peptides, or branched-chain amino acids. Deficiencies have not been reported in practical feeding situations. Type of carbohydrate (structural vs nonstructural) may also affect microbial maintenance requirements because of differences in rates of fermentation (microbial growth rate) and rates of passage and because of effects on rumen pH. Level of intake as it changes rate of passage and pH is important. Lipids provide little if any energy for ruminal microorganisms, and the energy obtained from protein fermentation is minimal (Nocek and Russell, 1988). Further, ensiled (fermented) forages may provide less energy for microorganisms than comparable fresh or dry feeds (Agricultural and Food Research Council, 1992), but this has not been documented for silages and high-moisture grains in the United States.

Carbohydrate digestion in the rumen is likely the most accurate predictor of BCP synthesis, and this mechanism is used in model level 2. However, for feedstuffs used for beef cattle few good data are available for rates of digestion and of passage of the different carbohydrates potentially digested in the rumen. More accurate values are available for TDN, and laboratory predictors of TDN can be used to estimate BCP synthesis. Therefore TDN is used as the indicator of energy availability in the rumen for level 1. The Agricultural and Food Research Council (1992) found that total tract digestible organic matter intake was the most precise indicator of BCP synthesis when nitrogen intake was adequate. Digestible organic matter and TDN are roughly equivalent in feedstuffs and diets.

The requirement for rumen degradable protein (including nonprotein nitrogen [NPN]) is considered equal to BCP synthesis. This assumes that the loss of ammonia from the rumen as a result of flushing to the duodenum and absorption through the rumen wall is equal to the amount of recycled nitrogen. A number of factors affect each of these fluxes of nitrogen (National Research Council, 1985) but rather complex modeling is needed (Russell et al., 1992) to account for them. Simply put, a deficiency of ruminal ammonia encourages recycling and an excess encourages absorption from the rumen. Therefore, a balance (rumen degradable protein in diet equal to BCP synthesis) minimizes both recycling and absorption. Few studies have attempted to titrate the need for rumen degradable protein. Karges (1990) found 10.9 percent of TDN as rumen degradable protein was needed to maximize gain in beef cows, presumably to maximize BCP synthesis; Hollingsworth-Jenkins (1994) found only 7.1 percent DIP was needed to maximize gain. These values are smaller than the value of 13 percent used in this publication to calculate BCP synthesis.

Optimum use of rumen degraded protein (including nonprotein nitrogen) would logically occur if protein and carbohydrate degradation in the rumen were occurring simultaneously. This is not the case in many diets. Protein degradation of many of the forages, for example, is rapid and degradation of energy-yielding components of NDF is much slower. With grains (for example, corn and sorghum) the opposite is true—slow protein degradation and rapid starch degradation. This results in low ruminal ammonia levels from high-grain diets postfeeding and high levels from forage diets, which is influenced by CP levels.

The ruminant compensates by recycling nitrogen. An excellent example of this is how cow performance is similar with protein supplementation either three times per week or once per day (Beaty et al., 1994). More basic studies with animals (Henning et al., 1993; Rihani et al., 1993) suggest little or no advantage to synchrony of energy availability and protein breakdown. Cattle also compensate by eating numerous meals per day such as in the feedlot.

Use of NPN is appropriate in high-grain diets (National Research Council, 1984, 1985; Sindt et al., 1993) because of the rapid rumen degradation of starch. The value of NPN in low-protein, high-forage diets is less clear (Rush and Totusek, 1975; Clanton, 1979). Reduced gains when using urea as opposed to a "natural" protein may be the result of insufficient UIP rather than the faster rate of ammonia release in the rumen. Until more information is available, it is advisable to use caution when using urea in low-protein, high-forage diets.

Russell et al. (1992) have demonstrated the need for amino acids and peptides for optimum BCP synthesis, and this concept is used in model level 2. A lack of amino acids or peptides is unlikely to be a problem in typical diets for beef cattle. Adequate MP in finishing diets can be accomplished by adding urea (Sindt et al., 1993). Fiber-digesting bacteria use primarily ammonia for BCP synthesis (Russell et al., 1992), so amino acids/peptides should not be limiting in the rumen. However, these fiber-digesting bacteria may require branched-chain volatile fatty acids (National Research Council, 1985), which would be supplied by amino acid degradation. A need for rumen degradable protein (other than NPN) might occur in diets containing mixtures of forage and grain such as "step-up" rations for finishing cattle (Sindt et al., 1993).

Digestibility of protein is important—for both BCP and UIP. In this publication, the value of 80 percent digestibility of BTP (National Research Council, 1985) is used. UIP digestibility may vary with the source; however, it is assumed that UIP is 80 percent digestible. National Research Council (1985) used 0.8 BCP = BTP because BCP contains approximately 20 percent nucleic acids. This value has been challenged by other MP systems (Agricultural and Food Research Council, 1992). Logically, the important measure is amino acid content (true protein) of BCP. These measures (Agricultural and Food Research Council, 1992) suggest a value of 0.75 rather than 0.8. However, the net absorption of amino acids is the important coefficient. Systems using lower BCP to BTP values used higher (0.85) digestibility values for BTP; therefore, these values compensate. Until more definitive data are available in the United States on digestible amino acid content of rumen bacteria, use of the value of 0.64, calculated as 0.8 BCP = BTP * 0.8 digestibility of BTP is suggested.

MP REQUIREMENTS

NRC requirements (1984, 1985) for MP were based on the factorial method. Factors included were metabolic fecal losses, urinary losses, scurf losses, growth, fetal growth, and milk. Metabolic fecal, urinary, and scurf losses represent the requirement needed for maintenance. It is difficult, however, to measure fecal and urinary losses independent of each other. It also is difficult to separate microbial (complete cells or cell walls) losses in the feces from true metabolic fecal losses. In the preceding edition of this report (National Research Council, 1984) metabolic fecal loss was calculated as a percentage of dry matter intake; in *Ruminant Nitrogen Usage* (National Research Council, 1985) metabolic fecal loss was calculated as a percentage of indigestible dry matter intake. Diet digestibility obviously affects the resulting calculated metabolic fecal losses. Most beef cows are fed diets containing 45 to 55 percent TDN during gestation. Consequently, for most beef cows, MP and CP requirements, using the calculation based on indigestible dry matter intake (National Research Council, 1985), are unrealistically high. The high requirement can be attributed to the fact that nitrogen is being excreted in the feces as microbial protein rather than as urea in the urine (National Research Council, 1985) as a result of microbial growth in the postruminal digestive tract.

The Institute National de la Recherche Agronomique (INRA) (1988), using nitrogen balance studies that included scurf, urinary, and metabolic fecal losses, determined that the maintenance requirement was 3.25 g MP/kg SBW$^{0.75}$. This system simplifies calculations and is based on metabolic body weight (BW$^{0.75}$), as are maintenance energy requirements, and is similar to the concept and value proposed by Smuts (1935). Assuming CP * 0.64 (CP converted to BCP: 80 percent true protein * 80 percent digestibility) = MP, Smuts (1935) calculated the requirement to be 3.52 g MP/kg BW$^{0.75}$. Wilkerson et al. (1993) estimated the maintenance requirement of 253 kg growing calves was 3.8 g MP/kg BW$^{0.75}$ using growth as the criteria. Their diets were high in roughage and were based on the assumption that 0.13 TDN = BCP. If actual BCP synthesis efficiency was less than 0.13, the estimate of the maintenance would be less than 3.8 g MP/kg BW$^{0.75}$. In this publication 3.8 g MP/kg BW$^{0.75}$ is used because the maintenance requirement estimated was based on animal growth rather than on nitrogen balance. However, recent nitrogen balance data reported by Susmel et al. (1993) do support the 3.8 g MP/kg BW$^{0.75}$ value.

CONVERSION OF MP TO NP

Studies by Armstrong and Hutton (1975) and Zinn and Owens (1983) reported that the average biological value of absorbed amino acids was reported to be 66 percent (National Research Council, 1984). A constant conversion of MP to net protein (NP) for gain of 0.5 and to NP for milk of 0.65 was assumed (National Research Council, 1985). These efficiency values are based on two components—the biological value of the protein and the efficiency of use of an "ideal mixture of amino acids" (Oldham, 1987). Oldham (1987) suggests that the efficiency value is 0.85 for all physiological functions. Biological values will vary with the source(s) of UIP in the diet. Biological value is defined herein and by Oldham (1987) as the relative amino acid balance. The biological value of microbial protein is quite high and strongly influences the biological value of the MP in many diets. Biological value will vary for different functions (Oldham, 1987)—for example, it is likely that the overall efficiency value for pregnancy and lactation are higher than for gain. Based on data for lactation and pregnancy (National Research Council, 1985), this subcommittee has chosen to use 0.65 (0.85 * 0.76; efficiency * biological value).

Efficiency of use for gain is not likely to be constant across body weights (maturity) and rates of gain. The INRA (1988) system assumes a decreasing efficiency as body weight increases. This was confirmed by Ainslie et al. (1993) and Wilkerson et al. (1993). Based on these data, the following equation is used:

If EQEBW ≤ 300 kg,
    percent efficiency of MP to NP =
83.4 − (0.114 * EQEBW), otherwise 49.2,

where EQSBW is equivalent shrunk body weight in kilograms.

This is the overall efficiency value (biological value * efficiency of use of ideal protein). This equation was developed by Ainslie et al. (1993) from data presented by INRA (Institut National de la Recherche Agronomique, 1988). The equation predicts a conversion efficiency of MP to NP of 66.3 percent for a 150-kg calf. A 300-kg steer has an efficiency of only 49.2 percent. The data of Ainslie et al. (1993) and Wilkerson et al. (1993) only cover the weight range from 150 to 300 kg. Therefore, these bounds have been placed on the conversion efficiency equation. Thus, for cattle weighing more than 300 kg, this maintains similar protein requirements to previous NRC publications (National Research Council, 1984, 1985) and recognizes the low CP requirements of cattle weighing more than 400 kg (Preston, 1982).

Validation

Few studies have been conducted that were designed either to validate protein requirement systems or to meet the requirements for validation. Most difficult to interpret are data where energy intake increases with protein supplementation because one does not know whether the increased gain is the result of increased MP or NE_g. Also, it is often difficult to determine whether the effect was caused by DIP or UIP. Karges (1990) maintained equal intakes in gestating cows and supplemented low-quality prairie hay with rumen degraded protein. He obtained a requirement of 608 g CP/day. Hollingsworth-Jenkins (1994) estimated a requirement of 605 g CP/day for gestating cows grazing winter range. The system proposed herein estimates the requirement to be 684 g CP/day. Based on predicted intake, the requirement is 725 g CP/day (National Research Council, 1984); at actual intake, the requirement was calculated to be 658 g CP/day. The calculation of 828 g CP/day (National Research Council, 1985) seems unreasonably high as a result of the high metabolic fecal protein value based on indigestible dry matter intake.

Validation data sets were developed for growing-finishing cattle (Wilkerson et al., 1993; Ainslie et al., 1993). Rates of gain varied from 0 to 1.5 kg/day. Diets ranged from 90 percent low-quality roughage to 90 percent concentrate. Generally, the cattle used were young because a deficiency in MP is difficult to demonstrate at heavier weights (Zinn, 1988; Ainslie et al, 1993; Sindt et al., 1993; Zinn and Owens, 1993). The data sets included 70 observations.

Prediction model level 1 had an r^2 of 0.80 and a bias of +20 percent, and level 2 had an r^2 of 0.67 and +18 percent bias. By comparison, gain predicted by ME intake had, in level 1, an r^2 of 0.90 and a +19 percent bias; level 2 had an r^2 of 0.95 and a bias of +13 percent. Gain limited by the first-limiting amino acid in level 2 had an r^2 of 0.74 and a +16 percent bias. Gain limited by the first-limiting nutrient (ME, MP, first-limiting amino acid) gave an r^2 of 0.81 and a bias of +12 in level 1 and an r^2 of 0.92 with 0 bias in level 2.

Validation is more difficult with cattle on high-grain finishing diets. Corn is the most common feed grain in the United States. It contains 8 to 10 percent protein, but approximately 60 percent of the protein escapes ruminal digestion. In diets that are 85 percent corn, this results in 4.0 to 5.3 percent of the diet being UIP. Shain et al. (1994) and Sindt et al. (1994) found that 4.6 percent UIP in addition to the BCP was sufficient to meet the needs of yearling cattle. In addition Shain et al. (1994), Milton and Brandt (1994) estimated the requirement for DIP for yearling cattle by feeding graded amounts of urea. Both found a response to urea that is consistent with the DIP requirement calculated herein (6.8 percent of dry matter). In the work of Shain et al. (1994), the UIP supplied was higher than the requirement (5.3 vs 3.6), and the CP required was 12 percent of dry matter because UIP was overfed. Presumably, the DIP requirement is needed to maximize microbial activity in the rumen because MP was in excess.

REFERENCES

Agricultural and Food Research Council. 1992. Nutritive requirements of ruminant animals: Protein. Nutr. Abstr. Rev. Ser. B 62:787–835.

Ainslie, S. J., D. G. Fox, T. C. Perry, D. J. Ketchen, and M. C. Barry. 1993. Predicting amino acid adequacy of diets fed to Holstein steers. J. Anim. Sci. 71:1312–1319.

Armstrong, D. G., and K. Hutton. 1975. Fate of nitrogenous compounds entering the small intestine. P. 432 in Digestion and Metabolism in the Ruminant, I. W. McDonald and A. C. I. Warner, eds. Armidale, NSW, Australia: The University of New England Publishing Unit.

Beaty, J. L., R. C. Cochran, B. A. Lintzenick, E. S. Vanzant, J. L. Morrill, R. T. Brandt, Jr., and D. E. Johnson. 1994. Effect of frequency of supplementation and protein concentration in supplements on performance and digestion characteristics of beef cows consuming low quality forages. J. Anim. Sci. 72:2475–2486.

Burroughs, W., A. H. Trenkle, and R. L. Vetter. 1974. A system of protein evaluation for cattle and sheep involving metabolizable protein (amino acids) and urea fermentation potential of feedstuffs. Vet. Med. Small Anim. Clin. 69:713–722.

Clanton, D. C. 1979. Nonprotein nitrogen in range supplements. J. Anim. Sci. 47:765–779.

Hannah, S. M., R. C. Cochran, E. S. Vanzant, and D. L. Harmon. 1991. Influence of protein supplementation on site and extent of digestion, forage intake, and nutrient flow characteristics in steers consuming dormant bluestem-range forage. J. Anim. Sci. 69:2624–2633.

Henning, P. H., D. G. Steyn, and H. H. Meissner. 1993. Effect of synchronization of energy and nitrogen supply on ruminal characteristics and microbial growth. J. Anim. Sci. 71:2516–2528.

Hollingsworth-Jenkins, K. J. 1994. Escape Protein, Rumen Degradable Protein, or Energy as the First Limiting Nutrient of Nursing Calves Grazing Native Sandhills Range. Ph.D. dissertation. University of Nebraska, Lincoln, Nebraska.

Institut National de la Recherche Agronomique. 1988. Alimentation des Bovins, Ovins, et Caprins. R. Jarrige, ed. Paris: Institut National de la Recherche Agronomique.

Karges, K. K. 1990. Effects of Rumen Degradable and Escape Protein on Cattle Response to Supplemental Protein on Native Pasture. M.S. thesis. University of Nebraska, Lincoln, Nebraska.

Krysl, J. J., M. E. Branine, A. U. Cheema, M. A. Funk, and M. L. Galyean. 1989. Influence of soybean meal and sorghum grain supplementation on intake, digesta kinetics, ruminal fermentation, site and extent of digestion and microbial protein synthesis in beef steers grazing blue grama rangeland. J. Anim. Sci. 67:3040–3051.

Lintzenick, B. A., R. C. Cochran, E. S. Vanzant, J. L. Beaty, R. T. Brandt, Jr., G. St. Jean, and T. G. Nagaraja. 1993. Influence of method of processing supplemental alfalfa on intake and utilization of dormant, bluestem-range forage by beef steers. J. Anim. Sci. 71(Suppl. 1):186.

Milton, C. T., and R. T. Brandt, Jr. 1994. Level of urea in high grain diets: Finishing steer performance. J. Anim. Sci. 72(Suppl. 1):231 (abstr.).

National Research Council. 1984. Nutrient Requirements of Beef Cattle, Sixth Revised Ed. Washington, D.C.: National Academy Press.

National Research Council. 1985. Ruminant Nitrogen Usage. Washington, D.C.: National Academy Press.

National Research Council. 1989. Nutrient Requirements of Dairy Cattle, Sixth Rev. Ed. Washington, D.C.: National Academy Press.

Nocek, J., and J. B. Russell. 1988. Protein and carbohydrate as an integrated system. Relationship of ruminal availability to microbial contribution and milk production. J. Dairy Sci. 71:2070–2107.

Oldham, J. D. 1987. Efficiencies of amino acid utilization. Pp. 171–186 in Feed Evaluation and Protein Requirement Systems for Ruminants, R. Jarrige and G. Alderman, eds. Luxembourg: CC.

Preston, R. L. 1982. Empirical value of crude protein systems for feedlot cattle. Pp. 201–217 in Protein Requirements of Cattle: Proceedings of an International Symposium, F. N. Owens, ed. MP-109. Stillwater, Okla.: Oklahoma State University, Division of Agriculture.

Rihani, N., W. N. Garrett, and R. A. Zinn. 1993. Influence of level of urea, and method of supplementation on characteristics of digestion of high-fiber diets by sheep. J. Anim. Sci. 71:1657–1665.

Rush, I. G., and R. Totusek. 1975. Effects of frequency of ingestion of high-urea winter supplements by range cattle. J. Anim. Sci. 41:1141–1146.

Russell, J. B., and R. J. Wallace. 1988. Energy yielding and consuming reactions. Pp. 185–216 in The Rumen Microbial Ecosystem, P. N. Hobson, ed. London: Elsevier Applied Science.

Russell, J. B., J. D. O'Connor, D. G. Fox, P. J. Van Soest, and C. J. Sniffen. 1992. A net carbohydrate and protein system for evaluating cattle diets: I. Ruminal fermentation. J. Anim. Sci. 70:3551–3561.

Shain, D. H., R. A. Stock, T. J. Klopfenstein, and R. P. Huffman. 1994. Level of rumen degradable nitrogen in finishing cattle diets. J. Anim. Sci. 72(Suppl. 1):923 (abstr.).

Sindt, M. H., R. A. Stock, T. J. Klopfenstein, and D. H. Shain. 1993. Effect of protein source and grain type on finishing calf performance and ruminal metabolism. J. Anim. Sci. 71:1047–1056.

Sindt, M. H., R. A. Stock, and T. J. Klopfenstein. 1994. Urea versus urea and escape protein for finishing calves and yearlings. Anim. Feed Sci. Tech. 49:103–117.

Smuts, D. 1935. The relation between the basal metabolism and the endogenous nitrogen metabolism, with particular reference to the maintenance requirement of protein. J. Nutr. 9:403–433.

Spicer, L. A., C. B. Theurer, J. Sorne, and T. H. Noon. 1986. Ruminal and post-ruminal utilization of nitrogen and starch from sorghum grain-, corn-, and barley-based diets by beef steers. J. Anim. Sci. 62:521–530.

Stokes, S. R., A. L. Goetsch, A. L. Jones, and K. M. Landis. 1988. Feed intake and digestion by beef cows fed prairie hay with different levels of soybean meal and receiving postruminal administration of antibiotics. J. Anim. Sci. 66:1778–1789.

Susmel, P., M. Spanghero, B. Stefano, C. R. Mills, and E. Plazzotta. 1993. Digestibility and allantoin excretion in cows fed diets differing in nitrogen content. Livest. Prod. Sci. 36:213–222.

Villalobos, G. 1993. Integration of Complementary Forages with Rangeland for Efficient Beef Production in the Sandhills of Nebraska. Ph.D. dissertation. University of Nebraska, Lincoln, Nebraska.

Wilkerson, V. A., T. J. Klopfenstein, R. A. Britton, R. A. Stock, and P. S. Miller. 1993. Metabolizable protein and amino acid requirements of growing beef cattle. J. Anim. Sci. 71:2777–2784.

Zinn, R. A. 1988. Crude protein and amino acid requirements of growing-finishing Holstein steers gaining 1.43 kg per day. J. Anim. Sci. 66:1755–1763.

Zinn, R. A., and F. N. Owens. 1983. Influence of feed intake level on site of digestion in steers fed a high concentrate diet. J. Anim. Sci. 56:471–475.

Zinn, R. A., and F. N. Owens. 1993. Ruminal escape protein for lightweight feedlot calves. J. Anim. Sci. 71:1677–1687.

3 Growth and Body Reserves

ENERGY AND PROTEIN REQUIREMENTS FOR GROWING CATTLE

Net energy for gain (NE_g) is defined herein as the energy content of the tissue deposited, which is a function of the proportion of fat and protein in the empty body tissue gain (Garrett et al., 1959; fat contains 9.367 kcal/g and nonfat organic matter contains an average of 5.686 kcal/g). Simpfendorfer (1974) summarized data from steers of British beef breeds from birth to maturity and found that within cattle of a similar mature size, 95.6 to 98.9 percent of the variation in the chemical components and empty body energy content was associated with the variation in weight (Figure 3-1 A and B). When energy does not limit growth, the empty body contains an increasingly smaller percentage of protein and an increasingly larger percentage of fat, and reaches chemical maturity when additional weight contains little additional protein. Figure 3-1A shows that steers in this data base contained little additional protein in the gain after an SBW of 750 kg. At SBW in excess of 200 to 300 kg, there appeared to be an influence of the effect of plane of nutrition, as evidenced by the scatter of points on the plot of body fat content (Figure 3-1A).

The energy content of weight gain across a wide range of ME intakes and rates of gain was described in equation form by Garrett (1980), equations that were adapted by the Subcommittee on Beef Nutrition for use in the preceding edition of this volume (National Research Council, 1984). This data set included 72 comparative slaughter experiments conducted at the University of California between 1960 and 1980 of approximately 3,500 cattle receiving various diets. The equation developed with British-breed steers describes the relationship between retained energy (RE) and empty body weight gain (EBG) for a given empty body weight (EBW);

$$RE = 0.0635 * EBW^{0.75} * EBG^{1.097}. \qquad \text{Eq. 3-1}$$

Because energy is retained as either protein or fat, the composition of the gain at different weights can be estimated from RE computed in Eq. 3-1 (Garrett, 1987);

$$\text{proportion of fat} = 0.122 * RE - 0.146, \text{ and} \qquad \text{Eq. 3-2}$$

$$\text{proportion of protein} = 0.248 - 0.0264 * RE. \qquad \text{Eq. 3-3}$$

Using these relationships, the relationship between stage of growth (percentage of mature weight), rate of gain, and composition of gain can be computed (Table 3-1). The resulting NE_g requirement in Table 3-1 for various shrunk body weights (SBW) and shrunk daily gains (SWG) are those presented in Table 1 of the 1984 edition of this volume for a medium-frame steer, except the last line shows requirements for 1.3 kg SWG rather than 1.2 kg SWG. These ranges in SWG represent those in that data base. Several relationships are shown in this table. First, energy content of the gain at a particular SWG increases with weight in a particular body size. Second, protein and fat content of the gain and expected body fat at a particular weight depend on rate of gain. Eqs. 3-1, 3-2, and 3-3 were used to compute the expected percentage of body fat at different SBW from the NE concentrations in the gain (Mcal/kg) when the 1984 National Research Council (NRC) medium-frame steer was grown from 200 kg SBW at 11.5 percent body fat at SWG of 1 kg/day (1.01 Mcal NE_g/kg diet) for the first 100 kg and 1.3 kg/day (1.35 Mcal NE_g/kg diet) to various SBW (Table 3-1). Eqs. 3-1 and 3-2 were used for the computations of protein and fat at various SWG, using constants of 0.891 and 0.956, respectively, for converting EBW and EBG to SBW and SWG (National Research Council, 1984). Table 3-1 shows the percentage body fat expected at various weights for the 1984 NRC medium-frame steer with typical two-phase feeding programs (grown on high-quality forage and finished on high-

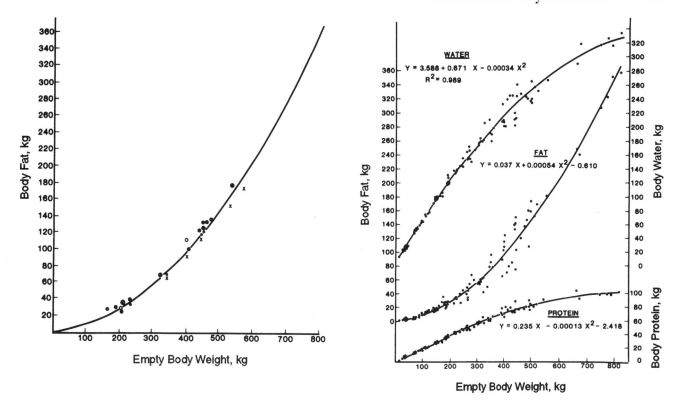

FIGURE 3-1 Relationship between empty body weight (kg) and body fat (kg) in male castrates of British beef breeds. A: From Simpfendorfer (1974). B: From Simpfendorfer (1974); superimposed points are from Lofgreen and Garrett (1968); Fox et al. (1972); Jesse et al. (1976); Crickenberger et al. (1978); Harpster (1978); Lomas et al. (1982); and Woody et al. (1983).

energy grain diets). Table 3-1 shows that even at low rates of gain and early stages of growth, some fat is deposited and both protein and fat are synthesized as rate of gain increases. Lightweight (90 kg) Holstein calves restricted to 0.23 to 0.53 kg ADG/day had 14.2 to 16.5 percent fat in the gain, respectively (Abdalla et al., 1988), which agrees with the values in Table 3-1. Phospholipids are required for cellular membrane growth (Murray et al., 1988). As energy intake above maintenance increases, protein synthesis rate becomes first limiting, and excess energy is deposited as fat; this dilutes body content of protein, ash, and water, which are deposited in nearly constant ratios to each other at a particular age (Garrett, 1987).

To predict NE_g required for SBW and SWG, EBW and EBG were converted to 4 percent shrunk liveweight gain with the following equations developed for use in the 1984 edition of this volume from the Garrett (1980) body composition data base:

$$EBW = 0.88 * SBW + 14.6 * NE_m \qquad \text{Eq. 3-4}$$
$$- 22.9 \ (r = 0.98) \text{ and}$$

$$EBG = 0.93 * SWG + 0.174 * NE_m \qquad \text{Eq. 3-5}$$
$$- 0.28 \ (r = 0.96)$$

or with constants of 0.891 * SBW and 0.956 * SWG.

These equations were rearranged to predict EBG and SWG;

$$EBG = 12.341 * (RE/EBW^{0.75})^{0.9116} \qquad \text{Eq. 3-6}$$
$$= 12.341 * EBW^{-0.6837} * RE^{0.9116}.$$

$$SWG = 13.91 * RE^{0.9116} * SBW^{-0.6837}. \qquad \text{Eq. 3-7}$$

In the rearranged equations, RE is equivalent to NE available for gain. Thus, if intake is known, the net energy required for gain (NEFG) may be calculated as (DMI − feed required for maintenance) * diet NE_g. NEFG can then be substituted into Eqs. 3-6 and 3-7 for RE to predict ADG.

Given the relationship between energy retained and protein content of gain, protein content of SWG is given as (National Research Council, 1984):

$$\text{protein retained} = SWG * (268 - \qquad \text{Eq. 3-8}$$
$$(29.4 * (RE/SWG))); r^2 = 0.96.$$

The weight at which cattle reach the same chemical composition differs depending on mature size and sex; hence, composition is different even when the weight is the same (Fortin et al., 1980; Figure 3-2 A and B). Each type reached 28 percent body fat (equivalent body composition) at different weights (Figure 3-2A). Figure 3-2B shows a similar plot for empty body protein, with the end

TABLE 3-1   Relationship of Stage of Growth and Rate of Gain to Body Composition, Based on NRC 1984 Medium-Frame Steer

Shrunk ADG, kg	Shrunk body weight, kg						
	200	250	300	350	400	450	500
	NE_g required, Mcal/d[a]						
0.6	1.68	1.99	2.28	2.56	2.83	3.09	3.34
0.8	2.31	2.73	3.13	3.51	3.88	4.24	4.59
1.0	2.95	3.48	4.00	4.49	4.96	5.42	5.86
1.3	3.93	4.65	5.33	5.98	6.61	7.22	7.81
	Protein in gain, percent[b]						
0.6	20.4	19.5	18.8	18.0	17.3	16.6	16.0
0.8	18.7	17.6	16.5	15.5	14.6	13.6	12.7
1.0	17.0	15.6	14.2	13.0	11.7	10.5	9.3
1.3	14.4	12.5	10.7	9.0	7.3	5.7	4.2
	Fat in gain, percent[c]						
0.6	5.9	9.7	13.2	16.6	19.9	23.1	26.2
0.8	13.6	18.7	23.6	28.2	32.8	37.1	41.4
1.0	21.4	27.9	34.1	40.1	45.6	51.5	56.9
1.3	22.3	29.0	35.4	41.5	47.4	53.2	58.7
	Body fat, percent						
0.6	11.6	10.8	10.9	11.5	12.3	13.4	14.5
0.8	11.6	12.5	13.9	15.6	17.5	19.4	21.4
1.0	11.6	14.2	17.0	19.9	22.8	25.6	28.5
1.3	11.6	14.4	17.4	20.4	23.4	26.4	29.3
1 then 1.3	11.6	14.2	17.0	20.1	23.1	26.1	29.1

[a]Computed from the 1984 NRC equation which was determined from 72 comparative slaughter experiments (Garrett, 1980); retained energy (RE) = 0.0635 EBW$^{0.75}$ EBG$^{1.097}$, where EBW is 0.891 SBW and EBG is .956 SBG.

[b]Computed from the equations of Garrett (1987), which were determined from the 1984 NRC data base; proportion of fat in the shrunk body weight gain = 0.122 RE − 0.146, and proportion of protein = 0.248 − 0.0264 RE. The proportion of fat and protein in the gain is for the body weight and ADG the RE is computed for.

[c]Percent body fat was determined when grown at 1 kg ADG to 300 kg and 1.3 kg ADG to each subsequent weight as described above.

of the line corresponding to the weight at 28 percent body fat. Weight at the same 12th rib lipid content varied 170 kg among steers of different biological types (Cundiff et al., 1981). The first NRC net energy system (National Research Council, 1976) used the Lofgreen and Garrett (1968) system to predict energy requirements, which was based on British breed steers given an estrogenic implant. From 1970 to 1990, larger mature-size European breed sires were increasingly used with the U.S. base British breed cow herd, resulting in the development of more diverse types of cows in the United States. This change, along with the use of sire evaluation programs that led to selection for larger body size to achieve greater absolute daily gain, resulted in an increase in average steer slaughter weights. The preceding edition of this volume (National Research Council, 1984) provided equations for medium- and large-frame cattle to adjust requirements for these changes. The current population of beef cattle in the United States varies widely in biological type and slaughter weight. By 1991, steers slaughtered averaged 542 kg, 48 percent choice with a weight range of 399 to 644 kg (M. Berwin, U.S. Department of Agriculture Market News data, Des Moines, IA, personal communication, 1992).

All systems developed since the NRC 1984 system use some type of size-scaling approach to adjust for differences in weight at a given composition. The Commonwealth Scientific and Industrial Research Organization (CSIRO) system (Commonwealth Scientific and Industrial Research Organization, 1990) uses one table of energy requirements for proportion of a standard reference weight, then gives a table of "standard reference weights" for different breed types. This standard reference weight is defined as the weight at which skeletal development is complete and the empty body contains 25 percent fat, which corresponds to a condition score 3 on a 0 to 5 scale. Oltjen et al. (1986) developed a mechanistic model to predict protein accretion from initial and mature DNA content, with the residual between net energy available for gain and that required for protein synthesis assumed to be deposited as fat. The animal's current weight as a proportion of mature weight is used to adjust for differences in mature size and use of implants.

The Institut National de la Recherche Agronomique (INRA) system (Institut National de la Recherche Agronomique, 1989) uses allometric relationships between the EBW and SBW, the weight of the chemical components, and the weight of the fat-free body mass to predict energy and protein requirements. Coefficients in the equations are

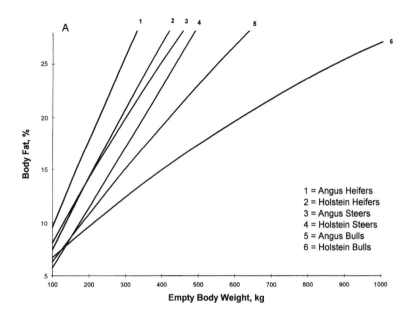

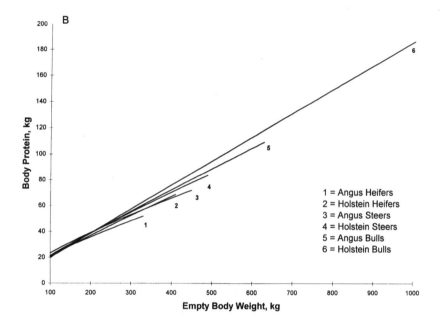

FIGURE 3-2 Relationship between empty body weight (kg) and body fat (%) in Angus and Holstein heifers, steers, and bulls; composition differs even when weight is the same. A: Each type reached 28 percent body fat (equivalent body composition) at different weights. B: A similar plot for empty body protein; the end of the line corresponds to the weight at 28 percent body fat.

parameters from the Gompertz equation (Taylor, 1968), which represents changes in liveweight with time. Initial and final weights with growth curve coefficients are given for six classes of bulls, two classes of steers, and two classes of heifers for finishing cattle, and two classes each for male and female growing cattle. The amount of lipids deposited daily is proportional to the daily liveweight gain raised to the power 1.8. Daily gain of protein is calculated from the gain in the fat-free body mass because protein content of fat free gain varies little with type of animal, growth rate, or feeding level (Garrett, 1987). Byers et al. (1989) developed an equation for steers similar to that of NRC (1984), except weight is replaced by proportional weight (current

weight/dam mature weight). A different exponent is used for "no growth regulator" (nonGR); the growth regulator (GR) equation assumes use of an estrogenic implant.

Fox et al. (1992) developed a system to interrelate the Beef Improvement Federation (BIF) frame-size system for describing breeding females and the USDA system for describing feeder cattle with energy and protein requirements. Dam mature weight is predicted from the BIF (1986) frame sizes of 1 to 9, which is assumed to be the same as the weight at which a similar frame size steer is 28 percent body fat (USDA low-choice grade). That weight is subsequently divided into the frame size of steer assumed to represent NRC 1984, the medium-framed steer equation, to obtain an adjustment factor that is used to compute the weight at which other frame sizes and sexes are equivalent in body composition. This approach is similar to the CSIRO (1990) standard reference weight system.

Based on input from industry specialists and land-grant university research and extension animal scientists, this subcommittee decided to use the NRC 1984 net energy system and the body weights and energy content of gain represented by the medium-frame steer equation as a standard reference base because of its widespread acceptance, success with its use, and the large body-composition data base underlying that system. The focus of this revision was on refining that system so that energy and protein requirements can be predicted for the wide ranges in body sizes of breeding and feeder cattle in North America, including both *Bos taurus* and *Bos indicus* types.

Because neither their actual composition nor mature weight is known, body composition and subsequent NE requirements must be predicted from estimated mature cow weight for breeding cattle or final weight and grade of feeder cattle. Because of the large number of breed types used, the widespread use of crossbreeding, anabolic implants, steers rather than bulls, feeding systems, and carcass grading systems used in North America, the European and CSIRO systems used to predict energy and protein requirements are not readily adaptable to North American conditions. Other proposed systems (Oltjen et al., 1986; Byers et al., 1989; Institut National de la Recherche Agronomique, 1989; Commonwealth Scientific and Industrial Research Orgainzation, 1990; Agricultural and Food Research Council, 1993) either did not account for as much of the variation with the validation data set described later or are not sufficiently complete to allow prediction of requirements from common descriptions of cattle and all conditions that must be taken into account in North America (bulls, steers, and heifers; various implant combinations; wide variations in body size, feeding systems, and final weights and grades).

The system developed for predicting energy and protein requirements for growing cattle assumes cattle have a similar body composition at the same degree of maturity, based on the evaluations presented previously. The NRC 1984 medium-frame steer equation (Eq. 3-1) is used as the standard reference base to compute the energy content of gain at various stages of growth and rates of gain for all cattle types. This is accomplished by adjusting the body weights of cattle of various body sizes and sexes to a weight at which they are equivalent in body composition to the steers in the Garrett (1980) data base, as described by Tylutki et al. (1994):

$$EQSBW = SBW * (SRW/FSBW); \qquad Eq. 3-9$$

EQSBW is weight equivalent to the 1984 NRC medium-frame size steer, SBW is shrunk body weight being evalu-

ated, SRW is standard reference weight for the expected final body fat (Table 3-2), and FSBW is final shrunk body weight at the expected final body fat (Table 3-2). These values were determined by averaging the percent body fat within all cattle in each of three marbling categories in the energy and protein retained validation data (Harpster, 1978; Danner et al., 1980; Lomas et al., 1982; Woody et al., 1983). Body fat percent averaged 27.8 (±3.4), 26.8 (±3), and 25.2 (±2.91) for those pens in the small, slight, or trace marbling categories, respectively. In comparison, the body fat data of Perry et al. (1991a,b) and Ainslie et al. (1992) averaged 28.4 percent (±4.1) for those in the small-marbling category. These steers had been selected to be a cross section of the current breed types and body sizes used in the United States. This variable SRW allows adapting the system to both U.S. and Canadian grading systems and determining SRW for marketing cattle at different end points. For breeding herd replacement heifers, FSBW is expected mature weight (MW). When computed as shown in Table 3-1 for heifers grown at 0.6 to 0.8 kg/day, accumulated fat content was 18 to 22 percent at the 28 percent fat steer SRW. Therefore, the SRW for breeding herd replacement heifers was assumed to be the same as the 1984 medium-frame steer fed to 28 percent fat. This approach is supported by a summary of the U.S. Meat Animal Research Center (MARC) data (Smith et al., 1976; Cundiff et al., 1981; Jenkins and Ferrell, 1984) in which mature weights of heifer mates averaged 10 percent more than implanted steer mates finished on high-energy diets

TABLE 3-2   Standard Reference Weights for Different Final Body Compositions

	Average Marbling Score		
	Traces	Slight	Small
Body fat, percent SE[a]	25.2 ± 2.9	26.8 ± 3.0	27.8 ± 3.4
Standard reference weight, kg[b]	435	462	478

[a]The means and standard errors (SE) shown for body fat in each marbling score category were determined by averaging the percentage body fat across all cattle in each of three marbling categories in the energy and protein retained validation data (Harpster, 1978; Danner et al., 1980; Lomas et al., 1982; and Woody et al., 1983). In a second comparison evaluation, the body fat data of Perry et al. (1991a, 1991b) and Ainslie et al. (1992) averaged 28.4 percent (±4.1) for those in the small marbling category. These relate to the current USDA and Canadian grading standards, respectively, as follows: traces, standard or A; slight, select or AA; and small, choice or AAA.

[b]The standard reference body weights (SBW basis) were determined from the NE_g concentrations in the gain (Mcal/kg) when the reference animal (1984 NRC medium frame steer) was grown from 200 kg SBW at 11.5 percent body fat at SWG of 1 kg/day (1.01 Mcal NE_g/kg diet) for the first 100 kg and 1.3 kg/day (1.35 Mcal NE_g/kg diet) until the percentage body fat in table 2 was reached. Eq. 1 and 2 were used for the computations, using constants of .891 and .956, respectively for converting EBW and EWG to SBW and SWG. The SRW and FSBW (mature weight) of replacement heifers (18 to 22 percent fat) is assumed to be the same as the 28 percent fat weight as implanted steer mates, based on the data of Smith et al. (1976), Cundiff et al. (1981), Jenkins and Ferrell (1984), and Harpster (1978) and accumulated fat content when heifers are grown at replacement heifer rates (Table 3-1). Breeding bulls are assumed to be 67 percent greater than cows, giving an SRW of 800 kg.

after weaning. Based on MARC data, breeding bulls are assumed to be 67 percent heavier at maturity than cows, giving an SRW of 800 kg, which is the mature weight of a bull with the same genotype as the 1984 NRC medium-frame steer.

The EQSBW computed from the SRW/FSBW multiplier is then used in Eq. 3-7 to compute the NE_g requirement. If Eq. 3-1 or 3-6 is used, SBW is adjusted to EBW with Eq. 3-4. Alternatively, the equation of Williams et al. (1992; EBW = full BW * [1-gut fill], where gut fill is 0.0534 + 0.329 * fractional forage NDF) can be used to predict EBW from unshrunk liveweight. Predicted gut fill is then corrected with multipliers for full BW, physical form of forage, and fraction of concentrates.

Because a table of requirements can be generated for any body size using the computer disk provided, only one example is shown (533 kg FSBW to represent the average steer in the United States). A similar table can be computed and printed for any body size with the computer disk containing the model. In this representative example, an FSBW change of 35 kg alters the NE_g requirement by approximately 5 percent. Heifers and bulls with similar parents as the steers represented in this table have 18 percent greater and lesser, respectively, NE_g requirements at the same weight as these steers. This system requires accurate estimation of FSBW. Most cattle feeders are experienced with results expected with feedlot finishing on a high-energy diet of backgrounded calves or yearlings that have received an estrogenic implant. Guidelines for other conditions are

- reduce FSBW 25 to 45 kg for nonuse of an estrogenic implant,
- increase FSBW 25 to 45 kg for use of an implant containing trenbolone acetate (TBA) plus estrogen,
- increase FSBW 25 to 45 kg for extended periods at slow rates of gain, and
- decrease FSBW 25 to 45 kg for continuous use of a high-energy diet from weaning.

Anabolic Agents

A variety of anabolic agents are available for use in steers and heifers destined for slaughter to enhance growth rate, feed efficiency, and lean tissue accretion. Trade names, active ingredients, and restrictions on animal use for products currently available in North America are given in Table 3-3. With the exception of melengestrerol acetate (MGA), which is added to the feed, these products are implanted into the ear. They have been approved for use by the Food and Drug Administration in the United States and the Bureau of Veterinary Drugs in Canada, although not all of the products listed in Table 3-3 are approved in both countries. The mode of action of anabolic agents is not completely understood but, in the final analysis, they enhance the rate of protein accretion in the body (National Research Council, 1994). Effects of these agents on growth, body, and carcass composition have also been reviewed (Galbraith and Topps, 1981; Unruh, 1986).

These products enhance rate of gain and feed intake. Rate of gain is usually enhanced more than intake, and feed efficiency is also improved. Their effect on nutrient utilization is minimal, so their impact on requirements can

---

Representative Example of Requirements

This example, a 320-kg steer with an FSBW of 600 kg (or herd replacement heifer with an MW of 600 kg) has an EQSBW of (478/600) * 320 = 255 kg. A 320-kg heifer with an FSBW of 480 has an EQSBW of (478/480) * 320 = 319 kg. The predicted SWG for the 320-kg steer consuming 5 Mcal NE_g is (Eq. 3-7); $13.91 * 5^{0.9116} * 255^{-0.6837}$ = 13.91 * 4.337 * 0.02263 = 1.365 kg/day. The SWG of the heifer consuming the same amount of energy will be $13.91 * 5^{0.9116} * 319^{-0.6837}$ = 1.17 kg/day. To compute NE_g requirement in this example 320-kg steer using Eq. 3-1 (0.891 * SBW to compute EBW and 0.956 * SWG to compute EBG): 255 * 0.891 = 227 kg EBW; 1.365 * 0.956 = 1.305 EBG; RE = $0.0635 * 227^{0.75} * 1.305^{1.097}$ = 0.0635 * 58.5 * 1.339 = 4.97 Mcal. Assuming NE_m requirement is 0.077 $SBW^{0.75}$, the NE_m requirement is (0.077 * $320^{0.75}$) = 5.83 Mcal/day. Net protein requirement for gain is then (Eq. 3-8); 268 − (29.4 * (5/1.365)) * 1.365 = 147 g/day. This value is then divided by the efficiency of use of absorbed protein to obtain the metabolizable protein required for gain (0.83 − (0.00114 * EQSBW)), which is added to the metabolizable protein required for maintenance (3.8 * $SBW^{0.75}$) to obtain the total metabolizable protein required. For the 320-kg steer, MP = 147/(0.83 − 0.00114 ((478/600) * 320)) + (3.8 * $320^{0.75}$) = 560 g.

TABLE 3-3 Anabolic Agents Used for Growing and Finishing Cattle in North America

Trade Name	Active Ingredients	Animal Use
Compudose	Estradiol	Steers over 270 kg
Finaplix	Trenbolone acetate	Steers or heifers
Forplix	Zeranol	Steers or heifers
	Trenbolone acetate	
Implus-H	Estradiol benzoate	Heifers
	Testosterone	
Implus-S	Estradiol benzoate	Steers
	Progesterone	
MGA	Melengesterol acetate	Heifers
Ralgro, Magnum	Zeranol	Steers or heifers
Revalor	Estradiol	Steers or heifers
	Trenbolone acetate	
Synovex—C	Estradiol benzoate	Suckling calves
	Progesterone	
Synovex—H	Estradiol benzoate	Heifers over 180 kg
	Testosterone	
Synovex—S	Estradiol benzoate	Steers over 180 kg
	Progesterone	

be accounted for by their effect on protein, fat, and energy accretion, which is taken into account by adjusting slaughter weight at constant finish. Effects on dry matter intake have also been quantified and are discussed in Chapter 7.

All anabolic implants that contain an estrogenic substance yield similar increases in performance when evaluated under similar conditions (Byers et al., 1989). Nearly all the increase in weight gain can be accounted for by an increased growth of lean tissue and skeleton (Trenkle, 1990). Recent studies (Trenkle, 1990; Perry et al., 1991b; Bartle et al., 1992) indicate that compared to not using an implant, estrogenic implants increase protein content of gain equivalent to a 35-kg change in FSBW, whereas estradiol and trenbolone acetate (TBA) combination implants alter the protein content of gain equivalent to a change of approximately 70 kg in FSBW. If the FSBW is reduced by 25 to 45 kg when no implant is given or is increased approximately 25 to 45 kg if TBA + estrogen is given, the NE_g requirement is changed by approximately 5 percent. This change is consistent with the 5 percent increase in net energy requirement when estrogenic implants were not in use (National Research Council, 1984) and the 4 percent adjustment in net energy requirement for nonuse of an anabolic implant in the model of Oltjen et al. (1986). Use of the two EBG exponents (GR and nonGR) in the equation of Byers et al. (1989) results in an 18 percent greater NE_g requirement at 750 g ADG and a 20 percent greater NE_g requirement at 1,500 g daily gain without an estrogenic implant compared with continuous use of an estrogenic implant. Solis et al. (1988) found, however, that the continuous use of an estrogenic implant in steers increased final weight at a similar composition by 25 kg as a result of 4.4 percentage units less fat in the gain over the growth period. These results are consistent with the recommendations given here for adjusting FSBW for the use of anabolic implants.

Ionophore Effects

Ionophores are polyether compounds included in diets of growing and finishing cattle to improve feed efficiency and animal health. Four products are currently licensed in North America, referred to by chemical name as lasalocid, laidlomycin propionate, monensin, and salinomycin. Lasalocid and monensin are licensed in both the United States and Canada, laidlomycin propionate is licensed in the United States, and salinomycin is licensed in Canada.

The ionophore's mechanisms of action are initiated by channeling ions through cell membranes (Bergen and Bates, 1984), and they have a marked effect on microbial cells in particular. There is a shift in volatile fatty acids produced in the rumen toward more propionate with corresponding reductions in acetate and butyrate. Measurements with rumensin in vivo have shown that it increases

propionate production by 49 and 76 percent for high-roughage and high-concentrate diets, respectively (Van Maanen et al., 1978). This magnitude of response implies a significant improvement in the capture of feed energy during ruminal fermentation with less methane produced. Thus, metabolizable and net energy values of feeds should increase when ionophores are consumed.

In a comparative slaughter trial, Byers (1980) found that the efficiency of energy use for maintenance was increased 5.7 percent by monensin with no effect on efficiency for gain. Delfino et al. (1988) made a similar observation with respect to lasalocid; they observed a 10 percent improvement in NE_m of the feed with no effect on NE_g. In a review of feedlot data, Raun (1990) reported that for cattle fed high-concentrate diets (average 15.7 percent forage), rumensin increased feed efficiency by 5.6 percent and gain by 1.8 percent but decreased dry matter intake by 4 percent. Simulations using the model in this publication (Chapter 10), with a 90 percent concentrate diet, showed that a 12 percent increase in NE_m concentration of the diet with a 4 percent reduction in intake gave a 5.3 and 1.5 percent improvement in feed efficiency and gain, respectively.

With lower energy rations (40 percent concentrate only), Goodrich et al. (1984) concluded that monensin increased feed efficiency and gain by 7.5 and 1.6 percent, respectively, with 6.4 percent lower intake. Simulation of these results using a 12 percent enhancement of ration NE_m with monensin gave 7.9 and 4.5 percent improvements in feed efficiency and gain, respectively. These simulations confirm observed results that the proportional response in feed efficiency and gain to including monensin decreases as ration energy level increases.

There are insufficient data available to develop individual recommendations for each ionophore and its effect on NE_m. Thus, for all ionophores it is recommended that the NE_m concentration of the diet be increased by 12 percent. Ionophores have characteristic effects on intake; and this is discussed in Chapter 7.

Several reports have suggested that ionophores can improve energetic efficiency in cows and breeding animals. However, data are inconsistent (for review, see Sprott et al., 1988).

Ionophores can have significant effects on nutrients other than energy. In general, they enhance absorption of nitrogen, magnesium, phosphorus, zinc, and selenium with inconsistent effects on calcium, potassium, and sodium. For further information see Chapter 5 and the review by Spears (1990).

From experimental data on the simultaneous use of anabolic agents and ionophores, the subcommittee has concluded that interaction is minimal. Thus, it is recommended that adjustments made to slaughter weight based on use of anabolic agents are independent of ionophore use and adjustments made to ration NE_m based on use of ionophores are independent of anabolic agents. Their effects on feed intake have been considered to be additive.

Previous Plane of Nutrition Effects

Energy intake above maintenance can vary considerably, depending on diet fed during early growth in stocker and backgrounding programs. Table 3-1 indicates that a reduced intake above maintenance results in a greater proportion of protein in the gain at a particular weight, which is supported by several studies (Fox and Black, 1984; Abdalla et al., 1988; Byers et al., 1989) and the model by Keele et al. (1992). When thin cattle are placed on a high-energy diet, however, compensatory fat deposition occurs. Most of the improved efficiency of gain results from a decreased maintenance requirement and increased feed intake (Fox and Black, 1984; Ferrell et al., 1986; Carstens et al., 1987; Abdalla et al., 1988). As discussed in the maintenance requirement section, it is assumed NE_m requirement is 20 percent lower in a very thin animal (CS 1), is increased 20 percent in a very fleshy animal (CS 9), and changes 5 percent per condition score. The NE_m adjustment for previous nutrition (COMP) is thus computed as

$$COMP = 0.8 + (CS - 1) * 0.05, \qquad Eq. \ 3\text{-}10$$

where CS is body condition score. The effect of plane of nutrition is taken into account by the rate of gain function (increased fat deposition with increased rate of gain) and EQSBW in the primary equations. Thus, the user determines the expected final weight and body fat, and the model computes EQSBW to use in computing NE_g required as shown in Eq. 3-9. The change in efficiency of energy utilization is accounted for by a reduced NE_m requirement and increased DMI above maintenance.

Effects of Special Dietary Factors

Diet composition and level of intake differences will cause the composition of the ME (ruminal volatile fatty acids, intestinally digested carbohydrate, and fat) to vary (Ferrell, 1988), which can affect the composition of gain (Fox and Black, 1984). Most of these effects will alter rate of gain, which is taken into account by the primary equations; however, fat distribution may be altered, which could affect carcass grade (Fox and Black, 1984).

Unique Breed Effects

Most of the unique breed effects on NE_g requirements are accounted for by differences in the weight at which different breeds reach a given chemical composition (Harp-

ster, 1978; Cundiff et al., 1986; Institut National de la Recherche Agronomique, 1989). Nonetheless, breeds can differ in fat distribution, which can influence carcass grade (Cundiff et al., 1986; Perry et al., 1991a).

Validation of Energy and Protein Requirement System

The standard reference weight (SRW) approach was validated and compared to the 1984 NRC system with three distinctly different data sets that were completely independent of those used to develop the NRC systems—the one presented in this publication and the one developed for the preceding edition of this volume (National Research Council, 1984). The Oltjen et al. (1986) model was also compared to the other two with the first two data sets. For the 1984 NRC system, cattle with frame sizes larger than 6 were considered large-framed. For this publication, the standard reference weight (478 kg) was divided by the pen mean weight at 28 percent body fat to obtain the body size adjustment factor, which was then applied to the actual weight for use in the standard reference equations to predict energy and protein retained.

Data set 1 (Harpster, 1978; Danner et al., 1980; Lomas et al., 1982; Woody et al., 1983) included 82 pen observations (65 pens of steers and 17 pens of heifers) with body composition determined by the same procedures used by Garrett (1980) in developing the NRC 1984 system. Included were FSBW representative of the range in cattle fed in North America; all silage to all corn-based diets; no anabolic implant, estrogen only or estrogen + TBA; and *Bos taurus* breed types representative of those fed in North America (British, European, Holstein, and their crosses).

Data set 2 included 142 serially slaughtered (whole body chemical analysis by component; Fortin et al., 1980; Anrique et al., 1990) nonimplanted steers, heifers, and bulls ranging widely in body size. A detailed description of these data sets, validation procedures, and results were published by Tylutki et al. (1994), except the SRW has been increased from 467 to 478 kg. In nearly every subclass, the system developed for this publication accounted for more of the variation and had less bias than did the other two systems. Nearly identical results were obtained between the 1984 NRC and present systems when energy retained was used to predict SWG in Eq. 3-7; this equation is the one most commonly used to predict ADG. Figure 3-3 shows the results when all subclasses were combined. The present model accounted for 94 percent of the variation with a 2 percent overprediction bias for retained energy and 91 percent of the variation in retained protein with a 2 percent underprediction bias. Figure 3-3 shows that use of the NRC 1984 medium-frame steer as a standard reference base results in accurate prediction of net energy requirements for growth across wide variations in cattle breed, body size, implant, and nutritional management systems.

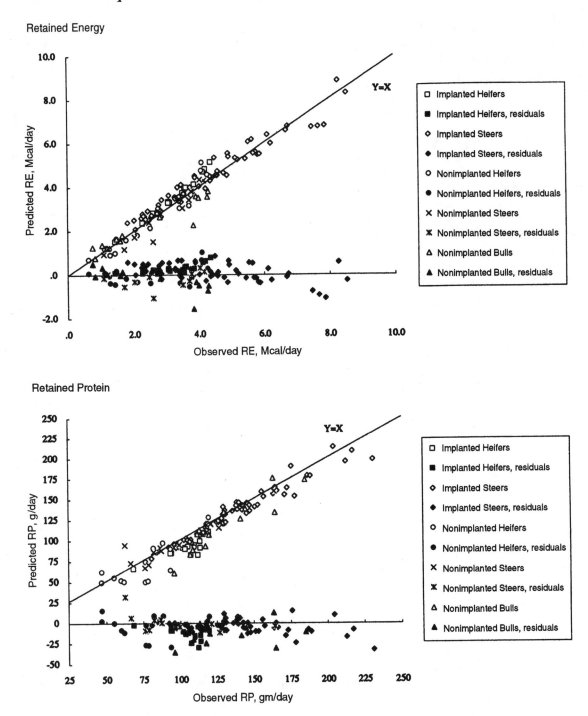

FIGURE 3-3   Use of the NRC 1984 medium-frame steer as a standard reference base results in accurate prediction of net energy requirements for growth across wide variations in cattle breed, body size, implant, and nutritional management systems.

Data set 3 included ADG predicted, using model level 2, from 63 different diets fed to a total of 687 *Bos indicus* bulls (Nellore breed) in which ME intake and body composition were determined (Lanna et al., 1994). FSBW was determined from final EBW fat content. For the NRC 1984 and the present systems, the r^2 was 0.51 and 0.67, and the bias was -21 percent and -1 percent for the respective systems.

These validations indicate that given the accuracies obtained, problems with predicting net energy and protein requirements and SWG are likely to include one of the following:

1. choosing the wrong FSBW
2. short-term, transitory effects of previous nutrition, gut fill, or anabolic implants,
3. variation in NE_m requirement,
4. variation in ME value assigned to the feed because of variations in feed composition and extent of ruminal or intestinal digestion,
5. variation in NE_m and NE_g derived from the ME because of variation in end products of digestion and their metabolizability, and
6. variations in gut fill.

AMINO ACID REQUIREMENTS

In recent studies, abomasal infusion of high-quality sources of amino acids significantly increased nitrogen balance in steers, despite the fact that they were fed diets balanced to optimize ruminal fermentation and to provide protein in excess of NRC requirements (Houseknecht et al., 1992; Robinson et al., 1994). These studies indicate that protein accretion was constrained by quantity and/or proportionality of amino acids absorbed.

Amino acid requirements for tissue growth are a function of the percentage of each amino acid in the net protein accretion and thus depend on the accuracy of prediction of protein retained. Ainslie et al. (1993) summarized various studies that have determined essential amino acid content of tissue protein in selected muscles (Hogan, 1974; Evans and Patterson, 1985), in daily accretion (Early et al., 1990), or in the whole empty body (Williams, 1978; Rohr and Lebzien, 1991; Ainslie et al., 1993). In a sensitivity analysis with model predicted vs first-limiting amino acid allowable gain, the average of the three whole empty body studies gave the least bias (Fox et al., 1995). These average values (average of Williams, 1978; Rohr and Lebzien, 1991; Ainslie et al., 1993) are as follows (g/100 g empty body protein); arginine, 3.3; histidine, 2.5; isoleucine, 2.8; leucine, 6.7; lysine, 6.4; methionine, 2.0; phenylalanine, 3.5; threonine, 3.9; and valine, 4.0. Tryptophan values were not given because of limitations in assay procedures.

A number of recent studies have evaluated tissue amino acid requirements by measuring net flux of essential amino acids across the hind limb of growing steers (Merchen and Titgemeyer, 1992; Byrem et al., 1993; Boisclair et al., 1994; Robinson et al., 1995). The proportionality of individual amino acid uptake did not markedly change when protein accretion was increased by infusing various compounds (bovine somatotropin, cimaterol, or casein). The proportions of the essential amino acids in the net flux in these studies followed the same trends as suggested by the tissue composition values listed above.

The above studies and the data previously cited in this section suggest that both quantity and proportionality of amino acid availability are important to achieve maximum energy allowable ADG. In a first NRC attempt to accomplish this for cattle, the model level 2, as described in Chapter 10, has been provided to allow the user to estimate both quantity and proportion of essential amino acids required by the animal and supplied by the diet. The critical steps involved are the prediction of microbial growth and composition; amount and composition of diet protein escaping ruminal degradation; intestinal digestion and absorption; and net flux of absorbed amino acids into tissue. Because of limitations in the ability to predict each of these components, the estimates of amino acid balances provided should be used only as a guide. The subcommittee has taken this step to provide a structure that is intended to stimulate research that will improve the ability to predict amino acid balances, which should lead to increased efficiency of energy and protein utilization in cattle.

Net daily tissue synthesis of protein represents a balance between synthesis and degradation (Oltjen et al., 1986; Early et al., 1990; Lobley, 1992). Lobley (1992) indicated that a 500-kg steer with a net daily protein accretion of 150 g actually degrades and resynthesizes at least another 2,550 g. Thus, balancing for daily net accretion accounts for only about 5.5 percent of the total daily protein synthesis. Protein metabolism is very dynamic, and a kinetic approach is needed to accurately predict amino acid requirements. Small changes in either the rate of synthesis or degradation can cause great alterations in the rate of gain, and the relative maintenance requirement changes with level of production. Lobley (1992), however, concluded that the precision of kinetic methods is critical; a 2 percent change in synthesis rate would alter net protein accretion 20 to 40 percent, and many of the procedures are not accurate within 4 to 5 percent. When combined with a system that has limitations in predicting absorbed amino acids from microbial and feed sources, errors could be greatly magnified with an inadequate mechanistic metabolism model. Given present knowledge, the subcommittee decided that protein and amino acids required for growth should be based on net daily accretion values that have been actually measured. Maintenance requirements for protein have been measured with metabolism trials (Institut National de la Recherche Agronomique, 1989) or in growth trials beginning at or slightly above maintenance (Wilkerson et al., 1993). Net daily protein and amino acid accretion have been measured and validated in the comparative slaughter studies reported here. However, this subcommittee recommends that models such as that of Oltjen et al. (1986) be developed, refined, and validated so that in the future this approach can be used to allow more accurate prediction of daily amino acid requirements.

ENERGY AND PROTEIN REQUIREMENTS FOR BREEDING HERD REPLACEMENTS

No rate of gain requirement has been given in previous NRC publications for growing cattle because they are used to market available forage at early stages of growth, which results in wide variations in rate of gain before feedlot finishing. However, replacement heifer growth rate that results in first parturition at 2 years of age is most economical (Gill and Allaire, 1976). In addition, inadequate size at first parturition may limit milk production and conception during first lactation. Excess energy intake, however, can have negative effects on mammary development. For example, excessive energy intake had a negative effect on mammary parenchyma (ductular epithelial tissue; Harrison et al., 1983; Foldager and Serjsen, 1987). Because puberty is associated with weight, parenchyma tissue growth, which is not linearly related to body growth, may be truncated before full ductal development as a result of excess energy intake before puberty (Van Amburgh et al., 1991). Excess energy intake, as evidenced by overconditioning from 2 to 3 months of age until after conception, should be avoided.

Numerous data are available to support the concept of a genetically determined threshold age and weight at which bulls or heifers attain puberty (for reviews, see Robinson, 1990; Ferrell, 1991; Dunn and Moss, 1992; Patterson et al., 1992; Schillo et al., 1992). Joubert (1963) proposed that heifers would not attain puberty until they reached a given degree of physiological maturity, which is similar to the "target weight" concept proposed by Lamond (1970). Simply stated, the concept is to feed replacement heifers to attain a preselected or target weight at a given age (Spitzer et al., 1975; Dziuk and Bellows, 1983; Wiltbank et al., 1985). In general, heifers of typical beef breeds (e.g., Angus, Charolais, Hereford, Limousin) are expected to attain puberty at about 60 percent of mature weight (Laster et al., 1972, 1976, 1979; Stewart et al., 1980; Ferrell, 1982; Sacco et al., 1987; Martin et al., 1992; Gregory et al., 1992). Heifers of dual purpose or dairy breeds (e.g., Braunvieh, Brown Swiss, Friesian, Gelbvieh, Red Poll) tend to attain puberty at a younger age and lower weight relative to mature weight (about 55 percent of mature weight) than those of beef breeds. Conversely, heifers of Bos indicus breeds (e.g., Brahman, Nellore, Sahiwal) generally attain puberty at older ages and heavier weight, and at a slightly higher percentage of mature weight (65 percent) as compared to European beef breeds.

The following model was developed to compute target weights and growth rates for breeding herd replacement heifers, using the data summarized in Chapter 4 (target breeding weights are 60 and 65 percent of mature weight for Bos taurus and Bos indicus, respectively). Then the equations described previously are used to predict net energy and protein requirements for growth. Based on the data summarized by Gregory et al. (1992) it is assumed that target first calving weights are 80 percent of mature weight, which is the 6 breed average for 2-year-old as a percentage of 6-year-old weight in this MARC data base. Target calving weight factors for 3 and 4 year olds (0.92 and 0.96, respectively) are from the model described by Fox et al. (1992).

Predicting target weights and rates of gain:

$TPW = MW * (0.60$ for Bos taurus; 0.65 for Bos indicus; and 0.55 for dual purpose or dairy)

$TCA = $ Target calving age in days

$TPA = TCA - 280$

$BPADG = (TPW - SBW)/(TPA - T_{age})$

$TCW1 = MW * 0.80$

$TCW2 = MW * 0.92$

$TCW3 = MW * 0.96$

$TCW4 = MW * 1$

$APADG = (TCW1 - TPW) / (TCA - TPA)$

$ACADG = (TCWxx - TCWx)/CI$

where:

MW is mature weight, kg;

LW is liveweight, kg;

TPW is target puberty weight, kg;

TCW1 is target calving weight, kg at 24 months;

TCW2 is target calving weight, kg at 36 months;

TCW3 is target calving weight, kg at 48 months;

TCW4 is target calving weight, kg at >48 months;

TCWx is current target calving weight, kg;

TCWxx is next target calving weight, kg;

TCA is target calving age in days;

TPA is target puberty age in days;

BPADG is prepubertal target ADG, kg/day;

APADG is postpubertal target ADG, kg/day;

ACADG is after calving target ADG, kg/day

T_{age} is heifer age, days;

CI is calving interval, days.

The equations in the previous section are used to compute requirements for the target ADG, and adjustments to reach these targets because of previous nutrition are made by determining ADG and NE_g requirements needed to achieve the targets. For pregnant animals, gain due to gravid uterus growth should be added to predicted daily gain (SWG), as follows:

$$ADG_{preg} = CBW * (0.3656 - 0.000523t) * e^{((0.0200 * t) - (0.000143 * t^2))};$$

where CBW is calf birth weight, kg. For pregnant heifers, weight of fetal and associated uterine tissue should be deducted from EQEBW to compute growth requirements. The conceptus weight (CW) can be calculated as follows:

$$CW = (CBW * (743.9/40699) * e^{((0.0200 * t) - (0.000143 * t^2))};$$

where, CW is conceptus weight, kg; and t is days pregnant.

Net energy requirement for optimal growth of breeding heifer replacements can be determined for these rates of growth with the primary net energy requirement equations, using expected mature weight as FSBW.

ENERGY AND PROTEIN RESERVES OF BEEF COWS

In utilizing available forage, beef cows usually do not consume the amount of energy that matches their requirements for maintenance, gestation, or milk production. Reserves are depleted when forage quality and (or) quantity declines because of weather, overstocking, or inadequate forage management, but are replenished when these conditions improve. In addition, most beef cows are not housed and must continually adjust energy balance for changes in environmental conditions.

Optimum management of energy reserves is critical to economic success with cows. Whether too fat or thin, cows at either extreme are at risk from metabolic problems and diseases, decreased milk yield, low conception rates, and difficult calving (Ferguson and Otto, 1989). Overconditioning is expensive and can lead to calving problems and lower dry matter intake during early lactation. Conversely, thin cows may not have sufficient reserves for maximum milk production and will not likely rebreed on schedule. To maintain a 12-month calving interval, cows must be bred by 83 days after calving (365 minus an average gestation length of 282 days). Dairy cows usually ovulate the first dominant follicle, but beef cows average three dominant follicles being produced before ovulation, depending on the suppressive effects of suckling, body condition, or energy intake (Roche et al., 1992). Both postcalving cow condition score and energy balance control ovulation (Wright et al., 1992). Conception rates reach near maximum at body condition score 5 (Wright et al., 1992). Ovulation occurs in dairy cattle 7 to 14 days after the energy balance nadir is reached during early lactation (Butler and Canfield, 1989). Beef cows in adequate body condition with adequate energy intake may have a similar response

because the negative effects of suckling may be offset by the lower energy demands of beef cows (W.R. Butler, Cornell University, personal communication, 1992). Allowing for three ovulations (assuming the first ovulation goes undetected), and allowing for two observed ovulations and inseminations for conception, the first ovulation must occur 41 days after calving. To allow this, the feeding program must be managed so that maximum negative energy balance during early lactation is reached by about 31 days after calving (41 days to first ovulation minus 10 days for ovulation after maximum negative energy balance). If the cow is too fat, intake will be lower and reserves will be used longer during early lactation, resulting in an extended time to maximum negative energy balance. Even if thin cows consume enough to meet requirements by 31 days, a feedback mechanism mediated through hormonal changes seems to inhibit ovulation if body condition is inadequate (Roche et al., 1992). Additional signals relative to the need for a given body condition before ovulation appear to occur in cows nursing calves.

In previous NRC publications, changes in energy reserves were accounted for by allowing for weight gain or loss. However, in practice, few producers weigh beef cows to determine if their feeding program is allowing for the appropriate energy balance. Energy reserves are more often managed by observing body condition changes, and all systems developed since the last NRC publication use condition scores (CS) to describe energy reserves. Body condition score is closely related to body fat and energy content (Wagner, 1984; Houghton et al., 1990; Fox et al., 1992; Buskirk et al., 1992). The CSIRO nutrient requirement recommendations (Commonwealth Scientific Industrial Research Organization, 1990) adapted the 0 to 5 body condition scoring system of Wright and Russel (1984a,b). In their system, a CS change of 1 contains 83 kg body weight change, which contains 6.4 Mcal/kg for British breeds and 5.5 Mcal/kg for large European breeds; this is equivalent to 55 kg and 330 Mcal/CS on a 1 to 9 scale. The INRA (1989) nutrient requirement recommendations use a 0 to 5 system also and assume 6 Mcal lost/kg weight loss, which is equivalent to 332 Mcal/CS on a 9-point scale.

The Oklahoma (Cantrell et al., 1982; Wagner, 1984; Selk et al., 1988) and Colorado groups (Whitman, 1975) developed a 9-point system for condition scoring. The Purdue group (Houghton et al., 1990) used a 5-point scale with minus, average, and plus within each point, which in effect approximates the dairy 1 to 5 system; both are similar to a continuous 9-point scale. Empty body lipid was 3.1, 8.7, 14.9, 21.5 and 27.2, respectively, for CS 1 to 5, which they proposed correspond to CS 2, 5, and 8 on the 1 to 9 scale. Empty body weights averaged 75 kg per increase in condition score, which is equivalent to 50 kg/CS on a 9-point system. The Texas group (Herd and Sprott, 1986) used a 9-point scale and reported 0, 4, 8, 12, 16, 24, 28, and 32 percent body fat, respectively, for CS 1-9. The Cornell group (Fox et al., 1992) used the Oklahoma 9-point scoring system and 14 studies of body composition in cows to develop a model to predict weight and energy lost or gained with changes in age, mature size, and condition score.

In a 455-kg vs a 682-kg mature cow with a CS of 5, a loss of 1 CS from 5 to 4 is associated with 30 kg and 167 Mcal vs 45 kg and 257 Mcal, respectively, which is 5.6 Mcal/kg. From CS 2 to CS 1, the weight lost contains 4.4 Mcal/kg. The Purdue group (Buskirk et al., 1992) predicted from body weight and CS changes energy content of tissue gain (or loss) at each CS to be 2.16, 2.89, 3.62, 4.34, 5.07, 5.8, 6.53, 7.26, and 7.98 Mcal/kg for CS 1 to 9, respectively. Their CS 5 value of 5.07 compares to the CSIRO (1990) value of 6.4 for British breeds and 5.5 for European breeds; the INRA (1989) value of 6; and the Fox et al. (1992) value of 5.6 Mcal/kg weight change at a CS of 5, which reaches a maximum of 5.7 at CS 9 and declines to 4.4 by CS 2, on a 1 to 9 scale. The Buskirk et al. (1992) system assumes a linear decline in energy content of gain as weight is lost, which implies proportional protein and fat in the gain or loss with changes in weight as occurs during growth. The other systems (Institut National de la Recherche Agronomique, 1989; Commonwealth Scientific Industrial Research Organization, 1990; National Research Council, 1989; Fox et al., 1992) assume a hierarchical loss of fat energy first in mature animals using and replenishing reserves. Another difference is that the Buskirk et al. (1992) system uses NE_g values of feeds to meet NE reserves requirements, whereas the CSIRO, INRA, and NRC systems as well as others (Moe, 1981; Fox et al., 1992) assume higher efficiencies of use of ME for energy reserves than for growth.

The model below was developed from a body composition data set provided by MARC (C.L. Ferrell, personal communication, 1995). Body condition score, body weight, and body composition are used to calculate energy reserves. The equations were developed from data on chemical body composition and visual appraisal of condition scores (1 to 9 scoring system) from 105 mature cows of diverse breed types and body sizes. Characteristics of the data set were EBW = 0.851 * SBW; mean EBW, 546 (range 302 to 757) kg; percentage empty body fat, 19.3 (range 4.03 to 31.2); percentage empty body protein, 15.3 (range 13.2 to 18.0); and body condition score, 5.56 (range 2.25 to 8.0). The developed equations were validated on an independent data set of 65 mature cows (data from C.L. Ferrell, MARC, personal communication, 1995). The validation data set consisted of 9 year old cows of diverse sire breeds and Angus or Hereford dams with mean EBW, 471 (range 338 to 619) kg; mean percentage empty body fat, 20.3 (range 8.5 to 31.3); mean percentage empty body protein, 18.2 (range 13.9 to 21.3); and mean condition score, 4.9 (range 3.0 to 7.5). The resulting best-fit equations to

describe relationships between CS and empty body percentage fat, protein water, and ash were linear (Figure 3-4). A zero intercept model was used to describe the relationship between percent empty body fat and CS. The mean SBW change associated with a CS change was computed as 44 kg. It is assumed that for a particular cow the ash mass does not change when condition score changes. In the validation of this model, CS accounted for 67, 52, and 66 percent of the variation in body fat, body protein, and body energy, respectively.

1. Body composition is computed for the current CS:
   $AF = 0.037683 * CS; r^2 = 0.67.$
   $AP = 0.200886 - 0.0066762 * CS; r^2 = 0.52.$
   $AW = 0.766637 - 0.034506 * CS; r^2 = 0.67.$
   $AA = 0.078982 - 0.00438 * CS; r^2 = 0.66.$
   $EBW = 0.851 * SBW$
   $TA = AA * EBW$

where:

   AF = proportion of empty body fat
   AP = proportion of empty body protein

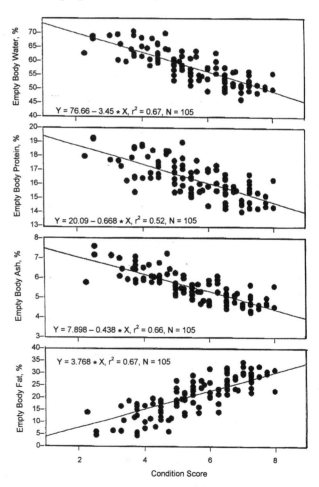

FIGURE 3-4   Relationship of empty body weight, protein, ash, and fat (as percentage) of body condition score in mature cows.

AW = proportion of empty body water
AA = proportion of empty body ash
SBW = shrunk body weight, kg
EBW = empty body weight, kg
TA = total ash, kg

2. For CS = 1 ash, fat, and protein composition are as follows:
   AA1 = 0.074602
   AF1 = 0.037683
   AP1 = 0.194208

where:

AA1 is proportion of empty body ash @ CS = 1
AF1 is proportion of empty body fat @ CS = 1
AP1 is proportion of empty body protein @ CS = 1

3. Assuming that ash mass does not vary with condition score, EBW and component body mass at condition score 1 is calculated:
   EBW1 = TA/AA1
   TF = AF * EBW
   TP = AP * EBW
   TF1 = EBW1 * AF1
   TP1 = EBW1 * AP1

where:

EBW1 is Calculated empty body weight at CS = 1, kg
TF is total fat, kg
TP is total protein, kg
TF1 is total body fat @ CS = 1, kg
TP1 is total body protein @ CS = 1, kg

4. Mobilizable energy and protein are computed:
   FM = (TF − TF1)
   PM = (TP − TP1)
   ER = 9.4FM + 5.7PM

where:

FM is mobilizable fat, kg
PM is mobilizable protein, kg
ER is energy reserves, Mcal

5. EBW, AF and AP are computed for the next CS to compute energy and protein gain or loss to reach the next CS:

   EBW = TA/AA

where:

EBW is EBW at the next score
TA is total kg ash at the current score
AA is proportion of ash at the next score
AF, AP, TF and TP are computed as in steps 1 and 3 for the next CS and FM, PM, and ER are computed as the difference between the next and current scores.

Table 3-4 gives CS descriptions and Table 3-5 shows the percentage composition and SBW change associated with each CS computed with this model. This model predicts energy reserves to be a constant 5.82 Mcal/kg liveweight loss, which compares to the 1989 NRC dairy value of 6 Mcal/kg, the CSIRO values of 6.4 for British breeds and 5.5 for European breeds, the INRA value of 6 and the AFRC value of 4.54. Protein loss is predicted to be 81 g/kg, compared to 117, 135, 138, and 160 g/kg weight loss for the Buskirk et al. (1992), CSIRO (1990), AFRC (1993), and NRC (1985) systems. SBW is predicted to be 76.5,

TABLE 3-4   Cow Condition Score

Condition Score	Body Fat, percent[a]	Appearance of Cow[b]
1	3.77	Emaciated—Bone structure of shoulder, ribs, back, hooks and pins sharp to touch and easily visible. Little evidence of fat deposits or muscling.
2	7.54	Very thin—Little evidence of fat deposits but some muscling in hindquarters. The spinous processes feel sharp to the touch and are easily seen, with space between them.
3	11.30	Thin—Beginning of fat cover over the loin, back, and foreribs. Backbone still highly visible. Processes of the spine can be identified individually by touch and may still be visible. Spaces between the processes are less pronounced.
4	15.07	Borderline—Foreribs not noticeable; 12th and 13th ribs still noticeable to the eye, particularly in cattle with a big spring of rib and ribs wide apart. The transverse spinous processes can be identified only by palpation (with slight pressure) to feel rounded rather than sharp. Full but straightness of muscling in the hindquarters.
5	18.89	Moderate—12th and 13th ribs not visible to the eye unless animal has been shrunk. The transverse spinous processes can only be felt with firm pressure to feel rounded—not noticeable to the eye. Spaces between processes not visible and only distinguishable with firm pressure. Areas on each side of the tail head are fairly well filled but not mounded.
6	22.61	Good—Ribs fully covered, not noticeable to the eye. Hindquarters plump and full. Noticeable sponginess to covering of foreribs and on each side of the tail head. Firm pressure now required to feel transverse process.
7	26.38	Very good—Ends of the spinous processes can only be felt with very firm pressure. Spaces between processes can barely be distinguished at all. Abundant fat cover on either side of tail head with some patchiness evident.
8	30.15	Fat—Animal taking on a smooth, blocky appearance; bone structure disappearing from sight. Fat cover thick and spongy with patchiness likely.
9	33.91	Very fat—Bone structure not seen or easily felt. Tail head buried in fat. Animal's mobility may actually be impaired by excess amount of fat.

[a]Based on the model presented in this chapter.
[b]Adapted from Herd and Sprott, 1986.

TABLE 3-5 Empty Body (EB) Chemical Composition at Different Condition Scores (CS)

| CS | Percent in EB | | | | SBW, percent of CS 5[a] |
	Fat	Protein	Ash	Water	
1	3.77	19.42	7.46	69.35	77
2	7.54	18.75	7.02	66.69	81
3	11.30	18.09	6.58	64.03	87
4	15.07	17.04	6.15	61.36	93
5	18.84	16.75	5.71	58.70	100
6	22.61	16.08	5.27	56.04	108
7	26.38	15.42	4.83	53.37	118
8	30.15	14.75	4.39	50.71	130
9	33.91	14.08	3.96	48.05	144

[a]Weight change from CS5 weight can be estimated from the difference between CS5 weight and CS5 weight * percent of CS5 weight for the CS in question. Net energy reserves provided, or required to change CS, is kg weight change * 5.82.

81.3, 86.7, 92.9, 108.3, 118.1, 129.9, and 144.3 percent of a CS 5 cow for CS 1, 2, 3, 4, 6, 7, 8, and 9, respectively. A 500 kg cow is predicted to weigh 465, 434, 407, and 383 kg at CS 4, 3, 2, and 1 with weight losses of 35, 31, 27, and 24 kg for CS 5, 4, 3 and 2, respectively. Corresponding values for a 650 kg cow are 604, 564, 528, and 585 kg SBW at CS 4, 3, 2 and 1 with weight losses per CS of 46, 40, 35 and 31 kg for CS 5, 4, 3 and 2, respectively.

Table 3-6 gives Mcal mobilized in moving to the next lower score, or required to move from the next lower score, to the one being considered for cows with different mature sizes. These cows are within the range included in the data base used to develop the regression equations (433 to 887 kg SBW). Diet NE_m replaced by mobilized reserves, or required to replenish reserves, are computed by assuming 1 Mcal of mobilized tissue will replace 0.8 Mcal of diet NE_m, and 1 Mcal of diet NE_m will provide 1 Mcal of tissue NE, based on Moe (1981) and NRC (1989). For example, a 500 kg cow at CS 5 will mobilize 207 Mcal in declining to a CS 4. If NE_m intake is deficient 3 Mcal/day, this cow will lose 1 CS in (207 * 0.8)/3 = 55 days. If consuming 3 Mcal NE_m above daily requirements, this cow will move back to a CS 5 in 207/3 = 69 days.

The weakest link in this model is the prediction of body weight change associated with each CS change. This is a critical step because it is used to compute total energy reserves available and energy required to replenish reserves. In this model, this calculation is based on the assumption that ash mass is constant. The weights and weight changes appear to agree well with other data at CS 5 and below, but appear to be high above CS 7. A reasonable alternative would be to use the weight change and energy reserves per CS computed for CS 5 for CS categories above a 5. Additional research is needed to be able to predict more accurately the body weights and weight changes associated with each condition score on diverse cattle types.

REFERENCES

Abdalla, H. O., D. G. Fox, and M. L. Thonney. 1988. Compensatory gain by Holstein calves after underfeeding protein. J. Anim. Sci. 66:2687–2695.

Agricultural and Food Research Council. 1993. Energy and protein requirements of ruminants. Wallingford, U.K.: CAB International.

Ainslie, S. J., D.G. Fox and T. C. Perry. 1992. Management systems for Holstein steers that utilize alfalfa silage and improve carcass value. J. Anim. Sci. 70:2643–2651.

Ainslie, S. J., D. G. Fox, T. C. Perry, D. J. Ketchen, and M. C. Barry. 1993. Predicting amino acid adequacy of diets fed to Holstein steers. J. Anim. Sci. 71:1312–1319.

Anrique, R. G., M. L. Thonney, and H. J. Ayala. 1990. Dietary Energy losses of cattle influenced by body type, size, sex and intake. Anim. Prod. 50:467–474.

Bartle, S. J., R. L. Preston, R. E. Brown, and R. J. Grant. 1992. Trenbolone acetate/estradiol combinations in feedlot steers: dose-response and implant carrier effects. J. Anim. Sci. 70:1326–1332.

Beef Improvement Federation. 1986. Guidelines for Uniform Beef Improvement Programs, Fifth Ed. Raleigh, N.C.: Beef Improvement Federation.

Bergen, W. G., and D. B. Bates. 1984. Ionophores: Their effect on production efficiency and mode of action. J. Anim. Sci. 58:1465–1483.

Boisclair, Y. R., D. E. Bauman, A. W. Bell, F. R. Dunshea, and M. Harkins. 1994. Nutrient utlization and protein turnover in the hindlimb of cattle treated with bovine somatotropin. J. Nutr. 124:664–673.

Buskirk, D. D., R. P. Lemenager, and L. A. Horstman. 1992. Estimation of net energy requirements (NE_m and NE) of lactating beef cows. J. Anim. Sci. 70:3867–3876.

TABLE 3-6 Energy Reserves for Cows with Different Body Sizes and Condition Scores

| CS | Mcal NE Required or Provided for Each CS[a] at CS 5 Mature Weight | | | | | | | | |
	400	450	500	550	600	650	700	750	800
2	112	126	140	154	168	182	196	210	223
3	126	141	157	173	189	204	220	236	251
4	144	162	180	198	217	235	253	271	289
5	165	186	207	227	248	269	289	310	331
6	193	217	242	266	290	314	338	362	386
7	228	267	285	314	342	371	399	428	456
8	275	309	343	378	412	446	481	515	549
9	335	377	419	461	503	545	587	629	670

[a]Represents the energy mobilized in moving to the next lower score, or required to move from the next lower score to this one. Each kg of SBW change contains 5.82 Mcal, and SBW at CS 1, 2, 3, 6, 7, 8, and 9 are 76.5, 81.3, 86.7, 92.9, 108.3, 118.1, 129.9, and 144.3 percent of CS 5 weight, respectively.

Butler, W. R. and R. W. Canfield. 1989. Interrelationships between energy balance and postpartum reproduction. Proc. Cornell Nutr. Conf:66.

Byers, F. M. 1980. Determining effects of monensin on energy value of corn silage diets for beef cattle by linear or semi-log methods. J. Anim. Sci. 51:158–169.

Byers, F. M., G. T. Schelling and L. W. Greene. 1989. Development of growth functions to describe energy density of growth in beef cattle. Energy Metab. Proc. Symp. 43:195–198.

Byrem, T. M., T. F. Robinson, A. W. Bell, and D. H. Beermann. 1993. The B-adrenergic agonist cimaterol enhances protein accretion during a close-arterial infusion into the hind limb of steers. FASEB J. 7(4):A645 (abstr. 3735).

Cantrell, J. A., J. R. Kropp, S. L. Armbruster, K. S. Lusby, R. P. Wettemann, and R. L. Hintz. 1982. The influence of postpartum nutrition and weaning age of calves on cow body condition, estrus, conception rate and calf performance of fall-calving beef cows. Oklahoma Agric. Exp. Sta. Res. Rep. MP-112:53.

Carstens, G. E., D. E. Johnson, and M. A. Ellenberger. 1987. The energetics of compensatory growth in beef cattle. J. Anim. Sci. 65(Suppl. 1):263 (abstr.).

Commonwealth Scientific and Industrial Research Organization. 1990. Feeding standards for Australian livestock. Ruminants. East Melbourne, Victoria, Australia: CSIRO Publications.

Crickenberger, R. G., D. G. Fox, and W. T. Magee. 1978. Effect of cattle size and protein level on the utilization of high corn silage or high grain rations. J. Anim. Sci. 46:1748–1758.

Cundiff, L. V., R. W. Koch, and G. M. Smith. 1981. Characterization of biological types of cattle—cycle II. IV. Post weaning growth and feed efficiency of steers. J. Anim. Sci. 53:332–346.

Cundiff, L. V., K. E. Gregory, R. H. Koch and G. E. Dickerson. 1986. Genetic diversity among cattle breeds and its use to increase beef production efficiency in a temperate climate. Proceedings of the Third World Congress on Genetics Applied to Beef Production. Lincoln, NE IX:271.

Danner, M. L., D. G. Fox, and J. R. Black. 1980. Effect of feeding system on performance and carcass characteristics of yearling steers, steer calves and heifer calves. J. Anim. Sci. 50:394–404.

Delfino, J., G. W. Mathison, and M. W. Smith. 1988. Effect of lasalocid on feedlot performance and energy partitioning in cattle. J. Anim. Sci. 66:136–150.

Dunn, T. G. and G. E. Moss. 1992. Effects of nutrient deficiencies and excesses on reproductive efficiency of livestock. J. Anim. Sci. 70:1580–1593.

Dziuk, P. J. and R. A. Bellows. 1983. Management of reproduction in beef cattle, sheep and pigs. J. Anim. Sci. 57(Suppl. 2):355.

Early, R. J., B. W. McBride, and R. O. Ball. 1990. Growth and metabolism in somatotropin-treated steers: III. Protein synthesis and tissue energy expenditures. J. Anim. Sci. 68:4153–4166.

Evans, E. H., and R. J. Patterson. 1985. Use of dynamic modelling seen as good way to formulate crude protein, amino acid requirements for cattle diets. Feedstuffs 57(42):24.

Ferguson, J. D., and K. A. Otto. 1989. Managing body condition in cows. Proc. Cornell Nutr. Conf. 75.

Ferrell, C. L. 1982. Effects of postweaning rate of gain on onset of puberty and productive performance of heifers of different breeds. J. Anim. Sci. 55:1272–1283.

Ferrell, C. L. 1988. Energy metabolism. In The Ruminant Animal—Digestive Physiology and Nutrition, D. Church, ed. Englewood Cliffs, N.J.: Prentice Hall.

Ferrell, C. L. 1991. Nutritional influences on reproduction. In Reproduction in Domestic Animals, Fourth Ed., P. T. Cupps, ed. Los Angeles: Academic.

Ferrell, C. L., L. J. Koong, and J. A. Nienaber. 1986. Effect of previous nutrition on body composition and maintenance energy costs of growing lambs. Br. J. Nutr. 56:595–605.

Foldager, J., and K. Serjsen. 1987. Research in Cattle Production: Danish Status and Perspectives. Tryk Denmark: Landhusholdningsselskabets Forlag.

Fortin, A., S. Simpfendorfer, J. T. Reid, H. J. Ayala, R. Anrique, and A. F. Kertz. 1980. Effect of level of energy intake and influence of breed and sex on the chemical composition of cattle. J. Anim. Sci. 51:604–614.

Fox, D. G., and J. R. Black. 1984. A system for predicting body composition and performance of growing cattle. J. Anim. Sci. 58:725–739.

Fox, D. G., R. R. Johnson, R. L. Preston, T. R. Dockerty, and E. W. Klosterman. 1972. Protein and energy utilization during compensatory growth in beef cattle. J. Anim. Sci. 34:310.

Fox, D. G., C. J. Sniffen, J. D. O'Connor, J. B. Russell, and P. J. Van Soest. 1992. A net carbohydrate and protein system for evaluating cattle diets. III. Cattle requirements and diet adequacy. J. Anim. Sci. 70:3578–3596.

Fox, D. G., M. C. Barry, R. E. Pitt, D. K. Roseler, and W. C. Stone. 1995. Application of the Cornell net carbohydrates and protein model for cattle consuming forages. J. Anim. Sci. 73:267.

Galbraith, H., and J. H. Topps. 1981. Effect of hormones on the growth and body composition of animals. Nutr. Abstr. Rev. Ser. B. 51:521–540.

Garrett, W. N. 1980. Energy utilization by growing cattle as determined in 72 comparative slaughter experiments. Energy Metab. Proc. Symp. 26:3–7.

Garrett, W. N. 1987. Relationship between energy metabolism and the amounts of protein and fat deposited in growing cattle. Energy Metab. Proc. Symp. 32:98–101.

Garrett, W. N., J. H. Meyer and G. P. Lofgreen. 1959. The comparative energy requirements of sheep and cattle for maintenance and gain. J. Anim. Sci. 18:528–547.

George, P. D. 1984. A Deterministic Model of Net Nutrient Requirements for the Beef Cow. Ph.D. dissertation. Cornell University, Ithaca, N.Y.

Gill, C. S., and F. R. Allaire. 1976. Relationship of age at first calving, days open, days dry, and herd life with a profile function for dairy cattle. J. Dairy Sci. 59:1131–1139.

Goodrich, R. D., J. E. Garrett, D. R. Ghast, M. A. Kirich, D. A. Larson, and J. C. Meiske. 1984. Influence of monesin on the performance of cattle. J. Anim. Sci. 58:1484–1498.

Gregory, K. E., L. V. Cundiff and R. M. Koch. 1992. Composite Breeds to Use Heterosis and Breed Differences to Improve Efficiency of Beef Production. Misc. Pub. Washington, D.C.: U.S..Department of Agriculture, Agricultural Research Service.

Haecker, T. L. 1920. Investigation in beef production. Minnesota Agr. Exp. Sta. Bull. 688.

Haigh, L. D., C. R. Moulton, and P. F. Trowbridge. 1920. Composition of the bovine at birth. Missouri Agr. Exp. Sta. Bull. 38.

Harpster, H. W. 1978. Energy requirements of cows and the effect of sex, selection, frame size, and energy level on performance of calves of four genetic types. Ph.D. dissertation. Michigan State University, East Lansing, Michigan.

Harrison, R. D., I. P. Reynolds and W. Little. 1983. A quantitative analysis of mammary glands of dairy heifers reared at different rates of liveweight gain. J. Dairy Res. 50:405–412.

Herd, D. B., and L. R. Sprott. 1986. Body condition, nutrition and reproduction of beef cows. Texas A&M Univ. Ext. Bull. 1526.

Hogan, J. P. 1974. Quantitative aspects of nitrogen utilization in ruminants. Symposium: Protein and amino acid nutrition in the high producing cow. J. Dairy Sci. 58:1164.

Houghton, P. L., R. P. Lemenager, G. E. Moss, and K. S. Hendrix. 1990. Prediction of postpartum beef cow body composition using weight to height ratio and visual body condition score. J. Anim. Sci. 68:1428.

Houseknecht, K. L., D. E. Baumn, D. G. Fox, and D. F. Smith. 1992.

Abomasal infusion of casein enhances nitrogen retention in somatotropin-treated steers. J. Nutr. 122:1717–1725.

Institut National de la Recherche Agronomique. 1989. Ruminant Nutrition. Montrouge, France: Libbey Eurotext.

Jenkins, T. G., and C. L. Ferrell. 1984. Output/input differences among biological types. Proceedings Beef Cow Efficiency Symposium pp 15–37. Michigan State Univ., East Lansing.

Jesse, G. W., G. B. Thompson, J. L. Clark, H. B. Hedrick, andvK. G. Weimer. 1976. Effects of ration energy and slaughter weight on composition of empty body and carcass gain of cattle. J. Anim. Sci. 43:418.

Joubert, D. M. 1963. Puberty in female farm animals. Anim. Breed. Abstr. 31:295–306.

Keele, J. W., C. B. Williams, and G. L. Bennett. 1992. A computer model to predict the effects of level of nutrition on composition of empty body gain in beef cattle: I. Theory and development. J. Anim. Sci. 70:841–857.

Lamond, D. R. 1970. The influence of undernutrition on reproduction in the cow. Anim. Breed. Abstr. 38:359–372.

Lanna, D. P., C. Boin, and D. G. Fox. 1994. Validation of the CNCPS and NRC (1984) for estimating nutritional requirements and performance of growing Zebu. Anais da XXXI Reunaio Anual da Sociedade Brasileira de Zootecnia, July 17–21, Maringa, PR, Brazil.

Laster, D. B., H. A. Glimp, and K. E. Gregory. 1972. Age and weight at puberty and conception in different breeds and breed crosses of beef heifers. J. Anim. Sci. 34:1031–1036.

Laster, D. B., G. M. Smith, and K. E. Gregory. 1976. Characterization of biological types of cattle. IV. Postweaning growth and puberty of heifers. J. Anim. Sci. 43:63–70.

Laster, D. B., G. M. Smith, L. V. Cundiff, and K. E. Gregory. 1979. Characterization of biological types of cattle (Cycle II). II. Postweaning growth and puberty of heifers. J. Anim. Sci. 48:500–508.

Lobley, G. E. 1992. Control of the metabolic fate of amino acids in ruminants: A review. J. Anim. Sci. 70:3264–3275.

Lofgreen, G. P., and W. N. Garrett. 1968. A system for expressing net energy requirements and feed values for growing and finishing cattle. J. Anim. Sci. 27:793–806.

Lomas, L. W., D. G. Fox, and J. R. Black. 1982. Ammonia treatment of corn silage. I. Feedlot performance of growing and finishing cattle. J. Anim. Sci. 55:909–923.

Lucas, H. L., Jr., W. W. G. Smart, Jr., M. A. Cipolloni, and H. D. Gross. 1961. Relations Between Digestibility and Composition of Feeds and Feeds, S-45 Report. Raleigh: North Carolina State College.

Martin, L. C., J. S. Brinks, R. M. Bourdon, and L. V. Cundiff. 1992. Genetic effects on beef heifer puberty and subsequent reproduction. J. Anim. Sci. 70:4006–4017.

Merchen, N. R., and E. C. Titgemeyer. 1992. Manipulation of amino acid supply to the growing ruminant. J. Anim. Sci. 70:3238–3247.

Moe, P. 1981. Energy metabolism of dairy cattle. J. Dairy Sci. 64:1120–1139.

Moulton, C. R., P. F. Trowbridge, and L. D. Haigh. 1922. Studies in animal nutriton. III. Changes in chemical composition on different planes of nutrition. Missouri Agr. Exp. Sta. Res. Bull. 55.

Murray, R. K., D. K. Granner, P. A. Mayes, and V. W. Rodwell. 1988. Harper's Biochemistry. Norwalk, Conn.: Appleton and Lange.

National Research Council. 1976. Nutrient Requirements of Beef Cattle. Washington, D.C.: National Academy of Sciences.

National Research Council. 1984. Nutrient Requirements of Beef Cattle, Sixth Revised Ed. Washington, D.C.: National Academy Press.

National Research Council. 1985. Ruminant Nitrogen Usage. Washington, D.C.: National Academy Press.

National Research Council. 1987. Predicting Feed Intake of Food-Producing Animals. Washington, D.C.: National Academy Press.

National Research Council. 1989. Nutrient Requirements of Dairy Cattle, Sixth Revised Ed. Update. Washington, D.C.: National Academy Press.

National Research Council. 1994. Metabolic Modifiers: Effects on the Nutrient Requirements of Food-Producing Animals. Washington, D.C.: National Academy Press.

Oltjen, J. W., A. C. Bywater, R. L. Baldwin, and W. N. Garrett. 1986. Development of a dynamic model of beef cattle growth and composition. J. Anim. Sci. 62:86–97.

Patterson, D. J., R. C. Perry, G. H. Kiracofe, R. A. Bellows, R. B. Staigmiller, and L. R. Corah. 1992. Management considerations in heifer development and puberty. J. Anim. Sci. 70:4018–4035.

Perry, T. C., D. G. Fox, and D. H. Beermann. 1991a. Influence of final weight, implants, and frame size on performance and carcass quality of Holstein and beef breed steers. Northeast Regional Agr. Eng. Serv. Bull. 44:142.

Perry, T. C., D. G. Fox, and D. H. Beermann. 1991b. Effect of an implant of trenbolone acetate and estradiol on growth, feed efficiency and carcass composition of Holstein and beef breed steers. J. Anim. Sci. 69:4696–4702.

Raun, A. P. 1990. Rumensin "then and now." Pp. A1–A20 in Rumensin in the 1990s. Indianapolis, Ind.: Elanco Animal Health.

Robinson, J. J. 1990. Nutrition in the reproduction of farm animals. Nutr. Res. Rev. 3:253–276.

Robinson, T. F., D. H. Beermann, T. M. Byrem, M. E. VanAmburgh, and D. A. Ross. 1994. Effects of abomasal casein infusion on nitrogen and amino acid absorption and nitrogen utilization in growing Holstein steers. J. FASEB 8(4):A176 (abstr.).

Robinson, T. F., D. H. Beermann, T. M. Byrem, D. E. Ross, and D. G. Fox. 1995. Effects of abomasal casein infusion on mesenteric drained viscera amino acid absorption, hindlimb amino acid net flux and whole body nitrogen balance in Holstein steers. J. Anim. Sci. 73(Suppl. 1):140.

Roche, J. F., M. A. Corwe, and M. P. Boland. 1992. Postpartum anestrus in dairy and beef cows. Anim. Reprod. Sci. 28:371–378.

Rohr, K., and P. Lebzien. 1991. Present knowledge of amino acid requirements for maintenance and production. In Proceedings of the Sixth International Symposium on Protein Metabolism and Nutrition. Herning, Denmark.

Sacco, R. E., J. F. Baker, and T. C. Cartwright. 1987. Production characters of primiparous females of a five-breed diallel. J. Anim. Sci. 64:1612–1618.

Schillo, K. K., J. B. Hall, and S. M. Hileman. 1992. Effect of nutrition and season on onset of puberty in the beef heifers. J. Anim. Sci. 70:3994–4005.

Selk, G. E., R. P. Wettemann, K. S. Lusby, J. W. Oltjen, S. L. Mobley, R. J. Rasby, and J. C. Garmendia. 1988. Relationships among weight change, body condition and reproductive performance of range beef cows. J. Anim. Sci. 66:3153–3159.

Simpfendorfer, S. 1974. Relationship of Body Type, Size, Sex, and Energy Intake to the Body Composition of Cattle. Ph.D. dissertation. Cornell University, Ithaca, N.Y.

Smith, G. M., D. B. Laster, and L. V. Cundiff. 1976. Characterization of biological types of cattle. II. Postweaning growth and feed efficiency of steers. J. Anim. Sci. 43:37–47.

Solis, J. C., F. M. Byers, G. T. Schelling, C. R. Long, and L. W. Green. 1988. Maintenance requirements and energetic efficiency of cows of different breeds. J. Anim. Sci. 66:764–773.

Spears, J. W. 1990. Ionophores and nutrient digestion and absorption in ruminants. J. Nutr. 120:632–638.

Spitzer, J. C., J. N. Wiltbank, and D. C. Lefever. 1975. Increasing beef cow productivity by increasing reproductive performance. Colorado State Univ. Exp. Sta., Gen. Ser. 949.

Sprott, L. R., T. B. Goehring, and J. R. Beverly, and L. R. Corah. 1988. Effects of ionophores on cow herd production: A review. J. Anim. Sci. 66:1340–1346.

Stewart, T.S., C. R. Long, and T. C. Cartwright. 1980. Characterization

[handwritten margin note: See front cover "Errata"]

of cattle of a five breed diallel. III. Puberty in bulls and heifers. J. Anim. Sci. 50:808–820.

Taylor, C. S. 1968. Time taken to mature in relation to mature weight for sexes, strains, and species for domesticated mammals and birds. Anim. Prod. 10:157–169.

Trenkle, A. 1990. Impact of implant strategies on performance and carcass merit of feedlot cattle. In Proceedings of the 1990 Southwest Nutrition Conference. Tempe, Ariz.

Trowbridge, P. F., C. R. Moulton, and L. D. Haigh. 1918. Effect of limited food supply on the growth of young beef animals. Missouri Agr. Exp. Sta. Res. Bull. 28.

Tylutki, T. P., D. G. Fox, and R. G. Anrique. 1994. Predicting net energy and protein requirements for growth of implanted and nonimplanted heifers and steers and nonimplanted bulls varying in body size. J. Anim. Sci. 72:1806–1813.

Unruh, J. A. 1986. Effects of endogenous and exogenous growth promoting compounds on carcass composition, meat quality and meat nutritional value. J. Anim. Sci. 62:1441–1448.

Van Amburgh, M., D. Galton, D. G. Fox, and D. E. Bauman. 1991. Optimizing heifer growth. Proc. Cornell Nutr. Conf:85.

Van Maanen, R. W., J. H. Herbein, A. D. McGilliard, and J. W. Young. 1978. Effects of monensin on in vivo rumen propionate production and blood glucose kinetics in cattle. J. Nutr. 108:1002–1007.

Van Soest, P. J. 1994. Nutritional Ecology of the Ruminant, Second Ed. London: Comstock Publishing Assoc.

Wagner, J. J. 1984. Carcass Composition in Mature Hereford Cows: Estimation and Influence on Metabolizable Energy Requirements for Maintenance during Winter. Ph.D. dissertation. Oklahoma State Univ., Stillwater, Oklahoma.

Whitman, R. W. 1975. Weight Change, Body Condition and Beef Cow Reproduction. Ph.D. dissertation. Colorado State University, Fort Collins, Colorado.

Wilkerson, V. A. 1993. Metabolizable protein and amino acid requirements of growing cattle. J. Anim. Sci. 71:2777–2784.

Wilkerson, V. A., T. J. Klopferstein, R. A. Button, R. A. STock, and P. S. Miller. 1993. Metabolizable protein and amino acid requirements of growing cattle. J. Anim. Sci. 71:2777–2784.

Williams, A. P. 1978. The amino acid, collagen and mineral composition of preruminant calves. J. Agric. Sci. Camb. 90:617.

Williams, C. B., J. W. Keele, and D. R. Waldo. 1992. A computer model to predict empty body weight in cattle from diet and animal characteristics. J. Anim. Sci. 70:3215–3222.

Wiltbank, J. N., S. Roberts, J. Nix, and L. Rowden. 1985. Reproductive performance and profitability of heifers fed to weight 272 or 318 kg at the start of the first breeding season. J. Anim. Sci. 60:25–34.

Woody, H. D., D. G. Fox, and J. R. Black. 1983. Effect of diet grain content on performance of growing and finishing cattle. J. Anim. Sci. 57:717–728.

Wright, V. A., and A. J. Russel. 1984a. Partition of fat, body composition, and body condition score in mature cows. Anim. Prod. 38:23–32.

Wright, V. A., and A. J. Russel. 1984b. Estimation in vivo of the chemical composition of the bodies of mature cows. Anim. Prod. 38:33–43.

Wright, I. A., S. M. Rhind, T. K. Whyte, and A. J. Smith. 1992. Effects of body condition at calving and feeding level after calving on LH profiles and the duration of the post-partum anoestrous period in beef cows. Anim. Prod. 55:41–46.

4 Reproduction

GESTATION

Meeting the nutrient requirements of pregnant female cattle is important to ensuring an adequate nutrient supply for proper growth and development of the fetus and to ensuring that the female is in adequate body condition to calve and lactate, to rebreed within 80 days after calving, and to provide, in the case of the 2- or 3-year-old heifer, adequate nutrients for continued growth. This section will concentrate on nutrient requirements for pregnancy—in particular, energy and protein—in cattle and some of the factors affecting those requirements.

For lack of information to the contrary, it is generally assumed that nutrient needs for pregnancy are proportional to birth weight of the calf. Thus, it is assumed that factors that affect calf birth weight have a proportional affect on nutrient requirements during pregnancy. Factors known to affect calf birth weight include breed of sire, breed of dam, heterosis, parity of the dam, number of fetuses, sex of the fetus, environmental temperature, and nutrition of the dam (Ferrell, 1991a).

Of the factors affecting calf birth weight, breed or genotype of the sire, dam, or calf generally has the greatest influence (Andersen and Plum, 1965). Typical birth weight of calves of various breeds are listed in Table 4-1. Birth weights of calves in one study differed by as much as 18 kg (Agricultural and Food Research Council, 1990). Ranges to 10 kg were reported for mean birth weights of calves of different breeds typically used for beef production in the United States (Beef Improvement Federation, 1990) or those of crossbred calves from Angus and Hereford dams (Gregory et al., 1982; Cundiff et al., 1988). Heterosis, resulting in increased birth weight, is generally about 6 to 7 percent when *Bos taurus* breeds are crossed, less (0 or negative) when *Bos taurus* sires are crossed on *Bos indicus* dams, but considerably higher (20 to 25 percent) when the reciprocal mating is made (Ellis et al., 1965; Long, 1980;

TABLE 4-1 Estimated Birth Weight of Calves of Different Breeds or Breed Crosses, kg

Breed	BIF	AFRC	MARC
Angus	31	26	35
Brahman	31	—	41
Braford	36	—	—
Brangus	33	—	—
Braunvieh	—	—	39
Charolais	39	43	40
Chianina	—	—	41
Devon	32	34	—
Galloway	—	—	36
Gelbvieh	39	—	39
Hereford	36	35	37
Holstein	—	43	—
Jersey	—	25	31
Limousin	37	38	39
Longhorn	—	—	33
Maine-Anjou	40	—	41
Nellore	—	—	40
Piedmontese	—	—	38
Pinzgauer	33	—	40
Polled Hereford	33	—	36
Red Poll	—	—	36
Sahiwal	—	—	38
Santa Gertrudis	33	—	—
Salers	35	—	38
Shorthorn	37	32	39
Simmental	39	43	40
South Devon	33	42	38
Tarentaise	33	—	38

NOTE: BIF, Beef Improvement Federation; AFRC, Agricultural and Food Research Council; MARC, Roman L. Hruska U.S. Meat Animal Research Center (USDA/ARS).

Sources: Beef Improvement Federation (1990), AFRC (1990), MARC, from data reported by Cundiff et al. (1988), and Gregory et al. (1982), which are from a particular sire breed on mature Angus and Hereford cows.

Gregory et al., 1992a). Weight of heifer calves average 7 percent less than bull calves at birth (Agricultural and Food Research Council, 1990; Beef Improvement Federation, 1990), and weight of calves born to 2-, 3-, and 4-year old cows average 8, 5, and 2 percent less than those born to

40

5- to 10-year-old cows (Beef Improvement Federation, 1990; Gregory et al., 1990). Birth weight of calves born as twins is 25 percent less, but the total weight of twins average 150 percent of the birth weight of calves born as singles (Gregory et al., 1990).

Severe energy or protein underfeeding has resulted in marked reductions of calf birth weight (Hight, 1966, 1968a,b; Tudor, 1972). Inadequate food intake during late pregnancy is also associated with weak labor, increased dystocia, reduced milk production and growth of progeny, and lowered rebreeding performance of the dam (Bellows and Short, 1978; Kroker and Cummins, 1979). Conversely, gross overfeeding during pregnancy can also result in reduced birth weight and subsequent decreased milk production, increased dystocia and neonatal death loss, and poor rebreeding performance (Arnett et al., 1971; Robinson, 1977). The relationship of calf birth weight to cow condition score is typified by data shown in Figure 4-1. Birth weight decreased as cow condition score decreased below 3.5 or increased above 7, but did not change within the range of cow condition scores of about 3.5 to 7. It is suggested that calf birth weight is not substantially influenced by cow nutritional status within a broad range, but may be reduced by extreme over- or underfeeding. In those situations, negative influences on rebreeding performance, dystocia, etc., are of greater concern than calf birth weight.

Effects of Temperature

Although this section is primarily concerned with factors affecting calf birth weight, it is important to note that high environmental temperature during or shortly after conception can significantly increase embryonic mortality in cattle as well as several other species (Bell, 1987). In addition, high environmental temperatures, particularly during early pregnancy, may result in a wide range of

congenital defects. Limited data are available from well-controlled studies of cattle to characterize the influence of elevated temperatures on calf birth weight (Collier et al., 1982) and, to this subcommittee's knowledge, no data are available from controlled experiments to characterize influences of chronic cold exposure, although these effects have been well documented in sheep (Alexander and Williams, 1971; Rutter et al., 1971, 1972; Cartwright and Thwaites, 1976; Thompson et al., 1982; Bell, 1987). Numerous data are available, however, to indicate that calves born in the spring are heavier than those born in the fall (McCarter et al., 1991a), calves born in the northern areas of the United States are heavier than those born in southern areas, and that genotype/environment interactions may have important influences on calf birth weight (Burns et al., 1979; Olson et al., 1991). The magnitude of response of calf birth weight to environmental temperature is influenced by severity, duration, and timing of exposure as well as genotype of the dam.

Factors Affecting Fetal Growth

Considerable progress has been made toward understanding how various factors affect fetal growth and the ensuing birth weight. Normal fetal growth follows an exponential pattern (Figure 4-2). In cattle, weight of uterine and placental tissues also increase exponentially (Ferrell et al., 1976a; Prior and Laster, 1979). Growth and development of the uterus and placental tissues precedes fetal growth. Development of those tissues is required to support subsequent fetal growth (Ferrell, 1991b,c). Growth of the fetus is a result of its genetic potential for growth, which is reflected in its demand for nutrients and constraints imposed by the maternal and placental systems in meeting that demand (Gluckman and Liggins, 1984; Ferrell, 1989). The potential of the maternal and placental systems to meet those demands are reflected in uterine blood flow or placental size and functional capacity. The influence of maternal nutrition on fetal development is complicated by the fact that the fetus can be undernourished in well-fed mothers when placental size or function is inadequate to meet fetal demands. Conversely, even though the mother is undernourished, the maternal and placental systems may compensate such that fetal malnutrition is minimal (Bassett, 1986, 1991). Weight and perfusion of uterine and placental tissues are reduced with heat (Alexander and Williams, 1971; Cartwright and Thwaits, 1976; Reynolds et al., 1985; Bell et al., 1987) and with twins as compared with single fetuses (Bellows et al., 1990; Ferrell and Reynolds, 1992). These variables are also influenced by genotype of sire, dam, or fetus (Ferrell, 1991c). Numerous other data are available to indicate that perfusion of uterine and placental tissues and functional capacity of the placenta have central roles in fetal growth (Alexander, 1964a,b; Owens et al., 1986).

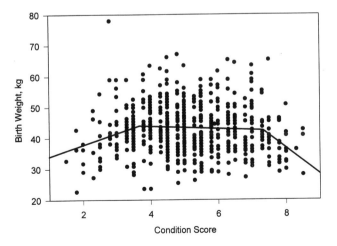

FIGURE 4-1　Relationship of calf birth weight to cow condition score in mature cows of nine breeds.

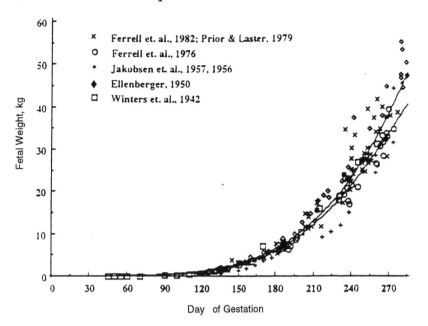

FIGURE 4-2　Relationship of fetal weight to day of gestation in cattle.

The Role of the Placenta

Functions of the placenta include exchange of metabolites, water, heat, and respiratory gasses. The placenta also serves as a site of synthesis and secretion of numerous hormones and extensive interconversion of nutrients and other metabolites (Munro et al., 1983; Battaglia, 1992). Placental transport of oxygen, glucose, amino acids, and urea and placental clearance of highly diffusible solutes increase during gestation as indicated by net fetal uptake or loss in both sheep and cattle (Bell et al., 1986; Reynolds et al., 1986). Because of the numerous metabolic functions of the uterus and placenta (uteroplacenta), oxidative metabolism is extensive throughout gestation. Even in late gestation when the fetus is several times larger than the placenta, energy consumption of the uteroplacenta is about equal to that of the fetus (Reynolds et al., 1986). Similarly, uteroplacental net use of glucose is at least 70 percent of gravid uterine glucose uptake, even in late gestation. Likewise, a major proportion of the net use of amino acids taken up from the uterine circulation is metabolized by the uteroplacenta (Reynolds et al., 1986; Ferrell, 1991b). An increase in maternal metabolism is also required to support the requirements of pregnancy. Thus, of the total increase in energy expenditure associated with pregnancy, about one-half may be attributed to metabolism of tissues of the gravid uterus and about one-fourth may be attributed to the fetus per se (Kleiber, 1961; Ferrell and Reynolds, 1987).

Energy Requirements

Energy accretion in the gravid uterus of Hereford heifers bred to Hereford bulls has been reported by Ferrell et al.

(1976a). The equation used to describe the relationship of energy content of the gravid uterus (Ye) vs day of gestation (t) in kcal, was

$$Ye, kcal = 69.73e^{(0.03233 - 0.0000275t)t}. \qquad \text{Eq. 4-1}$$

Similar values can be calculated from the data of Prior and Laster (1979) who used crossbred heifers bred to Brown Swiss bulls, and from the data of Jakobsen (1956) and Jakobsen et al. (1957) who used Red Danish cattle. Other information related to bovine fetal growth and weight change of the pregnant cow is available (Winters et al., 1942; Ellenberger et al., 1950; Eley et al., 1978; Silvey and Haycock, 1978).

Eq. 4-1 was associated with a predicted calf birth weight of 38.5 kg. Scaling Eq. 4-1 by birth weight yields the following equation (kcal):

$$Ye = \text{birth weight } (1.811)e^{(0.03233 - 0.0000275t)t}. \qquad \text{Eq. 4-2}$$

This equation may be differentiated with respect to t to estimate daily energy accretion in the tissues of the gravid uterus, yielding (kcal/day):

$$Ye = \text{birth weight } (0.05855 - 0.0000996t)\, e^{(0.03233 - 0.0000275t)t} \qquad \text{Eq. 4-3}$$

The gross efficiency of metabolizable energy (ME) use for accretion in the gravid uterus of cattle averaged 14 percent (Ferrell et al., 1976b). Other estimates with cattle and sheep average about 13 percent (Graham, 1964; Langlands and Southerland, 1968; Lodge and Heaney, 1970; Moe et al., 1970; Moe and Tyrrell, 1971; Sykes and Field, 1972; Rattray et al., 1974; Robinson et al., 1980). Some of the potential reasons for the low estimates of apparent gross efficiency have been discussed previously. Use of the average value of 13 percent efficiency results in the following equation to estimate the daily ME requirement for pregnancy in cattle:

$$ME = \text{birth weight } (0.4504 - 0.000766t)\, e^{(0.03233 - 0.0000275t)t} \qquad \text{Eq. 4-4}$$

Some evidence is available to indicate that efficiencies of ME use for maintenance and pregnancy vary similarly (Robinson et al., 1980). Values for efficiency of utilization of ME for maintenance (k_m) may be calculated from the equation of Garrett (1980a) as follows:

$$K_m = (1.37\, ME - 0.138\, ME^2 + 0.0105\, ME^3 - 1.12)/ME \qquad \text{Eq. 4-5}$$

or,

$$K_m = NE_m/ME;$$

where NE_m is net energy required for maintenance. The estimate of ME required for pregnancy may be converted to NE_m equivalent (kcal/day) by use of appropriate estimate of k_m as follows:

$$NE_m = k_m * \text{birth weight } (0.4504 - 0.000766)e^{(0.03233 - 0.0000275t)t}. \qquad \text{Eq. 4-6}$$

If it is assumed, for example, that cows typically consume primarily forage diets containing 2.0 Mcal ME/kg, k_m is expected to be 0.576. With this assumption, the NE_m required for pregnancy may be estimated from the following equation (kcal/day):

$$NE_m = 0.576 \text{ birth weight } (0.4504 - 0.000766t)e^{(0.03233 - 0.00275t)t}. \qquad \text{Eq. 4-7}$$

Estimates of the NE_m required for pregnancy, from this equation, are shown in Table 4-2. For comparison purposes, previous estimates from NRC (1984) and CSIRO (1990) are also shown.

Protein Requirements

Protein requirements for pregnancy may be estimated using the approach used with energy. Estimates of nitrogen (N) content of gravid uterine tissues at various stages of

gestation have been reported by Jakobsen (1956), Ferrell et al. (1976a), and Prior and Laster (1979). The equation derived by Ferrell et al. (1976a) to relate N (g) content of those tissues to day of gestation (t) was

$$N = 2.312e^{(0.0278 - 0.0000176t)t}. \qquad \text{Eq. 4-8}$$

As with energy, this relationship may be scaled by predicted calf birth weight (38.5 kg) to derive the following equation

$$N = \text{birth weight } (0.060)e^{(0.0278 - 0.0000176t)t}. \qquad \text{Eq. 4-9}$$

Daily accretion of N in gravid uterine tissues may be calculated by differentiation of Eq. 3-9 with respect to t as follows:

$$N = \text{birth weight } (0.001669 - 0.00000211t)e^{(0.0278 - 0.0000176t)t}. \qquad \text{Eq. 4-10}$$

Supplementary net protein required for pregnancy is estimated from daily N accretion in gravid uterine tissues as

$$\text{net protein} = N \text{ accretion, g/day} * 6.25. \qquad \text{Eq. 4-11}$$

Resulting values are shown in Table 4-3 for several stages of gestation. It should be noted that because of the high rate of metabolism of amino acids by uteroplacental and fetal tissues relative to accretion (Ferrell et al., 1983; Battaglia, 1992), as well as changes in extrareproductive tissue metabolism, these should be considered minimal estimates.

LACTATION

Milk production in the beef cow is difficult to assess. In contrast to the dairy cow, which is generally milked by machine two or more times daily, the beef cow is generally in a pasture or range environment and milk produced is consumed by the suckling calf. Numerous efforts have been made to assess milk production of beef cows with suckling calves with minimal disturbance of the normal routine of the cow and calf (Lampkin and Lampkin, 1960; Neville, 1962; Christian et al., 1965; Gleddie and Berg, 1968; Lamond et al., 1969; Deutscher and Whiteman,

TABLE 4-2 Estimates of NE_m (Mcal/day) Required for Pregnancy

Days of Gestation	This Report	NRC, 1984	CSIRO, 1990
130	0.327	0.199	0.280
160	0.634	0.505	0.509
190	1.166	1.083	0.923
220	2.027	1.952	1.673
250	3.333	2.916	3.029
280	5.174	3.518	5.478

NOTE: Estimates are based on calf birth weight of 38.5 kg.

TABLE 4-3 Estimates of Available Net Protein Required for Pregnancy by Beef Cows on Several Days of Gestation

Days of Gestation	Available Protein, g/day
130	9.1
160	17.5
190	32.2
220	56.0
250	95.2
280	156.1

NOTE: Estimates are based on calf birth weight of 38.5 kg.

1971; Totusek et al., 1973). The primary methods include hand milking with the calf nursing, machine or hand milking after oxytocin injection, and weighing the calf before and after (weigh-suckle-weigh) nursing (Kropp et al., 1973; Totusek et al., 1973; Cundiff et al., 1974; Holloway et al., 1975; Neidhardt et al., 1979; Boggs et al., 1980; Gaskins and Anderson, 1980; Chenette and Frahm, 1981; Hansen et al., 1982; Butson and Berg, 1984a,b; Jenkins and Ferrell, 1984; Holloway et al., 1985; McMorris and Wilton, 1986; Daley et al., 1987; Clutter and Nielson, 1987; Beal et al., 1990; McCarter et al., 1991; Hohenboken et al., 1992; Jenkins and Ferrell, 1992). Estimates of milk yield of grazing cows have been made at intervals varying from daily to twice during the entire lactation period. Time of separation of the calf from the cow has varied from 4 to 16 hours.

Under the above situations, milk yield estimates vary depending on the method used. In addition, milk yield estimates differ based on the genetic potential of the cow to produce milk, age and breed of the cow, capacity of the calf to consume milk (which is influenced by breed, size, age, and sex of the calf), nutritional status, thermal environment, and stage of lactation. The most commonly adapted procedure has been the weigh-suckle-weigh procedure, but several groups of researchers have used machine or hand milking. Many of the latter groups have reported composition of milk, as well as yield (Melton et al., 1967; Wilson et al., 1969; Kropp et al., 1973; Totusek et al., 1973; Cundiff et al., 1974; Holloway et al., 1975; Lowman et al., 1979; Rogers et al., 1979; Bowden, 1981; Chenette and Frahm, 1981; Mondragon et al., 1983; Butson and Berg, 1984a,b; McMorris and Wilton, 1986; Daley et al., 1987; Diaz et al., 1992; Masilo et al., 1992). It is important to note that composition as well as yield is variable. Some of the factors influencing milk composition include milk collection procedure, breed and age of cow, stage of lactation, and nutritional status.

Whereas numerous reports have included measures of milk yield of beef cows or cows with suckling calves, the primary emphasis has been to assess relative yields for breed group comparisons or to estimate the relative influence of milk yield on calf preweaning growth (Drewry et al., 1959; Christian et al., 1965; Notter et al., 1978; Reynolds et al., 1978; Robinson et al., 1978; Williams et al., 1979; Bartle et al., 1984; Marshall et al., 1984; Miller and Deutscher, 1985; Fiss and Wilton, 1989; Montano-Bermudez et al., 1990; Green et al., 1991; Freking and Marshall, 1992; Gregory et al., 1992a,b). Only a limited number of studies have reported data from which the shape of the lactation curve can be assessed (Deutscher and Whiteman, 1971; Kropp et al., 1973; Totusek et al., 1973; Grainger and Wilhelm, 1979; Neidhardt et al., 1979; Gaskins and Anderson, 1980; Chenette and Frahm, 1981; Jenkins and Ferrell, 1984; Holloway et al., 1985; Jenkins et al., 1986; Clutter and Neilson, 1987; Sacco et al., 1987; Mezzadra et al., 1989; McCarter et al., 1991; Hohenboken et al., 1992; Jenkins and Ferrell, 1992) (Figure 4-3). These studies, unlike those with dairy cows, generally include a limited number of data points for a given cow during lactation, largely because of logistical problems described previously.

The most widely applied equation for describing the lactation curve of dairy cattle has been that proposed and described by Wood (1967, 1969, 1976, 1979, 1980) of the form

$$Y_n = an^b e^{cn}$$

where the coefficients a, b, and c define the curve of production of a character Y at week n. Several other approaches have been proposed (Rowlands et al., 1982;

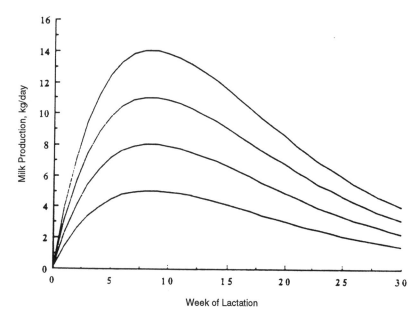

FIGURE 4-3 Generalized lactation curves for cows producing 5, 8, 11, or 14 kilograms of milk at peak milk production.

Elston et al., 1989; Morant and Granaskthy, 1989) but, as with the Woods' equation, their use with beef cattle milk production has been very limited because of the relatively large number of data points to fit the equation form. Jenkins and Ferrell (1984) proposed a similar equation form:

$$Y_n = n/ae^{kn} \qquad \text{Eq. 4-11}$$

where Y_n equals daily milk yield (kg/day) at week n postpartum, a and k are solution parameters, and e is the base of natural logarithms. This equation may be used to estimate the following values:

Time of peak milk yield $= 1/k$,   Eq. 4-12

Milk yield at time of peak $= 1/ake$ Eq. 4-13

Total yield for n weeks of lactation $=$
$$-7/ak * (ne^{-kn} + 1/ke^{-kn} - 1/k). \qquad \text{Eq. 4-14}$$

This equation form has been criticized (Hohenboken et al., 1992) but has an advantage over that of the Woods' equation in that it can be fit with a minimal number of data points. In addition, curve parameters may be estimated from published data with a minimum of information.

Available data (Deutscher and Whiteman, 1971; Jeffery et al., 1971; Kropp et al., 1973; Totusek et al., 1973; Grainger and Wilhelms, 1979; Neidhardt et al., 1979; Gaskins and Anderson, 1980; Chenette and Frahm, 1981; Jenkins and Ferrell, 1984; Holloway et al., 1985; Jenkins et al., 1986; Clutter and Neilson, 1987; Sacco et al., 1987; Lubritz et al., 1989; Mezzadra et al., 1989; McCarter et al., 1991; Hohenboken et al., 1992; Jenkins and Ferrell, 1992) indicate that peak lactation occurred at approximately 8.5 weeks postpartum in cows with suckling calves. Those data included a wide variety of breeds or breed crosses of cows, calves, milk yields, and sampling protocols. This value is somewhat later than generally observed for dairy cows and may reflect the influence of calf consumption capacity. Rearrangement of Eq. 4-12 yields

$$k = 1/8.5 \qquad \text{Eq. 4-15}$$
or
$$k = 0.1176.$$

Maximum or peak yield of cows with suckling calves is variable, as noted above. Reported values range from about 4 to 20 kg/day. The highest values have been reported for Holstein or Friesian cows. More typically, reported values for dual purpose or dairy × beef crossbred cows have rarely exceeded 14 kg/day. Therefore, for the purposes of this publication, NE_m and net protein requirements are for peak yield values of 5, 8, 11, and 14 kg/day for four types of cows typical of beef production enterprises (Tables 4-4 and 4-5). Rearrangement of Eq. 4-13 and solving for "a" yields estimates of 0.6257, 0.3911, 0.2844, and 0.2235 for cows having maximum yields of 5, 8, 11, and 14 kg/day at 8.5 weeks postpartum. Substitution of these values

TABLE 4-4   Net Energy (NE_m, Mcal/day) Required for Milk Production

Week of Lactation	Peak Milk Yield, kg/day			
	5	8	11	14
3	2.42	3.87	5.32	6.77
6	3.40	5.44	7.48	9.52
9	3.58	5.73	7.88	10.03
12	3.36	5.37	7.39	9.40
15	2.95	4.72	6.49	8.26
18	2.49	3.98	5.47	6.96
21	2.04	3.26	4.48	5.71
24	1.64	2.62	3.60	4.58
27	1.29	2.07	2.85	3.62
30	1.01	1.46	2.19	2.83

NOTE: Requirement assumes milk contains 4.0% fat, 3.4% protein, 8.3% SNF, and 0.72 Mcal/kg.

TABLE 4-5   Net Protein (g/day) Required for Milk Production

Week of Lactation	Peak Milk Yield, kg/day			
	5	8	11	14
3	115	183	252	321
6	161	258	354	451
9	170	272	373	475
12	159	254	350	445
15	140	223	307	391
18	118	188	259	330
21	97	154	212	270
24	68	124	170	217
27	61	98	135	172
30	48	77	105	134

NOTE: Requirement assumes milk contains 3.4% protein.

into Eq. 4-14 yields estimates of total milk yield over a 30-week lactation period of 701, 1,122, 1,543, and 1,963 kg. These values encompass nearly all reported values for total milk yield of beef cows with suckling calves. Expected maximum milk production is highly dependent on cow genotype and is about 26 and 12 percent lower for 2- and 3-year-old heifers, respectively, than for cows 4 years old or older (Gleddie and Berg, 1968; Gaskins and Anderson, 1980, Hansen et al., 1982; Butson and Berg, 1984a,b; Clutter and Nielson, 1987).

Insufficient data are available to fully characterize the effects of age and breed of cow, stage of lactation, nutritional status, etc., on milk composition in beef cows. Therefore, for general purposes, mean of composition values for beef cows (Melton et al., 1967; Wilson et al., 1969; Kropp et al., 1973; Totusek et al., 1973; Cundiff et al., 1974; Holloway et al., 1975; Lowman et al., 1979; Bowden, 1981; Chenette and Frahm, 1981; Grainger et al., 1983; Mondragon et al., 1983; Butson and Berg, 1984a,b; McMorris and Wilson, 1986; Daley et al., 1987; Diaz et al., 1992; Masilo et al., 1992) is assumed. The average (mean ± SD) value for milk fat was 4.03 ± 1.24 percent (18 studies), for milk protein was 3.38 ± 0.27 percent (10 studies), for solids not fat (SNF) was 8.31 ± 1.38 percent (10 studies), and for lactose was 4.75 ± 0.91 percent (5 studies). Energy content (E, Mcal/kg) of milk may be calculated as follows (Tyrrell and Reid, 1965):

$$E = 0.097 * \text{fat percent} + 0.361, \qquad \text{Eq. 4-16}$$

or

$$E = (0.092 * \text{fat percent}) + \qquad \text{Eq. 4-17}$$
$$(0.049 * \text{SNF percent}) - 0.0569.$$

Committees of the National Research Council (1984, 1989) concluded that ME is utilized for lactation and maintenance with similar efficiencies; thus the energy content of the milk produced is equivalent to the NE_m required for milk production (National Research Council, 1984). Data reported by Moe et al. (1970, 1972), Patle and Mudgal (1976), van der Honing (1980), Agricultural Research Council (1980), Garrett (1980b), Moe (1981), Munger (1991), Windisch et al. (1991), Gadeken et al. (1991), and Unsworth (1991), among others, support this conclusion. Although limited data are available, differences among breeds in efficiency of ME use for milk production appear to be minimal.

BREEDING PERFORMANCE

Beef cattle are managed under a wide variety of conditions. To a large extent their usefulness lies in their ability to harvest and utilize feed resources available under existing environmental conditions. The large variation in animal genotypes, environmental conditions, and available feed resources presents a challenge in determining and applying nutrient requirement guidelines. Providing nutrients to meet animal requirements is necessary for attainment of maximum production levels. However, it is frequently not economically advantageous to feed beef cattle in the breeding herd to meet their nutrient requirements throughout the year. Production levels to maximize net economic return vary based on interrelationships among numerous factors including, but not limited to, feed resources available, animal genotype, physiological state, costs of supplements, and environmental conditions. It should be recognized, however, that if the animals' nutrient requirements are not met during part of the year, deficits must be made up during other parts of the year if production is to be maintained.

In grazing, as in nongrazing situations, maximum efficiency of diet utilization is attained by providing nutritionally balanced diets. When energy is first limiting, for example, protein, minerals and vitamins are not efficiently utilized. Supplemental protein, in this case, will be used to meet energy needs until energy and protein are equally limiting. Conversely, if protein is first limiting, provision of additional energy will not improve performance and

may in fact depress performance. These concepts are applicable to other nutrients as well, i.e., performance is limited to that which is supported by the first-limiting nutrient. In the grazing animal, the quantity and quality of forages are of primary concern because they provide the nutrient base. The most limiting nutrients are especially difficult to establish for grazing cattle because the quantity and quality of the diets selected by the animal are difficult to assess. This is of less concern when minimal variation in forage quality results in limited opportunity for selectivity, such as occurs most commonly during spring and winter grazing.

The ultimate result of malnutrition of the beef herd is a reduction in the number of viable offspring produced. Influences of malnutrition are seen through effects on attainment of puberty, duration of the postpartum estrus, gametogenesis, conception rate, embryonic mortality, prenatal development, and sexual behavior. Some of these effects will be discussed briefly in subsequent sections. Readers are referred to recent reviews by Hurley and Doane (1989), Robinson (1990), Short et al. (1990), Ferrell (1991a), Dunn and Moss (1992), Schillo et al. (1992), and Patterson et al. (1992) for greater detail.

Heifer Development

Age at puberty is an important production trait in cattle because many of the currently used management systems require that heifers be bred, during a restricted breeding system, at 14- to 16-months-old to calve at 2 years old. Heifers that reach puberty early and have a number of estrous cycles prior to the breeding season have a higher conception rate and conceive earlier in the breeding season than ones that reach puberty later. In addition, heifers that conceive early in their first breeding season have a greater probability of weaning more and heavier calves during their productive lifetime.

EFFECTS OF FEEDING

Underfeeding, resulting in low growth rate of heifers, delays puberty in cattle; and the effects are more pronounced when applied in the early prenatal phase than when applied immediately prepubertal. In an extreme example, Rege et al. (1993) reported age at first calving in White Fulani cattle in Nigeria to be as late as 2,527 days (6.9 years). As an example of more typical conditions in temperate regions, Angus-Hereford crossbred heifers fed to gain 0.27, 0.45, or 0.68 kg/day reached puberty at an average age of 433, 411, and 388 days old, respectively (Short and Bellows, 1971). Although these differences are relatively small, pregnancy rates after a 60-day breeding season were 50, 86, and 87 percent, respectively.

EFFECTS OF MATURITY

Both age and weight at puberty differ substantially among breeds of cattle (Laster et al., 1972, 1976, 1979; Stewart et al., 1980; Sacco et al., 1987). Within beef breeds, those having larger mature size tend to reach puberty at a later age and heavier weight. *Bos indicus* heifers tend to reach puberty at an older age than *Bos taurus* heifers, and heifers from higher milk-producing breeds are generally younger at puberty than those from breeds having lower milk production. Some of those differences are likely the result of direct maternal effects expressed through higher rates of preweaning gain by calves from higher milk-producing breeds.

Numerous data are available that indicate that neither age nor weight is a reliable indicator of reproductive development but that threshold values for both age and weight must be reached before puberty can occur. This conclusion is similar to the "physiological maturity" concept proposed by Joubert (1963) and to the "target weight" concept proposed by Lamond (1970). These concepts have been used by Spitzer et al. (1975), Dziuk and Bellows (1983), and Wiltbank et al. (1985) to suggest that replacement heifers should be fed to reach a preselected or "target" weight at a given age. Heifers of most *Bos taurus* breeds of cattle are expected to reach puberty by 14 months old or younger, if fed adequately. However, threshold ages of some heifers of *Bos indicus* breeds may be older than 14 months. Generally, heifers of typical *Bos taurus* beef breeds (e.g., Angus, Charolais, Hereford, Limousin) are expected to reach puberty at about 60 percent of mature weight. Heifers of dual purpose or dairy breeds (e.g., Braunvieh, Brown Swiss, Friesian, Gelbvieh, Red Poll) tend to reach puberty at a younger age and lower weight, relative to mature weight (about 55 percent of mature weight) than those of beef breeds. Conversely, heifers of *Bos indicus* breeds (e.g., Brahman, Nellore, Sahiwal) generally reach puberty at older ages and heavier weights (about 65 percent of mature weight) than those of *Bos taurus* beef breeds (Laster et al., 1972, 1976, 1979; Stewart et al., 1980; Ferrell, 1982; Sacco et al., 1987; Martin et al., 1992; Gregory et al., 1992b; Vera et al., 1993).

Mature weight refers to weight reached at maturity by cows of the same genotype in a nonrestrictive environment (for example, mature weight as determined by genetic potential). In a restrictive environment (high environmental temperature, limited nutrition, parasite loads, etc.), mature weight of cows is often less than that of cows of similar genotype maintained in a less restrictive environment (Butts et al., 1971; Pahnish et al., 1983). Heifer weight at puberty is also reduced, but to a lesser extent than is mature weight. Thus, under those types of conditions, weight at puberty is generally a greater percentage of observed mature weight than described above (Vera et al., 1993).

If the target weight and age to reach puberty are established, and present age and weight are known, rates of gain needed to achieve the target weight and age can easily be calculated. Energy and protein needs to meet those rates of gain can be estimated by use of the previously described net energy and net protein equations for growing heifers. Excessive feeding should be avoided. In addition to increasing feed costs, overfeeding that results in excess fat accretion may have detrimental effects on expression of behavioral estrous, conception rate, embryonic and neonatal survival, calving ease, milk production, and productive life.

Weight and Condition Changes in Reproducing Females

Composition of weight change in growing and mature cattle has been discussed in other sections and will not be discussed in detail here. In the mature cow, weight change, with the exception of weight change associated with pregnancy or parturition, primarily reflects change in body condition. In the developing heifer, percentage of body fat and body condition may decrease, even though weight may continue to increase because of skeletal and muscle growth at the expense of body fat. In both the heifer and cow, weight gain associated with pregnancy and weight loss at parturition should not be construed as change in maternal weight or condition. Weight gain during pregnancy and loss at parturition is about 1.7 times calf birth weight and represents weight gain or loss of the fetus, fetal fluids, placenta, and uterus. For many practical purposes, subjective evaluation of body fatness by use of a visual condition scoring system (1 = thinnest, 9 = fattest) is frequently of benefit. More accurate methods are available for measuring body composition, but their use is generally limited to experimentation because of high costs or amount of labor required.

Death of calves perinatally represents a major production loss for beef cattle. Neonatal mortality is related to birth weight with the greatest losses occurring at low and high birth weights and lower mortality associated with moderate birth weights. Because dystocia, which is positively associated with birth weight, is a major cause of neonatal calf death (Laster and Gregory, 1973; Bellows et al., 1987), some cattle producers have attempted to reduce calf birth weight, particularly in first calf heifers, by underfeeding during the last trimester of pregnancy. As noted previously, malnutrition must be relatively severe to result in substantial reductions in calf birth weight. In nine studies reviewed by Dunn (1980), birth weight was reduced in all but one by severe underfeeding, but dystocia was reduced in only one (Dunn and Moss, 1992); but by underfeeding sufficiently to reduce birth weight, calf survival was reduced. In addition, numerous data (Short et al., 1990; Ferrell, 1991; Dunn and Moss, 1992) indicate the interval from calving to rebreeding is increased by underfeeding during late pregnancy. Inadequate prepartum nutrition is also associated with lower milk production and decreased calf weight at weaning (Hight, 1968; Corah et al., 1975; Bellows and Short, 1978). Negative effects of underfeeding during pregnancy are more severe in first-calf heifers than in more mature cows.

The interval from calving until conception cannot exceed approximately 80 days if an annual calving interval is to be maintained in beef cows. To have a high probability of conception by 80 days postcalving, the interval of postpartum anestrous should be 60 days or less. For this reason, numerous researchers have studied the period of postpartum anestrous (see Short et al., 1990). The duration of the postpartum anestrous period is increased in cows fed low concentrations of energy during late gestation or early lactation. However, response to low energy intake prepartum or weight change prepartum depends on body condition at calving. Cows that are in good body condition at calving (condition score ≥5) are minimally affected by either pre- or postpartum weight changes. Postpartum anestrous interval is increased by weight loss in cows that are in thin-to-moderate body condition (condition score ≤4) prior to calving. This problem is exacerbated by insufficient energy intake and weight loss postpartum. Effects of poor body condition at calving can be partly overcome by increased postpartum feeding. However, the postpartum period is a period of high metabolic demand because of the high nutrient requirements during early lactation. Thus, it is difficult to feed enough energy to cows during the early postpartum period to compensate for poor body condition at calving. This problem is intensified in heifers because of the additional nutrient needs for growth during the lactational period. Conversely, cows that are obese at calving have greater incidence of metabolic, infectious, digestive, and reproductive disorders than cows in moderate-to-good body condition.

The duration of the postpartum interval of anestrous is longer in suckled than in milked or nonlactating cows. The delay in initiation of estrous cycles postpartum appears to result primarily from calf contact rather than suckling or lactation per se. In addition, the calf stimulus interacts with the nutritional status of the cow such that postpartum interval of anestrous is increased to a greater extent in cows in poor body condition than in those in good condition. Early weaning of calves, short-term weaning, or partial weaning, such as once per day suckling, have reduced the postpartum interval in anestrous beef cows, but successful use of any of these approaches requires intensive management and other inputs.

It should be most efficient biologically to maintain cows in good condition throughout the year because of inefficiencies involved in depletion and repletion of body tissues. In addition, cows in good body condition are more tolerant of cold and other stresses. However, in many production situations, cows lose weight during early lactation when feed quantity is limited and quality is low and gain weight when higher quality feeds are more abundant or when nutrient demands are less. This cyclic loss and gain, although biologically less efficient, may be more efficient economically and may not be detrimental to total production, depending on the duration and severity of poor-feed conditions and the physiological status of the animals.

Males

Nutrient requirements for normal growth of young bulls have been discussed in previous sections, and estimates of requirements for maintenance and growth have been indicated. Details about nutritional influences on sexual development of young bulls as well as influences on sexual behavior, mating ability, and semen quantity and quality have been discussed in greater detail in reviews cited earlier in this section; thus it will be discussed only briefly here. Nutrient intakes below requirements result in reduced growth rates and delayed puberty in the male, as in the female and, if severe enough, can permanently impair sperm output (Bratton et al., 1959; VanDemark et al., 1964; Nolan et al., 1990). Inadequate nutrient intake is associated with reduced testicular weight, secretory output of the accessory sex glands, sperm motility and sperm concentration. Similarly, the reproductive potential of young males may also be impaired by overfeeding (Coulter and Kozub, 1984). Overfeeding has been associated with decreased scrotal circumference, epididymal sperm reserves, and seminal quality; however, it appears to be more likely to underfeed, particularly bulls of large breeds, than to overfeed (Pruitt and Corah, 1985). Negative influences of specific nutrient deficiencies have been discussed in detail by Hurley and Doane (1989).

Mating behavior is an important aspect of male reproductive function as it has a direct bearing on the number of females mated. Moderate energy or protein deficiency or excess seem to have little effect on mating behavior, spermatogenesis, or semen quality. Severe deficiencies may result in diminished libido, depression of endocrine testicular function, and arrest of growth and secretory activity of accessory sex glands. Prolonged severe malnutrition, particularly insufficient intake of energy, protein, or water can lead to reduction or cessation of spermatogenesis and a reduction in semen quality. These effects are accompanied by decreased size of the testes and accessory sex glands. Atrophy of the interstitial and Sertoli cell populations may accompany these changes. Nevertheless, overall, it is evident that unless males are severely deprived, there is minimal effect on the sexual responses and efficiency of the mating responses. Conversely, overfeeding and obesity may result in diminished sexual activity. Overly fat males may become less willing and able to inseminate females. Specific nutrient deficiencies may result in lowered physical ability to mate in addition to specific effects noted by Hurley and Doane (1989).

It should be noted that the negative effects of malnutrition are more evident in the young male than in older animals. The mature male is remarkably resistant to nutritional stress, and infertility problems of nutritional origin are not often encountered. Both young and mature males frequently lose weight during the breeding season resulting from both decreased food consumption and substantially increased physical activity. Thus, bulls should be in good body condition at the beginning of the breeding season to provide energy and protein reserves for use during breeding.

REFERENCES

Agricultural and Food Research Council. 1990. Nutrient Requirements of Ruminant Animals: Energy. Nutr. Abstr. Rev. Ser. B 60:729–804.

Agricultural Research Council. 1980. The Nutrient Requirements of Ruminant Livestock: Technical Review. Farnham Royal, U.K.: Commonwealth Agricultural Bureaux.

Alexander, G. 1964a. Studies on the placenta of the sheep (*Ovis aries* L.): Placental size. J. Reprod. Fertil. 7:289–305.

Alexander, G. 1964b. Studies on the placenta of the sheep (*Ovis aries* L.): Effect of surgical reduction of the number of caruncles. J. Reprod. Fertil. 7:307–322.

Alexander, G., and D. Williams. 1971. Heat stress and development of the conceptus in domestic sheep. J. Agric. Sci., Camb. 76:53–72.

Andersen, H., and M. Plum. 1965. Gestation length and birth weight in cattle and buffaloes: A review. J. Dairy Sci. 48:1224–1235.

Arnett, D.W., G. L. Holland, and R. Totusek. 1971. Some effects of obesity in beef females. J. Anim. Sci. 33:1129–1136.

Bartle, S. J., J. R. Males, and R. L. Preston. 1984. Effect of energy intake on the postpartum interval in beef cows and the adequacy of the cows milk production for calf growth. J. Anim. Sci. 58:1068–1074.

Bassett, J. M. 1986. Nutrition of the conceptus: Aspects of its regulation. Proc. Nutr. Soc. 45:1–10.

Bassett, J. M. 1991. Current perspectives on placental development and its integration with fetal growth. Proc. Nutr. Soc. 50:311–319.

Battaglia, F. C. 1992. New concepts in fetal and placental amino acid metabolism. J. Anim. Sci. 70:3258–3263.

Beal, W. E., D. R. Notter, and R. M. Akers. 1990. Techniques for estimation of milk yield in beef cows and relationships of milk yield to calf weight gain and postpartum reproduction. J. Anim. Sci. 68:937–943.

Beef Improvement Federation. 1990. Guidelines for Uniform Beef Improvement Programs, 6th Ed. Stillwater, Okla.: Oklahoma State University, Beef Improvement Federation.

Bell, A. W. 1987. Consequences of severe heat stress for fetal development. Pp. 313–333 in Heat Stress: Physical Exertion and Environment, J. R. S. Hales and D. A. B. Richards, eds. Amsterdam: Elsevier/North Holland.

Bell, A. W., J. M. Kennaugh, F. C. Battaglia, E. L. Makowski, and G. Meschia. 1986. Metabolic and circulatory studies of fetal lambs at midgestation. Am. J. Physiol. 250:E538–E544.

Bell, A. W., B. B. Wilkening, and G. Meschia. 1987. Some aspects of placental function in chronically heat stressed ewes. J. Develop. Physiol. 9:17–29.

Bellows, R. A., and R. E. Short. 1978. Effects of precalving feed level

on birth weight, calving difficulty and subsequent fertility. J. Anim. Sci. 46:1522–1528.

Bellows, R. A., D. J. Patterson, P. J. Burfening, and D. A. Phelps. 1987. Occurrence of neonatal and postnatal mortality in range beef cattle. II. Factors contributing to calf death. Theriogenology 28:573–586.

Bellows, R. A., R. E. Short, G. P. Kitto, R. B. Staigmiller, and M. D. MacNeil. 1990. Influence of sire, sex of fetus and type of pregnancy on conceptus development. Theriogenology 34:941–954.

Boggs, D. L., E. F. Smith, R. R. Shalles, B. E. Brent, L. R. Corah, and R. J. Pruitt. 1980. Effects of milk and forage intake on calf performance. J. Anim. Sci. 51:550–553.

Bowden, D. M. 1981. Feed utilization for calf production in the first lactation by 2-year-old F_1 crossbred beef cows. J. Anim. Sci. 51:304–315.

Bratton, R. W., S. D. Musgrove, H. O. Dunn, and R. H. Foote. 1959. Causes and prevention of reproductive failures in dairy cattle. II. Influence of underfeeding and overfeeding from birth to 80 weeks of age on growth, sexual development and semen production of Holstein bulls. Bull. No. 940. Ithaca, N.Y.: Cornell University Agricultural Experiment Station.

Burns, W. C., M. Koger, W. T. Butts, O. F Pahnish, and R. L. Blackwell. 1979. Genotype by environmental interaction in Hereford cattle. II. Birth and weaning traits. J. Anim. Sci. 49:403–409.

Butson, S., and R. T. Berg. 1984a. Lactation performance of range beef and dairy beef cows. Can. J. Anim. Sci. 64:253–265.

Butson, S., and R. T. Berg. 1984b. Factors influencing lactation performance of range beef and dairy-beef cows. Can. J. Anim. Sci. 64:267–277.

Butts, W. T., M. Koger, O. F. Pahnish, W. C. Burns, and E. J. Warwick. 1971. Performance of two lines of Hereford cattle in two environments. J. Anim. Sci. 33:923–932.

Cartwright, G. A., and C. J. Thwaites. 1976. Foetal stunting in sheep. 1. The influence of maternal nutrition and high ambient temperature on the growth and proportions of Merino fetuses. J. Agric. Sci. Camb. 86:573–580.

Chenette, C. G., and R. R. Frahm. 1981. Yield and composition of milk from various two-breed cross cows. J. Anim. Sci. 52:483–492.

Christian, L. L., E. R. Hauser, and A. B. Chapman. 1965. Association of preweaning and postweaning traits with weaning weight in cattle. J. Anim. Sci. 24:652–659.

Clutter, A. C., and M. K. Nielson. 1987. Effect of level of beef cow milk production on pre- and postweaning calf growth. J. Anim. Sci. 64:1313–1322.

Collier, R. J., S. G. Doelger, H. H. Head, W. W. Thatcher, and C. J. Wilcox. 1982. Effects of heat stress during pregnancy on maternal hormone concentrations, calf birth weight and postpartum milk yields in postpartum cows. J. Anim. Sci. 54:309–319.

Commonwealth Scientific and Industrial Research Organization. 1990. Feeding Standards for Australian Livestock: Ruminants. East Melbourne, Victoria, Australia: CSIRO Publications.

Corah, L. R., T. G. Dunn, and C. C. Kaltenbach. 1975. Influence of prepartum nutrition on the reproductive performance of beef females and the performance of their progeny. J. Anim. Sci. 41:819–824.

Coulter, G. H., and G. C. Kozub. 1984. Testicular development, epididymal sperm reserves and seminal quality in two-year-old Hereford and Angus bulls: Effects of two levels of dietary energy. J. Anim. Sci. 59:432–440.

Cundiff, L. V., R. M. Koch, and K. E. Gregory. 1988. Pp. 3–4 in Germ plasm evaluation in cattle. Beef Progress Report No. 3. USDA-ARS-71. Washington, D.C.: U.S. Department of Agriculture.

Cundiff, L. V., K. E. Gregory, F. J. Schwulst, and R. M. Koch. 1974. Effects of heterosis on maternal performance and milk production in Hereford, Angus, and Shorthorn cattle. J. Anim. Sci. 38:728–745.

Daley, D. R., A. McCuskey, and C. M. Bailey. 1987. Composition and yield of milk from beef-type *Bos taurus* and *Bos indicus* × *Bos taurus* dams. J. Anim. Sci. 64:373–384.

Deutscher, G. H., and J. V. Whiteman. 1971. Productivity as two-year-olds of Angus-Holstein crossbreds compared to Angus heifers under range conditions. J. Anim. Sci. 33:337–342.

Diaz, C., D. R. Notter, and W. E. Beal. 1992. Relationship between milk expected progeny differences of Polled Hereford sires and actual milk production of their crossbred daughters. J. Anim. Sci. 70:396–402.

Drewry, K. J., C. J. Brown, and R. S. Honea. 1959. Relationships among factors associated with mothering ability in beef cattle. J. Anim. Sci. 18:938–946.

Dunn, T. G. 1980. Relationship of nutrition to successful embryo transplantation. Theriogenology 13:27–39.

Dunn, T. G., and G. E. Moss. 1992. Effects of nutrient deficiencies and excesses on reproductive efficiency of livestock. J. Anim. Sci. 70:1580–1593.

Dziuk, P. J., and R. A. Bellows. 1983. Management of reproduction in beef cattle, sheep and pigs. J. Anim. Sci. 57(Suppl. 2):355.

Eley, R. M., W. W. Thatcher, F. W. Bazer, C. J. Wilcox, R. B. Becker, H. H. Head, and R. W. Adkinson. 1978. Development of the conceptus in the bovine. J. Dairy Sci. 61:467–473.

Ellenberger, H. G., J. A. Newlander, and C. H. Jones. 1950. Composition of the bodies of dairy cattle. Vermont Agric. Exp. Sta. Bull. 558:3–66.

Ellis, G. J., Jr., T. C. Cartwright, and W. E. Kruse. 1965. Heterosis for birth weight in Brahman-Hereford crosses. J. Anim. Sci. 24:93–96.

Elston, D. A., C. A. Glasbey, and D. R. Nielson. 1989. Non-parametric lactation curves. Anim. Prod. 48:331–339.

Ferrell, C. L. 1982. Effects of postweaning rate of gain on onset of puberty and productive performance of heifers of different breeds. J. Anim. Sci. 55:1272–1283.

Ferrell, C. L. 1989. Placental regulation of fetal growth. In Animal Growth Regulation, D. R. Campion, G. J. Hausman, and R. J. Martin, eds. New York: Plenum Press.

Ferrell, C. L. 1991a. Maternal and fetal influences on uterine and conceptus development in the cow. I. Growth of tissues of the gravid uterus. J. Anim. Sci. 69:1945–1953.

Ferrell, C. L. 1991b. Maternal and fetal influences on uterine and conceptus development in the cow. II. Blood flow and nutrient flux. J. Anim. Sci. 69:1954–1965.

Ferrell, C. L. 1991c. Nutrient requirements and various factors affecting fetal growth and development. Minnesota Nutr. Conf. 52:76–92.

Ferrell, C. L. 1991d. Nutritional influences on reproduction. In Reproduction in Domestic Animals, 4th Ed., P. T. Cupps, ed. Los Angeles: Academic Press.

Ferrell, C. L., and L. P. Reynolds. 1987. Oxidative metabolism of gravid uterine tissues of the cow. Energy Metab. Proc. Symp. 32:298–301.

Ferrell, C. L., and L. P. Reynolds. 1992. Uterine and umbilical blood flows and net nutrient uptake by fetuses and uteroplacental tissues of cows gravid with either single or twin fetuses. J. Anim. Sci. 70:426–433.

Ferrell, C. L., W. N. Garrett, and N. Hinman. 1976a. Growth, development and composition of the udder and gravid uterus of beef heifers during pregnancy. J. Anim. Sci. 42:1477–1489.

Ferrell, C. L., W. N. Garrett, N. Hinman, and G. Grichting. 1976b. Energy utilization by pregnant and nonpregnant heifers. J. Anim. Sci. 42:937–950.

Ferrell, C. L., S. P. Ford, R. K. Christenson, and L. Prior. 1983. Blood flow, steroid secretion and nutrient uptake of the gravid bovine uterus and fetus. J. Anim. Sci. 56:656–667.

Fiss, C. F., and J. W. Wilton. 1989. Effects of breeding system, cow weight and milk yield on reproductive performance in beef cattle. J. Anim. Sci. 67:1714–1721.

Freking, B. A., and D. M. Marshall. 1992. Interrelationships for heifer milk production and other biological traits with production efficiency to weaning. J. Anim. Sci. 70:646–655.

Gadeken, D., K. Rohr, and P. Lebzien. 1991. Effect of nitrogen fertilizing and of harvest season on net energy values of hay in dairy cows. Energy Metab. Proc. Symp. 58:321–324.

Garrett, W. N. 1980a. Energy utilization by growing cattle as determined in 72 comparative slaughter experiments. Energy Metab. Proc. Symp. 26:3–7.

Garrett, W. N. 1980b. Discussion Paper: Use of energy in reproduction and lactation. Energy Metab. Proc. Symp. 26:383–389.

Gaskins, C. T., and D. C. Anderson. 1980. Comparison of lactation curves in Angus-Hereford, Jersey-Angus and Simmental-Angus cows. J. Anim. Sci. 50:828–832.

Gleddie, V. M., and R. T. Berg. 1968. Milk production in range beef cows and its relationship to calf gains. Can. J. Anim. Sci. 48:323–333.

Gluckman, P. D., and G. C. Liggins. 1984. Regulation of fetal growth. Pp. 511–557 in Fetal Physiology and Medicine, R. W. Beard and P. W. Nathanielsz, eds. London: Butterworths.

Graham, N. M. 1964. Energy exchanges in pregnant and lactating ewes. Aust. J. Agric. Res. 15:127–141.

Grainger, C., and G. Wilhelm. 1979. Effect of pattern and duration of underfeeding in early lactation on milk production and reproduction of dairy cows. Aust. J. Exp. Agric. Anim. Husb. 19:395–401.

Grainger, C., C. W. Holmes, A. M. Bryant, and Y. Moore. 1983. A note on the concentration of energy in the milk of Friesian and Jersey cows. Anim. Prod. 36:307–308.

Green, R. D., L. V. Cundiff, G. E. Dickerson, and T. G. Jenkins. 1991. Output/input differences among nonpregnant, lactating *Bos indicus-Bos taurus* and *Bos taurus-Bos taurus* F_1 cross cows. J. Anim. Sci. 69:3156–3166.

Gregory, K. E., L. V. Cundiff, and R. M. Koch. 1982. Characterization of breeds representing diverse biological types: Preweaning traits. Beef Research Program, Progress Report No. 1. USDA-ARS, ARM-NC-21:7. Peoria, Ill.: Agricultural Research North Central Region, Science and Education Administration, U.S. Department of Agriculture.

Gregory, K. E., S. E. Echternkamp, G. E. Dickerson, L. V. Cundiff, R. M. Koch, and L. D. Van Vleck. 1990. Twinning in cattle: III. Effects of twinning on dystocia, reproductive traits, calf survival, calf growth and cow productivity. J. Anim. Sci. 68:3133–3144.

Gregory, K. E., L. V. Cundiff, and R. M. Koch. 1992a. Breed effects and heterosis on milk yield and 200-day weight in advanced generations of composite populations of beef cattle. J. Anim. Sci. 70:2366–2372.

Gregory, K. E., L. V. Cundiff, and R. M. Koch. 1992b. Composite breeds to use heterosis and breed differences to improve efficiency of beef production. Misc. Publ. Clay Center, Neb.: U.S. Department of Agriculture, Agricultural Research Service, USDA-ARS Roman L. Hruska U.S. Meat Animal Research Center.

Hansen, P. J., D. H. Baik, J. J. Rutledge, and E. R. Hauser. 1982. Genotype × environmental interactions on reproductive traits of bovine females. II. Postpartum reproduction as influenced by genotype, dietary regimen, level of milk production and parity. J. Anim. Sci. 55:1458–1472.

Hight, G. K. 1966. Effects of undernutrition in late pregnancy on beef cattle production. N. Z. J. Agric. Res. 9:479–490.

Hight, G. K. 1968a. Plane of nutrition effects in late pregnancy and during lactation on beef cows and their calves to weaning. N. Z. J. Agric. Res. 11:71–84.

Hight, G. K. 1968b. Comparison of the effects of three nutritional levels in late pregnancy on beef cows and their calves to weaning. N. Z. J. Agric. Res. 11:477–486.

Hohenboken, W. D., A. Dudley, and D. E. Moddy. 1992. A comparison among equations to characterize lactation curves in beef cows. Anim. Prod. 55:23–28.

Holloway, J. W., D. F Stephens, J. V. Whiteman, and R. Totusek. 1975. Performance of 3-year-old Hereford, Hereford × Holstein and Holstein cows on range and in drylot. J. Anim. Sci. 40:114–125.

Holloway, J. W., W. T. Butts, Jr., J. R. McCurley, E. E. Beaver, H. L.

Peeler, and W. L. Backus. 1985. Breed × nutritional environmental interactions for beef female weight and fatness, milk production and calf growth. J. Anim. Sci. 61:1354–1363.

Hurley, W. L., and R. M. Doane. 1989. Recent developments in the roles of vitamins and minerals in reproduction. J. Dairy Sci. 72:784–804.

Jakobsen, P. E. 1956. Protein requirements for fetus formation in cattle. Proc. 7th Intern. Cong. Anim. Husb. 6:115–126.

Jakobsen, P. E., P. H. Sorensen, and H. Larsen. 1957. Energy investigations as related to fetus formation in cattle. Acta Agric. Scand. 7:103–112.

Jeffery, H. B., R. T. Berg, and R. T. Hardin. 1971. Factors influencing milk yield of beef cattle. Can. J. Anim. Sci. 51:551–560.

Jenkins, T. G., and C. L. Ferrell. 1984. A note on the lactation curves of crossbred cows. Anim. Prod. 39:479–482.

Jenkins, T. G., and C. L. Ferrell. 1992. Lactation characteristics of nine breeds of cattle fed various quantities of dietary energy. J. Anim. Sci. 70:1652–1660.

Jenkins, T. G., C. L. Ferrell, and L. V. Cundiff. 1986. Relationship of components of the body among mature cows as related to size, lactation potential and possible effects on productivity. Anim. Prod. 43:245–254.

Joubert, D. M. 1963. Puberty in female farm animals. Anim. Breed Abstr. 31:295–306.

Kleiber, M. 1961. The Fire of Life. New York: Wiley & Sons.

Kroker, G. A., and L. J. Cummins. 1979. The effect of nutritional restriction on Hereford heifers in late pregnancy. Aust. Vet. J. 55:467–474.

Kropp, J. R., D. F. Stephens, J. W. Holloway, J. V. Whiteman, L. Knori, and R. Totusek. 1973. Performance on range and in drylot of two-year-old Hereford, Hereford × Holstein and Holstein females as influenced by level of winter supplementation. J. Anim. Sci. 37:1222–1232.

Lamond, D. R. 1970. The influence of undernutrition on reproduction in the cow. Anim. Breed. Abstr. 38:359–372.

Lamond, D. R., J. H. G. Holmes, and K. P. Haydock. 1969. Estimation of yield and composition of milk produced by grazing beef cows. J. Anim. Sci. 29:606–611.

Lampkin, K., and G. H. Lampkin. 1960. Studies on production of beef from Zebu cattle in East Africa. II. Milk production in suckled cows and its effect on calf growth. J. Agric. Sci., Camb. 55:233–239.

Langlands, J. P., and H. A. M. Southerland. 1968. An estimate of the nutrients utilized for pregnancy by Merino sheep. Br. J. Nutr. 22:217–227.

Laster, D. B., and K. E. Gregory. 1973. Factors affecting peri- and early postnatal calf mortality. J. Anim. Sci. 37:1092–1097.

Laster, D. B., H. A. Glimp, and K. E. Gregory. 1972. Age and weight at puberty and conception in different breeds and breed crosses of beef heifers. J. Anim. Sci. 34:1031–1036.

Laster, D. B., G. M. Smith, and K. E. Gregory. 1976. Characterization of biological types of cattle. IV. Postweaning growth and puberty of heifers. J. Anim. Sci. 43:63–70.

Laster, D. B., G. M. Smith, L. V. Cundiff, and K. E. Gregory. 1979. Characterization of biological types of cattle (Cycle II). II. Postweaning growth and puberty of heifers. J. Anim. Sci. 48:500–508.

Lodge, G. A., and D. P. Heaney. 1970. Energy cost of pregnancy in the ewe. Energy Metab. Proc. Symp. 13:109–112.

Long, C. R. 1980. Crossbreeding for beef production: Experimental results. J. Anim. Sci. 51:1197–1223.

Lowman, B. G., R. A. Edwards, S. H. Somerville, and G. M. Jolly. 1979. The effect of plane of nutrition in early lactation on the performance of beef cows. Anim. Prod. 29:293–303.

Lubritz, D. L., K. Forrest, and O. W. Robinson. 1989. Age of cow and age of dam effects on milk production of Hereford cows. J. Anim. Sci. 67:2544–2549.

Marshall, D. M., R. R. Frahm, and G. W. Horn. 1984. Nutrient intake

and efficiency of calf production by two-breed cross cows. J. Anim. Sci. 59:317–328.

Martin, L. C., J. S. Brinks, R. M. Bourdon, and L. V. Cundiff. 1992. Genetic effects on beef heifer puberty and subsequent reproduction. J. Anim. Sci. 70:4006–4017.

Masilo, B. S., J. S. Stevenson, R. R. Shalles, and J. E. Shirley. 1992. Influence of genotype and yield and composition of milk on interval to first partum ovulation in milked beef and dairy cows. J. Anim. Sci. 70:379–385.

McCarter, M. N., D. S. Buchanan, and R. R. Frahm. 1991a. Comparison of crossbred cows containing various proportions of Brahman in spring or fall calving systems. II. Milk production. J. Anim. Sci. 69:77–84.

McCarter, M. N., D. S. Buchanan, and R. R. Frahm. 1991b. Comparison of crossbred cows containing various proportions of Brahman in spring or fall calving systems: III. Productivity as three-, four-, and five-year olds. J. Anim. Sci. 69:2754–2761.

McMorris, M. R., and J. W. Wilton. 1986. Breeding systems, cow weight and milk yield effects on various biological variables in beef production. J. Anim. Sci. 63:1361–1372.

Melton, A. A., J. K. Riggs, L. A. Nelson, and T. C. Cartwright. 1967. Milk production, composition and calf gains of Angus, Charolais and Hereford cows. J. Anim. Sci. 26:804–809.

Mezzadra, C., R. Paciaroni, S. Vulich, E. Villarreal, and L. Melucci. 1989. Estimation of milk consumption curve parameters for different genetic groups of bovine calves. Anim. Prod. 49:83–87.

Miller, H. L., and G. H. Deutscher. 1985. Beef production of Simmental × Angus and Hereford × Angus cows under range conditions. J. Anim. Sci. 61:1364–1369.

Moe, P. W. 1981. Energy metabolism of dairy cattle. J. Dairy Sci. 64:1120–1139.

Moe, P. W., and H. F Tyrrell. 1971. Metabolizable energy requirements of pregnant dairy cows. J. Dairy Sci. 55:480–483.

Moe, P. W., H. F. Tyrrell, and W. P. Flatt. 1970. Partial efficiency of energy use for maintenance, lactation, body gain and gestation in the dairy cow. Energy Metab. Proc. Symp. 13:65–68.

Moe, P. W., W. P. Flatt, and H. F Tyrrell. 1972. Net energy value of feeds for lactation. J. Dairy Sci. 55:945–958.

Mondragon, I., J. W. Wilton, O. B. Allen, and H. Song. 1983. Stage of lactation effects: Repeatabilities and influences on weaning weights of yield and composition of milk in beef cattle. Can. J. Anim. Sci. 63:751–761.

Montano-Bermudez, M., M. K. Nielson, and G. H. Deutscher. 1990. Energy requirements for maintenance of crossbred beef cattle with different genetic potentials for milk. J. Anim. Sci. 68:2279–2288.

Morant, S. V., and A. Granaskthy. 1989. A new approach to the mathematical formulation of lactation curves. Anim. Prod. 49:151–162.

Munger, A. 1991. Milk production efficiency in dairy cows of different breeds. Energy Metab. Proc. Symp. 58:292–295.

Munro, H. N., S. J. Philistine, and M. E. Fant. 1983. The placenta in nutrition. Annu. Rev. Nutr. 3:97–124.

National Research Council. 1984. Nutrient Requirements of Beef Cattle, Sixth Revised Edition. Washington, D.C.: National Academy Press.

National Research Council. 1989. Nutrient Requirements of Dairy Cattle, Sixth Revised Edition. Washington, D.C.: National Academy Press.

Neidhardt, R., D. Plasse, J. H. Weniger, O. Verde, J. Beltran, and A. Benavides. 1979. Milk yield of Brahman cows in a tropical beef production system. J. Anim. Sci. 48:1–6.

Neville, W. E., Jr. 1962. Influence of the dam's milk production and other factors on 120- and 240-day weight of Hereford calves. J. Anim. Sci. 21:315–320.

Nolan, C. J., D. A. Nevendorff, R. W. Godfrey, P. G. Harms, T. H. Welsh, Jr., N. H. McArthur, and R. D. Randel. 1990. Influence of dietary energy intake on prepubertal development of Brahman bulls. J. Anim. Sci. 68:1087–1096.

Notter, D. R., L. V. Cundiff, G. M. Smith, D. B. Laster, and K. E. Gregory. 1978. Characterization of biological types of cattle. VII. Milk production in young cows and transmitted and maternal effects on preweaning growth of progeny. J. Anim. Sci. 46:908–921.

Olson, T. A., K. Euclides Fiho, L. V. Cundiff, M. Koger, W. T. Butts, Jr., and K. E. Gregory. 1991. Effects of breed group by location interaction on crossbred cattle in Nebraska and Florida. J. Anim. Sci. 69:104–114.

Owens, J. A., J. Falconer, and J. S. Robinson. 1986. Effects of restriction of placental growth on umbilical and uterine blood flows. Am. J. Physiol. 250:R427–R434.

Pahnish, O. F., M. Koger, J. J. Urick, W. C. Burns, W. T. Butts, and G. V. Richardson. 1983. Genotype × environmental interaction in Hereford cattle. III. Postweaning traits of heifers. J. Anim. Sci. 56:1039–1046.

Patle, B. R., and V. D. Mudgal. 1976. Utilization of diet energy for maintenance and milk production by lactating crossbred cows (Brown Swiss × Sahiwal) during their early stage of lactation. Ind. J. Anim. Sci. 46:1–7.

Patterson, D. J., R. C. Perry, G. H. Kiracofe, R. A. Bellows, R. B. Staigmiller, and L. R. Corah. 1992. Management considerations in heifer development and puberty. J. Anim. Sci. 70:4018–4035.

Prior, R. L., and D. B. Laster. 1979. Development of the bovine fetus. J. Anim. Sci. 48:1546–1553.

Pruitt, R. J., and L. R. Corah. 1985. Effect of energy intake after weaning on sexual development of beef bulls. 1. Semen characteristics and serving capacity. J. Anim. Sci. 61:1186–1193.

Rattray, P. V., W. N. Garrett, N. E. East, and N. Hinman. 1974. Efficiency of utilization of metabolizable energy during pregnancy in sheep. J. Anim. Sci. 38:383–393.

Rege, J. E. O., R. R. Von Kaufmann, W. N. M. Mwenya, E. O. Otchere, and R. I. Mani. 1993. On-farm performance of Bunaji (White Fulani) cattle. 2. Growth, reproductive performance, milk offtake and mortality. Durrant. 57:211–220.

Reynolds, L. P., C. L. Ferrell, J. A. Nienaber, and S. P. Ford. 1985. Effects of chronic environmental heat stress on blood flow and nutrient uptake of the gravid bovine uterus and foetus. J. Agric. Sci., Camb. 104:289–297.

Reynolds, L. P., C. L. Ferrell, D. A. Robertson, and S. P. Ford. 1986. Metabolism of the gravid uterus, foetus and utero-placenta at several stages of gestation in cows. J. Agric. Sci., Camb. 106:437–444.

Reynolds, W. L., T. M. DeRouen, and R. A. Bellows. 1978. Relationships of milk yield of dam to early growth rate of straightbred and crossbred calves. J. Anim. Sci. 47:584–594.

Robinson, J. J. 1977. The influence of maternal nutrition on ovine growth. Proc. Nutr. Soc. 36:9–16.

Robinson, J. J. 1990. Nutrition in the reproduction of farm animals. Nutr. Res. Rev. 3:253–276.

Robinson, J. J., I. McDonald, C. Fraser, and J. G. Gordon. 1980. Studies on reproduction in prolific ewes. 6. The efficiency of energy utilization for conceptus growth. J. Agric. Sci., Camb. 94:331–338.

Robinson, O. W., M. K. M. Yusuff, and E. U. Dillard. 1978. Milk production in Hereford cows. I. Means and correlations. J. Anim. Sci. 47:131–136.

Rogers, G. L., A. M. Bryant, K. E. Jury, and J. B. Hutton. 1979. Silage and dairy cow production. II. Milk yield and composition of cows fed pasture silage supplemented with pasture, maize silage and protein concentrates. N. Z. J. Agric. Res. 22:523–531.

Rowlands, G. J., S. Lucey, and A. M. Russell. 1982. A comparison of different models of the lactation curve of dairy cattle. Anim. Prod. 35:135–144.

Rutter, W., T. R. Laird, and P. J. Broadbent. 1971. The effects of clipping pregnant ewes at housing and of feeding different basal roughages. Anim. Prod. 13:329–336.

Rutter, W., T. R. Laird, and P. J. Broadbent. 1972. A note on the effects of clipping pregnant ewes at housing. Anim. Prod. 14:127–130.

Sacco, R. E., J. F. Baker, and T. C. Cartwright. 1987. Production characteristics of primiparous females of a five-breed diallel. J. Anim. Sci. 64:1612–1618.

Schillo, K. K., J. B. Hall, and S. M. Hileman. 1992. Effect of nutrition and season on onset of puberty in the beef heifers. J. Anim. Sci. 70:3994–4005.

Short, R. E., and R. A. Bellows. 1971. Relationships among weight gains, age at puberty and reproductive performance in heifers. J. Anim. Sci. 32:127–131.

Short, R. E., R. A. Bellows, R. B. Staigmiller, J. G. Berardinelli, and E. E. Custer. 1990. Physiological mechanisms controlling anestrous and infertility in postpartum beef cattle. J. Anim. Sci. 68:799–816.

Silvey, M. W., and K. P. Haycock. 1978. A note on live-weight adjustment for pregnancy in cows. Anim. Prod. 27:113–116.

Spitzer, J. C., J. N. Wiltbank, and D. C. Lefever. 1975. Increasing beef cow productivity by increasing reproductive performance. Gen. Ser. No. 949. Fort Collins, Colo.: Colorado State University Experiment Station.

Stewart, T. S., C. R. Long, and T. C. Cartwright. 1980. Characterization of cattle of a five breed diallel. III. Puberty in bulls and heifers. J. Anim. Sci. 50:808–820.

Sykes, A. R., and A. C. Field. 1972. Effect of dietary deficiencies of energy, protein and calcium on the pregnant ewe. III. Some observations on the use of biochemical parameters in controlling undernutrition during pregnancy and on the efficiency of energy utilization of energy and protein for fetal growth. J. Agric. Sci. Camb. 78:127–133.

Thompson, G. E., J. M. Bassett, D. E. Samson, and J. Lee. 1982. The effects of cold exposure of pregnant sheep on foetal plasma nutrients, hormones and birth weight. Br. J. Nutr. 48:59–64.

Totusek, R., D. W. Arnett, G. L. Holland, and J. V. Whiteman. 1973. Relation of estimation method, sampling interval and milk composition to milk yield of beef cows and calf gain. J. Anim. Sci. 37:153–158.

Tudor, G. D. 1972. The effect of pre- and post-natal nutrition on the growth of beef cattle. I. The effect of nutrition and parity of the dam on calf birth weight. Aust. J. Agric. Res. 23:389–385.

Tyrrell, H. F., and J. T. Reid. 1965. Prediction of the energy value of cow's milk. J. Dairy Sci. 48:1215–1223.

Unsworth, E. F. 1991. The efficiency of utilization of metabolizable energy for lactation from grass silage-based diets. Energy Metab. Proc. Symp. 58:329.

VanDemark, N. L., G. R. Fritz, and R. E. Marger. 1964. Effect of energy intake on reproductive performance of dairy bulls. II. Semen production and replenishment. J. Dairy Sci. 47:798–802.

van der Honing, Y. 1980. The utilization by high-yielding cows of energy from animal tallow or soya bean oil added to a diet rich in concentrates. Energy Metab. Proc. Symp. 26:315–318.

Vera, R. R., C. A. Ramirez, and H. Ayala. 1993. Reproduction in continuously underfed Brahman cows. Anim. Prod. 57:193–198.

Williams, J. H., D. C. Anderson, and D. D. Kress. 1979. Milk production in Hereford cattle. II. Physical measurements: Repeatabilities and relationship with milk production. J. Anim. Sci. 49:1443–1448.

Wilson, L. L., J. E. Gillooly, M. C. Rugh, C. E. Thompson, and H. R. Purdy. 1969. Effects of energy intake, cow body size and calf sex on composition and yield of milk by Angus-Holstein cows and preweaning growth rate of progeny. J. Anim. Sci. 28:789–795.

Wiltbank, J. N., S. Roberts, J. Nix, and L. Rowden. 1985. Reproductive performance and profitability of heifers fed to weight 272 or 318 kg at the start of the first breeding season. J. Anim. Sci. 60:25–34.

Windisch, W., M. Kirchgessner, and H. L. Muller. 1991. Effect of different energy supply on energy metabolism in lactating dairy cows after a period of energy restriction. Energy Metab. Proc. Symp. 58:304–307.

Winters, L. M., W. W. Green, and R. E. Comstock. 1942. The prenatal development of the bovine. Univ. Minn. Agric. Exp. Sta. Tech. Bull. No. 151:1.

Wood, P. D. P. 1967. Algebraic model of the lactation curve in cattle. Nature 216:164–165.

Wood, P. D. P. 1969. Factors affecting the shape of the lactation curve in cattle. Anim. Prod. 11:307–316.

Wood, P. D. P. 1976. Algebraic models of the lactation curves for milk, fat and protein production with estimates of seasonal variation. Anim. Prod. 22:35–40.

Wood, P. D. P. 1979. A simple model of lactation curves for milk yield, food requirement and body weight. Anim. Prod. 28:55–63.

Wood, P. D. P. 1980. Breed variation in the shape of the lactation curve of cattle and their implications for efficiency. Anim. Prod. 31:133–141.

5 Minerals

At least 17 minerals are required by beef cattle. This chapter presents information about not only mineral requirements but also the function, signs of deficiency, factors affecting requirements, sources, and toxicity of each essential mineral. Macrominerals required include calcium, magnesium, phosphorus, potassium, sodium and chlorine, and sulfur. The microminerals required are chromium, cobalt, copper, iodine, iron, manganese, molybdenum, nickel, selenium, and zinc. Others, including arsenic, boron, lead, silicon, and vanadium have been shown to be essential for one or more animal species, but there is no evidence that these minerals are of practical importance in beef cattle, and therefore are not discussed.

Calcium and phosphorus requirements discussed in the subsequent sections are included in the computer models. Requirements and maximum tolerable concentrations for other minerals are shown in Table 5-1. For certain miner-

als, requirements are not listed because research data are inadequate to determine requirements.

Many of the essential minerals are usually found in sufficient concentrations in practical feedstuffs. Other minerals are frequently insufficient in diets fed to cattle, and supplementation is necessary to optimize animal performance or health. Supplementing diets at concentrations in excess of requirements greatly increases mineral loss in cattle waste. Oversupplementation of minerals should be avoided to prevent possible environmental problems associated with runoff from waste or application of cattle waste to soil.

A number of elements that are not required (or at least required only in very small amounts) can cause toxicity in beef cattle. Maximum tolerable concentrations of several elements known to be toxic to cattle are given in Table 5-2. The maximum tolerable concentration for a mineral has been defined as "that dietary level that, when fed for

TABLE 5-1 Mineral Requirements and Maximum Tolerable Concentrations

Mineral	Unit	Requirement Growing and Finishing Cattle	Cows Gestating	Cows Early Lactation	Maximum Tolerable Concentration
Calcium	%	See Chapter 9			—
Chlorine	%	—	—	—	—
Chromium	mg/kg	—	—	—	1,000.00
Cobalt	mg/kg	0.10	0.10	0.10	10.00
Copper	mg/kg	10.00	10.00	10.00	100.00
Iodine	mg/kg	0.50	0.50	0.50	50.00
Iron	mg/kg	50.00	50.00	50.00	1,000.00
Magnesium	%	0.10	0.12	0.20	0.40
Manganese	mg/kg	20.00	40.00	40.00	1,000.00
Molybdenum	mg/kg	—	—	—	5.00
Nickel	mg/kg	—	—	—	50.00
Phosphorus	%	See Chapter 9			
Potassium	%	0.60	0.60	0.70	3.00
Selenium	mg/kg	0.10	0.10	0.10	2.00
Sodium	%	0.06–0.08	0.06–0.08	0.10	—
Sulfur	%	0.15	0.15	0.15	0.40
Zinc	mg/kg	30.00	30.00	30.00	500.00

TABLE 5-2 Maximum Tolerable Concentrations of Mineral Elements Toxic to Cattle

Element	mg/kg
Aluminum	1,000.00
Arsenic	50.00 (100.00 for organic forms)
Bromine	200.00
Cadmium	00.5
Fluorine	40.00 to 100.00
Lead	30.00
Mercury	2.00
Strontium	2,000.00

Source: Adapted from Table 1 in National Research Council. 1980. Mineral Tolerance of Domestic Animals. Washington, D.C.: National Academy of Sciences.

a limited period, will not impair animal performance and should not produce unsafe residues in human food derived from the animal" (National Research Council, 1980: p. 3).

MACROMINERALS

Calcium

Calcium is the most abundant mineral in the body; approximately 98 percent functions as a structural component of bones and teeth. The remaining 2 percent is distributed in extracellular fluids and soft tissues, and is involved in such vital functions as blood clotting, membrane permeability, muscle contraction, transmission of nerve impulses, cardiac regulation, secretion of certain hormones, and activation and stabilization of certain enzymes.

CALCIUM REQUIREMENTS

Estimated requirements for calcium were calculated by adding the available calcium needed for maintenance, growth, pregnancy, and lactation and correcting for the percentage of dietary calcium absorbed. Calcium requirements are similar to those in the previous edition of this volume (National Research Council, 1984) because new information is not sufficient to justify a change. The maintenance requirement was calculated as 15.4 mg Ca/kg body weight (Hansard et al., 1954, 1957). Retained needs in excess of maintenance requirements were calculated as 7.1 g Ca/100 g protein gain. Calcium content of gain was calculated from slaughter data (Ellenberger et al., 1950). The calcium requirement for lactation in excess of maintenance needs was calculated as 1.23 g Ca/kg milk produced. Fetal calcium content was assumed to be 13.7 g Ca/kg fetal weight. This requirement was distributed over the last 3 months of pregnancy and an average birth weight of 35 kg was assumed. Absolute calcium requirements were converted to dietary calcium requirements assuming a true absorption

for dietary calcium of 50 percent. Lower absorption values have been obtained in older cattle, but in many instances calcium intake may have exceeded dietary requirements in these animals (Hansard et al., 1954, 1957; Martz et al., 1990). Absorption of calcium is largely determined by requirement relative to intake. True calcium absorption is reduced when intake exceeds the animal's need. The Agricultural and Food Research Council (AFRC) recently used a value of 68 percent absorption to calculate calcium requirements of cattle (TCORN, 1991).

FACTORS AFFECTING CALCIUM REQUIREMENTS

Calcium is absorbed primarily from the duodenum and jejunum by both active transport and passive diffusion (McDowell, 1992). It should be noted that diets high in fat may decrease calcium absorption through the formation of soaps (Oltjen, 1975). Vitamin D is required for active absorption of calcium (DeLuca, 1979). The amount of calcium absorbed is affected by the chemical form and source of the calcium, the interrelationships with other nutrients, and the animal's requirement. Requirement is influenced by such factors as age, weight, and type and stage of production. In natural feedstuffs, calcium occurs in oxalate or phytate form. In alfalfa hay, 20 to 33 percent was present as insoluble calcium oxalate and apparently unavailable to the animal (Ward et al., 1979). True absorption of alfalfa calcium was much lower than absorption of corn silage calcium when fed to dairy cows (Martz et al., 1990). In cattle fed high-concentrate diets, dietary calcium in excess of requirements improved gain or feed efficiency in some studies (Huntington, 1983; Brink et al., 1984; Bock et al., 1991). Improvements in performance were likely the result of manipulation of digestive tract function and may not represent a specific calcium requirement. Increasing calcium from 0.25 to 0.40 or 1.11 percent reduced organic matter and starch digestion in the rumen but increased postruminal digestion of organic matter and starch (Goetsch and Owens, 1985). In finishing cattle fed a high-concentrate diet, increasing calcium more than 0.3 percent increased gain in one of two trials but did not affect calcium status based on bone calcium, bone ash, and plasma ionizable calcium concentrations (Huntington, 1983).

SIGNS OF CALCIUM DEFICIENCY

The skeleton stores a large reserve of calcium that can be utilized to maintain critical blood calcium concentrations. Depending on their age, cattle can be fed calcium-deficient diets for extended periods without developing deficiency signs if previous calcium intake was adequate. Calcium deficiency in young animals, however, prevents normal bone growth, thus causing rickets and retarding growth and development. Rickets can be caused by a deficiency

See front "Errata"

of calcium, phosphorus, or vitamin D. It is characterized by improper calcification of the organic matrix of bone, which results in weak, soft bones that may be easily fractured. Signs include swollen, tender joints, enlargement of the ends of bones, an arched back, stiffness of the legs, and development of beads on the ribs.

Osteomalacia is the result of demineralization of the bones of adult animals. Because calcium and phosphorus in bone are in a dynamic state, high demands on calcium and phosphorus stores, such as occur during pregnancy and lactation, may result in osteomalacia. This condition is characterized by weak, brittle bones that may break when stressed.

Blood calcium concentration is not a good indicator of calcium status because plasma calcium is maintained at between 9 and 11 mg/dL by homeostatic mechanisms. Parathyroid hormone is released in response to a lowering of plasma calcium. It stimulates the production of 1,25-dihydroxy cholecalciferol (vitamin D_3). The 1,25-dihydroxy cholecalciferol increases calcium absorption from the intestine and, in conjunction with parathyroid hormone, increases calcium resorption from bone. If plasma calcium concentrations become elevated, calcitonin is produced and parathyroid hormone production is inhibited. Thus, calcium absorption and bone resorption are decreased.

CALCIUM SOURCES

The calcium content in forage is affected by species, portion of plant consumed, maturity, quantity of exchangeable calcium in the soil, and climate (Minson, 1990). Forages are generally good sources of calcium, and legumes are higher in calcium content than grasses. Cereal grains are low in calcium, so high-grain diets require supplementation. Oilseed meals are much higher in calcium than grains. Sources of supplemental calcium include calcium carbonate, ground limestone, bone meal, dicalcium phosphate, defluorinated phosphate, monocalcium phosphate, and calcium sulfate. True absorption in young steers of calcium from different sources ranged from 45 percent for ground limestone to 64 percent for dibasic calcium phosphate (Hansard et al., 1957).

SIGNS OF CALCIUM TOXICITY

High concentrations of dietary calcium are tolerated well by cattle. Protein and energy digestibility were reduced when cattle were fed a diet containing 4.4 percent calcium (calcium carbonate) (Ammerman et al., 1963). High concentrations of dietary calcium may affect metabolism of phosphorus, magnesium, and certain trace elements, but the changes are relatively small (National Research Council, 1980; Alfaro et al., 1988).

Magnesium

More than 300 enzymes are known to be activated by magnesium (Wacker, 1980). Magnesium is essential, as the complex Mg-ATP, for all biosynthetic processes including glycolysis, energy-dependent membrane transport, formation of cyclic-AMP, and transmission of the genetic code. Magnesium also is involved in the maintenance of electrical potentials across nerve and muscle membranes and for nerve impulse transmission. Of the total percentage of magnesium in the body, 65 to 70 percent is in bone, 15 percent in muscle, 15 percent in other soft tissues, and 1 percent in extracellular fluid (Mayland, 1988).

MAGNESIUM REQUIREMENTS

Dietary requirements for magnesium vary depending on age, physiological state, and bioavailability from the diet. As a percentage of dry matter, recommended magnesium requirements are as follows:

- growing and finishing cattle, 0.10 percent;
- gestating cows, 0.12 percent; and
- lactating cows, 0.20 percent.

Absolute requirements for magnesium have been estimated as follows:

- replenishment of endogenous loss, 3 mg Mg/kg liveweight;
- growth, 0.45 g Mg/kg gain;
- lactation, 0.12 g Mg/kg milk; and
- pregnancy, 0.12, 0.21, and 0.33 g Mg/day for early, mid, and late pregnancy, respectively (Grace, 1983).

O'Kelly and Fontenot (1969, 1973) found that beef cows required 7 to 9 g Mg/day during gestation and 18 to 21 g Mg/day during lactation to maintain serum magnesium concentrations of 2.0 mg/dL. These daily quantities corresponded to 0.10 to 0.13 percent during gestation and 0.17 to 0.20 percent during lactation. In young calves fed milk, 12 to 16 mg Mg/kg body weight was adequate to maintain blood magnesium concentrations (Huffman et al., 1941; Blaxter and McGill, 1956).

SIGNS OF MAGNESIUM DEFICIENCY

Magnesium deficiency in calves results in excitability, anorexia, hyperemia, convulsions, frothing at the mouth, profuse salivation, and calcification of soft tissue (Moore et al., 1938; Blaxter et al., 1954). Grass tetany or hypomagnesemic tetany is characterized by low magnesium concentrations in plasma and cerebrospinal fluid and is a problem in lactating beef cows. Initial signs of grass tetany are nervousness, reduced feed intake, and muscular twitching around the face and ears. Animals are uncoordinated and walk with a stiff gait. In the advanced stages, cows go down on their side with their head back and go into convulsions. Death usually occurs unless the animal is treated intravenously or subcutaneously with a magnesium-salt solution.

Grass tetany is most common in lactating cows grazing lush spring pastures or fed harvested forages low in magnesium. With early spring pastures, the problem is more one of insufficient availability rather than low forage magnesium concentrations per se. Fertilizing pastures with fertilizers high in nitrogen and potassium is associated with increased incidence of grass tetany. Cows depend on a frequent supply of magnesium from the gastrointestinal tract to maintain normal blood magnesium concentrations because homeostatic mechanisms are not sufficient to regulate blood magnesium concentrations. Magnesium concentrations in bone are high, but mature animals lack the ability to mobilize large amounts of magnesium from bone (Rook and Storry, 1962). In young calves, at least 30 percent of the skeletal magnesium can be mobilized during magnesium deficiency (Blaxter et al., 1954).

FACTORS AFFECTING MAGNESIUM REQUIREMENTS

The rumen is the major site of magnesium absorption in ruminants (Grace et al., 1974; Greene et al., 1983). Magnesium absorption is high in young calves fed milk but decreases with age (Peeler, 1972). True absorption values for magnesium in mature ruminants fed hay and grass range from 10 to 37 percent (Agricultural Research Council, 1980). Magnesium in concentrates is more available than magnesium in forages (Peeler, 1972). A number of studies have shown that high-dietary potassium reduces magnesium absorption (Greene et al., 1983; Wylie et al., 1985). High dietary concentrations of nitrogen, organic acids (citric acid and trans-aconitate), long-chain fatty acids, calcium, and phosphorus also may reduce magnesium absorption or utilization (Fontenot et al., 1989). High-ruminal NH_3 concentrations have been associated with hypomagnesemia in cows grazing spring pastures high in crude protein (Martens and Rayssiguier, 1980). Magnesium absorption has been enhanced by feeding soluble carbohydrates or carboxylic ionophores (Fontenot et al., 1989; Spears et al., 1989). Evidence suggests that magnesium absorption from the rumen occurs by an active sodium-linked process (Martens and Rayssiguier, 1980), and sodium supplementation in a low-sodium diet increases magnesium absorption (Martens et al., 1987). It has also been reported that different breeds absorb magnesium differently (Greene et al., 1989). Excess magnesium absorbed is excreted primarily in the urine.

MAGNESIUM SOURCES

Cereal grains generally contain 0.11 to 0.17 percent magnesium; plant protein sources contain approximately twice this concentration (Underwood, 1981). Magnesium concentration in forages varies greatly depending on plant species, soil magnesium, stage of growth, season and environmental temperature (Minson, 1990). Legumes are usually higher in magnesium than are grasses. Magnesium oxide and magnesium sulfate are good sources of supplemental magnesium, but magnesium in magnesite and dolomitic limestone is poorly available (Gerken and Fontenot, 1967; Ammerman et al., 1972).

SIGNS OF MAGNESIUM TOXICITY

Magnesium toxicity is not a problem in beef cattle. Maximum tolerable concentrations have been estimated at 0.4 percent (National Research Council, 1980). Cows fed 0.39 percent magnesium showed no adverse effects (O'Kelly and Fontenot, 1969). Young calves fed 1.3 percent magnesium had lower feed intake and weight gain and diarrhea with mucus in feces (Gentry et al., 1978). Steers fed 2.5 or 4.7 percent magnesium exhibited severe diarrhea and a lethargic appearance, while 1.4 percent magnesium reduced dry matter digestibility (Chester-Jones et al., 1990).

Phosphorus

Phosphorus is often discussed in conjunction with calcium because the two minerals function together in bone formation; however, the effect of the calcium:phosphorus ratio on ruminant performance has been overemphasized in the past. Several studies (Dowe et al., 1957; Wise et al., 1963; Ricketts et al., 1970; Alfaro et al., 1988) have shown that dietary calcium to phosphorus ratios of between 1:1 and 7:1 result in similar performance, provided that phosphorus intake is adequate to meet requirements.

Approximately 80 percent of phosphorus in the body is found in bones and teeth with the remainder distributed in soft tissues. Phosphorus also functions in cell growth and differentiation as a component of DNA and RNA; energy utilization and transfer as a component of ATP, ADP, and AMP; phospholipid formation; and maintenance of acid-base and osmotic balance. Phosphorus is required by ruminal microorganisms for their growth and cellular metabolism.

PHOSPHORUS REQUIREMENTS

Requirements for phosphorus were calculated using the factorial method. Estimated requirements for maintenance, growth, pregnancy, and lactation were totaled and then corrected for the percentage of dietary phosphorus absorbed. The maintenance requirement for phosphorus was considered to be 16 mg P/kg body weight. This value is similar to fecal endogenous losses observed in cattle fed phosphorus concentrations at or near requirements (Tillman and Brethour, 1958; Tillman et al., 1959; Challa and Braithwaite, 1988; Challa et al., 1989). Slightly lower fecal endogenous losses were observed for dairy cows in negative phosphorus balance (Martz et al., 1990). Retained-phosphorus needs in excess of maintenance requirements were calculated as 3.9 g P/100 g protein gain. The phosphorus content of gain was calculated from data presented by Ellenberger et al. (1950). Phosphorus needs, during lactation, in excess of maintenance, were calculated as 0.95 g P/kg milk produced. Fetal phosphorus was assumed to be 7.6 g P/kg fetal weight. This requirement was distributed over the last 3 months of pregnancy, and a birth weight of 35 kg was assumed. *See "Errata" front cover*

FACTORS AFFECTING PHOSPHORUS REQUIREMENTS

A true absorption of 68 percent was assumed in converting absolute phosphorus requirements to dietary requirements. This value agrees well with most studies (Tillman and Brethour, 1958; Tillman et al., 1959; Challa et al., 1989; Martz et al., 1990) of cattle where true absorption has been measured. Absorption of phosphorus was much higher in young calves fed milk (Lofgreen et al., 1952). In their estimate of requirements, AFRC (TCORN, 1991) assumed an absorption coefficient of 64 percent for phosphorus in forages and 70 percent for phosphorus in concentrates.

In young calves with an initial weight of 96 kg, 0.22 percent phosphorus was adequate for maximum weight gains, but increasing phosphorus to 0.30 percent increased bone ash (Wise et al., 1958). A more recent study with dairy calves weighing approximately 70 kg indicated that 0.26 percent phosphorus was not adequate for maximum growth or bone ash (Jackson et al., 1988). Call et al. (1978) fed Hereford heifers (165 kg initial weight), beginning at approximately 7 months of age, diets containing 0.14 or 0.36 percent phosphorus for 2 years. No differences between the two groups were detected in growth, rib bone morphology and phosphorus content, age at puberty, conception rate, or calving interval.

In a second study, Hereford heifers were fed low-phosphorus diets from weaning through their fifth gestation and lactation (Call et al., 1986). The low-phosphorus group received 6 to 12.1 g P/day, while controls received 20.6 to 38.1 g P/day with phosphorus intake increased as the cattle grew larger. Females fed the low-phosphorus intake remained healthy, and growth and reproduction were similar to that observed in phosphorus supplemented animals. When phosphorus intake of 6 to 12.1 g P/day was reduced to 5.1 to 6.6 g P/day, clinical signs of deficiency occurred within 6 months (Call et al., 1986). Reproduction was not impaired until cows were fed the very low phosphorus diet for more than 1 year. It was concluded that 12 g P/day

throughout 1 production year was adequate for 450-kg Hereford cows (Call et al., 1986). No measurements of milk production or calf weaning weights were given in these papers (Call et al., 1978, 1986).

SIGNS OF PHOSPHORUS DEFICIENCY

In grazing livestock, phosphorus deficiency has been described as the most prevalent mineral deficiency throughout the world (McDowell, 1992). Studies in South Africa and Texas of cattle that grazed forages low in phosphorus showed large improvements in fertility and calf weaning weights with phosphorus supplementation (Dunn and Moss, 1992). Phosphorus deficiency results in reduced growth and feed efficiency, decreased appetite, impaired reproduction, reduced milk production, and weak, fragile bones (Underwood, 1981; Shupe et al., 1988). The skeleton provides a large reserve of phosphorus that can be drawn on during periods of inadequate phosphorus intake in mature animals. Skeletal reserves can subsequently be replaced during periods when phosphorus intake is high relative to requirements. Plasma phosphorus concentrations consistently below 4.5 mg/dL are indicative of a deficiency, but bone phosphorus is a more sensitive measure of phosphorus status (McDowell, 1992).

Phosphorus absorption occurs in the small intestine. The percentage absorbed is not greatly affected by amount of phosphorus intake (TCORN, 1991). Varying endogenous fecal excretion is an important homeostatic mechanism for controlling phosphorus in cattle. Endogenous fecal losses consist largely of unabsorbed salivary phosphorus (Challa et al., 1989). Salivary phosphorus is affected by plasma phosphorus concentration, which does depend on phosphorus intake as well as factors that affect salivary flow such as dry matter intake and physical form of the diet (TCORN, 1991). Thus, fecal endogenous loss of phosphorus may vary depending on intake and other factors that affect salivary phosphorus. In estimating the maintenance requirement, it is important that endogenous fecal excretion of phosphorus be measured in cattle fed approximately their phosphorus requirement. Urinary losses of phosphorus are generally lower but may increase in cattle fed high-concentrate diets (Reed et al., 1965).

PHOSPHORUS SOURCES

Phosphorus-deficient soils are widespread and forages produced on these soils are low in phosphorus. Drought conditions and increased forage maturity also can result in low forage-phosphorus concentrations. Cereal grains and oilseed meals contain moderate to high concentrations of phosphorus. Animal and fish products are high in phosphorus. In terms of availability, supplemental sources of phosphorus were ranked as follows: dicalcium phosphate, defluorinated phosphate, and bone meal (Peeler, 1972). More recent studies with calves have indicated that defluorinated phosphate (Miller et al., 1987) and monoammonium phosphate (Jackson et al., 1988) are equal in availability to dicalcium phosphate. Phytate phosphorus is not well utilized by nonruminants, but seems to be utilized by ruminants as readily as phosphorus from inorganic sources (McGillivray, 1974).

Potassium

Potassium is the third most abundant mineral in the body and the major cation in intracellular fluid. Potassium is important in acid-base balance, regulation of osmotic pressure, water balance, muscle contractions, nerve impulse transmission, and certain enzymatic reactions.

POTASSIUM REQUIREMENTS

Feedlot cattle require approximately 0.6 percent potassium. Studies conducted with potassium in cattle receiving no ionophore have been inconsistent. Roberts and St. Omer (1965) observed a response in gain with potassium supplementation of steer diets containing 0.50 to 0.56 percent potassium in only one of three trials. Devlin et al. (1969) noted improvements in steers' gain and feed intake when potassium was added to diets already containing 0.5 percent potassium. More recently, however, Kelley and Preston (1984) observed no improvement in steer performance when potassium was supplemented to a basal diet containing 0.4 percent potassium. Studies with feedlot cattle fed lasalocid (Ferrell et al., 1983; Spears and Harvey, 1987) or monensin (Brink et al., 1984) indicate that potassium requirement does not exceed 0.55 percent. Potassium requirements in young dairy calves not fed an ionophore also do not exceed 0.55 percent (Weil et al., 1988; Tucker et al., 1991). Because of the lower rates of gain observed in growing cattle in range conditions, potassium requirements for range cattle may be lower than those for feedlot cattle. Clanton (1980) concluded that growing cattle in range conditions require 0.3 to 0.4 percent potassium.

FACTORS AFFECTING POTASSIUM REQUIREMENTS

See "Errata" front cover

Potassium requirements of beef cows are not well defined. Clanton (1980) suggested that gestating beef cows require 0.5 to 0.7 percent potassium. Because of the relatively high secretion of potassium in milk (1.5 g/kg), requirements for potassium may be slightly higher in beef cows during lactation—for example, for cows producing 9 kg milk/day, approximately 13.5 g K/day or 0.13 percent of dry matter intake would be needed for milk production.

SIGNS OF POTASSIUM DEFICIENCY

A deficiency of potassium results in reduced feed intake and weight gain, pica, rough hair coat, and muscular weakness (Devlin et al., 1969). In beef cattle, a severe deficiency of potassium is unlikely. A marginal potassium deficiency results in decreased feed intake and retarded weight gain. Dietary potassium concentration is the best indicator of potassium status. Serum or plasma potassium is not a reliable indicator of potassium status. Reduced feed consumption appears to be an early indicator of marginal potassium deficiency, but the depression in feed intake is usually of relatively small magnitude, making it difficult to detect in field conditions.

Potassium is absorbed from the rumen and omasum as well as the intestine, and absorption is very high. The major route of potassium excretion is the urine. Body stores of potassium are small; therefore, a deficiency can occur rapidly (Ward, 1966).

POTASSIUM SOURCES

Forages are excellent sources of potassium, usually containing between 1 and 4 percent potassium. In fact, high potassium content in lush spring pastures seems to be a major factor associated with the occurrence of grass tetany in beef cows (Mayland, 1988).

As forages mature, the potassium content decreases, and low concentrations of potassium have been observed in range forage and in accumulated tall fescue during the winter (Clanton, 1980). Cereal grains are often deficient (<0.5 percent) in potassium, and high-concentrate diets may require potassium supplementation unless a high-potassium forage or protein supplement is included in the diet. Oilseed meals are good sources of potassium. Potassium can be supplemented to cattle diets as potassium chloride, potassium bicarbonate, potassium sulfate, or potassium carbonate. All forms are readily available.

SIGNS OF POTASSIUM TOXICITY

Increasing the potassium content of a liquid diet from 1.2 to 5.8 percent on a dry matter basis resulted in the deaths of 3 of 8 calves as a result of cardiac insufficiency (Blaxter et al., 1960). In calves, increasing dietary potassium from 2.77 to 6.77 percent reduced feed intake and retarded weight gain (Neathery et al., 1980). The maximum tolerable concentration of potassium has been set at 3 percent for cattle (National Research Council, 1980). Cattle grazing lush, spring pastures often consume more than 3 percent potassium, and other than reduced absorption of magnesium, no adverse effects have been reported.

Sodium and Chlorine

Sodium is the major cation, while chlorine is the major anion, in extracellular fluid. Both sodium and chlorine are involved in maintaining osmotic pressure, controlling water balance, and regulating acid-base balance. Sodium also functions in muscle contractions, nerve impulse transmission, and glucose and amino acid transport. Chlorine is necessary for the formation of hydrochloric acid in gastric juice and for the activation of amylase.

SODIUM AND CHLORINE REQUIREMENTS

Requirements for sodium in nonlactating beef cattle do not exceed 0.06 to 0.08 percent, while lactating beef cows require approximately 0.10 percent sodium (Morris, 1980). Ruminants have an appetite for sodium, and if it is provided ad libitum, they will consume more salt than they actually require. In a 2-year study with beef cows grazing forage containing from 0.012 and 0.055 percent sodium, providing salt ad libitum did not affect calf weaning weights or cow body weights (Morris et al., 1980). Chlorine requirements are not well defined but a deficiency of chlorine does not seem likely in practical conditions (Neathery et al., 1981). Young calves fed 0.038 percent chlorine performed similar to those fed 0.5 percent chlorine (Burkhaltor et al., 1979).

SIGNS OF SODIUM DEFICIENCY

Signs of deficiency of sodium are rather nonspecific and include pica and reduced feed intake, growth, and milk production (Underwood, 1981). When sodium intake is low, the body conserves sodium by increasing reabsorption of sodium from the kidney in response to aldosterone (McDowell, 1992). The sodium:potassium ratio in saliva has been used as an indicator of sodium status. This ratio is normally 20:1, and a production response to sodium supplementation is likely when the sodium:potassium ratio is less than 10:1 (Morris, 1980). Serum or plasma sodium concentration is not a reliable indicator of sodium status. Dietary sodium concentration is a good measure of sodium adequacy.

SODIUM AND CHLORINE SOURCES

Cereal grains and oilseed meals usually provide inadequate amounts of sodium for beef cattle. Animal products are much higher in sodium and chlorine than plant products (Meyer et al., 1950). The sodium content of forages varies considerably (Minson, 1990). Sodium can be supplemented as sodium chloride or sodium bicarbonate and both forms are highly available.

SIGNS OF SODIUM TOXICITY

High concentrations of salt have been used to regulate feed intake and cattle can tolerate high-dietary concentrations provided that an adequate supply of water is available. Growing cattle were able to tolerate 9.33 percent salt for 84 days without adverse effects (Meyer et al., 1955). However, Leibholz et al. (1980) reported that 6.5 percent salt decreased organic matter intake and growth in calves. The maximum tolerable concentration for dietary salt in cattle was estimated at 9.0 percent in *Mineral Tolerance of Domestic Animals* (National Research Council, 1980).

Salt is much more toxic when present in the drinking water of cattle. Growing cattle were able to tolerate 1.0 percent added salt in drinking water without adverse effects (Weeth et al., 1960; Weeth and Haverland, 1961); however, the addition of 1.25 to 2.0 percent salt resulted in anorexia, reduced weight gain or weight loss, reduced water intake and physical collapse (Weeth et al., 1960). In some areas of the western United States, soils are high in saline, resulting in groundwater that can cause saline water intoxication. Consumption of water with more than 7,000 mg Na/kg resulted in reduced feed and water intake, decreased growth, mild digestive disturbances, and diarrhea (Jenkins and Mackey, 1979).

Sulfur

Sulfur is a component of methionine, cysteine, and cystine, and the B-vitamins, thiamin and biotin, as well as a number of other organic compounds. Sulfate is a component of sulfated mucopolysaccharides and also functions in certain detoxification reactions in the body. All sulfur-containing compounds with the exception of biotin and thiamin can be synthesized from methionine. Ruminal microorganisms are capable of synthesizing all organic sulfur containing compounds required by mammalian tissue from inorganic sulfur (Block et al., 1951; Thomas et al., 1951). Sulfur is required also by ruminal microorganisms for their growth and normal cellular metabolism.

SULFUR REQUIREMENTS

Requirements of beef cattle for sulfur are not well defined. The recommended concentration in beef cattle diets is 0.15 percent. Sulfur supplementation increased gain in steers fed corn silage-corn-urea based diets containing 0.10 to 0.11 percent sulfur (Hill, 1985). In steers fed high-concentrate diets containing 0.14 percent sulfur, increasing dietary sulfur tended to reduce ruminal lactic acid accumulation and improve feed efficiency (Rumsey, 1978). Other studies have indicated that 0.11 to 0.12 percent sulfur was adequate for growing cattle (Bolsen et al., 1973; Pendlum et al., 1976). In Australia, sulfur supplementation increased gain by 12 percent in steers grazing sorghum × sudangrass containing 0.08 to 0.12 percent sulfur (Archer and Wheeler, 1978). The sulfur requirement of ruminants grazing sorghum × sudangrass may be increased because of the need for sulfur in the detoxification of cyanogenic glucoside found in most sorghum forages.

SIGNS OF SULFUR DEFICIENCY

Severe sulfur deficiency results in anorexia, weight loss, weakness, dullness, emaciation, excessive salivation, and death (Thomas et al., 1951; Starks et al., 1953). Marginal deficiencies of sulfur can reduce feed intake, digestibility, and microbial protein synthesis. A dietary limitation of sulfur can dramatically decrease microbial numbers as well as microbial digestion and protein synthesis. Supplementation to increase the sulfur content of hay from 0.04 to 0.075 percent increased counts of ruminal bacteria, protozoa, and sporangia of anaerobic fungi in sheep (Morrison et al., 1990). Impaired utilization of lactate by ruminal microorganisms, resulting in lactate accumulation in the rumen and blood, also can occur as a result of sulfur deficiency (Whanger and Matrone, 1966).

~~SULFUR SOURCES~~ "Factors Affecting Sulfur Requirements"

Most rumen bacteria are able to synthesize the sulfur-containing amino acids from sulfide (Goodrich et al., 1978). Ruminal sulfide is derived from the reduction of inorganic sulfur sources and from the degradation of sulfur-containing amino acids. Sulfide can be absorbed from the rumen and oxidized by tissues to sulfate, a less toxic form of sulfur. Sulfur is found in feedstuffs largely as a component of protein. Dietary sulfur requirements may be higher when diets high in rumen bypass protein are fed because of a limitation of sulfur for optimal ruminal fermentation. Most practical diets are adequate in sulfur. When urea or other nonprotein nitrogen sources replace preformed protein, sulfur supplementation may be needed. Mature forages, forages grown in sulfur-deficient soils, corn silage, and sorghum × sudangrass can be low in sulfur. Sorghum forages seem inherently low in sulfur relative to most forages, and the sulfur content of sorghum × sudangrass did not increase in response to sulfur fertilization (Wheeler et al., 1980). "Sulfur Sources"

Sulfur can be supplemented in ruminant diets as sodium sulfate, ammonium sulfate, calcium sulfate, potassium sulfate, magnesium sulfate, or elemental sulfur. Based on in vitro microbial protein synthesis, the availability of sulfur to ruminal microorganisms from different sources has been ranked from most to least available as L-methionine, calcium sulfate, ammonium sulfate, D,L-methionine, sodium

sulfate, sodium sulfide, elemental sulfur, and methionine hydroxy analog (Kahlon et al., 1975).

SIGNS OF SULFUR TOXICITY

Acute sulfur toxicity is characterized by restlessness, diarrhea, muscular twitching, dyspnea, and, in prolonged cases, inactivity followed by death (Coghlin, 1944). Concentrations of sulfur lower than those needed to cause clinical signs of toxicity can reduce feed intake and retard growth rate (Kandylis, 1984) and decrease copper status (Smart et al., 1986). Increasing dietary sulfur from 0.12 to 0.41 percent using ammonium sulfate reduced feed intake by 32 percent in steers fed high-concentrate diets containing urea (Bolsen et al., 1973). Consumption of water high in sulfate (5,000 mg/kg) reduced feed and water intake (Weeth and Hunter, 1971). The maximum tolerable concentration of dietary sulfur has been estimated at 0.40 percent (National Research Council, 1980).

MICROMINERALS

Chromium

Chromium functions as a component of the glucose tolerance factor that serves to potentiate the action of insulin (Mertz, 1992). The addition of chromium as 0.4 mg chromium picolinate/kg diet (Bunting et al., 1994), or chromium polynicotinate/kg diet (Kegley and Spears, 1995), for growing cattle increased glucose clearance rate following intravenous glucose administration. Adding low concentrations (0.2 to 1.0 mg/kg) of chromium also increased immune response and growth rate in stressed cattle (Chang and Mowat, 1992; Moonsie-Shageer and Mowat, 1993). These studies suggest that in some situations supplemental chromium may be needed.

Current information is not sufficient to determine chromium requirements. Based on studies with humans and laboratory animals, organic chromium is much more bioavailable than inorganic chromium. The maximum tolerable concentration of trivalent chromium in the chloride form was estimated to be 1,000 mg Cr/kg diet for cattle (National Research Council, 1980). No adverse effects were observed in steers fed 4.0 mg chromium polynicotinate complex/kg diet for 70 days (Claeys and Spears, unpublished data). Hexavalent chromium is much more toxic than the trivalent form (National Research Council, 1980).

Cobalt

Cobalt functions as a component of vitamin B_{12} (cobalamin). Cattle are not dependent on a dietary source of vitamin B_{12} because ruminal microorganisms are capable

See "Errata" - front cover

of synthesizing B_{12} from dietary cobalt. Measurements of the amount of dietary cobalt converted to vitamin B_{12} in the rumen have ranged from 3 to 13 percent of intake (Smith, 1987). Ruminal bacteria also produce a number of B_{12} analogues that are active in bacteria but apparently inactive in animal tissues (Bigger et al., 1976). Two vitamin B_{12}-dependent enzymes occur in mammalian tissues (Smith, 1987)—methylmalonyl CoA mutase is essential for the metabolism of propionate to succinate, as it catalyzes the conversion of L-methylmalonyl CoA to succinyl CoA; and 5-methyltetrahydrofolate homocysteine methyltransferase (methionine synthase) catalyzes the transfer of methyl groups from 5-methyltetrahydrofolate to homocysteine to form methionine and tetrahydrofolate. This reaction is important in the recycling of methionine following transfer of its methyl group.

COBALT REQUIREMENTS

The cobalt requirement of cattle is approximately 0.10 mg/kg dry matter diet (Smith, 1987). Cobalt concentrations between 0.07 and 0.11 percent have been reported to be adequate in various studies (Smith, 1987). Young, rapidly growing cattle seem more sensitive to cobalt deficiency than older cattle. Feeding a high-concentrate diet may depress ruminal synthesis of vitamin B_{12} and increase production of B_{12} analogues (Walker and Elliot, 1972; Halpin et al., 1984). However, MacPhearson and Chalmers (1984) found no evidence that cobalt requirements were higher when high-concentrate diets were consumed.

SIGNS OF COBALT DEFICIENCY

Decreased appetite and failure to grow or moderate weight loss are early signs of cobalt deficiency (Smith, 1987). If the deficiency is allowed to become severe, animals exhibit severe unthriftiness, rapid weight loss, fatty degeneration of the liver, and pale skin and mucous membranes as a result of anemia. Cobalt deficiency also has been reported to impair the ability of neutrophils to kill yeast and reduce disease resistance (MacPherson et al., 1989). Recent findings indicate that an inability by ruminal microorganisms to convert succinate to propionate is an early manifestation of cobalt deficiency (Kennedy et al., 1991). Ruminal and plasma succinate concentrations were greatly elevated in lambs fed cobalt-deficient diets. Liver vitamin B_{12} or cobalt concentrations can be used to assess cobalt status (Smith, 1987). Vitamin B_{12} concentrations in liver of 0.10 µg/g wet weight or less are indicative of cobalt deficiency. Measurement of serum B_{12} in cattle may be of limited value because of the presence of B_{12} analogues in bovine serum (Halpin et al., 1984).

FACTORS AFFECTING COBALT REQUIREMENTS

Soils deficient in cobalt occur in many areas of the world including the southeastern Atlantic coast of the United States (Ammerman, 1970). Legumes are generally higher in cobalt than grasses and availability of cobalt in soil is highly dependent on soil pH (Underwood, 1981). Increasing soil pH from 5.4 to 6.4 reduced the cobalt content of ryegrass from 0.35 to 0.12 mg/kg (Mills, 1981). Cobalt can be supplemented to the diet in free-choice mineral mixtures. Feed-grade sources of cobalt include cobalt sulfate and cobalt carbonate. It is unclear how these two forms of cobalt compare in terms of relative bioavailability for vitamin B_{12} synthesis. Pellets containing cobalt oxide and finely divided iron, and controlled-release glass pellets containing cobalt have been used in grazing ruminants. Both types of pellets remain in the reticulorumen and release cobalt over an extended period.

SIGNS OF COBALT TOXICITY

Cobalt toxicity is not likely to occur unless an error is made in formulating a mineral supplement. Cattle can tolerate approximately 100 times the dietary requirement for cobalt (National Research Council, 1980). Signs of chronic cobalt toxicity, with the exception of elevated liver cobalt, are similar to those of cobalt deficiency and include decreased feed intake and reduced body weight gain, anemia, emaciation, hyperchromia, debility, and increased liver cobalt (National Research Council, 1980). Young dairy calves given up to 66 mg Co/kg body weight for up to 28 weeks showed no adverse effects (Keener et al., 1949). The sulfate, carbonate, and chloride forms of cobalt were similar in terms of toxicity (Keener et al., 1949).

Copper

Copper functions as an essential component of a number of enzymes including lysyl oxidase, cytochrome oxidase, superoxide dismutase, ceruloplasmin, and tyrosinase (McDowell, 1992).

COPPER REQUIREMENTS

Requirements for copper can vary from 4 to more than 15 mg/kg depending largely on the concentration of dietary molybdenum and sulfur. The recommended concentration of copper in beef cattle diets is 10 mg Cu/kg diet. This amount should provide adequate copper if the diet does not exceed 0.25 percent sulfur and 2 mg Mo/kg diet. Less than 10 mg Cu/kg diet may meet requirements of feedlot cattle because copper is more available in concentrate diets than in forage diets. Copper requirements may be affected by breed. Simmental cattle excrete more copper in their bile than Angus (Gooneratne et al., 1994). Ward et al. (1995) reported that Simmental and Charolais cows and their calves were more susceptible to copper deficiency than Angus when fed the same diet.

Copper requirements are greatly increased by molybdenum and sulfur. The antagonistic action of molybdenum on copper metabolism is exacerbated when sulfur is also high. Considerable evidence suggests that molybdate and sulfide interact to form thiomolybdates in the rumen (Suttle, 1991). Copper is believed to react with thiomolybdates in the rumen to form insoluble complexes that are poorly absorbed. Some thiomolybdates are absorbed and affect systemic metabolism of copper (Gooneratne et al., 1989). Thiomolybdates can result in copper being tightly bound to plasma albumin and not available for biochemical functions, and they may directly inhibit certain copper-dependent enzymes. In cattle grazing pastures containing 3 to 20 mg Mo/kg, copper concentrations in the range of 7 to 14 mg/kg were inadequate (Thornton et al., 1972).

FACTORS AFFECTING COPPER REQUIREMENTS

Sulfur reduces copper absorption, perhaps via formation of copper sulfide in the gut, independent from its role in the molybdenum-copper interaction (Suttle, 1974). Reducing the sulfate content of drinking water high in sulfate from 500 to 42 mg/L by reverse osmosis increased the copper status of cattle (Smart et al., 1986). A copper concentration of 10 mg/kg was not adequate in cows receiving sulfated water, which resulted in total dietary sulfur of 0.35 percent (Smart et al., 1986). High concentrations of iron (Phillippo et al., 1987a) and zinc (Davis and Mertz, 1987) also reduce copper status and may increase copper requirements.

SIGNS OF COPPER DEFICIENCY

Copper deficiency is a widespread problem in many areas of the United States and Canada. Signs that have been attributed to copper deficiency include

- anemia,
- reduced growth,
- depigmentation and changes in the growth and physical appearance of hair,
- cardiac failure,
- bones that are fragile and easily fractured,
- diarrhea, and
- low reproduction characterized by delayed or depressed estrus (Underwood, 1981).

Achromotrichia or lack of hair pigmentation is generally the earliest clinical sign of copper deficiency. Copper deficiency also reduces the ability of isolated neutrophils to kill yeast (Boyne and Arthur, 1981); and copper deficiency in grazing lambs increased susceptibility to bacterial infec-

tions (Woolliams et al., 1986). As discussed in the molybdenum section, some of the abnormalities that have been attributed to copper deficiency may be caused by molybdenosis rather than copper per se.

Copper is poorly absorbed in ruminants with a developed rumen. Absorbed copper is excreted primarily via the bile with small amounts lost in the urine (Gooneratne et al., 1989). Considerable storage of copper can occur in the liver.

COPPER SOURCES

Forage copper concentrations are of limited value in assessing copper adequacy unless forage concentrations of copper antagonists such as molybdenum, sulfur, and iron are also considered. Liver copper concentrations less than 20 mg/kg on a dry matter basis or plasma concentrations less than 50 µg/dL are indicative of deficiency (Underwood, 1981). However, in the presence of high dietary molybdenum and sulfur, copper in liver and plasma may not accurately reflect copper status because the copper can exist in tightly bound forms unavailable for biochemical functions (Suttle, 1991). Forages vary greatly in copper content depending on plant species and available copper in the soil (Minson, 1990). Legumes are usually higher in copper than grasses. Milk and milk products are low in copper. Cereal grains generally contain 4 to 8 mg Cu/kg, and oilseed meals and leguminous seeds contain 15 to 30 mg Cu/kg.

Copper is usually supplemented to diets or ad libitum minerals in the sulfate, carbonate, or oxide forms. Recent studies indicate that copper oxide is very poorly available relative to copper sulfate (Langlands et al., 1989a; Kegley and Spears, 1994). In early studies, copper carbonate was at least equal to copper sulfate (Chapman and Bell, 1963). Various organic forms of copper also are available. In calves fed diets high in molybdenum, copper proteinate was more available than copper sulfate (Kincaid et al., 1986). However, Wittenberg et al. (1990) found similar availability of copper from copper proteinate and copper sulfate in steers fed high-molybdenum diets. Studies comparing copper lysine to copper sulfate have yielded inconsistent results. Ward et al. (1993) reported that copper lysine and copper sulfate were of similar bioavailability when fed to cattle; however, Nockels et al. (1993) found that copper lysine was more avaiable than copper sulfate.

Injectable forms of copper such as copper glycinate or copper EDTA have been given at 3- to 6-month intervals to prevent copper deficiency (Underwood, 1981). Although feed-grade copper oxide is largely unavailable, copper oxide needles, which remain in the gastrointestinal tract and slowly release copper over a period of months, have been used as a copper source for cattle (Cameron et al., 1989).

SIGNS OF COPPER TOXICITY

Copper toxicity can occur in cattle as a result of excessive supplementation of copper or the use of feeds that have been contaminated with copper from agricultural or industrial sources. The liver can accumulate large amounts of copper before signs of toxicity are observed. When copper is released from the liver in large amounts (hemolytic crisis), hemolysis, methemoglobinemia, hemoglobinuria, jaundice, icterus, widespread necrosis, and often death occur (National Research Council, 1980). The maximum tolerable concentration of copper for cattle has been estimated at 100 mg Cu/kg diet (National Research Council, 1980). The concentration of copper needed to cause toxicity will depend on the concentration of molybdenum, sulfur, and iron in the diet. Adult cattle are less susceptible to copper toxicity than young cattle. In young calves, feeding 115 mg Cu/kg for 91 days resulted in signs of toxicity (Shand and Lewis, 1957).

Iodine

Iodine functions as an essential component of the thyroid hormones thyroxine (T_4) and triiodothyronine (T_3), which regulate the rate of energy metabolism in the body. Between 70 and 80 percent of dietary iodine is absorbed as iodide from the rumen with considerable resecretion occurring in the abomasum (Miller et al., 1988). Iodide that is secreted into the abomasum is largely reabsorbed from the small and large intestine. Absorbed iodide is largely taken up by the thyroid gland for thyroid hormone synthesis or is excreted in the urine. In lactating cows, approximately 8 percent of dietary iodine is secreted in milk (Miller et al., 1988). When the thyroid hormones are catabolized, much of the iodine is reused by the thyroid gland.

IODINE REQUIREMENTS

Iodine requirements of beef cattle are not well established; 0.5 mg I/kg diet should be adequate unless the diet contains goitrogenic substances that interfere with iodine metabolism. Iodine requirements have been estimated by measuring thyroid hormone secretion rate (Agricultural Research Council, 1980). Miller et al. (1988) calculated the theoretical iodine requirement to be 0.6 mg/100 kg BW assuming

- a daily thyroxine secretion rate of 0.2 to 0.3 mg I/100 kg BW,
- 30 percent uptake of dietary iodine by the thyroid, and
- 15 percent recycling of thyroxine iodine.

This would correspond to 0.2 to 0.3 mg I/kg in the total diet, depending on feed intake.

FACTORS AFFECTING IODINE REQUIREMENTS

Goitrogenic substances in the feed may increase iodine requirements substantially (2- to 4-fold) depending on the amount and type of goitrogens present. The cyanogenetic goitrogens include the thiocyanate derived from cyanide in white clover and the glucosinolates found in some *Brassica* forages such as kale, turnips, and rape. They impair iodine uptake by the thyroid, and their effect can be overcome by increasing dietary iodine. Soybean meal and cottonseed meal also have a goitrogenic effect (Miller et al., 1975). The thiouracil goitrogens are found in *Brassica* seeds and inhibit iodination of tyrosine residues in the thyroid gland. The action of thiouracil goitrogens is more difficult to reverse with iodine supplementation.

SIGNS OF IODINE DEFICIENCY

The first sign of iodine deficiency is usually enlargement of the thyroid (goiter) in the newborn (Miller et al., 1988). Iodine deficiency may result in calves born hairless, weak, or dead; reduced reproduction in cows characterized by irregular cycling, low conception rate, and retained placenta; and decreased libido and semen quality in males (McDowell, 1992). Deficiency signs may not appear for more than a year after cattle are fed an iodine-deficient diet. Protein-bound iodine, thyroid weight in newborns, and milk iodine have been used to assess iodine status (Underwood, 1981).

IODINE SOURCES

The iodine content of feeds depends on the iodine available in the soil. In the United States, much of the Northeast, the Great Lakes, and Rocky Mountain regions are deficient in iodine (Underwood, 1981). Iodine is usually supplemented in diets or in free-choice minerals as calcium iodate or ethylenediamine dihydroiodide (EDDI), an organic form of iodine. Both forms are highly available and stable in mineral supplements and diets. Iodide forms such as potassium or sodium iodide are less stable and considerable losses can occur as a result of heat, moisture, light, and exposure to other minerals. EDDI has been widely used in cattle to prevent foot rot. The amount of EDDI fed to prevent foot rot is much higher than dietary requirements. At present, 10 mg I from EDDI is the maximum concentration that can be fed per head per day.

SIGNS OF IODINE TOXICITY

The maximum tolerable level of iodine is 50 mg/kg diet (National Research Council, 1980). In calves, 50 mg/kg of iodine as calcium iodate reduced weight gain and feed intake, and caused coughing and excessive nasal discharge

(Newton et al., 1974). Iodine in the form of EDDI has been fed at concentrations exceeding 50 mg/kg without adverse effects in calves and lactating cows (National Research Council, 1980).

Iron

Iron is an essential component of a number of proteins involved in oxygen transport or utilization. These proteins include hemoglobin, myoglobin, and a number of cytochromes and iron-sulfur proteins involved in the electron transport chain. Several mammalian enzymes also either contain iron or are activated by iron (McDowell, 1992). More than 50 percent of the iron in the body is present in hemoglobin, with smaller amounts present in other iron-requiring proteins and enzymes, and in protein-bound stored iron.

IRON REQUIREMENTS

The iron requirement is approximately 50 mg/kg diet in beef cattle. Studies with young calves fed milk diets have indicated that 40 to 50 mg Fe/kg is adequate to support growth and prevent anemia (Bremner and Dalgarno, 1973; Bernier et al., 1984). Iron requirements of older cattle are not well defined. Requirements in older cattle are probably lower than in young calves because considerable recycling of iron occurs when red blood cells turn over (Underwood, 1977), and in older animals blood volume is not increasing, or at least not to the extent that it is in young animals.

SIGNS OF IRON DEFICIENCY

A deficiency of iron results in anemia (hypochromic microcytic), listlessness, reduced feed intake and weight gain, pale mucus membranes and atrophy of the papillae of the tongue (Blaxter et al., 1957; Bremner and Dalgarno, 1973). Iron deficiency can occur in young calves fed exclusively milk, especially if they are housed in confinement. Most practical feedstuffs are more than adequate in iron, and iron deficiency is unlikely in other classes of cattle unless parasite infestations or diseases exist that cause chronic blood loss. In the absence of blood loss, only small amounts of iron are lost in the urine and feces (McDowell, 1992).

IRON SOURCES

Cereal grains normally contain 30 to 60 mg Fe/kg; oilseed meals contain 100 to 200 mg Fe/kg (Underwood, 1981). With the exception of milk and milk products, feeds of animal origin are high in iron, with meat and fish meal containing 400 to 500 mg Fe/kg; blood meal usually has more than 3,000 mg Fe/kg. The iron content of forages is highly variable but most forages contain from 70 to 500 mg Fe/kg. Much of the variation in forage iron is probably caused by soil contamination. Water and soil ingestion also can be significant sources of iron for beef cattle. Availability of iron from forages appears to be lower than from most supplemental iron sources (Thompson and Raven, 1959; Raven and Thompson, 1959). Iron from soil is probably of low availability; however, research by Healy (1972) indicated that a significant amount of iron from various soil types was soluble in ruminal fluid.

Iron is generally supplemented in diets as ferrous sulfate, ferrous carbonate, or ferric oxide. Availability of iron is highest for ferrous sulfate with ferrous carbonate being intermediate (Ammerman et al., 1967; McGuire et al., 1985). Ferric oxide is basically unavailable (Ammerman et al., 1967).

SIGNS OF IRON TOXICITY

Iron toxicity causes diarrhea, metabolic acidosis, hypothermia, and reduced gain and feed intake (National Research Council, 1980). The maximum tolerable concentration of iron for cattle has been estimated at 1,000 mg Fe/kg (National Research Council, 1980). Dietary iron concentrations as low as 250 to 500 mg/kg have caused copper depletion in cattle (Bremner et al., 1987; Phillippo et al., 1987a). In areas where drinking water or forages are high in iron, dietary copper may need to be increased to prevent copper deficiency.

Manganese

Manganese functions as a component of the enzymes pyruvate carboxylase, arginase, and superoxide dismutase and as an activator for a number of enzymes (Hurley and Keen, 1987). Enzymes activated by manganese include a number of hydrolases, kinases, transferases, and decarboxylases. Of the many enzymes that can be activated by manganese, only the glycosyltransferases are known to specifically require manganese.

MANGANESE REQUIREMENTS

The manganese requirement for growing and finishing cattle is approximately 20 mg Mn/kg diet. Skeletal abnormalities were noted in calves from cows fed diets containing 15.8 mg Mn/kg but were not present when diets were supplemented to contain 25 mg Mn/kg (Rojas et al., 1965). The quantity of manganese needed for maximum growth is less than that required for normal skeletal development. Manganese requirements for reproduction are higher than for growth and skeletal development, and the recommended concentration for breeding cattle is 40 mg/kg. Cows fed a diet containing 15.8 mg Mn/kg had lower conception rates than cows fed 25 mg Mn/kg (Rojas et al., 1965). Heifers fed 10 mg Mn/kg exhibited impaired reproduction (delayed cycling and reduced conception rate) compared to those fed 30 mg Mn/kg, but growth was similar for the two groups (Bentley and Phillips, 1951). Supplementing a corn silage-based diet containing 32 mg Mn/kg with 14 mg Mn/kg, from a manganese polysaccharide complex, reduced services per conception from 1.6 to 1.1 but did not affect overall conception rate in beef cows (DiCostanzo et al., 1986).

SIGNS OF MANGANESE DEFICIENCY

Inadequate intake of manganese in young animals results in skeletal abnormalities that may include stiffness, twisted legs, enlarged joints, and reduced bone strength (Hurley and Keen, 1987). In older cattle, manganese deficiency causes low reproductive performance characterized by depressed or irregular estrus, low conception rate, abortion, stillbirths, and low birth weights.

FACTORS AFFECTING MANGANESE REQUIREMENTS

Absorption of manganese from ^{54}MnCl in lactating dairy cows was less than 1 percent (Van Vruwaene et al., 1984) and little is known concerning dietary factors that may influence manganese absorption. Some evidence suggests that high dietary calcium and phosphorus may increase manganese requirements (Hawkins et al., 1955; Dyer et al., 1964; Lassiter et al., 1972). Biliary excretion of manganese plays an important role in manganese homeostasis but little excretion of manganese occurs via the urine (Hidiroglou, 1979).

MANGANESE SOURCES

The concentration of manganese in forages varies greatly depending on plant species, soil pH, and soil drainage (Minson, 1990). Forages generally contain adequate manganese, assuming that the manganese is available for absorption. Corn silage can be low, or at best marginal, in manganese content (Buchanan-Smith et al., 1974). Cereal grains usually contain between 5 and 40 mg Mn/kg with corn being especially low (Underwood, 1981). Plant protein sources normally contain 30 to 50 mg Mn/kg, whereas animal-protein sources only contain 5 to 15 mg Mn/kg. Manganese can be supplemented to ruminant diets as manganese sulfate, manganese oxide, or various organic forms (manganese methionine, manganese proteinate, manganese polysaccharide complex, or manganese amino acid chelate). Manganese sulfate is more available than manganese oxide (Wong-Ville et al., 1989; Henry et al., 1992).

Compared to manganese sulfate, relative availability of manganese from manganese methionine is approximately 120 percent (Henry et al., 1992).

SIGNS OF MANGANESE TOXICITY

In *Mineral Tolerances of Domestic Animals* (National Research Council, 1980), the maximum tolerable concentration of manganese was set at 1,000 mg/kg, at least on a short-term basis. Calves fed 1,000 mg Mn/kg for 100 days showed no adverse effects (Cunningham et al., 1966); >2,000 mg Mn/kg was required in this study to reduce growth and feed intake. In young calves fed milk replacer, 1,000 mg Mn/kg reduced weight gain and feed efficiency (Jenkins and Hidiroglou, 1991).

Molybdenum

Molybdenum functions as a component of the enzymes xanthine oxidase, sulfite oxidase, and aldehyde oxidase (Mills and Davis, 1987). Requirements for molybdenum, however, are not established. There is no evidence that molybdenum deficiency occurs in cattle under practical conditions, but molybdenum may enhance microbial activity in the rumen in some instances. The addition of 10 mg Mo/kg to a high-roughage diet containing 1.7 mg Mo/kg increased the rate of in situ dry matter disappearance from the rumen of cattle (Shariff et al., 1990). In situ dry matter disappearance was not improved by molybdenum supplementation when steers were fed a ground barley-based diet containing 1.0 mg Mo/kg (Shariff et al., 1990). Molybdenum added to a semipurified diet containing 0.36 mg Mo/kg improved growth and cellulose digestion in lambs (Ellis et al., 1958). In three subsequent studies with lambs fed semipurified or practical diets, no responses to added molybdenum were observed (Ellis and Pfander, 1970).

FACTORS AFFECTING MOLYBDENUM UTILIZATION

Metabolism of molybdenum is greatly affected by copper and sulfur with both minerals acting antagonistically. Sulfide and molybdate interact in the rumen to form thiomolybdates, resulting in decreased absorption and altered postabsorptive metabolism of molybdenum (Mills and Davis, 1987). Sulfate shares common transport systems with molybdate in the intestine and kidney, thus decreasing intestinal absorption and increasing urinary excretion of molybdate (Mills and Davis, 1987). It is well documented that relatively low dietary molybdenum can cause copper deficiency and that increasing dietary copper can overcome molybdenum toxicity.

SIGNS OF MOLYBDENUM TOXICITY

In cattle, high concentrations of molybdenum (20 mg Mo/kg or higher) can cause toxicity characterized by diarrhea, anorexia, loss of weight, stiffness, and changes in hair color (Ward, 1978). Providing large amounts of copper will usually overcome molybdenosis. The maximum tolerable concentration of molybdenum for cattle has been estimated to be 10 mg/kg (National Research Council, 1980). Molybdenum concentrations of less than 10 mg/kg can result in copper deficiency, depending on the length of time the cattle are exposed and the concentration of dietary copper. Recent studies suggest that a relatively low concentration of molybdenum may exert direct effects on certain metabolic processes independent of alterations in copper status. The addition of 5 mg Mo/kg to diets containing 0.1 mg Mo/kg caused copper depletion associated with reduced growth and feed efficiency, loss of hair pigmentation, changes in hair texture, and infertility in heifers (Bremner et al., 1987; Phillippo et al., 1987a,b). In these same studies, cattle fed high dietary iron had similar copper status—based on plasma copper, liver copper, and ceruloplasmin and superoxide dismutase activity—to heifers fed molybdenum but did not show clinical signs of copper deficiency. Supplementation with 5 mg Mo/kg starting at 13 to 19 weeks of age increased age at puberty and decreased liveweight of heifers at puberty and reduced conception rate (Phillippo et al., 1987b). Feeding beef cows and their calves an additional 5 mg Mo/kg reduced calf gains from birth to weaning by 28 percent, whereas calf gains were not affected by the addition of 500 mg Fe/kg (Gengelbach et al., 1994).

MOLYBDENUM SOURCES

Forages vary greatly in molybdenum concentration depending on soil type and soil pH. Neutral or alkaline soils coupled with high moisture and organic matter favor molybdenum uptake by forages (McDowell, 1992). Cereal grains and protein supplements are less variable in molybdenum than forages.

Nickel

Nickel deficiency has been produced experimentally in a number of animals (Nielson, 1987). However, the function of nickel in mammalian metabolism is unknown. Nickel is an essential component of urease in ureolytic bacteria (Spears, 1984). Supplementation of nickel to ruminant diets has increased ruminal urease activity in a number of studies (Spears, 1984; Oscar and Spears, 1988).

Research data are not sufficient to determine nickel requirements of beef cattle. The maximum tolerable concentration of nickel was estimated to be 50 mg/kg diet (National Research Council, 1980). Growing steers fed diets supplemented with 50 mg Ni/kg in the chloride form for 84 days showed no adverse effects (Oscar and Spears, 1988).

Selenium

In 1973, glutathione peroxidase was identified as the first known selenium metalloenzyme (Rotruck et al., 1973). Glutathione peroxidase catalyzes the reduction of hydrogen peroxide and lipid hydroperoxides, thus preventing oxidative damage to body tissues (Hoekstra, 1974). Recently, a second selenometalloenzyme, iodothyronine 5'-deiodinase, was identified (Arthur et al., 1990). This enzyme catalyzes the deiodination of thyroxine (T_4) to the more metabolically active triiodothyronine (T_3) in tissues.

SELENIUM REQUIREMENTS

Based on available research data, the selenium requirement of beef cattle can be met by 0.1 mg Se/kg. Clinical or subclinical signs of selenium deficiency have been reported in beef cows and calves receiving forages containing 0.02 to 0.05 mg Se/kg (Morris et al., 1984; Hidiroglou et al., 1985; Spears et al., 1986); however, calves housed in confinement have been fed semipurified diets containing 0.02 to 0.03 mg Se/kg for months without showing clinical signs of deficiency, despite very low activities of glutathione peroxidase (Boyne and Arthur, 1981; Siddons and Mills, 1981; Reffett et al., 1988). Even in the absence of clinical deficiency signs, calves have reduced neutrophil activity (Boyne and Arthur, 1981) and humoral immune response (Reffett et al., 1988).

FACTORS AFFECTING SELENIUM REQUIREMENTS

Factors that affect selenium requirements are not well defined. The function of vitamin E and selenium are interrelated, and a diet low in vitamin E may increase the amount of selenium needed to prevent certain abnormalities such as nutritional muscular dystrophy (white muscle disease) (Miller et al., 1988). High dietary sulfur has resulted in an increased incidence of white muscle disease in some but not all studies (Miller et al., 1988). In sheep, the occurrence of white muscle disease is higher when legume hay rather than nonlegume hay is consumed, even when selenium contents are similar (Whanger et al., 1972). Harrison and Conrad (1984) reported that selenium absorption in dairy cows was minimal at low (0.4 percent) and high (1.4 percent) calcium intakes and maximal when dietary calcium was 0.8 percent. In young calves, varying dietary calcium from 0.17 to 2.35 percent did not significantly affect selenium absorption (Alfaro et al., 1987). High concentrations of unsaturated fatty acids in the diet or various stressors (environmental or dietary) also may increase the requirement for selenium. Form of selenium may affect dietary requirements. Selenium is generally supplemented in animal diets as sodium selenite, while selenomethionine is the predominant form of selenium in most feedstuffs. Selenium from selenomethionine or a selenium-containing yeast was approximately twice as available as sodium selenite or cobalt selenite in growing heifers (Pehrson et al., 1989). Availability of selenium from sodium selenate was similar to sodium selenite (Podoll et al., 1992).

Selenium is absorbed primarily from the duodenum with little or no absorption from the rumen or abomasum. Absorption of selenium in ruminants is much lower than in nonruminants (Wright and Bell, 1966). The lower absorption of selenium is believed to relate to the reduction of selenite to insoluble forms in the rumen. Fecal excretion is greater than urinary excretion in mature ruminants. Pulmonary excretion of selenium is important when intakes of selenium are high (Ganther et al., 1966).

SIGNS OF SELENIUM DEFICIENCY

White muscle disease in young ruminants is a common clinical sign of selenium deficiency that results in degeneration and necrosis in both skeletal and cardiac muscle (Underwood, 1981). Affected animals may show stiffness, lameness, or even cardiac failure. Other signs of selenium deficiency that have been observed include unthriftiness (often times with weight loss and diarrhea; Underwood, 1981), anemia with presence of heinz bodies (Morris et al., 1984), and increased mortality and reduced calf weaning weights (Spears et al., 1986). Selenium-depleted cattle have shown reduced immune responses in a number of studies (Stabel and Spears, 1993). Arthur et al. (1988) reported that selenium-deficient cattle had increased T_4 and decreased T_3 concentrations in plasma relative to selenium-supplemented cattle. Depressed activity of iodothyronine 5′-deiodinase may explain the unthriftiness and poor growth often observed in selenium deficiency. Decreases in glutathione peroxidase activity associated with selenium deficiency can explain the occurrence of white muscle disease, heinz body anemia, and possibly other signs of selenium deficiency.

Selenium concentrations in plasma, serum, and whole blood, and glutathione peroxidase activities in plasma, whole blood, and erythrocytes, have been used to assess selenium status. Glutathione peroxidase activities indicative of a selenium deficiency can vary from one laboratory to another depending on assay conditions. Langlands et al. (1989b) concluded from a number of on-farm studies with cattle in Australia that selenium concentrations in whole blood and plasma were poor indicators of responsiveness to selenium supplementation unless unthriftiness was apparent.

SELENIUM SOURCES

Feedstuffs grown in many areas of the United States and Canada are deficient or at least marginally deficient in selenium. Selenium-deficient areas are located in the northwestern, northeastern, and southeastern parts of the United States. The selenium content of forages and other feedstuffs varies greatly depending on plant species and particularly the selenium content of the soil. Selenium can legally be supplemented in beef cattle diets to provide 3 mg/head/day or 0.3 mg/kg in the complete diet. Alternate methods of supplementing selenium include injecting selenium every 3 to 4 months or at critical production stages and using boluses retained in the rumen that release selenium over a period of months (Hidiroglou et al., 1985; Campbell et al., 1990).

SIGNS OF SELENIUM TOXICITY

Selenium toxicity may occur as a result of excessive selenium supplementation or consumption of plants naturally high in selenium. Many plant species of *Astragalus* and *Stanleya* grow primarily on seleniferous areas and can accumulate up to 3,000 mg Se/kg. Consumption of forages containing 5 to 40 mg Se/kg results in chronic toxicosis (alkali disease). Chronic toxicity signs include lameness, anorexia, emaciation, loss of vitality, sore feet, cracked, deformed and elongated hoofs, liver cirrhosis, nephritis, and loss of hair from the tail (Rosenfeld and Beath, 1964). Acute selenium toxicity (blind staggers) causes labored breathing, diarrhea, ataxia, abnormal posture, and death from respiratory failure (National Research Council, 1980). The maximum tolerable concentration of selenium has been estimated to be 2 mg/kg (National Research Council, 1980). The addition of 10 mg Se/kg to a milk replacer for 42 days reduced gain and efficiency in young calves, but supplemented selenium at 5 mg/kg caused no noticeable effects (Jenkins and Hidiroglou, 1986).

Zinc

Zinc functions as an essential component of a number of important enzymes. In addition, other enzymes are activated by zinc. Enzymes that require zinc are involved in nucleic acid, protein, and carbohydrate metabolism (Hambidge et al., 1986). Zinc also is important for normal development and functioning of the immune system.

ZINC REQUIREMENTS

The recommended requirement of zinc in beef cattle diets is 30 mg Zn/kg diet. This concentration should satisfy requirements in most situations. Pond and Oltjen (1988) reported no growth responses to zinc supplementation in medium- or large-framed steers fed corn silage-corn-based diets containing 22 to 26 mg Zn/kg. Growth responses to zinc supplementation were observed in two of four studies with finishing steers fed diets containing 18 to 29 mg Zn/kg (Perry et al., 1968). In later studies, zinc added to diets containing 17 to 21 mg Zn/kg improved gain in only one of seven experiments (Beeson et al., 1977). Other studies with growing and finishing cattle have indicated no response to zinc supplementation when diets contained 22 to 32 mg Zn/kg (Pringle et al., 1973; Spears and Samsell, 1984). Zinc requirements of beef cattle fed forage-based diets and requirements for reproduction and milk production are less well defined. Zinc supplementation increased gain in nursing calves grazing mature forages that contained 7 to 17 mg Zn/kg (Mayland et al., 1980).

SIGNS OF ZINC DEFICIENCY

Severe zinc deficiency in cattle results in reduced growth, feed intake, and feed efficiency; listlessness; excessive salivation; reduced testicular growth; swollen feet with open, scaly lesions; parakeratotic lesions that are most severe on the legs, neck, head, and around the nostrils; failure of wounds to heal; and alopecia (Miller and Miller, 1962; Miller et al., 1965; Ott et al., 1965; Mills et al., 1967). Thymus atrophy and impaired immune response have been observed in calves with a genetic disorder that causes impaired absorption of zinc, resulting in zinc deficiency (Perryman et al., 1989). Subclinical deficiencies of zinc can reduce weight gain (Mayland et al., 1980) and perhaps reproductive performance. Plasma or liver zinc concentrations may be used to diagnose severe zinc deficiencies, but plasma zinc determination is of little value in detecting marginal deficiencies. Stress or disease causes a redistribution of zinc in the body that can temporarily result in low plasma concentrations characteristic of a severe deficiency (Hambridge et al., 1986).

FACTORS AFFECTING ZINC REQUIREMENTS

Absorption of zinc occurs primarily from the abomasum and small intestine (Miller and Cragle, 1965). Zinc absorption is homeostatically controlled and cattle adjust the percentage of dietary zinc absorbed based on their need for growth or lactation (Miller, 1975). Milk contains 3 to 5 mg Zn/L, but the increased demand for milk production is likely met by increased absorption, provided that dietary zinc is present in a form that can be absorbed. Dietary factors that affect zinc requirements in ruminants are not understood. In contrast to nonruminants, high-dietary calcium does not appear to increase zinc requirements greatly in ruminants (Pond, 1983; Pond and Wallace, 1986). Phytate also does not affect zinc absorption in ruminants with a functional rumen. A relatively large portion of the zinc

in forages is associated with the plant cell wall (Whitehead et al., 1985), but it is not known whether zinc's association with fiber reduces absorption.

ZINC SOURCES

The zinc content of forages is affected by a number of factors including plant species, maturity, and soil zinc (Minson, 1990). Legumes are generally higher in zinc than grasses. Cereal grains usually contain between 20 and 30 mg Zn/kg, whereas plant protein sources contain 50 to 70 mg Zn/kg. Feed-grade sources of bioavailable zinc include zinc oxide, zinc sulfate, zinc methionine, and zinc proteinate. Based on available data, zinc in the sulfate and oxide form are of similar bioavailability in ruminants (Kincaid, 1979; Kegley and Spears, 1992). Absorption of zinc from zinc methionine is similar to zinc oxide, but zinc methionine appears to be metabolized differently following absorption (Spears, 1989).

SIGNS OF ZINC TOXICITY

The amount of zinc necessary to cause toxicity is much greater than requirements. The maximum tolerable concentration of zinc is 500 mg/kg (National Research Council, 1980). Decreased weight gain was reported in calves fed 900 mg Zn/kg for 12 weeks (Ott et al., 1966). Young calves fed milk replacer tolerated 500 mg Zn/kg for 5 weeks without adverse effects; but 700 mg/kg reduced gain, feed intake, and feed efficiency (Jenkins and Hidiroglou, 1991).

REFERENCES

Agricultural Research Council. 1980. The Nutrient Requirements of Ruminant Livestock. Slough, U.K.: Commonwealth Agricultural Bureaux.

Alfaro, E., M. W. Neathery, W. J. Miller, R. P. Gentry, C. T. Crowe, A. S. Fielding, R. E. Etheridge, D. G. Pugh, and D. M. Blackmon. 1987. Effects of varying the amounts of dietary calcium on selenium metabolism in dairy calves. J. Dairy Sci. 70:831–836.

Alfaro, E., M. W. Neathery, W. J. Miller, C. T. Crowe, R. P. Gentry, A. S. Fielding, D. G. Pugh, and D. M. Blackmon. 1988. Influence of a wide range of calcium intakes on tissue distribution of macroelements and microelements in dairy calves. J. Dairy Sci. 71:1295–1300.

Ammerman, C. B. 1970. Recent developments in cobalt and copper in ruminant nutrition: A review. J. Dairy Sci. 53:1097–1106.

Ammerman, C. B., L. R. Arrington, M. C. Jayaswal, R. L. Shirley, and G. K. Davis. 1963. Effect of dietary calcium and phosphorus levels on nutrient digestibility by steers. J. Anim. Sci. 22:248 (abstr.).

Ammerman, C. B., J. M. Wing, B. G. Dunavant, W. K. Robertson, J.P. Feaster, and L. R. Arrington. 1967. Utilization of inorganic iron by ruminants as influenced by form of iron and iron status of the animal. J. Anim. Sci. 26:404–410.

Ammerman, C. B., C. F. Chicco, P. E. Loggins, and L. R. Arrington. 1972. Availability of different inorganic salts of magnesium to sheep. J. Anim. Sci. 34:122–126.

Archer, K. A., and J. L. Wheeler. 1978. Response by cattle grazing sorghum to salt-sulfur supplements. Aust. J. Exp. Agric. Anim. Husb. 18:741–744.

Arthur, J. R., P. C. Morrice, and G. J. Becket. 1988. Thyroid hormone concentrations in selenium-deficient and selenium-sufficient cattle. Res. Vet. Sci. 45:122–123.

Arthur, J. R., F. Nicol, and G. J. Becket. 1990. Hepatic iodothyronine 5′-deiodinase. Biochem. J. 272:537–540.

Beeson, W. M., T. W. Perry, and T. D. Zurcher. 1977. Effect of supplemental zinc on growth and on hair and blood serum levels of beef cattle. J. Anim. Sci. 45:160–165.

Bentley, O. G., and P. H. Phillips. 1951. The effect of low manganese rations upon dairy cattle. J. Dairy Sci. 34:396–403.

Bernier, J. F., F. J. Fillion, and G. J. Brisson. 1984. Dietary fibers and supplemental iron in a milk replacer for veal calves. J. Dairy Sci. 67:2369–2379.

Bigger, G. W., J. M. Elliot, and T. R. Richard. 1976. Estimated ruminal production of pseudovitamin B_{12}, factor A and factor B in sheep. J. Anim. Sci. 43:1077–1081.

Blaxter, K. L., and R. F. McGill. 1956. Magnesium metabolism in cattle. Vet. Res. Ann. 2:35–39.

Blaxter, K. L., J. A. F. Rook, and A. M. MacDonald. 1954. Experimental magnesium deficiency in calves: Clinical and pathological observations. J. Comp. Pathol. Therap. 64:157–175.

Blaxter, K. L., G. A. M. Sarman, and A. M. MacDonald. 1957. Iron-deficiency anaemia in calves. Br. J. Nutr. 11:234–246.

Blaxter, K. L., B. Cowlishaw, and J. A. Rook. 1960. Potassium and hypomagnesemic tetany in calves. Anim. Prod. 2:1–10.

Block, R. J., J. A. Stekol, and J. K. Loosli. 1951. Synthesis of sulfur amino acids from inorganic sulfate by ruminants. II. Synthesis of cystine and methionine from sodium sulfate by the goat and by the macroorganisms of the rumen of the ewe. Arch. Biochem. Biophys. 33:353–363.

Bock, B. J., D. L. Harmon, R. T. Brandt, Jr., and J. E. Schneider. 1991. Fat source and calcium level effects on finishing steer performance, digestion, and metabolism. J. Anim. Sci. 69:2211–2224.

Bolsen, K. K., W. Woods, and T. Klopfenstein. 1973. Effect of methionine and ammonium sulfate upon performance of ruminants fed high corn rations. J. Anim. Sci. 36:1186–1190.

Boyne, R., and J. R. Arthur. 1981. Effects of selenium and copper deficiency on neutrophil function in cattle. J. Comp. Pathol. 91:271–276.

Bremner, I., and A. C. Dalgarno. 1973. Iron metabolism in the veal calf. 2. Iron requirements and the effect of copper supplementation. Br. J. Nutr. 30:61–76.

Bremner, I., W. R. Humphries, M. Phillippo, M. J. Walker, and P. C. Morrice. 1987. Iron-induced copper deficiency in calves: Dose-response relationships and interactions with molybdenum and sulfur. Anim. Prod. 45:403–414.

Brink, D. R., O. A. Turgeon, Jr., D. L. Harmon, R. T. Steele, T. L. Mader, and R. A. Britton. 1984. Effects of additional limestone of various types on feedlot performance of beef cattle fed high corn diets differing in processing method and potassium level. J. Anim. Sci. 59:791–797.

Buchanan-Smith, J. G., E. Evans, and S. O. Poluch. 1974. Mineral analysis of corn silage produced in Ontario. Can. J. Anim. Sci. 54:253–256.

Bunting, L. D., J. M. Fernandez, D. L. Thompson, and L. L. Southern. 1994. Influence of chromium picolinate on glucose usage and metabolic criteria in growing Holstein calves. J. Anim. Sci. 72:1591–1599.

Burkhalter, D. L., M. W. Neathery, W. J. Miller, R. H. Whitlock, and J. C. Allen. 1979. Effects of low chloride intake on performance, clinical characteristics, and chloride, sodium, potassium and nitrogen metabolism in diary calves. J. Dairy Sci. 62:1895–1901.

Call, J. W., J. E. Butcher, J. T. Blake, R. A. Smart, and J. L. Shupe. 1978. Phosphorus influence on growth and reproduction of beef cattle. J. Anim. Sci. 47:216–225.

Call, J. W., J. E. Butcher, J. L. Shupe, J. T. Blake, and A. E. Olson. 1986. Dietary phosphorus for beef cows. Am. J. Vet. Res. 47:475–481.

Cameron, H. J., R. J. Boila, L. W. McNichol, and N. E. Stranger. 1989. Cupric oxide needles for grazing cattle consuming low-copper, high-molybdenum forage and high-sulfate water. J. Anim. Sci. 67:252–261.

Campbell, D. T., J. Maas, D. W. Weber, O. R. Hedstrom, and B. B. Norman. 1990. Safety and efficacy of two sustained-release intracellular selenium supplements and the associated placental and colostral transfer of selenium in beef cattle. Am. J. Vet. Res. 51:813–817.

Challa, J., and G. D. Braithwaite. 1988. Phosphorus and calcium metabolism in growing calves with special emphasis on phosphorus homeostases. 1. Studies on the effect of changes in the dietary phosphorus intake on phosphorus and calcium metabolism. J. Agric. Sci. Camb. 110:573–581.

Challa, J., G. D. Braithwaite, and M. S. Dhanoa. 1989. Phosphorus homeostasis in growing calves. J. Agric. Sci. Camb. 112:217–226.

Chang, X., and D. N. Mowat. 1992. Supplemental chromium for stressed and growing feeder calves. J. Anim. Sci. 70:559–565.

Chapman, H. L., Jr., and M. C. Bell. 1963. Relative absorption and excretion by beef cattle of copper from various sources. J. Anim. Sci. 22:82–85.

Chester-Jones, H., J. P. Fontenot, and H. P. Veit. 1990. Physiological and pathological effects of feeding high levels of magnesium to steers. J. Anim. Sci. 68:4400–4413.

Clanton, D. C. 1980. Applied potassium nutrition in beef cattle. Pp. 17–32 in Proceedings, Third International Mineral Conference, January 17 and 18, 1980, Miami, Fla.

Coghlin, C. L. 1944. Hydrogen sulfide poisoning in cattle. Can. J. Comp. Med. 8:111–113.

Cunningham, G. N., M. B. Wise, and E. R. Barrick. 1966. Effect of high dietary levels of manganese on the performance and blood constituents of calves. J. Anim. Sci. 25:532–538.

Davis, G. K., and W. Mertz. 1987. Copper. Pp. 301–364 in Trace Elements in Human and Animal Nutrition, Vol. 1, W. Mertz, ed. New York: Academic Press.

DeLuca, H. F. 1979. The vitamin D system in the regulation of calcium and phosphorus metabolism. Nutr. Rev. 37:161–193.

Devlin, T. J., W. K. Roberts, and V. V. E. St. Omer. 1969. Effects of dietary potassium upon growth, serum electrolytes and intrarumen environment of finishing beef steers. J. Anim. Sci. 28:557–562.

DiCostanzo, A., J. C. Meiske, S. D. Plegge, D. L. Haggard, and K. M. Chaloner. 1986. Influence of manganese copper and zinc on reproductive performance of beef cows. Nutr. Rep. Int. 34:287–293.

Dowe, T. W., J. Matsushima, and V. H. Arthaud. 1957. The effects of adequate and excessive calcium when fed with adequate phosphorus in growing rations for beef calves. J. Anim. Sci. 16:811–820.

Dyer, I. A., W. A. Cassatt, and R. R. Rao. 1964. Manganese deficiency in the etiology and deformed calves. BioScience 14:31–32.

Dunn, T. G., and G. E. Moss. 1992. Effects of nutrient deficiencies and excesses on reproductive efficiency of livestock. J. Anim. Sci. 70:1580–1593.

Ellenberger, H. G., J. A. Newlander, and C. H. Jones. 1950. Composition of the bodies of dairy cattle. Vt. Agric. Exp. Sta. Bull. 558.

Ellis, W. C., and W. H. Pfander. 1970. Further studies on molybdenum as a possible component of the "alfalfa ash factor" for sheep. J. Anim. Sci. 19:1260 (abstr.).

Ellis, W. C., W. H. Pfander, M. E. Muhrer, and E. E. Pickett. 1958. Molybdenum as a dietary essential for lambs. J. Anim. Sci. 17:180–188.

Ferrell, M. C., F. N. Owens, and D. R. Gill. 1983. Potassium levels and ionophores for feedlot steers. Okla. Agr. Exp. Sta. Res. Rep. MP-114:54–60.

Fontenot, J. P., V. G. Allen, G. E. Bunce, and J. P. Goff. 1989. Factors influencing magnesium absorption and metabolism in ruminants. J. Anim. Sci. 67:3445–3455.

Ganther, H. E., O. A. Levander, and C. A. Baumann. 1966. Dietary control of selenium volatilization in the rat. J. Nutr. 88:55–60.

Gengelbach, G. P., J. D. Ward, and J. W. Spears. 1994. Effect of dietary copper, iron, and molybdenum on growth and copper status of beef cows and calves. J. Anim. Sci. 72:2722–2727.

Gentry, R. P., W. J. Miller, D. G. Pugh, M. W. Neathery, and J. B. Bynoum. 1978. Effects of feeding high magnesium to young dairy calves. J. Dairy Sci. 61:1750–1754.

Gerken, H. J., and J. P. Fontenot. 1967. Availability and utilization of magnesium from dolomitic and magnesium oxide in steers. J. Anim. Sci. 26:1404–1408.

Goetsch, A. L., and F. N. Owens. 1985. Effects of calcium source and level on site of digestion and calcium levels in the digestive tract of cattle fed high-concentrate diets. J. Anim. Sci. 61:995–1003.

Goodrich, R. D., T. S. Kahlon, D. E. Pamp, and D. P. Cooper. 1978. Sulfur in Ruminant Nutrition. Des Moines: National Feed Ingredient Association.

Gooneratne, S. R., W. T. Buckley, and D. A. Christensen. 1989. Review of copper deficiency and metabolism in ruminants. Can. J. Anim. Sci. 69:819–845.

Gooneratne, S. R., H. W. Symonds, J. V. Bailey, and D. A. Christensen. 1994. Effects of dietary copper, molybdenum and sulfur on biliary copper and zinc excretion in Simmental and Angus cattle. Can. J. Anim. Sci. 74:315–325.

Grace, N. D. 1983. Manganese. Pp. 80–83 in The Mineral Requirements of Grazing Ruminants, N. D. Grace, ed. Occasional Publ. No. 9. New Zealand: New Zealand Society of Animal Producers.

Grace, N. D., M. J. Wyatt, and J. C. Macrae. 1974. Quantitative digestion of fresh herbage by sheep. III. The movement of Mg, Ca, P, K, and Na in the digestive tract. J. Agric. Sci. 82:321–330.

Greene, L. W., K. E. Webbe, Jr., and J. P. Fontenot. 1983. Effect of potassium level on site of absorption of magnesium and other macroelements in sheep. J. Anim. Sci. 56:1214–1221.

Greene, L. W., J. F. Baker, and P. F. Hardt. 1989. Use of animal breeds and breeding to overcome the incidence of grass tateny: A review. J. Anim. Sci. 67:3463–3469.

Halpin, C. G., D. J. Harris, I. W. Caple, and D. S. Patterson. 1984. Contribution of cobalamin analogues to plasma vitamin B_{12} concentrations in cattle. Res. Vet. Sci. 37:249–253.

Hambidge, K. M., C. C. Casey, and N. F. Krebs. 1986. Zinc. Pp. 1–137 in Trace Elements in Human and Animal Nutrition, Vol. 2, W. Mertz, ed. New York: Academic Press.

Hansard, S. L., C. L. Comar, and M. P. Plumlee. 1954. The effects of age upon calcium utilization and maintenance requirements in the bovine. J. Anim. Sci. 13:25–36.

Hansard, S. L., H. M. Crowder, and W. A. Lyke. 1957. The biological availability of calcium in feeds for cattle. J. Anim. Sci. 16:437–443.

Harrison, J. H., and H. R. Conrad. 1984. Effect of dietary calcium on selenium absorption by the nonlactating dairy cow. J. Dairy Sci. 67:1860–1864.

Hawkins, G. E., Jr., G. H. Wise, G. Matrone, R. K. Waugh, and W. L. Lott. 1955. Manganese in the nutrition of young dairy cattle fed different levels of calcium and phosphorus. J. Dairy Sci. 38:536–547.

Healy, W. B. 1972. In vitro studies on the effects of soil on elements in ruminal, duodenal and ileal liquors from sheep. New Zealand J. Agric. Res. 15:289–305.

Henry, P. R., C. B. Ammerman, and R. C. Littell. 1992. Relative bioavailability of manganese from a manganese-methionine complex and inorganic sources for ruminants. J. Dairy Sci. 75:3473–3478.

Hidiroglou, M. 1979. Manganese in ruminant nutrition. Can. J. Anim. Sci. 59:217–236.

Hidiroglou, M., J. Proulx, and J. Jolette. 1985. Intraruminal selenium pellet for control of nutritional muscular dystrophy in cattle. J. Dairy Sci. 68:57–66.

Hill, G. M. 1985. The relationship between dietary sulfur and nitrogen metabolism in the ruminant. P. 37 in Proceedings of the Georgia Nutrition Conference, University of Georgia, Athens.

Hoekstra, W. G. 1974. Biochemical role of selenium. Pp. 61–77 in Trace Element Metabolism in Animals, No. 2, W. G. Hoekstra, J. W. Suttie, H. E. Ganther and W. Mertz, eds. Baltimore: University Park Press.

Huffman, D. F., C. L. Conley, C. C. Lightfoot, and C. W. Duncan. 1941. Magnesium studies in calves. II. The effect of magnesium salts and various natural feeds upon the magnesium content of the blood plasma. J. Nutr. 22:609–620.

Huntington, G. B. 1983. Feedlot performance, blood metabolic profile and calcium status of steers fed high concentrate diets containing several levels of calcium. J. Anim. Sci. 56:1003–1011.

Hurley, L. S., and C. L. Keen. 1987. Manganese. Pp. 185–223 in Trace Elements in Human and Animal Nutrition, Vol. 1, W. Mertz, ed. New York: Academic Press.

Jackson, J. A., Jr., D. L. Langer, and R. W. Hemken. 1988. Evaluation of content and source of phosphorus fed to dairy calves. J. Dairy Sci. 71:2187–2192.

Jenkins, K. J., and M. Hidiroglou. 1986. Tolerance of the preruminant calf for selenium in milk replacer. J. Dairy Sci. 69:1865–1870.

Jenkins, K. J., and M. Hidiroglou. 1991. Tolerance of the preruminant calf for excess manganese or zinc in milk replacer. J. Dairy Sci. 74:1047–1053.

Jensen, R., and D. R. Mackey. 1979. Diseases of Feedlot Cattle. Philadelphia: Lea and Febiger.

Kahlon, T. S., J. C. Meiske, and R. D. Goodrich. 1975. Sulfur metabolism in ruminants. 1. In vitro availability of various chemical forms of sulfur. J. Anim. Sci. 41:1147–1153.

Kandylis, K. 1984. Toxicology of sulfur in ruminants: Review. J. Dairy Sci. 67:2179–2187.

Keener, H. A., G. P. Percival, K. S. Morrow, and G. H. Ellis. 1949. Cobalt tolerance in young dairy cattle. J. Dairy Sci. 32:527–533.

Kegley, E. B., and J. W. Spears. 1992. Performance and mineral metabolism of lambs as affected by source (oxide, sulfate or methionine) and level of zinc. J. Anim. Sci. 70(Suppl. 1):302.

Kegley, E. B., and J. W. Spears. 1994. Bioavailability of feed grade copper sources (oxide, sulfate or lysine) in growing cattle. J. Anim. Sci. 72:2728–2734.

Kegley, E. B., and J. W. Spears. 1995. Immune response, glucose metabolism, and performance of stressed feeder calves fed inorganic or organic chromium. J. Anim. Sci. 73:2721–2726.

Kelley, W. K., and R. L. Preston. 1984. Effect of dietary sodium, potassium and the anion form of these cations on the performance of feedlot steers. J. Anim. Sci. 59(Suppl. 1):450.

Kennedy, D. G., P. B. Young, W. J. McCaugley, S. Kennedy, and W. J. Blanchflower. 1991. Rumen succinate production may ameliorate the effects of cobalt-vitamin B_{12} deficiency on methylmalonyl CoA mutase in sheep. J. Nutr. 121:1236–1242.

Kincaid, R. L. 1979. Biological availability of zinc from inorganic sources with excess calcium. J. Dairy Sci. 62:1081–1085.

Kincaid, R. L., R. M. Blauwiekel, and J. D. Cronrath. 1986. Supplementation of copper as copper sulfate or copper proteinate for growing calves fed forages containing molybdenum. J. Dairy Sci. 69:160–163.

Langlands, J. P., G. E. Donald, J. E. Bowles, and A. J. Smith. 1989a. Trace element nutrition of grazing ruminants. III. Copper oxide powder as a copper supplement. Aust. J. Agric. Res. 40:187–193.

Langlands, J. P., G. E. Donald, J. E. Bowles, and A. J. Smith. 1989b. Selenium concentrations in the blood of ruminants grazing in northern New South Wales. III. Relationship between blood concentration and the response in liveweight of grazing cattle given a selenium supplement. Aust. J. Agric. Res. 40:1075–1083.

Lassiter, J. W., W. J. Miller, F. M. Pate, and R. P. Gentry. 1972. Effect of dietary calcium and phosphorus on [54]Mn metabolism following a single tracer intraperitoneal and oral doses in rats. Proc. Soc. Exp. Biol. Med. 139:345–348.

Leibholz, J., R. C. Kellaway, and G. T. Hargreave. 1980. Effects of sodium chloride and sodium bicarbonate in the diet on the performance of calves. Anim. Feed. Sci. Technol. 5:309–314.

Lofgreen, G. P., M. Kleiber, and J. R. Luick. 1952. The metabolic fecal phosphorus excretion of the young calf. J. Nutr. 47:571–581.

MacPhearson, A., and J. S. Chalmers. 1984. Effect of dietary energy concentration on the cobalt/vitamin B$_{12}$ requirements of growing calves. P. 145 in Trace Element Metabolism in Animals, No. 5, C. F. Mills, I. Bremner, and J. K. Chesters, eds. Slough, U. K.: Commonwealth Agricultural Bureaux.

MacPhearson, A., G. Fisher, and J. E. Paterson. 1989. Effect of cobalt deficiency on the immune function of ruminants. Pp. 397–298 in Trace Elements in Man and Animals, No. 6, L. S. Hurley, B. Lonnerdal, C. L. Keen, and R. B. Rucker, eds. New York: Plenum Press.

Martens, H., and Y. Rayssiguier. 1980. Magnesium metabolism and hypomagnesaemia. Pp. 447–466 in Digestive Physiology and Metabolism in Ruminants, Y. Ruckebusch and P. Thivend, eds. Lancastor, U. K.: MTP Press Limited.

Martens, H., O. W. Kubel, G. Gabel, and H. Honig. 1987. Effects of low sodium intake on magnesium metabolism of sheep. J. Agric. Sci. Camb. 108:237–243.

Martz, F. A., A. T. Belo, M. F. Weiss, R. L. Belyea, and J. P. Gaff. 1990. True absorption of calcium and phosphorus from alfalfa and corn silage when fed to lactating cows. J. Dairy Sci. 73:1288–1295.

Mayland, H. F. 1988. Grass tetany. Pp. 511–522 in The Ruminant Animal-Digestive Physiology and Nutrition, D. C. Church, ed. Englewood Cliffs, N.J.: Prentice Hall.

Mayland, H. F., R. C. Rosenau, and A. R. Florence. 1980. Grazing cow and calf responses to zinc supplementation. J. Anim. Sci. 51:966–974.

McDowell, L. R. 1992. Minerals in Animal and Human Nutrition. New York: Academic Press.

McGillivray, J. J. 1974. Biological availability of phosphorus in feed ingredients. P. 51 in Proceedings of the Minnosota Nutrition Conference, University of Minnesota, St. Paul, Minn.

McGuire, S. O., W. J. Miller, R. D. Gentry, N. W. Neathery, S. Y. Ho, and D. M. Blackmon. 1985. Influence of high dietary iron as ferrous carbonate and ferrous sulfate on iron metabolism in young calves. J. Dairy Sci. 68:2621–2628.

Mertz, W. 1992. Chromium: History and nutritional importance. Biol. Trace Elem. Res. 32:3–8.

Meyer, J. H., R. R. Grunert, R. H. Grummer, P. H. Phillips, and G. Bohstedt. 1950. Sodium, potassium, and chlorine content of feedstuffs. J. Anim. Sci. 9:153–156.

Meyer, J. H., W. C. Weir, N. R. Ittner, and J. D. Smith. 1955. The influence of high sodium chloride intakes by fattening sheep and cattle. J. Anim. Sci. 14:412–418.

Miller, J. K., and W. J. Miller. 1962. Experimental zinc deficiency and recovery of calves. J. Nutr. 76:467–474.

Miller, J. K., and R. G. Cragle. 1965. Gastrointestinal sites of absorption and endogenous secretion of zinc in dairy cattle. J. Dairy Sci. 48:370–373.

Miller, J. K., E. W. Swanson, and G. E. Spalding. 1975. Iodine absorption, excretion, recycling, and tissue distributions in the dairy cow. J. Dairy Sci. 58:1578–1593.

Miller, J. K., N. Ramsey, and F. C. Madsen. 1988. The trace elements. Pp. 342–401 in The Ruminant Animal-Digestive Physiology and Nutrition, D. C. Church, ed. Englewood Cliffs, N.J.: Prentice-Hall.

Miller, W. J. 1975. New concepts and developments in metabolism and homeostatis of inorganic elements in dairy cattle: A review. J. Dairy Sci. 58:1549–1560.

Miller, W. J., W. J. Pitts, C. M. Clifton, and J. D. Morton. 1965. Effects of zinc deficiency per se on feed efficiency, serum alkaline phosphatase,

zinc in skin, behavior, greying, and other measurements in the Holstein calf. J. Dairy Sci. 48:1329–1334.

Miller, W. J., M. W. Neathery, R. P. Gentry, D. M. Blackmon, C. T. Crowe, G. O. Ward, and A. S. Fielding. 1987. Bioavailability of phosphorus from defluorinated and dicalcium phosphates and phosphorus requirements of calves. J. Dairy Sci. 70:1885–1892.

Mills, C. F. 1981. Cobalt deficiency and cobalt requirements of ruminants. Pp. 129–141 in Recent Advances in Animal Nutrition, W. Haresign, ed. Boston: Butterworths.

Mills, C. F., and G. K. Davis. 1987. Molybdenum. Pp. 429–463 in Trace Elements in Human and Animal Nutrition, Vol. 1, W. Mertz, ed. New York: Academic.

Mills, C. F., A. C. Dalgarno, R. B. Williams, and J. Quarterman. 1967. Zinc deficiency and zinc requirements of calves and lambs. Br. J. Nutr. 21:751–768.

Minson, D. J. 1990. Forages in Ruminant Nutrition. New York: Academic Press.

Moonsie-Shageer, S., and D. N. Mowat. 1993. Effect of level of supplemental chromium on performance, serum constituents, and immune status of stressed feeder calves. J. Anim. Sci. 71:232–238.

Moore, L. A., E. T. Hallman, and L. B. Sholl. 1938. Cardiovascular and other lesions in calves fed diets low in magnesium. Arch. Pathol. 26:820–838.

Morris, J. G. 1980. Assessment of sodium requirements of grazing beef cattle: A review. J. Anim. Sci. 50:145–152.

Morris, J. G., R. E. Delmas, and J. L. Hull. 1980. Salt (sodium) supplementation of range beef cows in California. J. Anim. Sci. 51:722–731.

Morris, J. G., W. S. Cripe, H. L. Chapman, Jr., D. F. Walker, J. B. Armstrong, J. D. Alexander, Jr., R. Miranda, A. Sanchez, Jr., B. Sanchez, J. R. Blair-West, and D. A. Denton. 1984. Selenium deficiency in cattle associated with heinz bodies and anemia. Science 223:491–493.

Morrison, M., R. M. Murray, and A. N. Boniface. 1990. Nutrition metabolism and rumen microorganisms in sheep fed a poor-quality tropical grass hay supplemented with sulfate. J. Agric. Sci. Camb. 115:269–275.

National Research Council. 1980. Mineral Tolerance of Domestic Animals. Washington, D.C.: National Academy of Sciences.

National Research Council. 1984. Nutrient Requirements of Beef Cattle, Sixth Revised Ed. Washington, D.C.: National Academy Press.

Neathery, N. W., D. G. Pugh, W. J. Miller, R. P. Gentry, and R. H. Whitlock. 1980. Effects of sources and amounts of potassium on feed palatability and on potassium toxicity in dairy calves. J. Dairy Sci. 63:82–85.

Neathery, N. W., D. M. Blackmon, W. J. Miller, S. Heinmiller, S. McGuire, J. M. Tarabula, R. P. Gentry, and J. C. Allen. 1981. Chloride deficiency in Holstein calves from a low chloride diet and removal of abomasal contents. J. Dairy Sci. 64:2220–2233.

Newton, G. L., E. R. Barrick, R. W. Harvey, and M. B. Wise. 1974. Iodine toxicity: Physiological effects of elevated dietary iodine on calves. J. Anim. Sci. 38:449–455.

Nielson, F. H. 1987. Nickel. Pp. 245–274 in Trace Elements in Human and Animal Nutrition, Vol. 1, W. Mertz, ed. New York: Academic Press.

Nockels, C. F., J. DeBonis, and J. Torrent. 1993. Stress induction affects copper and zinc balance in calves fed organic and inorganic copper and zinc sources. J. Anim. Sci. 71:2539–2545.

O'Kelley, R. E., and J. P. Fontenot. 1969. Effects of feeding different magnesium levels to drylot-fed lactating beef cows. J. Anim. Sci. 29:959–966.

O'Kelley, R. E., and J. P. Fontenot. 1973. Effects of feeding different magnesium levels to drylot-fed gestating beef cows. J. Anim. Sci. 36:994–1000.

Oltjen, R. R. 1975. Fats for ruminants-utilization and limitations, including value of protected fats. Pp. 31–40 in Proceedings of the Georgia Nutrition Conference, University of Georgia, Athens.

Oscar, T. P., and J. W. Spears. 1988. Nickel-induced alterations of in vitro and in vivo ruminal fermentation. J. Anim. Sci. 66:2313–2324.

Ott, E. A., W. H. Smith, M. Stab, H. E. Parker, and W. M. Beeson. 1965. Zinc deficiency syndrome in the young calf. J. Anim. Sci. 24:735–741.

Ott, E. A., W. H. Smith, R. B. Harrington, and W. M. Beeson. 1966. Zinc toxicity in ruminants. II. Effects of high levels of dietary zinc on gains, feed consumption and feed efficiency of beef cattle. J. Anim. Sci. 25:419–423.

Peeler, H. T. 1972. Biological availability of nutrients in feeds: Availability of major mineral ions. J. Anim. Sci. 35:695–712.

Pehrson, B., M. Knutson, and M. Gyllensward. 1989. Glutathionine peroxidase activity in heifers fed diets supplemented with organic and inorganic selenium compounds. Swedish J. Agric. Res. 19:53–56.

Pendlum, L. C., J. A. Boling, and N. W. Bradley. 1976. Plasma and ruminal constituents and performance of steers fed different nitrogen sources and levels of sulfur. J. Anim. Sci. 43:1307–1314.

Perry, T. W., W. M. Beeson, W. H. Smith, and M. T. Mohler. 1968. Value of zinc supplementation of natural rations for fattening beef cattle. J. Anim. Sci. 27:1674–1677.

Perryman, L. E., D. R. Leach, W. C. Davis, W. D. Mickelson, S. R. Heller, H. D. Ochs, J. A. Ellis, and E. Brummerstedt. 1989. Lymphocyte alterations in zinc-deficient calves with lethal trait A46. Vet. Immuno. Immunopathol. 21:239–245.

Phillippo, M., W. R. Humphries, and P. H. Garthwaite. 1987a. The effect of dietary molybdenum and iron on copper status and growth in cattle. J. Agric. Sci. Camb. 109:315–320.

Phillippo, M., W. R. Humphries, T. Atkinson, G. D. Henderson, and P. H. Garthwaite. 1987b. The effect of dietary molybdenum and iron on copper status, puberty, fertility and oestrous cycles in cattle. J. Agric. Sci. Camb. 109:321–336.

Podoll, K. L., J. B. Bernard, D. E. Ullrey, S. R. DeBar, P. K. Ku, and W. T. Magee. 1992. Dietary selenium versus selenite for cattle, sheep, and horses. J. Anim. Sci. 70:1965–1970.

Pond, W. G. 1983. Effect of dietary calcium and zinc levels on weight gain and blood and tissue mineral concentrations of growing Columbia- and Suffolk-sired lambs. J. Anim. Sci. 56:952–959.

Pond, W. G., and M. H. Wallace. 1986. Effects of gestation-lactation diet calcium and zinc levels and of parenteral vitamin A, D and E during gestation on ewe body weight and on lamb weight and survival. J. Anim. Sci. 63:1019–1025.

Pond, W. G., and R. R. Oltjen. 1988. Response of large and medium frame beef steers to protein and zinc supplementation of a corn silage-corn finishing diet. Nutr. Rep. Int. 38:737–743.

Pringle, W. L., W. K. Dawley, and J. E. Miltimore. 1973. Sufficiency of Cu and Zn in barley, forage, and corn silage rations as measured by response to supplements by beef cattle. Can. J. Anim. Sci. 53:497–502.

Raven, A. M., and A. Thompson. 1959. The availability of iron in certain grasses, clover and herb species. I. Perennial ryegrass, cocksfoot and timothy. J. Agric. Sci. 52:177–186.

Reed, W. D. C., R. C. Elliott, and J. H. Topps. 1965. Phosphorus excretion of cattle fed on high-energy diets. Nature 208:953–954.

Reffett, J. K., J. W. Spears, and T. T. Brown, Jr. 1988. Effect of dietary selenium on the primary and secondary immune response in calves challenged with infectious bovine rhinotracheitis virus. J. Nutr. 118:229–235.

Ricketts, R. E., J. R. Campbell, D. E. Weinman, and M. E. Tumbleson. 1970. Effect of three calcium:phosphorus ratios on performance of growing Holstein steers. J. Dairy Sci. 53:898–903.

Roberts, W. K., and V. V. E. St. Omer. 1965. Dietary potassium requirements of fattening steers. J. Anim. Sci. 24:902 (abstr.).

Rojas, M. A., I. A. Dryer, and W. A. Cassatt. 1965. Manganese deficiency in the bovine. J. Anim. Sci. 24:664–667.

Rook, J. A. F., and J. E. Storry. 1962. Magnesium in the nutrition of farm animals. Nutr. Abstr. Rev. 32:1055–1077.

Rosenfeld, I., and O. A. Beath, eds. 1964. Selenium. New York: Academic Press.

Rotruck, J. T., A. L. Pope, H. E. Ganther, A. B. Swanson, D. G. Hafeman, and W. G. Hoekstra. 1973. Selenium: Biochemical role as a component of glutathione peroxidase, Science 179:588–590.

Rumsey, T. S. 1978. Effects of dietary sulfur addition and synovex-S ear implants on feedlot steers fed an all-concentrate finishing diet. J. Anim. Sci. 46:463–477.

Shand, A., and G. Lewis. 1957. Chronic copper poisoning in young calves. Vet. Rec. 69:618–619.

Shariff, M. A., R. J. Boila, and K. M. Wittenberg. 1990. Effect of dietary molybdenum on rumen dry matter disappearance in cattle. Can. J. Anim. Sci. 70:319–323.

Shupe, J. L., J. E. Butcher, J. W. Call, A. E. Olson, and J. T. Blake. 1988. Clinical signs and bone changes associated with phosphorus deficiency in beef cattle. Am. J. Vet. Res. 49:1629–1636.

Siddons, R. C., and C. F. Mills. 1981. Glutathione peroxidase activity and erythrocyte stability in calves differing in selenium and vitamin E status. Br. J. Nutr. 46:345–355.

Smart, M. E., R. Cohen, D. A. Christensen, and C. M. Williams. 1986. The effects of sulfate removal from the drinking water on the plasma and liver copper and zinc concentrations of beef cows and their calves. Can. J. Anim. Sci. 66:669–680.

Smith, R. M. 1987. Cobalt. Pp. 143–183 in Trace Elements in Human and Animal Nutrition, W. Mertz, ed. New York: Academic Press.

Spears, J. W. 1984. Nickel as a "newer trace element" in the nutrition of domestic animals. J. Anim. Sci. 59:823–835.

Spears, J. W. 1989. Zinc methionine for ruminants: Relative bioavailability of zinc in lambs and effects on performance of growing heifers. J. Anim. Sci. 67:835–843.

Spears, J. W., and L. J. Samsell. 1984. Effect of zinc supplementation on performance and zinc status of growing heifers. J. Anim. Sci. 49(Suppl 1):407.

Spears, J. W., and R. W. Harvey. 1987. Lasalocid and dietary sodium and potassium effects on mineral metabolism, ruminal volatile fatty acids and performance of finishing steers. J. Anim. Sci. 65:830–840.

Spears, J. W., R. W. Harvey, and E. C. Segerson. 1986. Effect of marginal selenium deficiency on growth, reproduction and selenium status of beef cattle. J. Anim. Sci. 63:586–594.

Spears, J. W., B. R. Schricker, and J. C. Burns. 1989. Influence of lysocellin and monensin on mineral metabolism in steers fed forage-based diets. J. Anim. Sci. 67:2140–2149.

Stabel, J. R., and J. W. Spears. 1993. Role of selenium in immune responsiveness and disease resistance. Pp. 333–356 in Human Nutrition-A Comprehensive Treatise, Vol. 8, Nutrition and Immunology, D. M. Klurfeld, ed. New York: Plenum Press.

Starks, P. B., W. H. Hale, U. S. Garrigus, and R. M. Forbes. 1953. The utilization of feed nitrogen by lambs as affected by elemental sulfur. J. Anim. Sci. 12:480–491.

Suttle, N. F. 1974. Effects of organic and inorganic sulfur on the availability of dietary copper to sheep. Br. J. Nutr. 32:559–568.

Suttle, N. F. 1991. The interactions between copper, molybdenum and sulfur in ruminant nutrition. Annu. Rev. Nutr. 11:121–140.

TCORN. 1991. A reappraisal of the calcium and phosphorus requirements of sheep and cattle. Nutr. Abstr. Rev. Ser. B. 61:573–612.

Thomas, W. E., J. K. Loosli, H. H. Williams, and L. A. Maynard. 1951. The utilization of inorganic sulfates and urea nitrogen by lambs. J. Nutr. 43:515–523.

Thompson, A., and A. M. Raven. 1959. The availability of iron in certain grasses, clover and herb species. II. Alsike, broad red clover, Kent wild white clover, trefoil and lucerne. J. Agric. Sci. 53:224–229.

Thornton, I., G. F. Kershaw, and M. K. Davies. 1972. An investigation into copper deficiency in cattle in the Southern Pennines. II. Response to copper supplementation. J. Agric. Sci. Camb. 78:165–171.

Tillman, A. D., and J. R. Brethour. 1958. Dicalcium phosphate and phosphoric acid as phosphorus sources for beef cattle. J. Anim. Sci. 17:100–103.

Tillman, A. D., J. R. Brethour, and S. L. Hansard. 1959. Comparative procedures for measuring the phosphorus requirement of cattle. J. Anim. Sci. 18:249–255.

Tucker, W. B., J. A. Jackson, D. M. Hopkins, and J. F. Hogue. 1991. Influence of dietary sodium bicarbonate on the potassium metabolism of growing dairy calves. J. Dairy Sci. 74:2296–2302.

Underwood, E. J. 1977. Trace Elements in Human and Animal Nutrition, 4th Ed. New York: Academic Press.

Underwood, E. J. 1981. The Mineral Nutrition of Livestock, 2nd Ed. Slough, U.K.: Commonwealth Agricultural Bureaux.

Van Vruwaene, R., G. B. Gerber, R. Kirchmann, J. Colard, and J. Van Kerkom. 1984. Metabolism of ^{51}Cr, ^{54}Mn, ^{59}Fe and ^{60}Co in lactating dairy cows. Health Phys. 46:1069–1082.

Wacker, W. E. C. 1980. Magnesium and Man. Cambridge, Mass.: Harvard University Press.

Walker, C. K., and J. M. Elliot. 1972. Lactational trends in vitamin B$_{12}$ status on conventional and restricted-roughage rations. J. Dairy Sci. 55:474–479.

Ward, G. M. 1966. Potassium metabolism of domestic ruminants: A review. J. Dairy Sci. 49:268–276.

Ward, G. M. 1978. Molybdenum toxicity and hypocuprosis in ruminants: A review. J. Anim. Sci. 46:1078–1085.

Ward, G., L. H. Harbors, and J. J. Blaha. 1979. Calcium-containing crystals in alfalfa: Their fate in cattle. J. Dairy Sci. 62:715–722.

Ward, J. D., J. W. Spears, and E. B. Kegley. 1993. Effect of copper level and source (copper lysine versus copper sulfate) on copper status, performance, and immune response in growing steers fed diets with or without supplemental molybdenum and sulfur. J. Anim. Sci. 71:2748–2755.

Ward, J. D., J. W. Spears, and G. P. Gengelbach. 1995. Differences in copper status and copper metabolism among Angus, Simmental, and Charolais cattle. J. Anim. Sci. 73:571–577.

Weeth, H. J., and L. H. Haverland. 1961. Tolerance of growing cattle for drinking water containing sodium chloride. J. Anim. Sci. 20:518–521.

Weeth, H. J., and J. E. Hunter. 1971. Drinking of sulfate-water by cattle. J. Anim. Sci. 32:277–281.

Weeth, H. J., L. H. Haverland, and D. W. Cassard. 1960. Consumption of sodium chloride water by heifers. J. Anim. Sci. 19:845–851.

Weil, A. B., W. B. Tucker, and R. W. Hemken. 1988. Potassium requirements of dairy calves. J. Dairy Sci. 71:1868–1872.

Whanger, P. D., and G. Matrone. 1966. Effect of dietary sulfur upon the production and absorption of lactate in sheep. Biochem. Biophys. Acta 124:273–279.

Whanger, P. D., P. H. Weswig, J. E. Oldfield, P. R. Cheeke, and O. H. Muth. 1972. Factors influencing selenium and white muscle disease: Forage types, salts, amino acids, and dimethyl sulfoxide. Nutr. Rep. Int. 6:21–37.

Wheeler, J. L., D. A. Hedges, K. A. Archer, and B. A. Hamilton. 1980. Effect of nitrogen, sulfur and phosphorus fertilizer on the production, mineral content and cyanide potential of forage sorghum. Aust. J. Exp. Agric. Anim. Husb. 20:330–338.

Whitehead, D. C., K. M. Goulden, and R. D. Hartley. 1985. The distribution of nutrient elements in cell wall and other fractions of the herbage of some grasses and legumes. J. Sci. Food Agric. 36:311–318.

Wise, M. B., S. E. Smith, and L. L. Barnes. 1958. The phosphorus requirement of calves. J. Anim. Sci. 17:89–99.

Wise, M. B., A. L. Ordoreza, and E. R. Barrick. 1963. Influence of variations in dietary calcium:phosphorus ratio on performance and blood constituents of calves. J. Nutr. 79:79–84.

Wittenberg, K. M., R. J. Boila, and M. A. Shariff. 1990. Comparison of copper sulfate and copper proteinate as copper sources for copper-depleted steers fed high molybdenum diets. Can. J. Anim. Sci. 70:895–904.

Wong-Ville, J., P. R. Henry, C. B. Ammerman, and P. V. Rao. 1989. Estimation of the relative bioavailability of manganese sources for sheep. J. Anim. Sci. 67:2409–2414.

Woolliams, C., N. F. Suttle, J. A. Woolliams, D. G. Jones, and G. Wiener. 1986. Studies on lambs from lines genetically selected for low and high copper status. Anim. Prod. 43:293–301.

Wright, P. L., and M. C. Bell. 1966. Comparative metabolism of selenium and tellurium in sheep and swine. Am. J. Physiol. 211:6–10.

Wylie, M. J., J. P. Fontenot, and L. W. Greene. 1985. Absorption of magnesium and other macrominerals in sheep infused with potassium in different parts of the digestive tract. J. Anim. Sci. 61:1219–1229.

6 Vitamins and Water

Vitamins are unique among dietary nutrients fed to ruminants. In addition to being vital, vitamins are required in adequate amounts to enable animals to efficiently utilize other nutrients. Many metabolic processes are initiated and controlled by specific vitamins during various stages of life.

Calves from adequately fed mothers have minimal stores of vitamins at birth. Unlike the adult ruminant, a young calf does not have a fully functional rumen and active microflora, which typically contribute to vitamin synthesis. Colostrum is rich in vitamins, particularly vitamin A, provided that vitamins have been adequately supplied to the dam. Thus, a dietary supply of vitamins is typically provided to the newborn calf through colostrum. However, deficiencies of the B vitamins have been produced experimentally in calves prior to rumen development (Miller, 1979).

Intensive production systems have placed an increased emphasis on the importance of supplying adequate vitamin concentrations to meet animal requirements. Ruminants may become more susceptible to vitamin deficiencies in confinement feeding situations and when increased levels of production increase metabolic requirements for vitamins. Determining optimal vitamin concentrations—specific to age, breed, environment, and a multiplicity of other factors—facilitates management and production.

FAT-SOLUBLE VITAMINS

Vitamin A

Vitamin A is likely the vitamin of most practical importance in cattle feed. The function of vitamin A at the molecular level includes production of retinaldehyde in the chromophoric group of the visual pigment or a component of the visual purple required for dim light vision (Moore, 1939, 1941). Vitamin A is also essential for normal growth and reproduction, maintenance of epithelial tissues, and bone development.

Vitamin A does not occur, as such, in plant material; however, its precursors, carotenes or carotenoids, are present in plants in various forms (α-carotene, β-carotene, γ-carotene, and cryptoxanthin). Efficiency of conversion of carotenoids to retinol is variable in beef cattle and is generally lower than that for nonruminant animals (Ullrey, 1972). Retinyl acetate was degraded by ruminal fluid from concentrate-fed cattle more rapidly than from animals fed hay or straw (Rode et al., 1990).

Few grains, except for yellow corn, contain appreciable amounts of carotenoid; carotene is rapidly destroyed by exposure to sunlight and air, especially at high temperatures. Ensiling effectively preserves carotene but the availability of carotene from corn silage may be low (Jordan et al., 1963; Smith et al., 1964; Miller et al., 1967). High-quality forages provide carotenoid in large amounts but tend to be seasonal in availability.

The liver can store vitamin A, and these stores can serve to prevent vitamin-A deficiency. Unfortunately, liver stores are highly variable and cannot be assessed accurately without taking samples by biopsy. Furthermore, liver stores are in a dynamic state (Frey and Jensen, 1947; Hayes et al., 1967). Factors influencing deposition and removal are not well understood, but cattle exposed to drought, winter feeds of less than high-quality forage, or stresses such as high temperature or elevated nitrate intake are particularly susceptible. On a practical basis, no more than 2 to 4 months of protection from stored vitamin A can be expected, and cattle should be observed carefully for signs of deficiency whenever the diet is deficient.

A protective role for vitamin A and β-carotene against diseases has been demonstrated (Chew, 1987). It has also been suggested that mechanisms that require β-carotene protect the mammary gland from infection (Daniel et al., 1991). Furthermore, dietary vitamin A and β-carotene supplementation (53,000 IU vitamin A plus 400 mg β-carotene) to dairy cows 6 weeks before dry off and 2 weeks after dry off influence the responsiveness of bovine neutrophils and lymphocytes (Tjoelker et al., 1988a,b).

Beef cattle requirements for vitamin A are 2,200 IU/kg dry feed for beef feedlot cattle; 2,800 IU/kg dry feed for pregnant beef heifers and cows; and 3,900 IU/kg dry feed for lactating cows and breeding bulls (Guilbert and Hart, 1935; Jones et al., 1938; Guilbert et al., 1940; Madsen et al., 1948; Church et al., 1956; Chapman et al., 1964; Cullison and Ward, 1965; Perry et al., 1965, 1968; Swanson et al., 1968; Kohlmeier and Burroughs, 1970; Meacham et al., 1970; Kirk et al., 1971; Eaton et al., 1972). These requirements are the same as those given in the sixth edition of this report (National Research Council, 1984); there has been no new research to determine requirements since then. An IU is defined as 0.300 μg of trans-vitamin A alcohol (retinol) or 0.550 μg of retinyl palmitate.

SIGNS OF VITAMIN-A DEFICIENCY

Vitamin-A deficiency results in tissue changes associated primarily with vision, bone development, and epithelial structure and maintenance. Signs of deficiency may be specific for vitamin-A deficiency or the clinical signs may be general.

Vitamin-A deficiency is most likely to occur when cattle are fed

- high-concentrate diets;
- bleached pasture or hay grown during drought conditions;
- feeds that have received excess exposure to sunlight, air, and high temperature;
- feeds that have been heavily processed or mixed with oxidizing materials such as minerals; and
- feeds that have been stored for long periods of time.

Most susceptible are newborn calves deprived of colostrum and cattle unable to establish or maintain liver stores because of environmental or dietary stresses. Attempts to improve the vitamin-A status of the newborn calf by supplementing the dam's diet have been successful, but very high levels of vitamin-A or carotene have been necessary (Branstetter et al., 1973). Deficiencies can be corrected by increasing carotene intake by adding to the diet fresh, leafy, high-quality forages, which contain large amounts of vitamin-A precursors and vitamin E, or by supplying vitamin-A supplements in the feed or by injection. Since inefficient conversion of carotene to vitamin A is often a part of the problem, administering preformed vitamin A is preferred when deficiencies are present. Injected vitamin A is more efficiently utilized than vitamin A provided in the diet (Perry et al., 1967; Schelling et al., 1975), possibly because of extensive destruction of the vitamin in the

rumen and abomasum (Keating et al., 1964; Klatte et al., 1964; Mitchell et al., 1967).

Signs of vitamin-A deficiency include reduced feed intake, rough hair coat, edema of joints and brisket, lacrimation, xerophthalmia, night blindness, slow growth, diarrhea, convulsive seizures, improper bone growth, blindness, low conception rates, abortion, stillbirths, blind calves, abnormal semen, and other infections (Guilbert and Hart, 1935; Jones et al., 1938; Guilbert et al., 1940; Guilbert and Rochfort, 1940; Hart, 1940; Madsen and Earle, 1947; Madsen et al., 1948; Moore, 1957; Mitchell, 1967); however, only night blindness has proven unique to vitamin-A deficiency (Moore, 1939, 1941). Vitamin-A deficiency should be suspected when several of these symptoms are present. Clinical verification may include ophthalmoscopic examination, liver biopsy and assay, blood assay, testing spinal fluid pressure, conjunctival smears, and response to vitamin-A therapy.

SIGNS OF VITAMIN-A TOXICITY

Vitamin A has a wide margin of safety for use in ruminant animals. Ruminants appear to have a relatively high tolerance for vitamin A, presumably due in part to microbial degradation of vitamin A in the rumen (Rode et al., 1990). Extremely high concentrations of vitamin A can be toxic; however, toxicity is rarely a problem in livestock, unless unreasonably high concentrations are fed inadvertently (National Research Council, 1987).

Vitamin D

As a general term, vitamin D encompasses a group of closely related antirachitic compounds. There are two primary forms of vitamin D: ergocalciferol (vitamin D_2), which is derived from the plant steroid, ergosterol; and cholecalciferol (vitamin D_3), which is derived from the precursor 7-dehydrocholesterol and is found only in animal tissues or products.

Vitamin D is required for calcium and phosphorus absorption, normal mineralization of bone, and mobilization of calcium from bone. In addition, a regulatory role in immune cell function of vitamin D (1,25-dihydroxy D) has been suggested (Reinhardt and Hustmyer, 1987). Research in laboratory animals (DeLuca, 1974) indicates that before serving these functions, vitamin D must be metabolized to active forms.

Vitamin D is absorbed from the diet in the intestinal tract in association with lipids and the presence of bile salts. Once in the liver, one metabolite (25-hydroxy-vitamin-D_3) is formed, which is about four times as active as vitamin D. This major circulating metabolite of vitamin D is then transported to the kidney, where another vitamin D metabolite (1,25-dihydroxy-vitamin-D_3) is formed. This form is about five times as active as 25-hydroxy-vitamin-D_3 (Horst and Reinhardt, 1983). How vitamin D is degraded in the rumen (Parakkasi et al., 1970; Sommerfeldt et al., 1979) may be of practical significance when considering how the vitamin D should be administered. Sommerfeldt et al. (1983) indicated that orally administered tritium-labeled vitamin D_2 has one-third to one-half the activity of tritium-labeled vitamin D_3.

The vitamin D requirement of beef cattle is 275 IU /kg dry diet. The IU is defined as 0.025 μg of cholecalciferol (D_3) or its equivalent. Ergocalciferol (D_2) also is active in cattle. Unlike aquatic species that store appreciable amounts of vitamin D in the liver, most land mammals, including ruminants, do not maintain body stores of vitamin D. However, because vitamin D is synthesized by beef cattle exposed to sunlight or fed sun-cured forages, these animals rarely require vitamin D supplementation.

SIGNS OF VITAMIN-D DEFICIENCY

The most clearly defined sign of vitamin-D deficiency in calves is rickets, caused by the failure of bone to assimilate and use calcium and phosphorus normally. Accompanying evidence frequently includes a decrease in calcium and inorganic phosphorus in the blood, swollen and stiff joints, anorexia, irritability, tetany, and convulsions. In older animals with a vitamin-D deficiency, bones become weak and easily fractured and posterior paralysis may accompany vertebral fractures. Calves may be born dead, weak, or deformed (Rupel et al., 1933; Wallis, 1944; Warner and Sutton, 1948; Stillings et al., 1964). General clinical signs of vitamin-D deficiency include decreased appetite and growth rate, digestive disturbances, labored breathing, and weakness.

SIGNS OF VITAMIN-D TOXICITY

Intakes of excessive amounts of vitamin D can result in a variety of effects. Most commonly, blood calcium concentration becomes abnormally high as a result of increased bone resorption and increased intestinal absorption of calcium. This can result in widespread calcification of soft tissues and bone demineralization. Other signs of vitamin-D toxicity include loss of appetite and weight loss (National Research Council, 1987).

Vitamin E

Vitamin E occurs naturally in feedstuffs as α-tocopherol. Other forms exist such as β, γ, δ, ε, ζ, and ή; and all may occur in feedstuffs isolated from the oils of plants. Of the several compounds that have vitamin E activity, the naturally occurring compound having the highest vitamin E activity is *RRR*-α-tocopherol (formerly D-α-tocopherol),

with a biopotency equivalent to 1.36 moles of all-*rac*-α-tocopherol (U.S. Pharmacopeia, 1985). All-*rac*-α-tocopherol is a synthetic mixture of eight stereoisomers. Tocopherul acetate does not occur naturally, but is often used in animal diets. The alcohol group linked to the acetate prevents the tocopherol from being destroyed in the diet and, when consumed, the ester is hydrolyzed in the intestine to make the tocopherol available for absorption. Terms for expressing vitamin E activity have changed over the years. The current preferred expression of vitamin E activity is in molar concentration and conversion equivalents for IU expression (now obsolete) are presented below:

1 mg all-*rac*-α-tocopheryl acetate = 1 International Unit
0.74 mg *RRR*-α-tocopheryl acetate = 1 International Unit
0.91 mg all-*rac*-α-tocopherol = 1 International Unit
0.67 mg *RRR*-α-tocopherol = 1 International Unit

Determining vitamin E requirements of ruminants is difficult because of this vitamin's interrelationships with other dietary components. Vitamin E requirements depend on concentrations of antioxidants, sulfur-containing amino acids, and selenium in the diet. In addition, high dietary concentrations of polyunsaturated fatty acids present in unsaturated oils such as corn oil, linseed oil, and soybean oil can significantly increase vitamin E requirements. Detrimental effects of polyunsaturated fatty acids may be somewhat reduced in the ruminant animal because ruminal microorganisms are capable of fatty acid saturation; however, some polyunsaturated fatty acids may escape ruminal hydrogenation (McMurray et al., 1980).

Vitamin E is not stored in the body in large concentrations. In general, vitamin E may be found in many tissues, with the highest amounts found in liver and adipose tissue. Thymus, muscle, kidney, lung, spleen, heart, and adrenal tissues increase concentration of vitamin E when high concentrations of vitamin E are in the diet. When 300 IU vitamin E/day was fed for 266 days to finishing steers, less discoloration of the muscle tissue occurred during refrigeration storage. A short-term feeding regimen (67 days of 1,266 IU vitamin E/day or 30 days of 1,317 IU vitamin E/day) resulted in similar improvements (Arnold et al., 1992). D-α sources of tocopherol in plasma and tissues were increased after feeding 1,000 IU of either D or DL sources of acetate or alcohol for 28 days (Hidiroglou et al., 1988).

Vitamin E serves various functions including its role as an inter- and intracellular antioxidant and in the formation of structural components of biological membranes. The role of vitamin E as a biological antioxidant and a free radical scavenger in the immune system and in disease resistance has been documented (Tappel, 1972; Hoekstra, 1975; McCay and King, 1980). Jersey steers fed 1,000 IU of vitamin E as DL-α-tocopherol acetate for 6 months had higher interleukin-1 in the cells than did other steers. (See also Chapter 8.)

Vitamin E functions as an antioxidant in cellular membranes and has been widely used to protect and facilitate the uptake and storage of vitamin A (Perry et al., 1968). Its action in metabolism is not clearly defined but is linked closely with selenium (Muth et al., 1958; Hoekstra, 1975). Vitamin E functions in the maintenance of structural and functional integrity of skeletal muscle, cardiac muscle, smooth muscle, and the peripheral vascular system.

There are many factors that influence the stability of vitamin E in feeds—heat, oxygen, moisture, unsaturated fatty acids, trace minerals, and high nitrates (Bunyan et al., 1961). Physical changes during storage also influence the stability of vitamin E in feeds; with natural drying, corn may lose 15 to 25 percent of vitamin E (Pond et al., 1971; Young et al., 1975; Bauernfeind, 1980). Also, high-moisture feeds lose vitamin E more rapidly than dry feed (Adams, 1982; Harvey and Bieber-Wlaschny, 1988). Adequate amounts of vitamin E may not be available from feedstuffs; thus, formulating diets to ensure adequate concentrations of vitamin E is more difficult.

The vitamin E requirement for beef cattle has not been established but is estimated to be between 15 and 60 IU/kg dry diet for young calves. Even diets very low in vitamin E did not affect growth, reproduction, or lactation when fed to cattle for four generations (Gullickson and Calverley, 1946). A depletion and refeeding study was conducted with vitamin E, and the data indicate that the requirement for optimum growth of growing finishing steers was 50 to 100 units of vitamin E added in the feed daily (Hutcheson and Cole, 1985).

SIGNS OF VITAMIN-E DEFICIENCY

Vitamin-E deficiencies can be precipitated or accentuated by the intake of unsaturated fats. Signs of deficiencies in young calves are characteristic of white-muscle disease; they include general muscular dystrophy, weak leg muscles, crossover walking, impaired suckling ability caused by dystrophy of tongue muscles, heart failure, paralysis, and hepatic necrosis (Stafford et al., 1954; Muth et al., 1958).

Animals exhibiting deficiency signs, particularly white-muscle disease, may respond to either selenium or vitamin E or may require both. Vitamin E supplements the action of glutathione peroxidase, a selenium-containing enzyme. (Vitamin E and selenium interactions are discussed in the selenium section in Chapter 5.)

SIGNS OF VITAMIN-E TOXICITY

Vitamin-E toxicity has not been demonstrated in ruminants and there seems to be a wide margin of safety regarding the use of vitamin E in most animals. Of the major fat-soluble vitamins, the risk of toxicity is less with vitamin E than with vitamins A and D (National Research Council, 1987).

Vitamin K

The term *vitamin K* is used to describe a group of quinone fat-soluble compounds that have characteristic antihemorrhagic effects. Vitamin K is required for the synthesis of plasma clotting factors prothrombin (factor II), proconvertin (factor VII), Christmas factor (factor IX), and Stuart-Prower factor (factor X). Two major natural sources of vitamin K are the phylloquinones (vitamin K_1), found in plant sources, and the menaquinones (vitamin K_2), which are produced by bacterial flora. For ruminants, vitamin K_2 is the most significant source of vitamin K, since it is synthesized in large quantities by bacterial flora in the rumen. Vitamin K_1 is abundant in pasture and green roughages. Both forms possess similar biological activity and function in blood clotting.

SIGNS OF VITAMIN-K DEFICIENCY

The only sign of deficiency to be reported in cattle is the "sweet clover disease" syndrome. This results from the metabolic antagonistic action of dicoumarol that occurs when an animal consumes moldy or improperly cured sweet clover hay. Consumption of dicoumarol, a fungal metabolite produced from substrates in sweet clover hay, leads to prolonged blood clotting and has caused death from uncontrolled hemorrhages. It is important to note that dicoumarol passes through the placenta, and thus, the fetus of pregnant animals may be affected.

The initial appearance and severity of signs associated with dicoumarol poisoning are directly related to the dicoumarol content of the hay consumed. If low levels are consumed, then clinical signs may not appear for several months. Mild cases can be treated effectively with vitamin K (McElroy and Goss, 1940a; Link, 1959).

SIGNS OF VITAMIN-K TOXICITY

Few systematic studies of the effects of excess vitamin K have been conducted in ruminant animals. Toxicity associated with excessive oral intake of phylloquinone or menadione has not been demonstrated in beef cattle. The toxic dietary level of menadione is at least 1,000 times the dietary requirement (National Research Council, 1987).

WATER-SOLUBLE VITAMINS

Vitamin B_{12}

Vitamin B_{12} is a generic descriptor for a group of compounds that have vitamin B_{12} activity. One of the unique features of vitamin B_{12} is that it contains 4.5 percent cobalt. The naturally occurring forms of vitamin B_{12} are adenosylcobalamin and methylcobalamin and these are found in plant and animal tissues. Cyanocobalamin, an artificially produced form of vitamin B, is used extensively because it is relatively stable and readily available. The primary functions of vitamin B_{12} involve metabolism of nucleic acids and proteins, in addition to metabolism of fats and carbohydrates. Specifically, this vitamin plays a role in purine and pyrimidine synthesis, transfer of methyl groups, protein formation, and metabolism of fats and carbohydrates. Vitamin B_{12} is of special interest in ruminant nutrition because of its role in propionate metabolism (Marston et al., 1961) and the practical incidence of vitamin-B_{12} deficiency as a secondary result of cobalt deficiency. The ruminant's requirement for vitamin B_{12} is higher than the nonruminant's requirement and is associated with the requirement for cobalt, since this trace mineral is a component of vitamin B_{12}. Cobalt content of the diet is the primary limiting factor for ruminal microorganism synthesis of vitamin B_{12}. Substantial areas of the United States, Australia, and New Zealand have soils without enough cobalt to produce adequate concentrations in plants to support optimum vitamin B_{12} synthesis in the rumen (Ammerman, 1970). (For additional information on cobalt, see Chapter 5.)

SIGNS OF VITAMIN-B_{12} DEFICIENCY

A vitamin-B_{12} deficiency is difficult to distinguish from a cobalt deficiency. The signs of deficiency may not be specific and can include poor appetite, retarded growth, and poor condition. In severe deficiencies, muscular weakness and demyelination of peripheral nerves occurs. In young ruminant animals, vitamin-B_{12} deficiency can occur when rumen microbial flora are not yet fully developed.

Thiamin

Thiamin functions in all cells as a coenzyme cocarboxylase. Thiamin is the coenzyme responsible for all enzymatic carboxylations of α-keto acids in the tricarboxylic acid cycle, which provides energy to the body. Thiamin also plays a key role in glucose metabolism, as a coenzyme in the pentose phosphate pathway.

Thiamin antimetabolites have been found in raw fish products and bracken fern (Somogyi, 1973). Polioencephalomalacia (PEM), a central nervous system disorder, in grain-fed cattle and sheep has been linked to thiaminase activity or production of a thiamin antimetabolite in the rumen (Loew and Dunlop, 1972; Sapienza and Brent 1974). Affected animals have responded to intravenous administration of thiamin (2.2 mg/kg BW). Thiamin analogs produced in the rumen by thiaminase I in the presence of a cosubstrate appeared to be responsible for PEM (Brent

and Bartley, 1984). Supplementation of high-concentrate diets with thiamin, however, yield inconsistent results (Grigat and Mathison, 1982, 1983).

Synthesis of thiamin by rumen microflora makes it difficult to establish a ruminant requirement. Animals with a functional rumen can generally synthesize an adequate amount of thiamin. However, the synthesis of thiamin is subject to dietary factors including levels of carbohydrate and nitrogen. In addition, high sulfur diets have been associated with thiamin deficiency and PEM, a laminar softening or degeneration of brain gray matter in steers (Gould et al., 1991). Animal size, genetic factors, and physiological status also influence thiamin requirements.

SIGNS OF THIAMIN DEFICIENCY

In all species, a thiamin deficiency results in central nervous system disorders, since thiamin is an important component of the biochemical reactions that break down glucose to supply energy to the brain. Other signs of thiamin deficiency include weakness, retracted head, and cardiac arrhythmia. As with other water-soluble vitamins, deficiencies can result in slowed growth, anorexia, and diarrhea.

Niacin

Niacin functions in carbohydrate, protein and lipid metabolism as a component of the coenzyme forms of nicotinamide, nicotinamide adenine dinucleotide (NAD), and nicotinamide adenine dinucleotide phosphate (NADP). Niacin is particularly important in ruminants because it is required for liver detoxification of portal blood NH_3 to urea and liver metabolism of ketones in ketosis.

Niacin has been reported to enhance protein synthesis by ruminal microorganisms (Riddell et al., 1980, 1981). Niacin synthesis in the rumen seemed adequate when no niacin was added to the diet; however, when 6 g was added per day, an increase in niacin flow from the rumen occurred (Riddell et al., 1985). Supplemental niacin was more effective in increasing microbial protein synthesis with urea than soybean meal (Brent and Bartley, 1984). Responses to supplemental niacin of feedlot cattle have been variable.

Niacin is supplied to the ruminant from three primary sources: dietary niacin, conversion of tryptophan to niacin, and ruminal synthesis. Although niacin is normally synthesized in adequate quantities in the rumen, there are several factors that can influence ruminant niacin requirements (Olentine, 1984). These factors include protein (amino acid) balance, dietary energy supply, dietary rancidity, de novo synthesis, and availability of niacin in feeds. Excess leucine, arginine, and glycine increase the niacin requirement; whereas increasing dietary tryptophan decreases the niacin requirement. High-energy diets and the use of particular antibiotics can increase the requirement for niacin.

SIGNS OF NIACIN DEFICIENCY

Young ruminants are most susceptible to niacin deficiencies, and a dietary source of niacin or tryptophan is required until the rumen is fully developed. The first signs of niacin deficiency in most species are loss of appetite, reduced growth, general muscular weakness, digestive disorders, and diarrhea. The skin may also be affected with a scaly dermatitis. Often, these signs are followed by a microcytic anemia.

Choline

Choline is essential for building and maintaining cell structure throughout the body and for the formation of acetylcholine, the compound responsible for transmission of nerve impulses. Abnormal accumulation of fat is prevented by the lipotropic actions of choline, and labile methyl groups are furnished by choline for formation of methionine. All naturally occurring fats contain choline, but little information is available on the biological availability of choline in feeds.

Unlike most vitamins, choline can be synthesized by most animal species. Because ruminants synthesize choline, a requirement has not been determined; however, it has been recommended that milk-fed calves receive supplementation of 0.26% choline in milk replacers.

Choline from dietary sources is only of value to adult animals if it can escape rumen degradation. Rumsey (1985) determined that for choline-supplemented steers fed an all-concentrate diet, supplementation did not affect feedlot performance, carcass measurements, acidosis, or products of rumen fermentation. However, increasing dietary rumen protected choline (0.24 percent) produced a linear increase in milk production for lactating dairy cows (Erdman and Sharma, 1991).

SIGNS OF CHOLINE DEFICIENCY

Calves fed a synthetic milk diet containing 15 percent casein exhibited apparent signs of choline deficiency. Within a week, calves developed extreme weakness, labored breathing, and were unable to stand. Supplementation with 260 mg choline/L milk replacer alleviated the signs of choline deficiency.

Summary

B vitamins are abundant in milk and many other feeds, and synthesis of B vitamins by ruminal microorganisms is extensive (McElroy and Goss, 1940a,b; 1941a,b; Wegner

et al., 1940, 1941; Hunt et al., 1943) and begins very soon after the introduction of dry feed into the diet (Conrad and Hibbs, 1954). As the concentration in the diet increases, thiamin results in a net loss; whereas niacin increases substantially in the rumen, while the duodenal concentration of thiamin, niacin, riboflavin, and biotin does not change (Miller et al., 1986a,b). Niacin decreases in the duodenum and ileum when monensin is added (22 mg/kg diet), while thiamin, riboflavin, and biotin are not affected.

Signs of insufficient intake of B complex vitamins have been clearly demonstrated for thiamin (Johnson et al., 1948), riboflavin (Wiese et al., 1947), pyridoxine (Johnson et al., 1950), pantothenic acid (Sheppard and Johnson, 1957), biotin (Wiese et al., 1946), nicotinic acid (Hopper and Johnson, 1955), vitamin B_{12} (Draper et al., 1952; Lassiter et al., 1953), and choline (Johnson et al., 1951) in young calves. The established metabolic functions of B vitamins are important and consequently, a physiological need for most B vitamins can be assumed for cattle of all ages.

Attempts to obtain responses to other B vitamins are numerous, but the overall results are considered inconclusive. Although B vitamin synthesis is altered by diet, considerable change is possible without producing signs of deficiency (Hayes et al., 1966; Clifford et al., 1967).

Supplemental riboflavin, niacin, folic acid, B_{12}, and ascorbic acid are degraded and/or absorbed anterior to the small intestine, while biotin and pantothenic acid primarily escape the rumen (Zinn et al., 1987). As a result, practical vitamin-B deficiency is limited to young animals with immature rumen development and situations in which an antagonist is present or ruminal synthesis is limited by lack of precursors.

WATER

Water constitutes approximately 98 percent of all molecules in the body. Water is needed for regulation of body temperature as well as for growth, reproduction, and lactation; digestion; metabolism; excretion; hydrolysis of protein, fat, and carbohydrates; regulation of mineral homeostasis; lubrication of joints; nervous system cushioning; transporting sound; and eyesight. Water is an excellent solvent for glucose, amino acids, mineral ions, water-soluble vitamins, and metabolic waste transported in the body.

Water intake from feeds plus that consumed ad libitum as free water is approximately equivalent to the water requirements of cattle. Water requirement is influenced by several factors, including rate and composition of gain, pregnancy, lactation, activity, type of diet, feed intake, and environmental temperature.

Restriction of water intake reduces feed intake (Utley et al., 1970), which results in lower production. However, water restriction also tends to increase apparent digestibility and nitrogen retention.

The minimum requirement of cattle for water is a reflection of that needed for body growth and for fetal growth or lactation and that lost by excretion in the urine, feces, or sweat or by evaporation from the lungs or skin. Anything influencing these needs or losses will influence the minimum requirement.

Cattle lose water from the body through excretion from the kidney as urine and from the gastrointestinal tract as feces, as sweat, and by water vapor from skin and lungs. The amount of urine produced daily varies with the activity of the animal, air temperature, and water consumption as well as certain other factors. The antidiuretic hormone vasopressin controls reabsorption of water from the kidney tubules and ducts; thus, it affects excretion of urine. Under conditions of restricted water intake, the body may resorb a greater amount of water than usual, thus concentrating urine. Although this capacity to concentrate urine solutes is limited, it can reduce water requirements by a small amount. Water requirements can increase when a diet is high in protein, salt, minerals, or diuretic substances.

The amount of water lost in the feces depends largely on the diet. Succulent diets and diets with high mineral content contribute to more water in the feces.

The amount of water lost through evaporation from the skin or lungs is important and may even exceed that lost in the urine. If temperature and/or physical activity increase, water loss through evaporation and sweating increases.

Because feeds themselves contain some water and the oxidation of certain nutrients in feeds produces water, not all water must be provided by drinking. Feeds such as silage, green chop, or growing pasture forage are usually very high in moisture, while grains, hays, and dormant pasture forage are low in moisture. High-energy feeds produce much metabolic water; low-energy feeds produce a lesser amount. These are obvious complications in the matter of assessing water requirements. Fasting animals or those fed a low-protein diet may form water from the destruction of body protein or fat, but this is of minor significance.

The results of water requirement studies conducted under various conditions imply that thirst is a result of need and that animals drink to fill this need. The need results from an increase in the electrolyte concentration in the body fluids, which activates the thirst mechanism.

As this discussion suggests, water requirements are affected by many factors, and it is impossible to list specific requirements with accuracy. A water equation for feedlot steers has been developed by Hicks et al. (1988):

$$\text{Water intake (1 L/day)} = -19.76 + (0.4202 * MT) + (0.1329 * DMI) - (6.5966 * PP) - (1.1739 * DS).$$

See "Errata"—front cover

TABLE 6-1 Approximate Total Daily Water Intake of Beef Cattle[a]

Weight		Temperature in °F (°C)[b]											
		40	(4.4)	50	(10.0)	60	(14.4)	70	(21.1)	80	(26.6)	90	(32.2)
kg	lb	Liter	Gal	Liter	Gal	Liter	Gal	Liter	Gal	Liter	Gal	Liter	Gal
Growing heifers, steers, and bulls													
182	400	15.1	4.0	16.3	4.3	18.9	5.0	22.0	5.8	25.4	6.7	36.0	9.5
273	600	20.1	5.3	22.0	5.8	25.0	6.6	29.5	7.8	33.7	8.9	48.1	12.7
364	800	23.0	6.3	25.7	6.8	29.9	7.9	34.8	9.2	40.1	10.6	56.8	15.0
Finishing cattle													
273	600	22.7	6.0	24.6	6.5	28.0	7.4	32.9	8.7	37.9	10.0	54.1	14.3
364	800	27.6	7.3	29.9	7.9	34.4	9.1	40.5	10.7	46.6	12.3	65.9	17.4
454	1,000	32.9	8.7	35.6	9.4	40.9	10.8	47.7	12.6	54.9	14.5	78.0	20.6
Wintering pregnant cows[c]													
409	900	25.4	6.7	27.3	7.2	31.4	8.3	36.7	9.7	—	—	—	—
500	1,100	22.7	6.0	24.6	6.5	28.0	7.4	32.9	8.7	—	—	—	—
Lactating cows[d]													
409	900	43.1	11.4	47.7	12.6	54.9	14.5	64.0	16.9	67.8	17.9	61.3	16.2
Mature bulls													
636	1,400	30.3	8.0	32.6	8.6	37.5	9.9	44.3	11.7	50.7	13.4	71.9	19.0
727	1,600+	32.9	8.7	35.6	9.4	40.9	10.8	47.7	12.6	54.9	14.5	78.0	20.6

[a]Winchester and Morris (1956).

[b]Water intake of a given class of cattle in a specific management regime is a function of dry matter intake and ambient temperature. Water intake is quite constant up to 40 °F (4.4 °C).

[c]Dry matter intake has a major influence on water intake. Heavier cows are assumed to be higher in body condition and to require less dry matter and, thus, less water intake.

[d]Cows larger than 409 kg (900) lbs are included in this recommendation.

MT is the maximum temperature in degrees fahrenheit, DMI is dry matter intake in kg fed daily, PP is precipitation in cm/day, DS is the percent of dietary salt. However, the major influences on water intake in beef cattle fed typical rations are dry matter intake, environmental temperature, and stage and type of production. Table 6-1 has been designed as a guide only, and it must be used with respect to the influences of water intake.

Water quality is important in maintaining water consumption of cattle. Cattle consume water from surface water sources such as ponds, lakes, and streams and from ground water sources such as wells. Beef cattle requirements for water are a function of different metabolic priorities. Restricting water intake to less than the animal's requirement will reduce cattle performance.

For more detailed information on toxic substances in water, refer to the National Research Council publication *Nutrients and Toxic Substances in Water for Livestock and Poultry* (National Research Council, 1974).

REFERENCES

Adams, C. R. 1982. Feedlot cattle need supplemental vitamin E. Feedstuffs 54(18):24–26.

Ammerman, C. B. 1970. Recent developments in cobalt and copper in ruminant nutrition: A review. J. Dairy Sci. 53:1097–1107.

Arnold, R. N., K. K. Scheller, S. C. Arp, S. N. Williams, D. R. Buege, and D. M. Schaefer. 1992. Effect of long- or short-term feeding of alpha-tocopherol acetate to Holstein and crossbred beef steers on performance, carcass characteristics and beef stability. J. Anim. Sci. 70:3055–3065.

Bauernfeind, J. 1980. Tocopherols in food. Pp. 99–168 in Vitamin E: A Comprehensive Treatis, L. J. Machlin, ed. New York: Marcel Dekker.

Branstetter, R. F., R. E. Tucker, G. E. Mitchell, Jr., J. A. Boling, and N. W. Bradley. 1973. Vitamin A transfer from cows to calves. Int. J. Vit. Nutr. Res. 2:142–146.

Brent, B. E., and E. E. Bartley. 1984. Thiamin and niacin in the rumen. J. Anim. Sci. 59:813–822.

Bunyan J., D. McHale, J. Green, and S. Marcinkiewicz. 1961. Forms of vitamin E and their activity. Br. J. Nutr. 15:253.

Chapman, H. L., Jr., R. L. Shirley, A. Z. Palmer, C. E. Haines, J. W. Carpenter, and T. J. Cunha. 1964. Vitamins A and E in steer fattening rations on pasture. J. Anim. Sci. 23:669–673.

Chew, B. P. 1987. Symposium: Immune function: Relationship of nutrition and disease control: Vitamin A and beta-carotene on host defense. J. Dairy Sci. 70:2732–2743.

Church, D. C., L. S. Pope, and R. MacVicar. 1956. Effect of plane of nutrition of beef cows on depletion of liver vitamin A during gestation and on carotene requirements during lactation. J. Anim. Sci. 15:1078–1088.

Clifford, A. J., R. D. Goodrich, and A. D. Tillman. 1967. Effects of supplementing ruminant all-concentrate and purified diets with vitamins of the B complex. J. Anim. Sci. 26:400–403.

Conrad, H. R., and J. W. Hibbs. 1954. A high roughage system for raising calves based on early rumen development. Synthesis of thiamin and riboflavin in the rumen as influenced by ratio of hay to grain fed and initiation of dry feed consumption. J. Dairy Sci. 37:512–522.

Cullison, A. E., and C. S. Ward. 1965. Coastal Bermuda grass hay as a source of vitamin A for beef cattle. J. Anim. Sci. 24:969–972.

Daniel, L. R., B. P. Chew, T. S. Tanaka, and L. W. Tjoelker. 1991. In vitro effects of β-carotene and vitamin A on peripartum bovine peripheral blood mononuclear cell proliferation. J. Dairy Sci. 74:911–915.

DeLuca, H. F. 1974. Vitamin D: The vitamin and the hormone. Fed. Proc. 33:2211–2219.

Draper, H. H., J. T. Sime, and B. C. Johnson. 1952. A study of vitamin B₁₂ deficiency in the calf. J. Anim. Sci. 11:332–340.

Eaton, H. D., J. E. Rousseau, Jr., R. C. Hall, Jr., H. I. Frier, and J. J. Lucas. 1972. Reevaluation of the minimum vitamin A requirement of Holstein male calves based upon elevated cerebrospinal fluid pressure. J. Dairy Sci. 55:232–237.

Erdman, R. A., and B. K. Sharma. 1991. Effect of dietary rumen-protected choline in lactating dairy cows. J. Dairy Sci. 74:1641–1647.

Frey, R. R., and R. Jensen. 1947. Depletion of vitamin A reserves in the livers of cattle. Science 105:313.

Gould, D.H., M.M. McAllister, J.C. Savage, and D.W. Hamar. 1991. High sulfide concentrations in rumen fluid associated with nutritionally induced polioencephalomalacia in calves. Am. J. Vet. Res. 52:1164–1167.

Grigat, G. A., and G. W. Mathison. 1982. Thiamin supplementation of all-concentrate diet for feedlot steers. Can. J. Anim. Sci. 62:807–819.

Grigat, G. A., and G. W. Mathison. 1983. Thiamin and magnesium supplementation diets for feedlot steers. Can. J. Anim. Sci. 63:117–131.

Guilbert, H. R., and G. H. Hart. 1935. Minimum vitamin A with particular reference to cattle. J. Nutr. 10:409–427.

Guilbert, H. R., and L. H. Rochfort. 1940. Beef production in California. Calif. Agric. Exp. Stn. Circ. 11.

Guilbert, H. R., C. E. Howell, and G. H. Hart. 1940. Minimum vitamin A and carotene requirement of mammalian species. J. Nutr. 19:91–103.

Gullickson, T. W., and C. W. Calverley. 1946. Cardiac failure in cattle on the vitamin E-free rations as revealed by electrocardiograms. Science 104:312.

Hart, G. H. 1940. Vitamin A deficiency and requirements of farm mammals. Nutr. Abstr. Rev. 10:261.

Harvey, J. D., and M. Bieber-Wlaschny. 1988. Vitamin E availability to livestock varies dramatically. Feedstuffs:15–17.

Hayes, B. W., G. E. Mitchell, Jr., C. O. Little, and N. W. Bradley. 1966. Concentrations of B vitamins in ruminal fluid of steers fed different levels and physical forms of hay and grain. J. Anim. Sci. 25:539–542.

Hayes, B. W., G. E. Mitchell, Jr., C. O. Little, and H. B. Sewell. 1967. Turnover of liver vitamin A in steers. J. Anim. Sci. 26:855–857.

Hicks, R. B., F. N. Owens, D. R. Gill, J. J. Martin, and C. A. Strasia. 1988. Water intake by feedlot steers. Okla. Anim. Sci. Rpt. Mp-125:208.

Hidiroglou, N., L. F. Laflamme, and L. R. McDowell. 1988. Blood plasma and tissue concentrations of vitamin E in beef cattle as influenced by supplementation of various tocopherol compounds. J. Anim. Sci. 66:3227–3234.

Hoekstra, W. G. 1975. Biochemical function of selenium and its relation to vitamin E. Fed. Proc. 34:2083–2089.

Hopper, J. H., and B. C. Johnson. 1955. The production and study of an acute nicotinic acid deficiency in the calf. J. Nutr. 56:303–310.

Horst, R. L., and T. A. Reinhardt. 1983. Vitamin D metabolism in ruminants and its relevance to the periparturient cow. J. Dairy Sci. 66:661–678.

Hunt, C. H., E. W. Burroughs, R. M. Bethke, A. F. Schalk, and P. Gerlaugh. 1943. Further studies on riboflavin and thiamin in the rumen contents of cattle. II. J. Nutr. 25:207–216.

Hutcheson, D. P., and N. A. Cole. 1985. Vitamin E and selenium for yearling feedlot cattle. Fed. Am. Soc. Exp. Biol. (abst.) 69:807.

Johnson, B. C., T. S. Hamilton, W. B. Nevens, and L. E. Boley. 1948. Thiamin deficiency in the calf. J. Nutr. 35:137–145.

Johnson, B. C., J. A. Pinkos, and K. A. Burke. 1950. Pyridoxine deficiency in the calf. J. Nutr. 40:309–322.

Johnson, B. C., H. H. Mitchell, and J. A. Pinkos. 1951. Choline deficiency in the calf. J. Nutr. 43:37–48.

Jones, J. H., J. K. Riggs, G. S. Fraps, J. M. Jones, H. Schmidt, R. E. Dickson, P. E. Howe, and W. H. Black. 1938. Carotene requirements for fattening beef cattle. Proc. Am. Soc. Anim. Prod. 94:102.

Jordan, H. A., G. S. Smith, A. L. Neumann, J. E. Zimmerman, and G. W. Breniman. 1963. Vitamin A nutrition of beef cattle fed corn silage. J. Anim. Sci. 22:738–745.

Keating, E. K., W. H. Hale, and F. Hubbert, Jr. 1964. In vitro degradation of vitamin A and carotene by rumen liquor. J. Anim. Sci. 23:111–117.

Kirk, W. G., R. L. Shirley, J. F. Easley, and F. M. Peacock. 1971. Effect of carotene-deficient rations and supplemental vitamin A on gain, feed utilization and liver vitamin A of calves. J. Anim. Sci. 33:476–480.

Klatte, F. J., G. E. Mitchell, Jr., and C. O. Little. 1964. In vitro destruction of vitamin A by abomasal and ruminal contents. J. Agric. Food Chem. 12:420–421.

Kohlmeier, R. H., and W. Burroughs. 1970. Estimation of critical of plasma and liver vitamin A levels in feedlot cattle with observation upon influences of body stores and daily dietary requirements. J. Anim. Sci. 30:1012–1018.

Lassiter, C. A., G. M. Ward, C. F. Huffman, C. W. Duncan, and H. D. Webster. 1953. Crystalline vitamin B_{12} requirement of the young dairy calf. J. Dairy Sci. 36:997–1005.

Link, K. P. 1959. The discovery of dicumarol and its sequels. Circulation 19:97.

Loew, F. M., and R. H. Dunlop. 1972. Induction of thiamin inadequacy and polioencephalomalacia in adult sheep with Amprolium. Am. J. Vet. Res. 33:2195.

Madsen, L. L., and I. P. Earle. 1947. Some observations on beef cattle affected with generalized edema or anasarca due to vitamin A deficiency. J. Nutr. 34:603–619.

Madsen, L. L., O. N. Eaton, L. Memmstra, R. E. Davis, C. A. Cabell, and B. Knapp, Jr. 1948. Effectiveness of carotene and failure of ascorbic acid to increase sexual activity and semen quality of vitamin A deficient beef bulls. J. Anim. Sci. 7:60–69.

Marston, H. R., S. H. Allen, and R. M. Smith. 1961. Primary metabolic defect supervening on vitamin B_{12} deficiency in sheep. Nature 190:1085–1091.

McCay, P. B., and M. M. King. 1980. Vitamin E: Its role as a biologic free radical scavenger and its relationship to the microsomal mixed-function oxidase system. Pp. 289–317 in Vitamin E: Comprehensive Treatise, L. J. Machlin, ed. New York: Marcel Dekker.

McElroy, L. W., and H. Goss. 1940a. A quantitative study of vitamins in the rumen contents of sheep and goats fed vitamin-low diets. I. Riboflavin and vitamin K. J. Nutr. 20:527–540.

McElroy, L. W., and H. Goss. 1940b. A quantitative study of vitamins in the rumen contents of sheep and goats fed vitamin-low diets. II. Vitamin B_6 (pyridoxine). J. Nutr. 20:541–550.

McElroy, L. W., and H. Goss. 1941a. A quantitative study of vitamins in the rumen contents of sheep and cows fed vitamin-low diets. III. Thiamin. J. Nutr. 21:163–173.

McElroy, L. W., and H. Goss. 1941b. A quantitative study of vitamins in the rumen contents of sheep and cows fed vitamin-low diets. IV. Pantothenic acid. J. Nutr. 21:405–409.

McMurray, C.H., D.A. Rice, and W.J. Blanchflower. 1980. Changes in plasma levels of linoleic and linolenic acids in calves recently introduced to spring pasture. Proc. Nutr. Soc. 39:65.

Meacham, T. N., K. P. Bovard, B. M. Priode, and J. P. Fontenot. 1970. Effect of supplemental vitamin A on the performance of beef cows and their calves. J. Anim. Sci. 31:428–433.

Miller, B. L., J. C. Meiske, and R. D. Goodrich. 1986a. Effects of grain source and concentrate level on B-vitamin production and absorption in steers. J. Anim. Sci. 62:473–483.

Miller, B. L., J. C. Meiske, and R. D. Goodrich. 1986b. Effects of grain source and concentrate level on B-vitamin production and absorption in steers. J. Anim. Sci. 62:484–496.

Miller, R. W., R. W. Hemken, D. R. Waldo, and L. H. Moore. 1967. Vitamin A metabolism in steers fed corn silage or alfalfa hay pellets. J. Dairy Sci. 50:997.

Miller, W. J. 1979. Dairy cattle feeding and nutrition. Academic Press, New York.

Mitchell, G. E., Jr. 1967. Vitamin A nutrition of ruminants. J. Am. Vet. Med. Assoc. 151:430.

Mitchell, G. E., Jr., C. O. Little, and B. W. Hayes. 1967. Pre-intestinal destruction of vitamin A by ruminants fed nitrates. J. Anim. Sci. 26:827–829.

Moore, L. A. 1939. Carotene intake, level of blood plasma carotene, and the development of papillary edema and nyctalopia in calves. J. Dairy Sci. 22:803.

Moore, L. A. 1941. Some ocular changes and deficiency manifestations in mature cows fed a ration deficient in vitamin A. J. Dairy Sci. 24:893.

Moore, T. 1957. Vitamin A. Amsterdam: Elsevier.

Muth, O. H., J. E. Oldfield, L. F. Remmert, and J. P. Schubert. 1958. Effect of selenium and vitamin E on white muscle disease. Science 128:1090.

National Research Council. 1974. Nutrients and Toxic Substances in Water for Livestock and Poultry. Washington, D.C.: National Academy Press.

National Research Council. 1987. Vitamin Tolerance of Animals. Washington, D.C.: National Academy Press.

National Research Council. 1984. Nutrient Requirements of Beef Cattle, Sixth Revised Ed., Washington, D.C.: National Academy Press.

Olentine, C. B vitamins for ruminants. In Feed Manag. 35(4):3775.

Parakkasi, A., G. E. Mitchell, Jr., R. E. Tucker, C. O. Little, and A. W. Young. 1970. Pre-intestinal disappearance of vitamin D_2 in ruminants. J. Anim. Sci. 35:273 (abst.).

Perry, T. W., W. H. Smith, W. M. Beeson, and M. T. Mohler. 1965. Value of supplemental vitamin A for fattening beef cattle on pasture. J. Anim. Sci. 25:814–816.

Perry, T. W., W. M. Beeson, W. H. Smith, and M. T. Mohler. 1967. Injectable vs oral vitamin A for fattening steer calves. J. Anim. Sci. 26:115–118.

Perry, T. W., W. M. Beeson, W. H. Smith, R. B. Harrington, and M. T. Mohler. 1968. Interrelationships among vitamins A, E, and K when added to the rations of fattening beef cattle. J. Anim. Sci. 27:190–194.

Pond, W. G., W. H. Allaway, E. F. Walker, Jr., and K. Krook. 1971. Effects of corn selenium content and drying temperature and of supplemenntal vitamin E on growth, liver selenium, and blood vitamin E of chicks. J. Anim. Sci. 33:996–1000.

Rakes, A. H., M. P. Owens, J. H. Britt, and L. W. Whitlow. 1985. Effects of adding beta-carotene to rations of lactating cows consuming different forages. J. Dairy Sci. 68:1732–1737.

Reinhardt, T.A., and F.G. Hustmyer. 1987. Role of vitamin D in the immune system. J. Dairy Sci. 66:1520.

Riddell, D. O., E. E. Bartley, and A. D. Dayton. 1980. Effect of nicotinic acid on rumen fermentation in vitro and in vivo. J. Dairy Sci. 63:1429–1436.

Riddell, D. O., E. E. Bartley, and A. D. Dayton. 1981. Effect of nicotinic acid on microbial protein synthesis in vitro and on dairy cattle growth and milk production. J. Dairy Sci. 64:782–791.

Riddell, D. O., E. E. Bartley, J. J. Arambel, T. G. Nagaraja, A. D. Dayton, and G. W. Miller. 1985. Effect of niacin supplementation on ruminal niacin synthesis and degradation in cattle. Nutr. Rpt. Int. 31:407.

Rode, L. M., T. A. Mc Allister, and K. J. Cheng. 1990. Microbial degradation of vitamin A in rumen fluid from steers fed concentrate, hay, or straw diets. Can. J. Anim. Sci. 70:227–233.

Rumsey, T. S. 1985. Effect of choline in all concentrate diets of feedlot steers and on rumen acidosis. Can. J. Anim. Sci. 65:135–146.

Rupel, I. W., W. G. Bohstedt, and E. B. Hart. 1933. Vitamin D in the nutrition of the dairy calf. Wisc. Agric. Exp. Stn. Bull. 115.

Sapienza, D. A., and B. E. Brent. 1974. Ruminal thiaminases vs. concentrate adaptation. J. Anim. Sci. 39:251 (abstr.).

Schelling, G. T., L. W. Chittenden, T. C. Jackson, Jr., S. A. Koch, R. L. Stuart, R. E. Tucker, and G. E. Mitchell, Jr. 1975. Liver storage of oral and intramuscular vitamin A. J. Anim. Sci. 40:199 (abstr.).

Sheppard, J. J., and B. C. Johnson. 1957. Pantothenic acid deficiency in growing calf. J. Nutr. 61:195–205.

Smith, G. W., W. M. Durdle, J. E. Zimmerman, and A. L. Neumann. 1964. Relationships of carotene intake, thyro-active substances and soil fertility to vitamin A depletion of feeder cattle fed corn silages. J. Anim. Sci. 23:625–632.

Sommerfeldt, J. L., R. L. Horst, E. T. Littledike, and D. C. Beitz. 1979. In vitro degradation of cholecalciferol in rumen fluid. J. Dairy Sci. 62(Suppl. 1):192 (abstr.).

Sommerfeldt, J. L., J. L. Napoli, E. T. Littledike, D. C. Beitz, and R. L. Horst. 1983. Metabolism of orally administered [³H]ergocalciferol and [³H]cholecalciferol by dairy calves. J. Nutr. 113:2595–2600.

Somogyi, J. C. 1973. Antivitamins. Pp. 254–275 in Toxicants Occurring Naturally in Foods. Washington, D.C.: National Academy Press.

Stabel, J. R., T. A. Reinhart, M. A. Stevens, M. E. Kehrli, Jr., and B. J. Nonnecke. 1992. Vitamin E effects on in vitro immunoglobulin M and interleukin-1β production and transcription in dairy cattle. J. Dairy Sci. 75:2190–2198.

Stafford, J. W., K. F. Swingle, and H. Marsh. 1954. Experimental tocopherol deficiency in young calves. Am. J. Vet. Res. 15:373.

Stillings, B. R., J. W. Bratzler, L. F. Marriott, and R. C. Miller. 1964. Utilization of magnesium and other minerals by ruminants consuming low and high nitrogen-containing forages and vitamin D. J. Anim. Sci. 23:1148–1154.

Swanson, E. W., G. G. Martin, F. E. Pardue, and G. M. Gorman. 1968. Milk production of cows fed diets deficient in vitamin A. J. Anim. Sci. 27:541–548.

Tappel, A. L. 1972. Vitamin E and free radical peroxidation of lipids. Vitamin E and its role in cellular metabolism. Ann. N.Y. Acad. Sci. 203:12–23.

Tjoelker, L. W., B. P. Chew, T. S. Tanaka, and L. R. Daniel. 1988. Bovine vitamin A and β-carotene intake and lactational status. 1. Responsiveness of peripheral blood polymorphonuclear leukocytes to vitamin A and β-carotene challenge in vitro. J. Dairy Sci. 75:3112–3119.

Tjoelker, L. W., B. P. Chew, T. S. Tanaka, and L. R. Daniel. 1988. Bovine vitamin A and β-carotene intake and lactational status. 2. Responsiveness of mitogen-stimulated peripheral blood lymphocytes to vitamin A and β-carotene challenge in vitro. J. Dairy Sci. 75:3120–3127.

U.S. Pharmacopeia. 1985. U.S. Pharmacopeia, 21st Rev. XXI:1118–1119.

Ullrey, D. E. 1972. Symposium-Biological Availability of Nutrients in Feeds. Biological availability of fat-soluble vitamins: Vitamin A and carotene. J. Anim. Sci. 35:648–657.

Utley, P. R., N. W. Bradley, and J. A. Boling. 1970. Effect of restricting water intake on feed intake, nutrient digestibility and nitrogen metabolism in steers. J. Anim. Sci. 31:130–135.

Wallis, G. C. 1944. Vitamin D deficiency in dairy cows: Symptoms, causes, and treatment. S. Dak. Agric. Exp. Stn. Bull. 372.

Warner, R. G., and T. S. Sutton. 1948. The nutrition of the newborn dairy calf. III. The response to a photolyzed milk diet. J. Dairy Sci. 31:976.

Wegner, M. I., A. N. Booth, C. A. Elvehjem, and E. B. Hart. 1940. Rumen synthesis of the vitamin B complex. Proc. Soc. Exp. Biol. Med. 45:769.

Wegner, M. I., A. N. Booth, C. A. Elvehjem, and E. B. Hart. 1941. Rumen synthesis of the vitamin B complex on natural rations. Proc. Soc. Exp. Biol. Med. 47:9.

Wiese, A. C., B. C. Johnson, and W. B. Nevens. 1946. Biotin deficiency in the dairy calf. Proc. Soc. Exp. Biol. Med. 63:521.

Wiese, A. C., B. C. Johnson, H. H. Mitchell, and W. B. Nevens. 1947. Riboflavin deficiency in the dairy calf. J. Nutr. 33:263–270.

Winchester, C. F., and M. J. Morris. 1956. Water intake rates of cattle. J. Anim. Sci. 15:722–740.

Young, L. G., A. Lun, J. Pos, R. P. Forshaw, and D. Edmeades. 1975. Vitamin E stability in corn and mixed feed. J. Anim. Sci. 40:495–499.

Zinn, R. A., F. N. Owens, R. L. Stuart, J. R. Dunbar, and B. B. Norman. 1987. B-vitamins supplementation of diets for feedlot calves. J. Anim. Sci. 65:267–277.

7 Feed Intake

FACTORS AFFECTING FEED INTAKE

Factors that regulate dry matter intake (DMI) by ruminants are complex and not understood fully. Nevertheless, accurate estimates of feed intake are vital to predicting rate of gain and to the application of equations for predicting nutrient requirements of beef cattle, as provided in *Predicting Feed Intake for Food-Producing Animals* (National Research Council, 1987). Previous research has established relationships between dietary energy concentration and DMI by beef cattle based on the concept that consumption of less digestible, low-energy (often high-fiber) diets is controlled by physical factors such as ruminal fill and digesta passage, whereas consumption of highly digestible, high-energy (often low-fiber, high-concentrate) diets is controlled by the animal's energy demands and by metabolic factors (National Research Council, 1987). This model of intake regulation, however, is not fully compatible with existing data. Ketelaars and Tolkamp (1992a) used data from voluntary intake and digestibility of 831 roughages to evaluate the relationship between organic matter digestibility (OMD) and organic matter intake (OMI). Across a range of 30 to 84 percent OMD, OMI and OMD were related linearly. If intake of highly digestible feeds is regulated by energy demand, OMI (or digestible OMI) would be expected to plateau with increasing OMD. Also difficult to reconcile with the theory that ruminal fill of indigestible residues controls intake are large increases in intake during lactation periods and cold stress and decreases often observed with advancing pregnancy (Ketelaars and Tolkamp, 1992a). This disparity led Tolkamp and Ketelaars (1992) to hypothesize that ruminants do not simply eat as much as they can, but rather eat an amount that will optimize the cost and benefits of oxygen consumption; in effect, ad libitum intake in the model corresponds to the point at which net energy (NE) intake per unit of oxygen consumption is maximized. The approach of Tolkamp and Ketelaars (1992) resulted in accurate predictions of ad libitum intake by roughage-fed sheep. These authors further hypothesized that optimum intake was linked to an optimum metabolic acid load (Ketelaars and Tolkamp, 1992b). Additional research will be needed to develop this hypothesis fully; however, for further discussion of intake regulation theories and comparisons of intake predicted from various models, readers are referred to the thorough review by Mertens (1994).

Because the factors regulating intake by ruminants are not completely understood, models for predicting intake are empirical by nature. Intake prediction equations given in the preceding edition of *Nutrient Requirements of Beef Cattle* (National Research Council, 1984) and in *The Nutrient Requirements of Ruminant Livestock* (Agricultural Research Council, 1980) relate feed intake to dietary energy concentration (NE_m and ME, respectively). Based on such equations, energy concentration probably accounts, in part, for effects on feed intake attributed to gastrointestinal fill, energy demands, and potential effects of absorbed nutrients. These equations, however, do not account directly for the numerous physiological, environmental, and management factors that alter feed intake. Clearly, the methods of predicting feed intake described are intended to provide general guidelines. No single, general equation applies in all production situations. Optimally, beef cattle producers should develop intake prediction equations specific to given production situations; such equations should account for a greater percentage of the variation in intake than would be possible with a generalized equation.

Physiological Factors Affecting Feed Intake

Body composition, especially percentage of body fat, seems to affect feed intake (National Research Council, 1987). As animals mature, adipose tissue may, in some way, have a feedback role in controlling feed intake (National Research Council, 1987). Regardless of the mechanism, the percentage of body fat is often considered in equations to predict feed intake by beef cattle. Fox et al. (1988) suggested that DMI decreases 2.7 percent for each 1 percent increase in body fat over the range of 21.3 to 31.5 percent body fat. As a result of the relationship between feed intake and body fat, careful monitoring of feed intake can be a useful management tool to determine when cattle have reached appropriate slaughter condition.

Sex (steer vs heifer) seems to have limited effects on feed intake (Agricultural Research Council, 1980; National Research Council, 1987). Intake differences attributable to sex may be evident at certain times; Ingvartsen et al. (1992a) reported that at body weights (BW) less than 250 kg, heifers had greater intake capacity than steers or bulls. In the previous edition of *Nutrient Requirements of Beef Cattle* (National Research Council, 1984), the Subcommittee on Beef Cattle Nutrition suggested that predicted DMI should be decreased by 10 percent for medium-framed heifers. At a given BW, heifers are proportionally more mature (fatter) than steers; hence, Fox et al. (1988) in their equation for predicting DMI use a frame-equivalent weight adjustment instead of a direct adjustment for sex.

The age of an animal when it is placed on feed can affect feed intake. Older animals (e.g., yearlings vs calves) typically consume more feed per unit BW than younger ones. Presumably, the greater ratio of age to body weight (age relative to proportion of mature body composition) for yearling cattle prompts greater feed intake. This effect has been likened to increased feed intake by cattle experiencing compensatory growth (National Research Council, 1987). Assuming that cattle started on feed at heavier BW are generally older cattle, age-related effects on feed intake are partly responsible for the positive relationship between initial weight on feed and DMI (National Research Council, 1987). The 1984 subcommittee (National Research Council, 1984) and Fox et al. (1988) suggested a 10 percent increase in predicted DMI by cattle started on feed as yearlings compared with cattle started on feed as calves. Before more accurate predictions of feed intake are possible, designed studies are needed in which independent effects of age and body weight or body composition on feed intake can be quantified.

The animal's physiological state can markedly alter feed intake. Lactating animals can increase feed intake by 35 to 50 percent compared with that of nonlactating animals of the same BW fed the same diet (Agricultural Research Council, 1980). For forages, Minson (1990) reported a mean increase in DMI of 30 percent during lactation. Based on data from dairy cows, the Agricultural Research Council (ARC) (1980) and National Research Council (NRC) (1987) reports suggested that DMI increases by 0.2 kg/kg fat-corrected milk. Hence, beef cows bred for greater milk-producing ability would be expected to have greater feed intakes per unit BW during lactation. Advancing pregnancy has an adverse affect on feed intake, most notably during the last month (Agricultural Research Council, 1980; National Research Council, 1987). Ingvartsen et al. (1992a) noted a 1.5 percent decrease per week during the last 14 weeks of pregnancy in Danish Black and White heifers fed diets predominantly of roughage; this value agrees fairly well with the decrease of 2 percent per week during the last month of pregnancy suggested in the NRC (1987) report.

Frame size varies considerably in beef cattle. The 1984 NRC subcommittee (National Research Council, 1984) factored frame size into intake predictions, whereas Fox et al. (1988) suggested predictions could be adjusted by scaling frame sizes to an equivalent mature weight (frame-equivalent weight). However, Holstein and Holstein × beef crosses may consume more feed relative to body weight than beef breeds (National Research Council, 1987). Fox et al. (1988) suggested that intake predictions should be increased 8 percent for Holsteins and 4 percent for Holstein × British breed crosses relative to British-breed cattle. In addition to possible breed-specific effects, in the NRC (1987) report it was noted that genetic selection for feed efficiency could produce animals with increased feed intake potential, suggesting that genetic potential for growth (or increased production demands) may affect feed intake.

Environmental Factors Affecting Feed Intake

Considerable research has been conducted to evaluate effects of ambient temperature on feed intake and digestive function, and the topic has been reviewed extensively (Kennedy et al., 1986; Minton, 1986; Young, 1986; Young et al., 1989). In experimental situations, feed intake has been shown to increase as the temperature falls below the thermoneutral zone and decrease above that zone. With cold stress, ruminal motility and digesta passage increase before changes in intake occur, prompting Kennedy et al. (1986) to conclude that the digestive tract response may be essential to accommodating greater feed intake. As noted by Young (1986), however, this general response to temperature can vary with thermal susceptibility of the animal, acclimation, and diet. Behavioral responses to thermal stress (e.g., decreased grazing time) are restricted by some experimental conditions that could heighten the effects of thermal stress on feed intake. For example, acute cold stress decreased forage intake by as much as 47 percent in grazing cattle (Adams, 1987); however, for thermally adapted grazing cows, Beverlin et al. (1989) reported only small changes in forage intake with temperature deviations of 8° to −16° C. Similarly, feed intake by confined beef cattle fed finishing diets did not generally increase during cold stress and was often less during winter than during other seasons (Stanton, 1995).

Other adverse environmental conditions (wind, precipitation, mud, and so on) can accentuate the effects of ambient temperature. Fox et al. (1988) suggested multiplicative correction factors to adjust intake predictions for various environmental effects. Duration of adverse conditions seems important, and because effects caused by environmental conditions are variable, feed intake in a variable environment is difficult to predict (National Research Council, 1987). Regardless of the variable nature of its effects, thermal stress can markedly alter energetic efficiency of ruminants as evidenced by the effects of cold stress on energy utilization by beef cattle (Delfino and Mathison, 1991).

Seasonal or photoperiod (day length) effects on feed intake are understood less fully than are thermal effects, and photoperiod has been suggested as a potentially important factor influencing feed intake by beef cattle (National Research Council, 1987). Ingvartsen et al. (1992b) evaluated effects of day length on voluntary DMI capacity of Danish Black and White bulls, steers, and heifers. Voluntary DMI increased 0.32 percent per hour increase in day length; the range in the literature reviewed by the authors was −0.6 to 1.5 percent. Based on the deviation from the voluntary intake at 12 hours of daylight, voluntary intake would be expected to be 1.5 to 2 percent greater in long-day months (July in the northern hemisphere) and 1.5 to 2 percent less in short-day months (January). Hicks et al. (1990) grouped intake data into four seasons and thereby accounted for much of the seasonal pattern in feed intake. Nevertheless, temperature, photoperiod, animal, and perhaps management differences contribute to seasonal patterns, and separate effects are difficult to delineate.

Management and Dietary Factors Affecting Feed Intake

With grazing cattle, quantity of forage available can affect feed intake. The authors of the NRC (1987) report reviewed data summarized by Rayburn (1986) and concluded that grazed forage intake was maximized when forage availability was approximately 2,250 kg dry matter/ha or a forage allowance of 40 g organic matter/kg BW. Intake decreased rapidly to 60 percent of maximum when forage allowance was 20 g organic matter/kg BW (450 kg/ha; National Research Council, 1987). Minson (1990) noted that bite size decreased with forage mass of less than 2,000 kg dry matter/ha; this decrease was only partially compensated for by increased grazing time, resulting in decreased forage intake. The break point at which intake of grazed forage was decreased with decreasing forage allowance seemed to lie between 30 and 50 g dry matter/kg BW. Relationships may vary with forage type and sward structure. McCollum et al. (1992) evaluated effects of forage availability on cattle grazing annual winter wheat pasture and noted that peak intake of digestible organic matter was predicted at 1,247 kg dry matter/ha or an allowance of approximately 300 g dry matter/kg BW. The data base for determining the relationship between forage availability and forage intake is derived largely from studies with actively growing pastures. As noted by Minson (1990), gain by sheep is related more closely to green (growing) forage allowance than to total forage dry matter offered. Similarly, Bird et al. (1989) reported that body weight gain by grazing cattle could be modeled more effectively from green pasture mass than from total pasture mass. Selective grazing of growing forage may increase in pastures with both growing and senescent material. Cattle eat only small amounts of senescent forage when some growing forage is available (Minson, 1990). Hence, effects of forage availability on intake should be considered in light of pasture composition and the potential for selective grazing.

Growth-promoting implants tend to increase feed intake. In two trials with beef steers fed a 60 percent concentrate diet, administering an estradiol benzoate/progesterone implant increased DMI from 4 to 16 percent, depending on when the implant was administered relative to slaughter (Rumsey et al., 1992). Fox et al. (1988) suggested that predicted feed intake should be decreased 8 percent for nonimplanted cattle.

Monensin, the ionophore feed additive, typically decreases feed intake. Fox et al. (1988) suggested that feed intake decreases by 10 and 6 percent with 33 and 22 mg monensin/kg diet respectively. With beef steers fed a 90 percent concentrate diet, Galyean et al. (1992) noted a 4 percent decrease in feed intake when animals were fed 31 mg monensin/kg dietary dry matter. Lasalocid, another ionophore approved for use in beef cattle, seems to have limited effects on feed intake. Fox et al. (1988) suggested that feed intake is decreased 2 percent by lasalocid, regardless of dietary concentration. Malcolm et al. (1992) found that feed intake increased approximately 4 percent with 85 percent concentrate diets that contained 33 mg lasalocid/kg diet compared with a nonionophore, control diet. Fewer data are available regarding effects of laidlomycin propionate, an ionophore approved for confined growing and finishing cattle, on feed intake. However, a summary of available data (Vogel, 1995) suggests that laidlomycin propionate has minimal effect on feed intake.

A dietary nutrient deficiency, particularly protein, can decrease feed intake. With low-nitrogen, high-fiber forage, nitrogen deficiency is common, and provision of supplemental nitrogen often increases DMI substantially (Galyean and Goetsch, 1993). Forage intake responses to protein are most typical when forage crude protein content is less than 6 to 8 percent (National Research Council, 1987). Supplementing forages with grain-based concentrates often decreases forage intake, such effects typically being greater with high- than with low-quality forages (Galyean and Goetsch, 1993).

Grinding feeds can affect intake, but effects depend on the type of feed. With forages, fine grinding can increase intake, presumably through effects on digesta passage (Galyean and Goetsch, 1993). With concentrates, fine grinding often decreases feed intake. Adjustments to intake predictions for finely processed diets as a function of dietary NE_m concentration have been suggested (National Research Council, 1987). Fermentation of feeds by ensiling generally has little effect on DMI unless the silage is unusually wet or dry and undesirable fermentation has occurred (National Research Council, 1987). Intake of wilted grass silages is usually greater than that of direct-cut silage, but reasons for the decrease with direct-cut silages are not fully understood (Minson, 1990).

PREDICTION OF FEED INTAKE BY BEEF CATTLE

The approach used to develop prediction equations for feed intake involved reevaluating relationships suggested in the previous edition of *Nutrient Requirements of Beef Cattle* (National Research Council, 1984). Equations presented in the previous edition have been used extensively in practice; however, description of the data base used and statistical validation of the equations were inadequate. Hence, efforts will be made to fully describe the approach used to develop prediction equations for growing and finishing cattle and beef cows. No attempt was made to develop prediction equations for intake by nursing calves; readers are referred to *Predicting Feed Intake for Food-Producing Animals* (National Research Council, 1987) for a proposed equation. It also should be noted that the focus of prediction in each case was average DMI over an extended feeding period. Although prediction of feed intake for shorter periods is highly desirable, no data base exists from which to develop such prediction equations for the wide variety of production situations and feeds available to beef cattle producers.

Growing and Finishing Cattle: Dietary Energy Concentration

As noted previously, the *Nutrient Requirements of Beef Cattle* (National Research Council, 1984) provided an equation to predict DMI by growing and finishing beef cattle. This equation describes DMI as a function of dietary NE_m concentration, with adjustments for frame size or sex. The base NRC 1984 equation is

$$DMI = SBW^{0.75} * (0.1493 * NE_m - 0.046 * NE_m^2 - 0.0196)$$ Eq. 7-a

where DMI is expressed in kg/day, SBW is expressed in kg, and NE_m concentration is expressed as Mcal/kg dietary dry matter. Data from the published literature were used to reevaluate the relationship between dietary NE_m concentration and DMI by growing and finishing beef cattle (Figure 7-1).

Data were obtained from experiments conducted with growing and finishing beef cattle and published in the *Journal of Animal Science* from 1980 to 1992. Each of 185 data points extracted from the literature represented a treatment mean for average DMI throughout a feeding period. Feeding periods varied from 56 to 212 days. Approximately 48 percent of the cattle were implanted with a growth-promoting implant, and approximately 50 percent were fed an ionophore. Information on frame size (small, medium, or large), sex (steer, heifer, or bull), age (calf or yearling), and initial and final SBW was recorded. Because this data contained a mix of full and shrunk body weights, the subcommittee assumed SBW in developing these equations. Dietary NE_m concentration was calculated from tabular values (National Research Council, 1984); however, actually determined NE_m values were used, when available. Because of the limited number of observations, bulls were classed as large-frame steers and large-frame heifers were classed as medium-framed yearling heifers. Total NE_m intake was calculated as the product of DMI and dietary NE_m concentration. Total NE_m intake was then divided by average metabolic body weight (average $SBW^{0.75}$ in kg). The intake of NE_m per unit $SBW^{0.75}$ was analyzed by stepwise regression procedures (SAS Institute, Inc., 1987) with dietary NE_m concentration, NE_m^2, length of the feeding period, and dummy variables used to account for effects of sex and frame classes as possible independent selections.

The relationship between NE_m intake per unit $SBW^{0.75}$

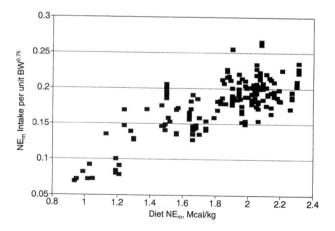

FIGURE 7-1   Relationship of dietary NE_m concentration to NE_m intake by beef cattle. Data points were obtained from published literature and represent treatment means for average intake during a feeding period.

and dietary NE_m concentration is shown in Figure 7-1. A regression equation that included NE_m, NE_m^2, and an intercept adjustment for yearling cattle accounted for 69.87 percent of the variation in NE_m intake per unit $SBW^{0.75}$. Expressed as total NE_m intake (Mcal/day), this equation is

$$NE_m \text{ intake} = SBW^{0.75} * (0.2435 * \qquad \text{Eq. 7-1}$$
$$NE_m - 0.0466 * NE_m^2 - 0.1128)$$

The intercept adjustment terms for medium-framed yearling steers and medium-framed yearling heifers differed slightly, but the standard errors of these adjustments overlapped. Hence, the mean value of the two intercept adjustments was used, resulting in one intercept term for both yearling steers and heifers of -0.0869 instead of -0.1128. DMI (kg/day) can be calculated from Eq. 7-1 by dividing total NE_m intake (Mcal/day) by dietary NE_m concentration (Mcal/kg).

DMI predicted from Eq. 7-1 and from Eq. 7-a were regressed on actual DMI for the 185 data points. The intake predicted from Eq. 7-a accounted for 62.35 percent of the variation in DMI, with a bias of -2.2 percent (under prediction). DMI predicted from Eq. 7-1 accounted for 72.85 percent of the variation in actual DMI, with a bias of -1.86 percent.

A comparison of the DMI predicted from the Eq. 7-a (with adjustments for frame size) and Eq. 7-1 is shown in Figure 7-2. In this example, DMI was predicted for a 410-kg average SBW, medium-frame steer (300 and 520 kg initial and final SBW, respectively) over a range in NE_m concentrations of 1 to 2.35 Mcal/kg. At low dietary NE_m concentrations, both equations yielded similar estimates of DMI. Eq. 7-1 predicted lesser intakes in the middle of the energy range and greater intakes at the upper end of the energy range than did Eq. 7-a.

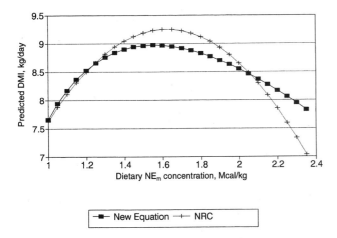

FIGURE 7-2 Comparison of dry matter intake predictions for a medium-frame (410-kg average BW) steer using Eq. 7-a (National Research Council, 1984) and Eq. 7-1, the equation developed from a literature data set.

As noted previously, Eq. 7-1 was developed to predict average DMI throughout a feeding period. Hence, the SBW term in the equation would be calculated as the average of initial and final SBW for a feeding period on a given diet. In practice, one would generally know the initial SBW and project the final SBW for the feeding period (e.g., estimated SBW at low-choice grade).

Because feed intake can vary greatly with environmental conditions, management factors, cattle type, and dietary factors, any equation should be viewed as providing a guideline rather than an absolute prediction of intake. Feedlot managers, nutritionists, and beef producers should combine such guidelines with their own data bases to develop more accurate predictions for specific situations. Hicks et al. (1990) reported that inclusion of feed intake data from the early portion of the feeding period (days 8 to 28) increased the coefficient of determination for prediction of mean DMI. Similarly, Oltjen and Owens (1987) used a statistical technique to adjust subsequent intake predictions for intake earlier in the feeding period. As noted by Hicks et al. (1990), it may be possible to use intake data obtained early in the feeding period of cattle to detect groups of cattle with particularly low or high feed intakes and thereby initiate appropriate management actions.

Growing and Finishing Cattle: Initial Weight on Feed

As discussed earlier, several factors other than dietary energy concentration can affect feed intake. Combined with data from cattle fed mostly high-energy diets, initial weight on feed seems to have predictive value (National Research Council, 1987). Hence, the relationship between initial body weight and DMI was evaluated in data obtained from the published literature. In addition, data from commercial feedlots were used to evaluate the relationship within a narrower range of dietary energy concentrations.

The data used in a preliminary analysis were the 185 data points described in the preceding section on dietary energy concentration. Dietary NE_m concentration ranged from slightly less than 1.0 to approximately 2.4 Mcal/kg. DMI (kg/day) was analyzed by stepwise regression procedures (SAS, Institute, Inc., 1987) with initial BW and dummy variables used to adjust the intercept and slope for effects of sex and frame classes as possible independent selections. The relationship between initial BW (kg) and DMI (kg/day) for the 185 data points taken from the literature is shown in Figure 7-3. Initial weight, with adjustments to the intercept for certain frame size/sex/age classes accounted for 59.78 percent of the variation in DMI. The equation is

$$DMI = 1.8545 + 0.01937 * iBW, \qquad \text{Eq. 7-2}$$

where iBW is initial BW in kg. For large-frame steer calves,

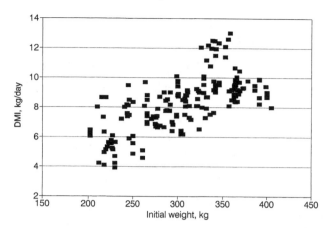

FIGURE 7-3 Relationship between initial BW and DMI for growing and finishing cattle. Data points were obtained from published literature and represent treatment means for average intake during a feeding period.

the intercept was 2.477, whereas for large-frame heifer calves and medium-frame yearling heifers, the intercept was 3.212. For medium-frame yearling steers, the intercept was 3.616. These equations were used to predict DMI, and predicted intake was regressed on actual intake for the 185 data points. Predicted intake accounted for 57.82 percent of the variation in actual DMI, with a bias of −2.1 percent (under prediction). As noted earlier, the intake equation in the preceding edition of this report (National Research Council, 1984) accounted for 62.35 percent of the variation in actual intake, with a bias of −2.2 percent.

The published data used to evaluate the relationship of initial BW to DMI covered a wide range in NE_m concentrations. In an effort to examine the relationship within a narrower range of energy concentrations that might be typical of beef feedlots, the subcommittee used initial weight and DMI data obtained from commercial feedlots. The first set of commercial data was collected from feedlots in Texas, Arizona, and California and included 929 pen means for DMI by crossbred steers and heifers. Average initial weight of cattle in this data set ranged from approximately 76 to 454 kg. Most cattle in this data set had some degree of Brahman breeding. The second data set included 732 pen means for DMI by crossbred steers collected from one feedlot in Kansas. Initial weight of cattle for this second data set ranged from to 201 to 528 kg. The degree of Brahman breeding was minimal in this second data set. Diets fed in both data sets were typical growing and finishing diets (NE_g ranged from approximately 1.1 to 1.59 Mcal/kg). Cattle in the first data set were typically on feed longer than those in the Kansas data set and, as a result, lower energy growing diets made up a greater proportion of the DMI than in the Kansas data set. For both commercial data sets, simple linear regression equations were developed with initial BW as an independent variable to predict

DMI. Results are shown in Table 7-1. For the first data set, which included both steer and heifer data, sex was not a significant factor, so the overall equation is presented.

The similarity of the relationship between initial weight and DMI in these two sets of commercial feedlot data are somewhat remarkable. The slope of both equations in Table 7-1 differs somewhat from the slope derived from the preliminary analysis of the literature data set, which might reflect the narrower range in dietary energy concentrations in the commercial data sets. For simplicity, the average values for the two equations shown in Table 7-1 can be used for a general prediction equation based on initial BW. Hence, Eq. 7-2 is revised as

$$DMI = 4.54 + 0.0125 * iBW, \qquad \text{Eq. 7-2}$$

where iBW is initial BW in kg. As with Eq. 7-1 described previously, it should be noted that Eq. 7-2 is designed to predict average feed intake throughout a feeding period.

These results suggest that initial BW when cattle are started on feed is related linearly to average DMI during a feeding period. This finding confirms previous research (National Research Council, 1987); thus feedlot managers and nutritionists should be able to use their own data bases to derive equations to predict DMI from initial BW. Other factors, like management, environment, and cattle type could be factored into such equations for individual production situations. Although Eq. 7-2 could be useful in practice, as noted previously, no single equation is likely to be effective in all production situations.

Validation of Prediction Equations

Three data sets were used as independent tests of Eq. 7-1, Eq. 7-2, and Eq. 7-a (with frame size adjustments). The first data set came from Cornell University (D. G. Fox, Cornell University, personal communication, 1995) (54 data points; average DMI by small-, medium-, and large-framed steers and heifers; NE_m [Mcal/kg] ranged from approximately 1.4 to 2.1; length of the feeding period was 100 days or longer). This data set was used to test the equations with diets in the middle-to-upper range of dietary NE_m concentrations. The second data set came from the University of Guelph, Ontario, Canada (J. G.

TABLE 7-1 Relationship Between Initial Weight on Feed and Dry Matter Intake by Beef Cattle in Two Sets of Commercial Feedlot Data

Data Set	Intercept	Slope	$S_{y \cdot x}$	r^2
A	4.4498	0.01081	0.6217	0.571
B	4.6346	0.01422	0.6048	0.452

NOTE: Data set A was collected from commercial feedlots in Texas, Arizona, and California, and included both steer and heifer data (n = 929). Data set B was collected from one feedlot in Kansas and included only steer data (n = 732).

Buchanan-Smith, personal communication, 1995) (38 data points; average DMI by medium- and large-frame steers and heifers fed mostly alfalfa/grass silage-based diets; length of the feeding period ranged from 16 to 24 weeks; NE_m [Mcal/kg] ranged from 1.12 to 1.95). This second data set was used to evaluate the equations in the lower-to-middle range of dietary NE_m concentrations. The third data set was taken from a summary of intake and digestion data compiled at the University of Alberta (Mathison et al., 1986). This data set included 139 observations with beef cattle fed all-forage diets. Dietary NE_m concentrations were calculated from the reported ME values of the diet (range in NE_m of 0.69 to 1.71 Mcal/kg). Grasses, legumes, and grass-legume mixtures, as well as crop residues, were included in the data set.

For each of the data sets, DMI predicted from the three equations was regressed on actual DMI. The r^2, $S_{y \cdot x}$, and bias for each prediction equation are shown in Table 7-2. Bias was also calculated by fitting the model with a forced intercept of 0 and expressing the deviation of the slope from this model as a percentage change from an ideal value of 1.0. Initial SBW data were not available for the Alberta data set.

For the Cornell data set, Eq. 7-1 accounted for approximately the same percentage of variation in actual DMI as Eq. 7-a but had less overprediction bias (Table 7-2). The rather simple Eq. 7-2 accounted for approximately 55 percent of the variation in actual DMI but tended to underestimate intake. With the Guelph data set, Eq. 7-a (with adjustments for frame size) and Eq. 7-1 yielded similar results (Table 7-2). Once again, however, the over-prediction bias of the NRC 1984 equation, Eq. 7-a, was corrected by Eq. 7-1. If frame adjustments were not made to the NRC 1984 predictions, the r^2 was 80.1 percent with a bias of 0.1 percent. Hence, the tendency for overprediction noted in this data set with Eq. 7-a was most likely a function of

use of the frame size adjustments. Eq. 7-2 accounted for approximately 35 percent of the variation in DMI and, in contrast to results with the Cornell data set, tended to overpredict DMI. Both Eq. 7-a and Eq. 7-1 yielded similar results when applied to the Alberta data set, accounting for approximately 30 percent of the variation in actual DMI and underpredicting DMI by approximately 8 percent.

Results of these independent tests were in agreement with the comparison of Eq. 7-a and Eq. 7-1 shown in Figure 7-2. The Guelph and Alberta data sets represented a range in dietary NE_m concentrations for which both equations predict similar DMI, whereas the Cornell data set included NE_m concentrations in the range for which predictions from the two equations are most divergent. Further testing of Eq. 7-1 with independent data sets will be required to determine whether it is a superior predictive tool than Eq. 7-a. For the three independent data sets evaluated, Eq. 7-1 seemed to decrease slightly the overprediction bias of Eq. 7-a . Evaluation of these data sets affirms the validity of the intake prediction equation, but also raises questions about the value of the suggested frame-size adjustments, in the National Research Council (1984) report.

The failure of both Eq. 7-1 and Eq. 7-a to accurately predict DMI of beef cattle fed all-forage diets (Alberta data set) raises some concerns. Specific considerations for all-forage diets will be dealt with in a subsequent section.

Adjustments to Predictions

Fox et al. (1988, 1992) reported on various factors that can affect feed intake, factors that can be used to adjust feed intake predictions of Eqs. 7-1 and 7-2 and Eq. 7-a. Some caution should be applied in making these adjustments, however, because of the possibility of double accounting. For example, the data base used to derive equations to predict intake includes intake data from cattle under a variety of management systems and an array of environmental conditions. Hence, the equations derived from the data base developed by this subcommittee may reflect partial adjustments for many of the factors suggested by Fox et al. (1988, 1992).

Three specific adjustments need to be addressed. First, as noted previously, approximately 50 percent of the 185 data points used to develop Eq. 7-1 represented cases in which cattle were fed an ionophore. Statistical evaluation of these data, however, suggested no basis for adjustments to intake predictions as a result of ionophore use. Nonetheless, based on field experience, this subcommittee believes considerable evidence suggests that monensin will typically decrease feed intake, whereas lasalocid and laidlomycin propionate have little effect on feed intake. As a result, the subcommittee suggests that predicted DMI be decreased by 4 percent if monensin is fed at concentrations

TABLE 7-2 Results of Regressing Predicted Dry Matter Intake on Actual Dry Matter Intake by Growing and Finishing Beef Cattle for Three Validation Data Sets

Data Set[a]	Equation[b]	Observations, n	r^2	$S_{y \cdot x}$	Bias, %[c]
Cornell	7-1	54	0.7647	0.3431	+0.16
	7-2	54	0.5481	0.3559	−6.49
	7-a (NRC, 1984)	54	0.7624	0.5498	+5.88
Guelph	7-1	38	0.7930	0.3731	−0.49
	7-2	38	0.3529	0.3330	+4.54
	7-a (NRC, 1984)	38	0.7827	0.5581	+8.34
Alberta	7-1	139	0.3078	0.7144	−8.40
	7-a (NRC, 1984)	139	0.3102	0.7028	−7.90

[a] See text for description of the data sets.

[b] Eq. 7-1 = NE_m intake = $BW^{0.75} * (0.02435 * NE_m - 0.0466 * NE_m^2 - 0.1128)$; Eq. 7-2 = DMI = $4.54 + 0.0125 * iBW$; and Eq. 7-a = DMI = $BW^{0.75} * (0.1493 * NE_m - 0.046 * NE_m^2 - 0.0196)$.

[c] Bias was calculated as the percentage deviation of the slope from a theoretical value of 1.0 when the predicted DMI was regressed on actual DMI with a zero-intercept model.

of 27.5 to 33 mg/kg dietary dry matter and that predicted DMI not be adjusted when lasalocid or laidlomycin propionate are added to the diet.

The second case relates to adjustments for use or nonuse of growth-promoting implants. As with ionophores, statistical evaluation of the 185 data points indicated no basis for adjustments to predicted DMI if growth-promoting implants were used. On the other hand, considerable research and field evidence suggest that such implants increase feed intake. Hence, the subcommittee suggests that values suggested by Fox et al. (1992) be used as a guideline for adjustments to predicted DMI in cases where implants are not used (e.g., 6 percent decrease in predicted DMI when implants are not used).

The third case deals with effects of forage allowance. Data presented in *Predicting Feed Intake of Food-Producing Animals* (National Research Council, 1987) relative to forage availability were reevaluated by Rayburn (1992). He constructed a quadratic regression of relative DMI on available forage mass. The resulting regression equation was

$$\text{percent of relative DMI} = 0.17 * FM - 0.000074 * FM^2 + 2.4,$$

where FM is available forage mass $\leq 1,150$ kg/ha. The FM value of 1,150 kg/ha represents the maxima of the quadratic equation (first derivative), and relative DMI is assumed to be 100 percent for FM greater than this maxima. For application to grazing situations, the subcommittee suggests that this relationship be used in two steps. First, the daily forage allowance (FA) should be determined

$$FA = (FM * 1,000 * \text{grazing unit})/ (SBW * \text{days of grazing}),$$

where grazing unit is the pasture size in hectares and SBW is in kg. If FM is $\geq 1,150$ kg/ha, or FA is four times the predicted DMI (expressed as g/kg SBW), no adjustment should be made to the predicted DMI. If neither of these conditions is true, relative DMI should be calculated from the equation shown above, and the predicted DMI should be multiplied by the relative DMI (expressed as a decimal) to adjust predicted DMI for the effects of limited FM. This adjustment procedure should be applied to all types of grazing systems; however, rotational or other intensive grazing systems with heavy stocking rates will result in more rapid changes in FM than continuous systems with lower stocking rates. This necessitates careful attention to FM in intensive grazing systems and frequent reevaluation of relative DMI.

BEEF COWS: DIETARY ENERGY CONCENTRATION

The preceding edition of *Nutrient Requirements of Beef Cattle* (National Research Council, 1984) includes an equation for feed intake by breeding beef females similar in form to an equation for growing and finishing beef cattle; DMI is described as a function of $SBW^{0.75}$, and linear and quadratic effects of dietary NE_m concentration (DMI, kg/day = $SBW^{0.75} * [0.1462 * NE_m - 0.0517 * NE_m^2 - 0.0074]$). As with the growing and finishing equation, the description of how this equation was developed was inadequate in that publication. *Predicting Feed Intake of Food-Producing Animals* (National Research Council, 1987) provides an alternative equation for beef cows that described DMI as a linear function of dietary NE_m concentration:

$$DMI = SBW^{0.75} * (0.0194 + 0.0545 * NE_m). \quad \text{Eq. 7-b}$$

To further evaluate the relationship between dietary NE_m concentration and intake by beef cows, an approach similar to that described previously for growing and finishing cattle was used. Treatment means for DMI were obtained from a variety of sources. Data were obtained from articles in the *Journal of Animal Science* (1979 through 1993), unpublished theses, and unpublished data that were solicited from individual scientists. The 153 data points used in the analysis represented treatment or breed × year means for DMI by nonpregnant beef cows or by cows during the middle and last one-third of pregnancy. As with growing and finishing beef cattle data, the beef cow data base contained a mix of full and shrunk body weights; the subcommittee assumed SBW in developing these equations. The data base was not sufficiently detailed to allow incorporation of information about body condition scores or frame sizes of the cows; and for some data points, only information on dietary NE_m concentration and DMI per unit $SBW^{0.75}$ was available. Dietary NE_m concentration (range = 0.76 to 2.08 Mcal/kg) was taken from the data source or calculated based on tabular values (National Research Council, 1984) for feeds. Total NE_m intake was calculated as the product of DMI and dietary NE_m concentration and expressed per unit $SBW^{0.75}$ (average $SBW^{0.75}$ during the intake measurement period). Data were then subjected to stepwise regression analysis (SAS Institute, Inc., 1987), with dummy variables included to account for the specific physiological stage of the cow.

It should be noted that data points were not included in the regression analysis when an obvious nutrient deficiency existed. This exclusion primarily impacted data points from beef cows fed low-quality forages that were deficient in crude protein. In such cases, only data from protein-supplemented cows were included in the data set. Hence, the resulting equation would not be applicable when the user wants to predict intake of a protein-deficient forage. Alternatively, the resulting equation would be applicable when the user wants to estimate total intake (e.g., forage plus supplement).

The relationship between dietary NE_m concentration

and NE_m intake is depicted in Figure 7-4. In contrast to the quadratic relationship noted for growing and finishing beef cattle (Figure 7-1), intake of NE_m by beef cows was relatively linear with dietary NE_m concentration. The regression equation that provided the best fit to the data included NE_m^2 and an intercept adjustment for nonpregnant cows. For pregnant cows,

$$NE_m \text{ intake} = SBW^{0.75} * (0.04997 * \qquad \text{Eq. 7-3}$$
$$NE_m^2 + 0.04631)$$

the intercept for nonpregnant cows = 0.03840. Eq. 7-3 accounted for 75.94 percent of the variation in NE_m intake. When Eq. 7-3 was used to predict DMI per unit $SBW^{0.75}$ for the 153 data points, the r^2 was 15.47 percent with a prediction bias of −2.2 percent. The relatively low r^2 resulted from the fact that a large proportion of the data points for middle-to-late pregnancy (breed × year means obtained from Pfennig, 1992) were for cows fed diets with a narrow range in dietary NE_m concentration (approximately 1.15 to 1.4 Mcal/kg). Compared with Eq. 7-3, Eq. 7-a for breeding females accounted for only 0.99 percent of the variation in actual DMI with a bias of −10 percent. Intake predicted from the NRC 1987 equation, Eq. 7-b, for breeding females accounted for 12.06 percent of the variation in actual DMI with a bias of −10.3 percent. The greater similarity in predictions between Eq. 7-3 and Eq. 7-b vs Eq. 7-a may reflect the fact that data points used to construct Eq. 7-b were included in the data set used to derive Eq. 7-3. Overall, these results seem to indicate that Eq. 7-3 provided a superior fit to these data than either Eq. 7-a or Eq. 7-b.

Predicted DMI by a 500-kg cow fed diets with varying NE_m concentration for Eq. 7-3, Eq. 7-a, and Eq. 7-b is

shown in Figure 7-5. Compared with Eq. 7-b, Eq. 7-3 predicted greater intakes at lower NE_m concentrations and lesser intakes at higher NE_m concentrations.

As with Eq. 7-1 for growing and finishing beef cattle, DMI is calculated from Eq. 7-3 by dividing the predicted NE_m intake by dietary NE_m concentration. Because of the mathematical form of this equation, predicted DMI will increase substantially for NE_m values less than approximately 0.95 Mcal/kg. This increase in predicted DMI results from division by a fraction and is not biologically realistic. Based on results that will be described in a subsequent section on all-forage diets, the subcommittee recommends that for feeds with NE_m concentrations less than 1.0 Mcal/kg, the user apply Eqs. 7-a or 7-b for breeding females, or, with Eq. 7-3, use a constant value of 0.95 for the NE_m concentration of the diet. The subcommittee further suggests that adjustments to predicted intake for effects of ionophores, implants, available forage mass, and other adjustments suggested by Fox et al. (1992) for growing and finishing cattle also be applied to intake predictions for beef cows.

VALIDATION OF THE BEEF COW EQUATION

Beef cows are not typically given ad libitum access to feed in production situations. As a result, obtaining data for both development and validation of intake prediction equations is difficult. In contrast to the equations derived for growing and finishing beef cattle, only one fully independent data set was available for validation of the beef cow equation. This data set, supplied by R. H. Pritchard (South Dakota State University, personal communication, 1995) included 36 pen observations of DMI by nonpregnant beef cows fed a high-concentrate diet (NE_m = 2.06 Mcal/kg). Cows were either implanted (Finaplix-H) or not

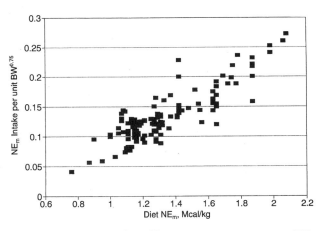

FIGURE 7-4   Relationship of dietary NE_m concentration to NE_m intake by beef cows (nonpregnant, and middle and last third of pregnancy). Data points were obtained from published and unpublished literature and represent treatment and breed × year means for average intake during a feeding period.

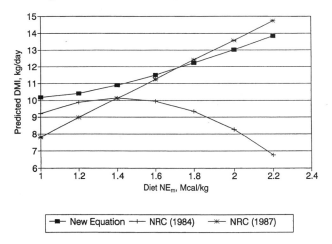

FIGURE 7-5   Comparison of dry matter intake (DMI) by a 500-kg, pregnant beef cow predicted from Eqs. 7-a (National Research Council, 1984) and 7-b (National Research Council, 1987) with that predicted from Eq. 7-3.

implanted and were classed in thin or nonthin body condition categories. Monensin was fed at 29.5 mg/kg diet. Cows were serially slaughtered, such that length of the feeding period ranged from 44 to 100 days.

Eq. 7-3 for nonpregnant beef cows, and Eqs. 7-a and 7-b for breeding females, were used to predict DMI. Predicted DMI was then regressed on actual DMI for all three equations. The r^2 values for Eq. 7-3 and Eqs. 7-a and 7-b were identical (36.25 percent), reflecting the fact that all cows were fed the same diet; only average SBW differed among observations. Bias was +7.57 percent for Eq. 7-3, −34.3 percent for Eq. 7-a, and +16.49 percent for Eq. 7-b. If predicted intakes for Eq. 7-3 were decreased by 4 percent for feeding monensin, the bias was +3.26 percent. Hence, in this particular set of validation data, Eq. 7-a for breeding females grossly underpredicted DMI, whereas Eq. 7-b overpredicted DMI, a situation that was partially corrected for use of Eq. 7-3. Further testing of these equations with independent data sets is desirable and is required to determine their relative predictive value.

SPECIAL CONSIDERATIONS FOR ALL-FORAGE DIETS

As noted previously, validation tests of Eq. 7-1 and Eq. 7-a, intake prediction equations for growing and finishing beef cattle, indicated that neither equation yielded accurate predictions of DMI by beef cattle fed all-forage diets in the Alberta (Mathison et al., 1986) data set. Because forages constitute all or most of the diet in many production situations, an accurate prediction equation for all-forage diets is critical to practical application of nutrient requirement data. Consequently, the Alberta data set was used to determine whether a specific equation for all-forage diets could be developed that would provide more accurate predictions of DMI by growing and finishing cattle and beef cows than either Eq. 7-1, Eq. 7-3, or Eqs. 7-a and 7-b. The Alberta data set consisted of 139 observations of ad libitum DMI by beef cattle consuming forages in three classes: grasses (65), legumes (39), and grass/legume mixtures (35). After first determining that SBW$^{0.75}$ accounted for significant (P <0.0001) variation in DMI, DMI expressed as kg/kg of SBW$^{0.75}$ was evaluated by stepwise regression with dietary crude protein (CP), neutral detergent fiber (NDF), and acid detergent fiber (ADF) concentrations, dietary ME concentration, and intercept adjustment terms for forage class. The resulting best-fit equation included terms for CP and ADF and an intercept adjustment term for grass/legume mixtures. Because the intercept adjustment term resulted in only a slight increase in r^2 and a slight decrease in S$_{y\cdot x}$, this term was deleted from the model, yielding the following equation (hereafter referred to as **CP__ADF**):

$$DMI \text{ kg/kg SBW}^{0.75} = 0.002774*CP$$
$$\text{percent in the forage} − 0.000864*ADF \quad \text{Eq. CP__ADF}$$
$$\text{percent in the forage} + 0.09826,$$

where percentages of CP and ADF in the forage are expressed on a DMI basis. The r^2 for this equation was 39.31 percent with an S$_{y\cdot x}$ of 0.0149. When this equation was used to predict DMI (kg/day) in the Alberta data set, and predicted intake was regressed on actual intake for the 139 data points, r^2 was 57.13 percent, S$_{y\cdot x}$ was 0.7501, and the bias was −2.93 percent.

The predicted vs actual DMI values for Eq. CP__ADF derived from the Alberta data set represent a considerable improvement relative to Eq. 7-a and Eq. 7-1 for this data set (see Table 7-2). Nonetheless, conclusions relative to the merit of various equations must be based on independent validation tests. Hence, two validation data sets were used to compare predicted vs actual intake among equations. The first data set was obtained from experiments with grazing beef steers and heifers. Data from Funk et al. (1987), Krysl et al. (1987), Pordomingo et al. (1991), Gunter (1993), and Gunter et al. (1993) were compiled and used to test Eq. CP__ADF developed from the Alberta data set, Eq. 7-1, and Eq. 7-a. Cattle in these experiments freely grazed native rangelands, and intake of organic matter was determined by marker-based methods. Dietary in vitro organic matter disappearance (IVOMD) was used to calculate dietary NE$_m$ concentration, assuming that IVOMD was equal to total digestible nutrient TDN and that 1 kg TDN was equal to 3.62 Mcal ME (National Research Council, 1984). Calculated NE$_m$ values for this data set ranged from 0.88 to 1.74 Mcal/kg.

The second data set was derived from the experiment of Vona et al. (1984), in which beef cows were fed warm-season grass hays. Dietary NE$_m$ concentration was calculated from in vivo DM digestibility in the same manner as described for IVOMD in the first data set. One data point with an extremely low DM digestibility (30.3 percent) relative to DMI was deleted from this data set. Calculated NE$_m$ concentrations for this data set ranged from 0.76 to 1.78 Mcal/kg. Because two of the predicted NE$_m$ values were less than 1.0 Mcal/kg, DMI was calculated from Eq. 7-3 by using a constant divisor of 0.95 for those two data points. Eqs. CP__ADF, 7-a, and 7-b for breeding females also were used to predict DMI for this data set.

Results for the two validation data sets are shown in Table 7-3. For the grazing steer and heifer data, Eq. CP__ADF accounted for less variation than either Eq. 7-a or Eq. 7-1. All three equations underpredicted DMI; however, underprediction bias was lowest for Eq. 7-1 and higher for Eq. CP__ADF. Standard errors of prediction were least for Eq. 7-a and Eq. CP__ADF and greatest for Eq. 7-1. For the beef cow data of Vona et al. (1984), Eq. 7-a accounted for considerably less variation in DMI than the other equations. Prediction bias was greatest for Eq. CP__ADF, and similar among Eqs. 7-a and 7-b and Eq.

TABLE 7-3 Results of Regressing Predicted Dry Matter Intake on Actual Dry Matter Intake for Two Validation Data Sets with Growing Beef Steers and Heifers and Beef Cows Fed All-Forage Diets

Data Set[a]	Equation	Obser-vations, n	r^2	$S_{y\cdot x}$	Bias, %[b]
Steers/heifers	CP_ADF[c]	38	0.4750	1.384	−9.71
	7-1[d]	38	0.5997	1.673	−0.93
	7-a (NRC, 1984)[e]	38	0.6800	1.322	−5.49
Cows	CP_ADF[c]	34	0.4357	0.6721	−10.83
	7-3[f]	34	0.5049	0.8005	+2.12
	7-a (NRC, 1984)[e]	34	0.1101	0.6496	−3.65
	7-b (NRC, 1987)[g]	34	0.6203	1.0536	−0.75

[a]See text for description of the data sets.

[b]Bias was calculated as the percentage deviation of the slope from a theorectical value of 1.0 when the predicted DMI was regressed on actual DMI with a zero-intercept model.

[c]CP_ADF = Mathison et al. (1986).

[d]Eq. 7-1 = NE_m intake (Mcal/day) = $BW^{0.75} * (0.02435 * NE_m - 0.0466 * NE_m^2) - 0.1128$.

[e]Eq. 7-a = the equation for either growing and finishing beef cattle or breeding females (National Research Council, 1984).

[f]Eq. 7-3 and for forages with calculated NE_m values of less than 0.95 Mcal/kg, a constant value of .95 was used as the divisor to calulate DMI from NE_m intake predicted by Eq. 7-3.

[g]Eq. 7-b = the equation for breeding females (National Research Council, 1987).

7-3. Standard error of prediction was least for Eq. 7-a and greatest for Eq. 7-b.

Neither of these two validation data sets is optimal. For the steer and heifer data set, the use of marker-based estimates of intake, and organic matter intake and digestibility rather than DM-based values, no doubt introduced some bias. For the cow data set, some caution should be used in interpreting the validation tests because Eq. 7-b was actually derived from this data set, and all the observations from this data set were included in the 153 data points used to develop Eq. 7-3. Despite these caveats, the validation tests indicate that Eqs. 7-1 and 7-3 and Eq. 7-a for growing and finishing cattle and breeding females generally yield estimates of DMI that are similar to those predicted from an empirical equation based on CP and ADF concentrations of the forage. Perhaps the similarity in predictions from these different approaches reflects the fairly high correlation between dietary energy metabolizability and fiber (NDF) concentration (Mertens, 1994). Further research is needed to develop more accurate means of predicting intake by beef cattle fed all-forage diets; but until such equations or models are developed, this subcommittee concludes that reasonable estimates of DMI can be obtained from Eqs. 7-1 and 7-3, as well as Eqs. 7-a and 7-b.

REFERENCES

Adams, D. C. 1987. Influence of winter weather on range livestock. Pp. 23–29 in Proceedings of the Grazing Livestock Nutrition Conference. Laramie: University of Wyoming.

Agricultural Research Council. 1980. The Nutrient Requirements of Ruminant Livestock. Technical Review. Farnham Royal, U. K.: Commonwealth Agricultural Bureaux.

Beverlin, S. K., K. M. Havstad, E. L. Ayers, and M. K. Petersen. 1989. Forage intake responses to winter cold exposure of free-ranging beef cows. Appl. Anim. Behav. Sci. 23:75–85.

Bird, P. R., M. J. Watson, and J. W. D. Cayley. 1989. Effect of stocking rate, season and pasture characteristics on liveweight gain of beef steers grazing perennial pastures. Aust. J. Agric. Res. 40:1277–1291.

Delfino, J. G., and G. W. Mathison. 1991. Effects of cold environment and intake level on the energetic efficiency of feedlot steers. J. Anim. Sci. 69:4577–4587.

Fox, D. G., C. J. Sniffen, and J. D. O'Connor. 1988. Adjusting nutrient requirements of beef cattle for animal and environmental variations. J. Anim. Sci. 66:1475–1495.

Fox, D. G., C. J. Sniffen, J. D. O'Connor, J. B. Russell, and P. J. Van Soest. 1992. A net carbohydrate and protein system for evaluating cattle diets. III. Cattle requirements and diet adequacy. J. Anim. Sci. 70:3578–3596.

Funk, M. A., M. L. Galyean, M. E. Branine, and L. J. Krysl. 1987. Steers grazing blue grama rangeland throughout the growing season. I. Dietary composition, intake, digesta kinetics and ruminal fermentation. J. Anim. Sci. 65:1342–1353.

Galyean, M. L., and A. L. Goetsch. 1993. Utilization of forage fiber by ruminants. Pp. 33–71 in Forage Cell Wall Structure and Digestibility, H. G. Jung, D. R. Buxton, R. D. Hatfield, and J. Ralph, eds. Madison, Wis.: ASA-CSSA-SSSA.

Galyean, M. L., K. J. Malcolm, and G. C. Duff. 1992. Performance of feedlot steers fed diets containing laidlomycin propionate or monensin plus tylosin, and effects of laidlomycin propionate concentration on intake patterns and ruminal fermentation in beef steers during adaptation to a high-concentrate diet. J. Anim. Sci. 70:2950–2958.

Gunter, S. A. 1993. Nutrient Intake and Digestion by Cattle Grazing Midgrass Prairie Rangeland and Plains Bluestem Pasture. Ph.D. dissertation, Oklahoma State University, Stillwater, Oklahoma.

Gunter, S. A., F. T. McCollum III, R. L. Gillen, and L. J. Krysl. 1993. Forage intake and digestion by cattle grazing midgrass prairie rangeland or sideoats grama/sweetclover pasture. J. Anim. Sci. 71:3432–3441.

Hicks, R. B., F. N. Owens, D. R. Gill, J. W. Oltjen, and R. P. Lake. 1990. Dry matter intake by feedlot beef steers: Influence of initial weight, time on feed and season of the year received in yard. J. Anim. Sci. 68:254–265.

Ingvartsen, K. L., H. R. Andersen, and J. Foldager. 1992a. Effect of sex and pregnancy on feed intake capacity of growing cattle. Acta Agric. Scand. (Sect. A). 42:40–46.

Ingvartsen, K. L., H. R. Andersen, and J. Foldager. 1992b. Random variation in voluntary dry matter intake and the effect of day length on feed intake capacity in growing cattle. Acta Agric. Scand. (Sect. A). 42:121–126.

Kennedy, P. M., R. J. Christopherson, and L. P. Milligan. 1986. Digestive responses to cold. Pp. 285–306 in Control of Digestion and Metabolism in Ruminants, L. P. Milligan, W. L. Grovum, and A. Dobson, eds. Englewood Cliffs, N.J.: Prentice-Hall.

Ketelaars, J. J. M. H., and B. J. Tolkamp. 1992a. Toward a new theory of feed intake regulation in ruminants 1. Causes of differences in voluntary feed intake: Critique of current views. Livestock Prod. Sci. 30:269–296.

Ketelaars, J. J. M. H., and B. J. Tolkamp. 1992b. Toward a new theory of feed intake regulation in ruminants 3. Optimum feed intake: In search of a physiological background. Livestock Prod. Sci. 31:235–258.

Krysl, L. J., M. L. Galyean, M. B. Judkins, M. E. Branine, and R. E. Estell. 1987. Digestive physiology of steers grazing fertilized and nonfertilized blue grama rangeland. J. Range Manage. 40:493–501.

Malcolm, K. J., M. E. Branine, and M. L. Galyean. 1992. Effects of ionophore management programs on performance of feedlot cattle. Agri-Practice 13:7–16.

Mathison, G. W., L. P. Milligan, and R. D. Weisenburger. 1986. Ruminant feed evaluation unit: Nutritive value of some Alberta feeds for cattle and sheep. Pp. 55–57 in The 65th Annual Feeders' Day Report: Agriculture and Forestry Bulletin, Special Issue. Edmonton: University of Alberta, Canada.

McCollum, F. T., M. D. Cravey, S. A. Gunter, J. M. Mieres, P. B. Beck, R. San Julian, and G. W. Horn. 1992. Forage availability affects wheat forage intake by stocker cattle. Oklahoma Agric. Exp. Sta. MP-136: 312–318.

Mertens, D. R. 1994. Regulation of forage intake. Pp. 450–493 in Forage Quality, Evaluation, and Utilization, G. C. Fahey, ed. Madison, Wis.: ASA-CSSA-SSSA.

Minson, D. J. 1990. Forage in Ruminant Nutrition. San Diego: Academic.

Minton, J. E. 1986. Effects of heat stress on feed intake of beef cattle. Pp. 325–327 in Symposium Proceedings: Feed Intake by Beef Cattle, MP-121, F. N. Owens, ed. Stillwater, Okla.: Oklahoma Agricultural Experiment Station.

National Research Council. 1984. Nutrient Requirements of Beef Cattle, Sixth Rev. Ed. Washington, D.C.: National Academy Press.

National Research Council. 1987. Predicting Feed Intake of Food-Producing Animals. Washington, D.C.: National Academy Press.

Oltjen, J. W., and F. N. Owens. 1987. Beef cattle feed intake and growth: Empirical Bayes derivation of the Kalman filter applied to a nonlinear dynamic model. J. Anim. Sci. 65:1362–1370.

Pfennig, M. 1992. Factors influencing feed intake in beef cow-calf pairs. M. S. thesis, Purdue University, West Lafayette, Indiana.

Pordomingo, A. J., J. D. Wallace, A. S. Freeman, and M. L. Galyean. 1991. Supplemental corn grain for steers grazing native rangeland during summer. J. Anim. Sci. 69:1678–1687.

Rayburn, E. B. 1986. Quantitative aspects of pasture management. Seneca Trail RC&D Technical Manual. Franklinville, N.Y.: Seneca Trail RC&D.

Rayburn, E. B. 1992. Modeling the effect of forage availability on the forage intake of grazing cattle. Paper presented at the American Society of Agronomy Northeastern Branch Meetings, University of Connecticut, Storrs, Conn., June 28–July 1, 1992.

Rumsey, T. S., A. C. Hammond, and J. P. McMurtry. 1992. Response to reimplanting beef steers with estradiol benzoate and progesterone: Performance, implant absorption pattern, and thyroxine status. J. Anim. Sci. 70:995–1001.

SAS Institute, Inc. 1987. Procedures Guide for Personal Computers, Version 6 Ed. Cary, N.C.: SAS Institute Inc.

Stanton, T. L. 1995. Damage control strategies for cattle exposed to cold stress. Pp. 289–298 in Symposium: Intake by Feedlot Cattle, P-942 F. N. Owens, ed. Stillwater: Oklahoma Agric. Exp. Sta.

Tolkamp, B. J., and J. J. M. H. Ketelaars. 1992. Toward a new theory of feed intake regulation in ruminants 2. Costs and benefits of feed consumption: An optimization approach. Livestock Prod. Sci. 30:297–317.

Vogel, G. 1995. The effect of ionophores on feed intake by feedlot cattle. Pp. 281–288 in Symposium: Intake by Feedlot Cattle, F. N. Owens, ed. Stillwater: Oklahoma Agricultural Experimental Station.

Vona, L. C., G. A. Jung, R. L. Reid, and W. C. Sharp. 1984. Nutritive value of warm-season grass hays for beef cattle and sheep: Digestiblity, intake and mineral utilization. J. Anim. Sci. 59:1582–1593.

Young, B. A. 1986. Food intake of cattle in cold climates. Pp. 328–340 in Symposium Proceedings: Feed Intake by Beef Cattle, MP-121, F. N. Owen, ed. Stillwater: Oklahoma Agricultural Experiment Station.

Young, B. A., B. Walker, A. E. Dixon, and V. A. Walker. 1989. Physiological adaptation to the environment. J. Anim. Sci. 67:2426–2432.

8 Implications of Stress

Stress in regard to beef cattle is defined as a nonspecific response of the body to any demand from the environment (Frazer et al., 1975; Selye, 1976). Stress can alter the steady state of the body and challenge physiological adaptive processes. Nutrition and stress are interactive and consequential in that stress can produce or aggravate nutritional deficiencies and nutritional deficiencies can produce a stress response. The major stresses observed in beef cattle are feed and water deprivation in the market system or during drought, weaning, crowding, and exposure to disease. Other stresses encountered by cattle are weather changes and castration, dehorning, vaccination, dipping, deworming, and other processing procedures. All these stresses can influence nutrient requirements of beef cattle; and because nutrition and stress interrelate as continuous processes, they should be considered as such. Management of stress in cattle has two major components: (1) management of the cause of stress and (2) management of the effects of stress—the quantified changes seen in animals.

One of the first stresses the animal encounters is weaning, a physical stress that is impossible to eliminate; however, preweaning and preconditioning management techniques have been used to decrease weaning stress (Cole, 1982). Though effective, these techniques often cannot be implemented because of cost and/or lack of adequate facilities.

During the marketing process, when the animal is deprived of feed and water, ruminal fermentation processes and capacity are significantly decreased and remain depressed for a few days after refeeding (Cole and Hutcheson, 1985a). Other changes include increased ruminal pH, serum osmolality, glucose, and urea nitrogen; however, once deprivation ceases, these variables return to predeprivation levels within 24 hours (Cole and Hutcheson, 1985b, 1987a). The number of ruminal protozoa and bacteria is lower in steers subjected to fasting and transit stress than in control animals (Galyean et al., 1980; Cole and Hutcheson,

1981), and the number increases more slowly when fasting occurs in conjunction with transit than when fasting is the only stressor. Baldwin (1967) suggests that the number of ruminal protozoa and bacteria decreases sharply following stress such as transportation. These ruminal changes tend to decrease appetite, thereby leading to decreased feed intake.

Feed intake decreases by more than 50 percent in cattle with respiratory disease and fever (Chirase et al., 1991). After the onset of bovine respiratory disease complex (BRDC), it takes as long as 10 to 14 days before feed intake returns to normal; consequently, nutrient demands for maintenance and growth are difficult to meet during periods of disease stress. The findings of a 7-year study of healthy and diseased calves newly arrived at feedlots (Hutcheson and Cole, 1986) are shown in Table 8-1.

ENERGY

Energy deficiency in cattle can severely depress the immune system (Nockles, 1988); however, excess dietary energy can also have detrimental effects. Calves newly arrived at a feedlot and fed a high-energy diet (75 percent concentrate) experienced increased performance, but inci-

TABLE 8-1 Dry Matter Feed Intake of Newly Arrived Calves (% of body weight)

Age, days	Healthy (SD)	Diseased (SD)
0–7	1.55 (0.51)	0.90 (0.75)
0–14	1.90 (0.50)	1.43 (0.70)
0–28	2.71 (0.50)	1.84 (0.66)
0–56	3.03 (0.43)	2.68 (0.68)

NOTE: SD, standard deviation.
SOURCE: Hutcheson, D. P., and N. A. Cole. 1986. Management of transit-stress syndrome in cattle: Nutritional and environmental effects. J. Anim. Sci. 62:555–560.

dence of disease was 57 percent compared with 47 percent when a 25 percent concentrate diet was used (Preston and Kunkle, 1974; Preston and Smith, 1974). Supplementing high-energy diets with hay for 3 to 7 days can overcome the adverse health effects of the high-energy diet (Lofgreen et al., 1981; Lofgreen, 1983, 1988). The source of grain type—corn, grain sorghum, barley or wheat—used in starter and receiving diets did not affect calf health or performance (Smith et al., 1988).

Grain type used in receiving diets did not affect calf health or performance. In fact, a better rate of gain was obtained with a mixture of grains (Brethour and Duitsman, 1972; Addis et al., 1975, 1978); however, highly stressed calves seem to have low tolerance for added fat, thus fat should probably not exceed 4 percent of dietary dry matter in receiving diets (Cole and Hutcheson, 1987b). Stressed calves prefer a dry diet compared to a diet high in corn silage, but they adapt to high amounts of corn silage in the diet after 7 to 14 days (Preston and Smith, 1973, 1974; Preston and Kunkle, 1974; Koers et al., 1975; Davis and Caley, 1977).

PROTEIN

Protein requirements of stressed calves do not seem to be different than those of nonstressed calves. Stressed calves, however, generally decrease their feed intake; therefore the concentration of protein in the diet should be increased for stressed or diseased calves (Cole and Hutcheson, 1990; Hutcheson et al., 1993). Protein concentrations of 13.5 to 14.5 percent on a dry matter basis in receiving diets meet the protein requirements of stressed calves (Embry, 1977; Bartle et al., 1988; Cole and Hutcheson, 1988; Eck et al., 1988; Cole and Hutcheson, 1990). Diseased calves exhibit a hypermetabolic response with increased excretion of nitrogen (Cole et al., 1986). The nitrogen kinetics of virus-infected calves are affected by shifts in the rates of protein metabolism (Orr et al., 1989). Figure 8-1 represents the differences in nitrogen (N) rate constants for infectious bovine rhinotracheitis virus calves. When fed increased protein, hyperurinary excretion of nitrogen during disease is partially alleviated (Boyles et al., 1989).

Stressed calves have a lower tolerance for nonprotein nitrogen (urea) than do nonstressed calves. Urea intakes of 30 gm/day or less seem to be tolerated by newly arrived or stressed calves during the first 2 weeks of feeding (Preston and Kunkle, 1974; Gates and Embry, 1975; Cole et al., 1984).

Feeding undigestible intake protein (UIP) to stressed calves resulted in increased performance (Preston and Kunkle, 1974; Preston and Smith, 1974; Grigsby, 1981; Phillips, 1984). UIP as 5.4 percent of dietary dry matter,

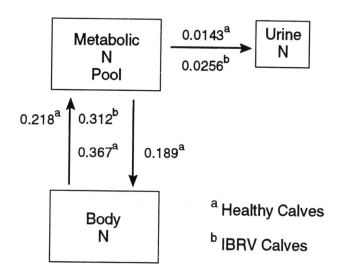

FIGURE 8-1 Changes in nitrogen (N) rate constants for calves with infectious bovine rhinotracheitis virus (IBRV). Infected calves fed an increased amount of protein experienced partial alleviation of hyperurinary excretion of nitrogen (Boyles et al., 1989).

at 45 percent of total protein, resulted in increased daily gains and dry matter intake (Preston and Bartle, 1990; Gunter et al., 1993; Hutcheson et al., 1993; Fluharty and Loerch, 1995).

MINERALS

Research indicates that, in general, mineral requirements for stressed cattle are not different than those for nonstressed cattle (Orr et al., 1990); however, decreased feed intake of stressed cattle suggests that higher concentrations of minerals should be formulated into their diets (Hutcheson, 1987, 1990). Cattle subjected to the stresses of marketing and shipping lose weight—primarily from loss of water from the digestive tract and, subsequently, from body cells. When intracellular water is lost, cellular deficiencies of potassium (K) and sodium (Na) can occur (Hutcheson, 1980). The potassium requirement of stressed calves is 20 percent more than that of nonstressed calves (Hutcheson et al., 1984). Data suggest that 1.2 to 1.4 percent potassium in the diet for 2 weeks is the optimum concentration for newly arrived, stressed calves. Additional potassium may not increase gain response if cattle shrink 2 to 4 percent; but with shrinkage of 7 or more percent, a significant effect may be observed with added potassium. Increasing dietary potassium allows the electrolyte and water balance to return to normal. When potassium is added as potassium chloride (KCl), however, care should be taken to limit salt (NaCl) to 0.25 percent of dietary dry matter so as not to increase chloride intake.

Many factors affect immune system response (Nockels, 1988; Hutcheson, 1990). On the other hand, during disease states trace mineral requirements may be affected by immune system response (Hutcheson, 1990). High concentrations of zinc have been shown to be beneficial to the animal's health during disease (Chirase et al., 1991), and zinc (Zn), copper (Cu), selenium (Se), and iron (Fe) seem to be necessary for immunocompetence (Chandra and Dayton, 1982; Brandt and Hutcheson, 1987; Drobe and Loerch, 1989; Erskine et al., 1989, 1990).

VITAMINS

Adding B vitamins to receiving rations for stressed calves increased their performance and feed intake in one (Overfield et al., 1976) but not all (Cole et al., 1979, 1982) experiments. Niacin added at 125 ppm seemed to increase average daily gain by healthy calves (Hutcheson and Cummins, 1984); however, diseased calves receiving niacin at 250 ppm seemed to have the best average daily gain. The most significant gains were observed when the cattle received 271 mg/cwt/day (Hutcheson and Cummins, 1984).

Vitamin E has been shown to be involved in immune system response; lymphocyte-stimulation indices were highest for calves fed 227.5 mg (250 IU) all-*rac*-α-tocopherol compared to controls (Cipriano et al., 1982). Increasing vitamin E intake during disease or infection produced varying results, but in general the data indicate that vitamin E is necessary for optimal functioning of the immune system. Vitamin E fed at 400 IU/day in receiving and starting diets of newly arrived feeder calves decreased disease and number of sick days and increased gain (Hicks, 1985).

Vitamin E fed at 450 IU/day to cattle that experienced more than 10 percent shrink increased gain (Lee et al., 1985). Vitamin E should be fed between 400 and 500 IU per head per day during the receiving and starting period. Calves receiving 125 mg/day (125 IU/day) of all-*rac*-α-tocopherol acetate consumed more than calves that did not receive additional vitamin E or 500 mg/day (500 IU/day) (Reddy et al., 1985).

Table 8-2 gives the suggested nutrient concentrations for receiving diets of stressed cattle. Many of the nutrients are based on the subcommittee's calculations; some are based on published data (Hutcheson, 1990). Decreased intake during disease stress is the single most common observation. Nutrient amounts recommended in Table 8-2 are for the first 2 weeks after arrival or until the cattle are consuming feed, on a dry matter basis, of 2 percent of body weight or more. Table 8-2 also gives nutrient amounts that would be consumed per day when suggested amounts are calculated: 1.55 percent of body weight, the average amount of feed consumed during the first week; and 1.90 percent of body weight, the average amount of feed consumed during the first 2 weeks—that is, the average of the 2 weeks.

REFERENCES

Addis, D. G., G. P. Lofgreen, J. G. Clark, J. R. Dunbar, and C. Adams. 1975. Barley vs milo in receiving rations. Calif. Feeders Day Rept. 14:53.

Addis, D. G., G. P. Lofgreen, J. G. Clark, C. Adams, F. Prigge, J. R. Dunbar, and B. Norman. 1978. Barley vs wheat in receiving rations for new calves. Calif. Feeders Day Rept. 17:54.

TABLE 8-2 Suggested Nutrient Concentrations for Stressed Calves (dry matter basis)

Nutrient	Unit	Suggested Range	Unit/day	Daily Nutrient Intake for 250-kg Calf[a] 0–7 days	0–14 days
Dry matter	%	80.0–85.0	kg	3.875	4.750
Crude protein	%	12.5–14.5	kg	0.48–0.56	0.59–0.69
Net energy of maintenance	Mcal/kg	1.3–1.6	Mcal	4.84	4.84
Net energy of gain	Mcal/kg	0.8–0.9	Mcal	0.01–0.8	0.6–1.6
Calcium	%	0.6–0.8	g	23.0–31.0	29.0–38.0
Phosphorus	%	0.4–0.5	g	16.0–19.0	19.0–24.0
Potassium	%	1.2–1.4	g	47.0–54.0	57.0–67.0
Magnesium	%	0.2–0.3	g	8.0–12.0	10.0–14.0
Sodium	%	0.2–0.3	g	8.0–12.0	10.0–14.0
Copper	~~ppm~~ *mg/kg*	10.0–15.0	mg	~~388.0–581.0~~ *39–58*	~~475.0–713.0~~ *47–71*
Iron	ppm "	100.0–200.0	mg	~~3,875.0–7,750.0~~ *388–775*	~~4,750.0–9,500.0~~ *475–950*
Manganese	ppm "	40.0–70.0	mg	~~1,550.0–2,713.0~~ *155–271*	~~1,900.0–3,325.0~~ *190–332*
Zinc	ppm "	75.0–100.0	mg	~~2,906.0–3,875.0~~ *290–387*	~~3,563.0–4,750.0~~ *356–475*
Cobalt	ppm "	0.1–0.2	mg	~~4.0–8.0~~ *.4–.8*	~~5.0–10.0~~ *.5–1.0*
Selenium	ppm "	0.1–0.2	mg	~~4.0–8.0~~ *.4–.8*	~~5.0–10.0~~ *.5–1.0*
Iodine	ppm "	0.3–0.6	mg	~~12.0–23.0~~ *1.2–2.3*	~~14.0–29.0~~ *1.4–2.9*
Vitamin A	IU/kg	4,000.0–6,000.0	IU	15,500.0–23,250.0	19,000.0–28,500.0
Vitamin E	IU/kg	~~400.0–500.0~~ *75–100*	IU	~~47.0–291.0~~ *–388,*	~~475.0–356.0~~ *–475*

[a]Intake levels are based on 1.55% for days 0 through 7 and 1.90% for days 0 through 14 from Table 8-1.

See "Errata" – front cover

Baldwin, R. L. 1967. Effect of starvation and refeeding upon rumen fermentation. California Feeders Day Rept. 7:7.

Bartle, S. J., K. J. Karr, J. G. Ross, D. L. Hancock, and R. L. Preston. 1988. Supplemental protein sources and dietary crude protein level for receiving feedlot cattle. Tex. Tech. Univ. Anim. Sci. Res. Rept. T-5-251:61.

Boyles, D. W., C. R. Richardson, C. S. Cobb, M. D. Miller, and N. A. Cole. 1989. Influence of protein status on the severity of the hypermetabolic response of calves with infectious bovine rhinotracheitis virus. Tex. Tech. Univ. Anim. Sci. Res. Rept. T-5-263:14–15.

Brandt, R. T., and D. P. Hutcheson. 1987. Trace minerals on the immune system. Kan. Formula Feed Conf. 42:12.

Brethour, J. R., and W. W. Duitsman. 1972. Wheat or dehydrated alfalfa in starting rations. Kan. St. Roundup Bull. 556:38.

Chandra, R. K., and D. H. Dayton. 1982. Trace element regulation of immunity and infection. Nutr. Res. 2:721–733.

Chirase, N. K., D. P. Hutcheson, and G. B. Thompson. 1991. Feed intake, rectal temperature, and serum mineral concentrations of feedlot cattle fed zinc oxide or zinc methionine and challenged with infectious bovine rhinotracheitis virus. J. Anim. Sci. 69:4137–4145.

Cipriano, J. E., J. L. Morrill, and N. V. Anderson. 1982. Effect of dietary vitamin E on immune responses of calves. J. Dairy Sci. 65:2357–2365.

Cole, N. A. 1982. Nutrition-Health interactions of newly arrived feeder cattle. Symp. Mgmt. Food Prod. Anim. II:683–701.

Cole, N. A., and D. P. Hutcheson. 1981. Influence on beef steers of two sequential short periods of feed and water deprivation. J. Anim Sci. 53:907–915.

Cole, N. A., and D. P. Hutcheson. 1985a. Influence of prefast feed intake on recovery from feed and water deprivation by beef steers. J. Anim. Sci. 60:772–780.

Cole, N. A., and D. P. Hutcheson. 1985b. Influence of realimentation diet on recovery of rumen activity and feed intake in beef steers. J. Anim. Sci. 61:692–701.

Cole, N. A., and D. P. Hutcheson. 1987a. Influence of pre-fast dietary roughage content on recovery from feed and water deprivation in beef steers. J. Anim. Sci. 65:1049–1057.

Cole, N. A., and D. P. Hutcheson. 1987b. Influence of receiving diet fat level on the health and performance of feeder calves. Nutr. Rep. Int. 36:965–970.

Cole, N. A., and D. P. Hutcheson. 1988. Influence of protein concentration in prefast and postfast diets on feed intake of steers and nitrogen and phosphorus metabolism of lambs. J. Anim. Sci. 66:1764–1777.

Cole, N. A., and D. P. Hutcheson. 1990. Influence of dietary protein concentrations on performance and nitrogen and repletion in stressed calves. J. Anim. Sci. 68:3488–3497.

Cole, N. A., M. R. Irwin, and J. B. Mclaren. 1979. Influence of pretransit feeding regimen and posttransit B-vitamin supplementation on stressed feeder steers. J. Anim. Sci. 49:310–317.

Cole, N. A., J. B. McLaren, and D. P. Hutcheson. 1982. Influence of preweaning and B-vitamin supplementation of the feedlot receiving diet on calves subjected to marketing and transit stress. J. Anim. Sci. 54:911–917.

Cole, N. A., D. P. Hutcheson, J. B. McLaren, and W. A. Phillips. 1984. Influence of pretransit zeranol implant and receiving diet protein and urea levels on performance of yearling steers. J. Anim. Sci. 58:527–534.

Cole, N. A., D. D. Delaney, J. M. Cummins, and D. P. Hutcheson. 1986. Nitrogen metabolism of calves inoculated with bovine adenovirus-3 or with infectious bovine rhinotracheitis virus. Am. J. Vet. Res. 47:1160–1164.

Davis, G. V. and H. K. Caley. 1977. Influence of management and rations on the performance of stress calves. Rep. Prog. Kans. Agric. Exp. Stn. Kans. State Coll. Agric. Appl. Sci. 288:16–29.

Drobe, E. A., and S. C. Loerch. 1989. Effects of parenteral selenium

and vitamin E on performance, health and humoral immune response of steers new to the feedlot environment. J. Anim. Sci. 67:1350–1359.

Eck, T. P., S. J. Bartle, R. L. Preston, R. T. Brandt, and C. R. Richardson. 1988. Protein source and level for incoming feedlot cattle. J. Anim. Sci. 66:1871–1876.

Embry, L. B. 1977. Feeding and management of new feedlot cattle. S. Dak. Cattle Feeders Day Rept. 1977:47.

Erskine, J. R., R. J. Eberhart, P. J. Grasso, and R. W. Scholz. 1989. Induction of *Escherichia coli* mastitis in cows fed selenium deficient or selenium-supplemented diets. Am. J. Vet. Res. 50:2093–2100.

Erskine, J. R., R. J. Eberhart, and R. W. Scholz. 1990. Experimentally induced *Staphylococcus aureus* mastitis in selenium-deficient and selenium-supplemented dairy cows. Am. J. Vet. Res. 51:1107–1111.

Fluharty, F. L., and S. C. Loerch. 1995. Effects of protein concentration and protein source on performance of newly arrived feedlot steers. Ohio Beef Cattle Research and Industry report 57–61.

Frazer, D., J. S. D. Ritchie, and A. F. Frazer. 1975. The term "stress" in a veterinary context. Br. Vet. J. 131:653–658.

Galyean, M. L., R. W. Lee, and M. E. Hubbert. 1980. Influence of fasting and transit stress on rumen fermentation in beef steers. New Mexico University Research 426.

Gates, R. N., and L. B. Embry. 1975. Soybean meal or urea during feedlot adaptation and growing calves. S. Dak. Cattle Feeders Day Rept. 7525:27.

Grigsby, M. E. 1981. Protein supplementation for stressed calves. New Mex. St. Univ. Clayton Res. Cen. Rept. 1981:25.

Gunter, S. A., M. L. Galyean, K. J. Malcolm-Callis, and D. R. Garcia. 1993. Effects of origin of cattle and supplemental protein source on performance of newly received feeder cattle. New Mex. St. Univ. Clayton Res. Cen. Rept. 1993:85.

Hicks, R. B. 1985. Effects of nutrition, medical treatments, and management practices of newly received stocker cattle. M.S. thesis. Oklahoma State University. Stillwater, Oklahoma.

Holdeman, L. V., I. J. Good, and W. E. C. Moore. 1976. Human fecal flora: variation in bacterial composition within individuals and possible effect of emotional stress. Appl. Environ. Microbiol. 31:359–367.

Hutcheson, D. P. 1980. The role of potassium in stress. International Minerals and Chemicals Seminar, Denver, Colorado.

Hutcheson, D. P. 1987. Minerals for feedlot cattle. Agri-Practice 8:3–6.

Hutcheson, D. P. 1990. Effect of nutrition on stress of ruminants. Liquid Feed Symposium Minneapolis, Mn. 1990:20.

Hutcheson, D. P., and N. A. Cole. 1986. Management of transit-stress syndrome in cattle: Nutritional and environmental effects. J. Anim. Sci. 62:555–560.

Hutcheson, D. P., and J. M. Cummins. 1984. The use of niacin in receiving diets for feeder calves. Pp. 120–121 in Proceedings of the Annual Meeting of the American Society of Animal Science, Western Section. Champaign, Ill.: American Society of Animal Science.

Hutcheson, D. P., N. A. Cole, and J. B. McLaren. 1984. Effects of pretransit diets and post-transit potassium levels for feeder calves. J. Anim. Sci. 58:700–707.

Hutcheson, J. P., T. L. Stanton, C. F. Nockels, and O. Robertson. 1993. The effect of three protein sources on feedlot performance of receiving calves. Colorado St. Univ. Beef Prg. Rept. 1993:67.

Koers, W. C., J. C. Parrott, L. B. Sherrod, R. H. Klett, and W. G. Sheldon. 1975. Receiving and sick pen rations for stressed calves. Tex. Tech. Res. Rept. 25:73.

Lee, R. W., R. L. Stuart, K. R. Perryman, and K. R. Ridenhour. 1985. Effect of vitamin supplementation on the performance of stress beef calves. J. Anim. Sci. 61(Suppl.1):425.

Lofgreen, G. P. 1983. Nutrition and management of stressed beef calves. Vet. Clin. N. Am. Large Anim. Pract. 5:87–101.

Lofgreen, G. P. 1988. Nutrition and management of stressed beef calves an update. Vet. Clin. N. Am. Food Anim. Pract. 4:509–522.

Lofgreen, G. P., L. H. Stinocher , and H. E. Kiesling. 1980. Effect of dietary energy, free choice alfalfa hay and mass medication on calves subjected to marketing and shipping stresses. J. Anim. Sci. 50:590–596.

Lofgreen, G. P., A. E. Eltayeb, and H. E. Kiesling. 1981. Millet and alfalfa hays alone and in combination with a high-energy diet for receiving stressed calves. J. Anim. Sci. 52:959–968.

National Research Council. 1984. Nutrient Requirements of Beef Cattle, Sixth Revised Ed. Washington, D.C.: National Academy Press.

Nockels, C. F. 1988. The role of vitamins in modulating disease resistance. Vet. Clin. N. Am. Food Anim. Pract. 4:531–542.

Orr, C., D. P. Hutcheson, J. M. Cummins, and G. B. Thompson. 1989. Nitrogen kinetics of infectious bovine rhinotracheitis stressed calves. J. Anim. Sci. 66:1982–1989.

Orr, C. L., D. P. Hutcheson, R. B. Grainger, J. M. Cummins, and R. E. Mock. 1990. Serum copper, zinc, calcium, and phosphorus concentrations of bovine respiratory disease and infectious bovine rhinotracheitis-stressed calves. J. Anim. Sci. 68(7).

Overfield, J. R., D. L. Hixon, and E. E. Hatfield. 1976. Effect of nutritional treatment of stressed feeder calves. Univ. Ill. Cattle Feeders Day Rept. AS–6721:28.

Phillips, W. A. 1984. The effect of protein source on the poststress performance of steer and heifer calves. Nutr. Rept. Int. 30:853–858.

Preston, R. L., and S. J. Bartle. 1990. Quantification of ruminal escape protein and amino acids for new feedlot cattle. Tex. Tech. Anim. Res. Rept. T-5-283:17.

Preston, R. L., and W. E. Kunkle. 1974. Role of roughage source in the receiving ration of yearling feeder steers. Res. Summ. Ohio Agr. Res. Dev. Cent. 77:51–53.

Preston, R. L., and C. K. Smith. 1973. Feedlot response of new feeder calves following a creep feeding period on the same or different protein source as that fed immediately after shipment. Res. Summ. Ohio Agr. Res. Dev. Cent. 68:29–31.

Preston, R. L., and C. K. Smith. 1974. Role of protein level, protected soybean protein, and roughage on the performance of new feeder calves. Res. Summ. Ohio Agr. Res. Dev. Cent. 77:47–50.

Preston, R. L., F. M. Byers, P. E. Moffitt, and C. E. Parker. 1975. Soybean meal and urea as sources of supplemental protein for newly received feeder calves. Ohio Agr. Res. Dev. Cent. Res. Rept. 1975:6.

Reddy, P. G., J. L. Morrill, R. A. Frey, M. B. Morrill, H. C. Minocha, S. J. Galitzer, and A. D. Dayton. 1985. Effects of supplemental vitamin E on the performance and metabolic profiles of dairy calves. J. Dairy Sci.68:2259–2270.

Seyle, H. 1976. The Stress of Life. New York: McGraw-Hill Book Company.

Smith, R. A., V. S. Hayes, and D. R. Gill. 1988. Management of stockers. Agric-Practice 9:8–14.

9 Tables of Nutrient Requirements

This seventh revised edition of *Nutrient Requirements of Beef Cattle* attempts to predict beef cattle requirements and performance under specific animal, environmental, and dietary conditions. Many variables (e.g., maintenance, growth, milk, microbial growth) are continuous and interact with the effects of feed composition. With this edition, the computer model described in Chapter 10 is provided on disk to calculate the effects of these variables. Because of all of the complex interactions accounted for in these models, the model tables differ from the tables of nutrient requirements in previous National Research Council (NRC) publications. Tables of nutrient requirements are, nevertheless, useful and instructive for some applications, so a computer program has been developed that uses model level 1 to compute and print nutrient requirement tables. This program allows determination of requirements for any body size and level of production of growing and finishing cattle, breeding bulls, bred heifers, and beef cows. No environmental stress is assumed. This chapter includes an example of each type of table for each of these classes of cattle, using the estimated U.S. average body size of finished steer and mature cow (533 kg). Simplified versions of these tables are provided at the end of the User's Guide to be used as guidelines.

Two types of tables can be computed and printed. The first type, daily nutrient requirements, computes a table of daily nutrient requirements for the body size and production level specified. The second type, diet evaluator, allows the user to determine the concentration of protein, calcium (Ca), and phosphorus (P) required in a diet under specific conditions. The diet evaluator computes energy allowable production for specified diets, balances for DIP, UIP, and MP, and Ca and P needed in the diet to support the diet energy allowable production. The CP requirement is determined by adjusting diet CP and DIP until DIP and UIP requirements are met.

In addition to determining nutrient density require-

ments, the diet evaluator allows the user to see how well a particular diet meets requirements of cattle in a feeding group with the range of weights specified for growing cattle or at each of the 12 months of the reproductive cycle for beef cows. In most beef production situations, cattle are fed in groups that vary in stage of growth or reproduction. Each group is usually fed to appetite either available forage (stocker, backgrounding, cow-calf) or high-energy based diets (growing and finishing cattle) and are provided supplements as needed to support the energy allowable production. The objective in diet formulation for high-forage diets is to determine supplemental energy, protein, and minerals needed to meet target levels of production. The objective in high-energy diets is typically to determine the protein and minerals needed to support the energy allowable ADG. In all situations, the user attempts to develop a "best fit" diet, considering the variation of animals in a feeding group.

To use the diet evaluator, the user enters the diet TDN, CP, and percent of CP that is DIP. Diet CP and CP degradability must be entered because the relationships between CP, DIP, and UIP vary, depending on diet and animal interactions. Diet NE_m, NE_g, DMI, ADG, or energy balance, DIP, UIP, and MP balances (g/day) are predicted for each of the diets over a range of body weights for the body size specified for growing cattle or for each month of the reproductive cycle for breeding cattle. Next, the predicted DMI and diet NE values can be modified with adjusters until DMI and animal production level agree with observed values. Diet concentration of CP and DIP can then be altered until the requirement for the observed energy allowable production is met. The DIP balance can be increased by increasing diet CP percentage and/or increasing DIP as a percentage of CP. The UIP balance can be improved by increasing percent of CP and/or reducing DIP as a percentage of the CP.

Diet TDN is used to predict diet NE_m and NE_g. Diet

NE$_m$ and NE$_g$ can only be changed by adjusting diet TDN because the relationship between these energy values must be kept consistent. Diet TDN is used to predict microbial growth, which must be consistent with the energy value used to predict NE$_m$ available to meet maintenance, pregnancy, and lactation requirements and the energy value used to predict NE$_g$ allowable ADG. To get the diet NE value desired, the user adjusts TDN until the desired NE value is predicted. The subcommittee recognizes that the relationship between TDN, ME, NE$_m$, and NE$_g$ may vary because of differences in amount of intake, rates of digestion and passage, and end products of digestion in the ME and their metabolizability. However, the relationships between them, as described in the preceding edition of this volume published in 1984, have also been used here for the reasons discussed in Chapters 1 and 10.

The concentration of nutrients needed for a given level of production depends on the actual DMI of the diet being fed to support the observed level of performance in a particular production setting. The DMI predictions are from equations developed from experimental feeding period averages as reported in published feeding trials involving wide variations in cattle type and stage of growth, as discussed in Chapter 7. Thus, predicted and observed values often differ in a specific production setting. Cattle fed feedlot finishing rations will typically consume 0 to 25 percent more early in the feeding period than predicted by these equations, which is compensated for by DMI of 0 to 25 percent less late in the feeding period. Further, as discussed in Chapter 7, concerning feed intake, most DMI prediction equations account for only 50 to 60 percent of the variation, leaving 40 to 50 percent to be accounted for by variations in local conditions such as feeding management, cattle type, and environment. The DMI adjusters allow the user to change the predicted DMI until it agrees with observed DMI; then the NE adjuster can be changed until predicted and observed performance agree.

Many factors can influence the NE derived from a diet for production, including variation in maintenance requirements, rates of digestion and passage, and metabolizability. If only DMI is adjusted, predicted and observed performance may not agree. For example, unrealistically high rates and efficiencies of gain may be predicted for calves consuming high-energy rations. Conversely, when these animals approach choice grade at the end of the finishing period, unrealistically low ADG may be predicted if only DMI is adjusted. Given these problems of prediction early and late in growth, limits were set on the weight ranges in the diet density tables at 55 percent of finished weight for the lightest weight and 80 percent of finished weight for the heaviest weight.

The primary use of these tables is intended to be for teaching the interactions of body size, stage of growth, diet energy density, and energy and protein requirements. The diet densities for CP and DIP may not be practical because the CP may have to be overfed to meet both DIP and UIP requirements. The user is encouraged to use the model with actual feed ingredients available for computing requirements for specific conditions. Despite their limitations as discussed in this section, simple guideline tables with diet nutrient concentration requirements for different classes of cattle are all that are needed in many situations and are provided at the end of the User's Guide.

EXAMPLE TABLES FOR GROWING AND FINISHING CATTLE

Tables 9-1 and 9-2 show daily requirements (Table 9-1) and diet evaluations (Table 9-2) for growing and finishing cattle. Inputs for Table 9-1 are for a 533-kg finished weight at 28 percent fat, a weight range of 200 to 450 kg, an ADG range of 0.50 to 2.50 kg, and breed code 1. Table 9-1 shows NE$_m$, NE$_g$, MP, Ca, and P required daily for maintenance and gain at six shrunk body weights, which represent six different stages of growth. All these requirements can be used directly to formulate dietary requirements for the specified level of performance, except the diet CP, DIP, and UIP required to meet the MP requirement. The CP intake needed can be estimated by dividing the total MP requirement in this table by 0.67, which is based on 80 percent of the MP from MCP and 20 percent from UIP. This approach was used in developing the guideline tables at the end of the User's Guide. However, this assumes that the nitrogen difference between the diet CP and MP requirement will meet microbial requirements for DIP and tissue requirements for UIP. This approach, which was used in the preceding edition of this volume to compute CP requirements, has major limitations. For this edition, the dietary CP intake needed is computed in the model level 1 as a sum of the DIP needed for microbial growth plus the UIP needed above the MP required for maintenance plus gain not met by microbial protein. These variables are not directly accounted for when the CP required is determined as MP/0.67.

Table 9-2 shows the evaluation of five diets (rations A through E) with the diet evaluator for the same animal used in Table 9-1 between 55 and 80 percent of final weight. The diet concentration of eNDF, TDN, and CP and DIP as a percentage of CP were entered for each of the five diets, and all DMI and NE adjusters were set at 100 percent. The eNDF values are used to adjust microbial protein yield and are affected only when diet eNDF drops below 20 percent of diet DM. The feed eNDF values in Appendix Table 1 (the feed library) can be used to determine eNDF in the diet. The program first computed diet NE$_m$ and NE$_g$ values, DMI, energy allowable ADG, MP, Ca, and P required for that ADG, MCP from the TDN

TABLE 9-1   Nutrient Requirements for Growing and Finishing Cattle

Wt @ Small marbling		533 kg					
Weight range		200-450 kg					
ADG range		0.50-2.50 kg					
Breed Code		1 Angus					

Body Weight, kg		200	250	300	350	400	450
Maintenance Requirements							
NE$_m$	Mcal/d	4.1	4.84	5.55	6.23	6.89	7.52
MP	g/d	202	239	274	307	340	371
Ca	g/d	6	8	9	11	12	14
P	g/d	5	6	7	8	10	11
Growth Requirements							
(ADG)				*NE$_g$ required for gain, Mcal/d*			
0.5	kg/d	1.27	1.50	1.72	1.93	2.14	2.33
1.0	kg/d	2.72	3.21	3.68	4.13	4.57	4.99
1.5	kg/d	4.24	5.01	5.74	6.45	7.13	7.79
2.0	kg/d	5.81	6.87	7.88	8.84	9.77	10.68
2.5	kg/d	7.42	8.78	10.06	11.29	12.48	13.64
				MP required for gain, g/d			
0.5	kg/d	154	155	158	157	145	133
1.0	kg/d	299	300	303	298	272	246
1.5	kg/d	441	440	442	432	391	352
2.0	kg/d	580	577	577	561	505	451
2.5	kg/d	718	712	710	687	616	547
				Calcium required for gain, g/d			
0.5	kg/d	14	13	12	11	10	9
1.0	kg/d	27	25	23	21	19	17
1.5	kg/d	39	36	33	30	27	25
2.0	kg/d	52	47	43	39	35	32
2.5	kg/d	64	59	53	48	43	38
				Phosphorus required for gain, g/d			
0.5	kg/d	6	5	5	4	4	4
1.0	kg/d	11	10	9	8	8	7
1.5	kg/d	16	15	13	12	11	10
2.0	kg/d	21	19	18	16	14	13
2.5	kg/d	26	24	22	19	17	15

intake, and DIP required for the MCP produced and UIP required with the equations presented in Chapter 10 for level 1.

All five diets were then balanced for UIP and DIP for the 300-kg body weight category by changing both CP and DIP until both UIP and DIP were balanced. The DIP is balanced for all other weights for each diet because MCP yield stays constant at 13 percent of TDN. The UIP would be deficient at lighter weights because the animal tissue requirement for protein at the energy allowable ADG exceeds the MCP and UIP provided by the diet. At weights less than 300 kg, the UIP deficiency would increase with the high-energy diets compared to low-energy diets because their lower eNDF results in a lower rumen pH, which reduces microbial growth as described in Chapter 2. This deficiency can be overcome by increasing the CP

and lowering the DIP, *but not to exceed that needed to balance DIP*, until the UIP requirement is met. In practical diets, this means substituting sources of DIP in the supplement with sources of UIP. At weights more than 300 kg, the diet UIP provided exceeds the MP required because of less protein in the ADG as the cattle increase in weight. The UIP excess can be decreased by lowering the CP while increasing the DIP as needed to keep the DIP balanced. The only practical way to accomplish this in the diet formula is to replace sources of UIP with sources of DIP until the CP and DIP reach a level provided by the grain and forage plus urea.

If actual data were available, predicted DMI would have been adjusted until it agreed with observed DMI, then the NE adjusters would have been used to adjust feed NE values until predicted and observed performance agree.

TABLE 9-2   Diet Evaluation for Growing and Finishing Cattle

| Wt @ Small Marbling | 533 kg | | | | | | | |
| Breed Code | 1 Angus | | | | | | | |

Ration	eNDF % DM	TDN % DM	NE$_m$ Mcal/kg	NE$_g$ Mcal/kg	CP % DM	DIP % CP	Weight Class	NE Adjuster
A	57	50	1.00	0.45	7.4	88	325	100%
B	43	60	1.35	0.77	10.0	78	350	100%
C	30	70	1.67	1.06	12.6	72.4	375	100%
D	5	80	1.99	1.33	14.4	48.5	400	100%
E	3	90	2.29	1.59	16.6	44.2	425	100%

Body Weight, kg	DMI Adjuster	DMI kg/d	ADG kg/d	DIP	UIP	MP	Ca	P
				--- balances, g/d ---			-- requirements, % of DM --	
300—A	100%	7.9	0.32	1	0	0	0.22%	0.13%
—B	100%	8.4	0.89	0	0	0	0.35%	0.18%
—C	100%	8.2	1.36	2	0	0	0.48%	0.24%
—D	100%	7.7	1.69	1	2	1	0.60%	0.29%
—E	100%	7.1	1.90	1	2	1	0.71%	0.34%
325—A	100%	8.4	0.32	1	14	11	0.21%	0.13%
—B	100%	8.9	0.89	0	38	30	0.33%	0.18%
—C	100%	8.7	1.36	2	57	46	0.45%	0.22%
—D	100%	8.2	1.69	1	73	58	0.55%	0.27%
—E	100%	7.6	1.90	1	82	66	0.65%	0.31%
350—A	100%	8.9	0.32	1	27	22	0.20%	0.13%
—B	100%	9.4	0.89	0	75	60	0.31%	0.17%
—C	100%	9.2	1.36	2	114	91	0.42%	0.21%
—D	100%	8.7	1.69	1	143	114	0.51%	0.25%
—E	100%	8.0	1.90	1	160	128	0.60%	0.29%
375—A	100%	9.4	0.32	1	40	32	0.20%	0.13%
—B	100%	9.9	0.89	0	111	89	0.30%	0.16%
—C	100%	9.7	1.36	2	169	135	0.39%	0.20%
—D	100%	9.1	1.69	1	212	169	0.48%	0.24%
—E	100%	8.4	1.90	1	238	190	0.56%	0.28%
400—A	100%	9.8	0.32	1	53	43	0.19%	0.12%
—B	100%	10.4	0.89	0	147	118	0.28%	0.16%
—C	100%	10.2	1.36	2	223	178	0.37%	0.19%
—D	100%	9.6	1.69	2	279	223	0.44%	0.23%
—E	100%	8.8	1.90	1	314	251	0.52%	0.26%
425—A	100%	10.3	0.32	1	66	53	0.19%	0.12%
—B	100%	10.9	0.89	0	182	146	0.27%	0.15%
—C	100%	10.6	1.36	2	276	221	0.35%	0.19%
—D	100%	10.0	1.69	2	346	277	0.42%	0.22%
—E	100%	9.3	1.90	1	388	311	0.48%	0.25%

TABLE 9-3   Nutrient Requirements for Growing Bulls

Wt @ Maturity	890 kg					
Weight Range	300–800 kg					
ADG Range	0.50–2.50 kg					
Breed Code	1 Angus					

Body Weight, kg		300	400	500	600	700	800
Maintenance Requirements							
NE$_m$	Mcal/day	6.38	7.92	9.36	10.73	12.05	13.32
MP	g/d	274	340	402	461	517	572
Ca	g/d	9	12	15	19	22	25
P	g/d	7	10	12	14	17	19
Growth Requirements							
ADG				*NE$_g$ Required for Gain, Mcal/d*			
0.5	kg/d	1.72	2.13	2.52	2.89	3.25	3.59
1.0	kg/d	3.68	4.56	5.39	6.18	6.94	7.67
1.5	kg/d	5.74	7.12	8.42	9.65	10.83	11.97
2.0	kg/d	7.87	9.76	11.54	13.23	14.85	16.41
2.5	kg/d	10.05	12.47	14.74	16.90	18.97	20.97
				MP Required for Gain, g/d			
0.5	kg/d	158	145	122	100	78	58
1.0	kg/d	303	272	222	175	130	86
1.5	kg/d	442	392	314	241	170	102
2.0	kg/d	577	506	400	299	202	109
2.5	kg/d	710	617	481	352	228	109
				Calcium Required for Gain, g/d			
0.5	kg/d	12	10	9	7	6	4
1.0	kg/d	23	19	16	12	9	6
1.5	kg/d	33	27	22	17	12	7
2.0	kg/d	43	35	28	21	14	8
2.5	kg/d	53	43	34	25	16	8
				Phosphorus Required for Gain, g/d			
0.5	kg/d	5	4	3	3	2	2
1.0	kg/d	9	8	6	5	4	2
1.5	kg/d	13	11	9	7	5	3
2.0	kg/d	18	14	11	8	6	3
2.5	kg/d	22	17	14	10	6	3

EXAMPLE TABLES FOR BREEDING BULLS

Tables 9-3 and 9-4 are example nutrient requirement (Table 9-3) and diet evaluation (Table 9-4) tables for growing bulls, using an 890-kg mature weight. Diet inputs for Table 9-4 were made as described for Table 9-2, with different diet TDN values. Weight ranges were set as 55 to 80 percent of the 28 percent fat weight of a steer of the same genotype (bull mature SBW * 0.6). (See Chapter 3 for the biological basis for computing bull requirements.) Diet CP, DIP, and UIP were balanced as described for Table 9-2 for 300 kg, except for diet A, for which upper bound of 80 percent DIP was used. The interpretations and applications are as described for Table 9-2.

TABLE 9-4 Diet Evaluation for Growing Bulls

Wt @ Maturity	890 kg
Breed Code	1 Angus

Ration	eNDF % DM	TDN % DM	NE$_m$ Mcal/kg	NE$_g$ Mcal/kg	CP % DM	DIP % CP	Weight Class	NE Adjuster
A	43	50	1.00	0.45	8.2	80	325	100%
B	37	65	1.51	0.92	10.9	78	350	100%
C	30	70	1.67	1.06	12.0	76	375	100%
D	20	75	1.83	1.20	13.4	73	400	100%
E	5	80	1.99	1.33	13.8	51	425	100%

Body Weight, kg	DMI Adjuster	DMI kg/d	ADG kg/d	DIP	UIP	MP	Ca	P
				--------------- balances, g/d ----------------			-- requirements, % of DM --	
300—A	100%	7.9	0.22	5	103	83	0.18%	0.12%
—B	100%	8.3	1.02	4	8	6	0.39%	0.20%
—C	100%	8.2	1.23	2	−3	−2	0.45%	0.23%
—D	100%	8.0	1.41	3	10	8	0.51%	0.25%
—E	100%	7.7	1.56	5	−2	−2	0.56%	0.27%
325—A	100%	8.4	0.22	5	119	95	0.18%	0.12%
—B	100%	8.8	1.02	5	51	41	0.36%	0.19%
—C	100%	8.7	1.23	2	49	39	0.42%	0.21%
—D	100%	8.5	1.41	3	70	56	0.47%	0.24%
—E	100%	8.2	1.56	6	63	51	0.52%	0.26%
350—A	100%	8.9	0.22	5	134	107	0.18%	0.12%
—B	100%	9.4	1.02	5	94	75	0.34%	0.18%
—C	100%	9.2	1.23	2	100	80	0.39%	0.20%
—D	100%	9.0	1.41	3	129	103	0.44%	0.22%
—E	100%	8.7	1.56	6	128	102	0.48%	0.24%
375—A	100%	9.4	0.22	6	149	119	0.18%	0.12%
—B	100%	9.8	1.02	5	136	109	0.32%	0.17%
—C	100%	9.7	1.23	2	150	125	0.37%	0.19%
—D	100%	9.4	1.41	3	187	149	0.41%	0.21%
—E	100%	9.1	1.56	6	191	153	0.45%	0.23%
400—A	100%	9.8	0.22	6	161	131	0.17%	0.12%
—B	100%	10.3	1.02	5	177	142	0.31%	0.17%
—C	100%	10.2	1.23	2	199	159	0.35%	0.19%
—D	100%	9.9	1.41	3	244	195	0.39%	0.20%
—E	100%	9.6	1.56	7	253	202	0.42%	0.22%
425—A	100%	10.3	0.22	6	169	143	0.17%	0.12%
—B	100%	10.8	1.02	6	218	174	0.29%	0.16%
—C	100%	10.6	1.23	2	247	198	0.33%	0.18%
—D	100%	10.4	1.41	3	300	240	0.36%	0.19%
—E	100%	10.0	1.56	7	314	251	0.40%	0.21%

TABLE 9-5   Nutrient Requirements of Pregnant Replacement Heifers

Mature Weight	533 kg
Calf Birth Weight	40 kg
Age @ Breeding	15 months
Breed Code	1 Angus

	Months since conception								
	1	2	3	4	5	6	7	8	9
NE$_m$ required, Mcal/d									
Maintenance	5.98	6.14	6.30	6.46	6.61	6.77	6.92	7.07	7.23
Growth	2.29	2.36	2.42	2.48	2.54	2.59	2.65	2.71	2.77
Pregnancy	0.03	0.07	0.16	0.32	0.64	1.18	2.08	3.44	5.37
Total	8.31	8.57	8.87	9.26	9.79	10.55	11.65	13.23	15.37
MP required, g/d									
Maintenance	295	303	311	319	326	334	342	349	357
Growth	118	119	119	119	119	117	115	113	110
Pregnancy	2	4	7	18	27	50	88	151	251
Total	415	425	437	457	472	501	545	613	718
Minerals									
Calcium required, g/d									
Maintenance	10	11	11	11	12	12	12	13	13
Growth	9	9	9	8	8	8	8	8	8
Pregnancy	0	0	0	0	0	0	12	12	12
Total	19	19	20	20	20	20	33	33	33
Phosphorus required, g/d									
Maintenance	8	8	8	9	9	9	10	10	10
Growth	4	4	3	3	3	3	3	3	3
Pregnancy	0	0	0	0	0	0	7	7	7
Total	12	12	12	12	12	13	20	20	20
ADG, kg/d									
Growth	0.39	0.39	0.39	0.39	0.39	0.39	0.39	0.39	0.39
Pregnancy	0.03	0.05	0.08	0.12	0.19	0.28	0.40	0.57	0.77
Total	0.42	0.44	0.47	0.51	0.58	0.67	0.79	0.96	1.16
Body weight, kg									
Shrunk body	332	343	355	367	379	391	403	415	426
Gravid uterus mass	1	3	4	7	12	19	29	44	64
Total	333	346	360	375	391	410	432	459	491

EXAMPLE TABLES FOR PREGNANT REPLACEMENT HEIFERS

Tables 9-5 and 9-6 contain requirements (Table 9-5) and diet evaluations (Table 9-6) for pregnant heifers. As with the preceding table sets, these two tables are related in that the animal described in the requirements table is then used in the diet evaluator. The program computes energy and protein balances expected for each of the three diets (rations A through C) entered as well as percent Ca and P needed in the diet DM to meet requirements. Animal descriptions entered were 533 kg mature weight, 40 kg expected birth weight, 15 month age at breeding, and breed code 1. Table 9-5 shows predicted NE$_m$, MP, Ca, and P required daily for maintenance, growth, and pregnancy and target ADG, SBW, and expected gravid uterus weight used to compute requirements for each of 9 months of gestation, using the equations presented in Chapter 10. As described previously, all can be used directly to formulate dietary requirements for the specified level of performance, except diet CP intake to meet the MP requirement, which can be computed as described for Table 9-2.

Table 9-6 shows diet evaluations for this same heifer.

The diet concentration of TDN and CP and DIP as a percentage of CP were entered for each of the three diets and the intake multiplier was set at 100 percent. All DIP values were then set at 80 percent, and diet CP was adjusted until DIP requirement was approximately met. Predicted DMI increased as pregnancy progressed because of increasing predicted SBW (shown in Table 9-5). As with the growing and finishing cattle, the DIP balance was constant over gestation for a given diet because microbial requirement is a constant proportion of TDN. However, the UIP balance changes with composition of the ADG (reduced protein content of ADG with increasing weight) and conceptus requirements. The CP, DIP, and UIP requirements are determined as described for growing and finishing cattle. Diet A (50 percent TDN) does not supply enough energy to support target heifer growth during any month. Diet B (60 percent TDN) exceeds target energy allowable ADG in all but the last month of pregnancy and exceeded UIP requirements for the energy allowable ADG in all but the first month. Diet C (70 percent TDN) exceeded target ADG in all months, but UIP was deficient for the energy allowable ADG in all but months 7 and 8.

TABLE 9-6 Diet Evaluation for Pregnant Replacement Heifers

Mature Weight	533 kg					
Calf Birth Weight	40 kg					
Age @ Breeding	15 months					
Breed Code	1 Angus					

Ration	TDN % DM	NE_m Mcal/kg	NE_g Mcal/kg	CP % DM	DIP % DM	DMI Factor
A	50	1.00	0.45	8.2	80	100%
B	60	1.35	0.77	9.8	80	100%
C	70	1.67	1.06	11.4	80	100%

		Months Since Conception								
		1	2	3	4	5	6	7	8	9
	NE_m Req. Factor	100%	100%	100%	100%	100%	100%	100%	100%	100%
A	DM, kg	8.5	8.8	9.0	9.2	9.4	9.7	9.9	10.1	10.3
	NE allowed ADG	0.35	0.34	0.33	0.31	0.28	0.22	0.12	0.00	0.00
	DIP Balance, g/d	5	5	5	6	6	6	6	6	6
	UIP Balance, g/d	75	79	83	87	90	92	90	66	−53
	MP Balance, g/d	60	63	67	69	72	74	72	52	−42
	Ca % DM	0.22%	0.21%	0.21%	0.20%	0.19%	0.18%	0.28%	0.25%	0.25%
	P % DM	0.17%	0.17%	0.16%	0.16%	0.15%	0.14%	0.19%	0.16%	0.16%
B	DM, kg	9.0	9.3	9.5	9.7	10.0	10.2	10.4	10.7	10.9
	NE allowed ADG	0.96	0.96	0.95	0.92	0.88	0.82	0.71	0.54	0.30
	DIP Balance, g/d	4	4	4	4	4	4	4	4	4
	UIP Balance, g/d	5	14	22	30	38	49	54	46	18
	MP Balance, g/d	4	11	18	24	31	40	43	37	14
	Ca % DM	0.36%	0.35%	0.33%	0.32%	0.31%	0.29%	0.38%	0.34%	0.29%
	P % DM	0.27%	0.27%	0.26%	0.26%	0.25%	0.23%	0.27%	0.24%	0.20%
C	DM, kg	8.8	9.1	9.3	9.5	9.8	10.0	10.2	10.4	10.7
	NE allowed ADG	1.47	1.46	1.45	1.42	1.38	1.31	1.19	1.02	0.77
	DIP Balance, g/d	2	2	2	2	2	2	2	2	2
	UIP Balance, g/d	−66	−54	−43	−32	−19	−1	10	8	−18
	MP Balance, g/d	−53	−43	−34	−26	−15	−1	8	6	−14
	Ca % DM	0.48%	0.47%	0.45%	0.43%	0.41%	0.39%	0.48%	0.43%	0.38%
	P % DM	0.37%	0.36%	0.35%	0.35%	0.33%	0.31%	0.35%	0.32%	0.28%

NOTE: Requirements are for NE allowed ADG and target weight. NE allowed ADG is ADG independent of conceptus gain.

Table 9-7   Nutrient Requirements of Beef Cows

Mature Weight	533 kg	Milk Fat	4.0 %			
Calf Birth Weight	40 kg	Milk Protein	3.4 %			
Age @ Calving	60 months	Calving Interval	12 months			
Age @ Weaning	30 weeks	Time Peak	8.5 weeks			
Peak Milk	8 kg	Milk SNF	8.3 %			
Breed Code	1 Angus					

	Month since Calving											
	1	2	3	4	5	6	7	8	9	10	11	12
NE$_m$ Req. Factor	100%	100%	100%	100%	100%	100%	100%	100%	100%	100%	100%	100%
NE$_m$ required, Mcal/d												
Maintenance	10.25	10.25	10.25	10.25	10.25	10.25	8.54	8.54	8.54	8.54	8.54	8.54
Growth	0.00	0.00	0.00	0.00	0.00	0.00	0.00	0.00	0.00	0.00	0.00	0.00
Lactation	4.78	5.74	5.17	4.13	3.10	2.23	0.00	0.00	0.00	0.00	0.00	0.00
Pregnancy	0.00	0.00	0.01	0.03	0.07	0.16	0.32	0.64	1.18	2.08	3.44	5.37
Total	15.03	15.99	15.43	14.41	13.42	12.64	8.87	9.18	9.72	10.62	11.98	13.91
MP required, g/d												
Maintenance	422	422	422	422	422	422	422	422	422	422	422	422
Growth	0	0	0	0	0	0	0	0	0	0	0	0
Lactation	349	418	376	301	226	163	0	0	0	0	0	0
Pregnancy	0	0	1	2	4	7	14	27	50	88	151	251
Total	770	840	799	724	651	591	436	449	471	510	573	672
Calcium required, g/d												
Maintenance	16	16	16	16	16	16	16	16	16	16	16	16
Growth	0	0	0	0	0	0	0	0	0	0	0	0
Lactation	16	20	18	14	11	8	0	0	0	0	0	0
Pregnancy	0	0	0	0	0	0	0	0	0	12	12	12
Total	33	36	34	31	27	24	16	16	16	29	29	29
Phosphorus required, g/d												
Maintenance	13	13	13	13	13	13	13	13	13	13	13	13
Growth	0	0	0	0	0	0	0	0	0	0	0	0
Lactation	9	11	10	8	6	4	0	0	0	0	0	0
Pregnancy	0	0	0	0	0	0	0	0	0	5	5	5
Total	22	24	23	21	19	17	13	13	13	18	18	18
ADG, kg/d												
Growth	0.00	0.00	0.00	0.00	0.00	0.00	0.00	0.00	0.00	0.00	0.00	0.00
Pregnancy	0.00	0.00	0.02	0.03	0.05	0.08	0.12	0.19	0.28	0.40	0.57	0.77
Total	0.00	0.00	0.02	0.03	0.05	0.08	0.12	0.19	0.28	0.40	0.57	0.77
Milk kg/d	6.7	8.0	7.2	5.8	4.3	3.1	0.0	0.0	0.0	0.0	0.0	0.0
Body weight, kg												
Shrunk Body	533	533	533	533	533	533	533	533	533	533	533	533
Conceptus	0	0	1	1	3	4	7	12	19	29	44	64
Total	533	533	534	534	536	537	540	545	552	562	577	597

EXAMPLE TABLES FOR BEEF COWS

Tables 9-7 and 9-8 contain requirements (Table 9-7) and diet evaluations (Table 9-8) for beef cows. As with the bred heifers, these two tables are related; the animal described in the requirements table is used in the diet evaluator. It computes energy and protein balances expected for each of the three diets (rations A through C) entered and percent Ca and P needed in the diet DM to meet requirements. Animal descriptions entered were 533 kg mature weight, breed code 1, 40 kg expected birth weight, 60 months age, the breed default peak milk (8 kg), the default values for milk composition (4 percent fat, 3.4 percent protein, 8.3 percent solids not fat), 8.5 weeks at peak milk, and 30 months duration of lactation. Table

9-7 shows predicted NE$_m$, NE$_g$, MP, Ca, and P required daily for maintenance, growth, lactation, and pregnancy as well as predicted target ADG, SBW, daily milk production, and expected gravid uterus weight used to compute the requirements for each of the 12 months of the reproductive cycle using the equations presented in Chapter 10. As described previously, all can be used directly to formulate dietary requirements for the specified level of performance, except diet CP intake to meet DIP and UIP requirements, which can be computed as described for Table 9-2.

Table 9-8 shows diet evaluations for this same cow. The diet concentration of TDN and CP and DIP as a percentage of CP were entered for each of the three diets and the

TABLE 9-8   Diet Evaluation for Beef Cows

Mature Weight	533 kg	Milk Fat	4.0 %
Calf Birth Weight	40 kg	Milk Protein	3.4 %
Age @ Calving	60 months	Calving Interval	12 months
Age @ Weaning	30 weeks	Time Peak	8.5 weeks
Peak Milk	8 kg	Milk SNF	8.3 %
Breed Code	1 Angus		

Ration	TDN % DM	ME Mcal/kg	NE$_m$ Mcal/kg	CP % DM	DIP % CP	DMI Factor
A	50	1.84	1.00	7.9	82.5	100%
B	60	2.21	1.35	7.8	100.0	100%
C	70	2.58	1.67	9.1	100.0	100%

		Months since Calving											
		1	2	3	4	5	6	7	8	9	10	11	12
	NE$_m$ Req. Factor	100%	100%	100%	100%	100%	100%	100%	100%	100%	100%	100%	100%
A	Milk kg/d	6.7	8.0	7.2	5.8	4.3	3.1	0.0	0.0	0.0	0.0	0.0	0.0
	DM, kg	11.14	11.40	12.12	11.83	11.54	11.30	10.68	10.68	10.68	10.68	10.68	10.68
	Energy Balance, Mcal/d	−3.90	−4.59	−3.31	−2.58	−1.88	−1.34	1.81	1.50	0.95	0.06	−1.30	−3.24
	DIP Balance, g/d	7	7	7	7	7	7	6	6	6	6	6	6
	UIP Balance, g/d	−201	−270	−169	−96	−24	34	175	170	142	93	14	−110
	MP Balance, g/d	−161	−216	−136	−77	−19	27	149	136	113	75	11	−88
	Ca % DM	0.65%	0.70%	0.62%	0.57%	0.52%	0.47%	0.34%	0.34%	0.34%	0.59%	0.59%	0.59%
	P % DM	0.20%	0.21%	0.19%	0.18%	0.16%	0.15%	0.12%	0.12%	0.12%	0.17%	0.17%	0.17%
	Reserves Flux/mo, Mcal	−148	−174	−126	−98	−71	−51	55	46	29	2	−50	−123
B	DM, kg	11.96	12.23	12.72	12.43	12.14	11.90	11.28	11.28	11.28	11.28	11.28	11.28
	Energy Balance, Mcal	1.07	0.47	1.69	2.32	2.92	3.38	6.32	6.00	5.46	4.56	3.20	1.27
	DIP Balance, g/d	5	5	5	5	5	5	5	5	5	5	5	5
	UIP Balance, g/d	18	−47	44	114	182	233	221	221	221	221	209	85
	MP Balance, g/d	14	−38	35	91	146	189	304	291	269	230	167	68
	Ca % DM	0.27%	0.30%	0.27%	0.25%	0.22%	0.20%	0.15%	0.15%	0.15%	0.25%	0.25%	0.25%
	P % DM	0.19%	0.20%	0.18%	0.17%	0.16%	0.14%	0.11%	0.11%	0.11%	0.16%	0.16%	0.16%
	Reserves Flux/mo, Mcal	32	14	51	71	89	103	192	183	166	139	97	39
C	DM, kg	13.16	13.42	13.79	13.50	13.21	12.97	12.35	12.35	12.35	12.35	12.35	12.35
	Energy Balance, Mcal/d	6.99	6.48	7.65	8.18	8.69	9.07	11.80	11.49	10.95	10.05	8.69	6.76
	DIP Balance, g/d	3	3	3	3	3	3	2	2	2	2	2	2
	UIP Balance, g/d	295	233	314	308	301	296	282	282	282	282	282	282
	MP Balance, g/d	236	187	256	308	360	401	509	496	473	435	371	272
	Ca % DM	0.25%	0.27%	0.25%	0.23%	0.20%	0.19%	0.13%	0.13%	0.13%	0.23%	0.23%	0.23%
	P % DM	0.17%	0.18%	0.17%	0.15%	0.14%	0.13%	0.10%	0.10%	0.10%	0.14%	0.14%	0.14%
	Reserves Flux/mo, Mcal	212	197	233	249	264	276	359	349	333	306	264	205

intake multiplier was set at 100 percent. All DIP values were then set at 80 percent, and diet CP was adjusted until DIP requirements were close to being balanced. Predicted DMI varies with daily milk production and forage quality. The CP required to meet diet DIP required for microbial growth is constant for a given diet but increased as diet TDN increased because microbial growth is a constant proportion of TDN. However, the UIP balance changes with milk and pregnancy requirements.

Diet A (50 percent TDN) met energy and UIP requirements in months 7 to 10 (cows just dry), became deficient in energy in month 11, and deficient in both energy and UIP in month 12. Diet B (60 percent TDN) is adequate in energy in all months and UIP in all but month 2 of lactation. Diet C (70 percent TDN) exceeded energy and UIP requirements in all months.

The energy reserves flux (Mcal/mo) is given for each month of the reproductive cycle for each diet evaluated.

Appendix Table 13 can be used to estimate days for a CS change by dividing the Appendix Table 13 value by the predicted daily energy balance. To reduce a negative energy balance, 1 Mcal diet NE$_m$ will substitute for 1 Mcal negative energy balance. To utilize energy reserves, 1 Mcal diet NE$_m$ can be replaced by 0.8 Mcal tissue energy.

TABLE OF ENERGY RESERVES FOR BEEF COWS

Appendix Table 13 gives Mcal mobilized in moving to the next lower CS, or required to move from the next lower CS to the one being considered, for cows with different mature weights. For example, a 500-kg cow at CS 5 will mobilize 207 Mcal in declining to a CS 4. If NE$_m$ intake is deficient 3 Mcal/day, this cow will lose 1 CS in (207 ∗ 0.8)/3 = 55 days. If consuming 3 Mcal NE$_m$ above daily requirements, this cow will move from a CS 4 to a CS 5 in 207/3 = 69 days. The equations developed for computation of energy reserves are discussed in Chapter 3.

TABLE OF MAINTENANCE REQUIREMENT MULTIPLIERS FOR ENVIRONMENTAL CONDITIONS

The program used to develop the tables of requirements does not adjust for environmental conditions. Appendix Table 14 gives multipliers developed from the computer model level 1 that can be used to adjust NE_m requirements for environmental stress.

10 Prediction Equations and Computer Models

The National Research Council's (NRC) Nutrient Requirement Series is used in many ways—teaching, research, and practical diet formulation. The level of solution needed depends on the intended use, information available, knowledge of the user, and risk of use. As the complexity of the information desired and the completeness of prediction of animal responses increases, the information and knowledge needed also increases. A computer program containing two levels of equations was developed to (1) predict requirements and energy and protein allowable production from the dietary ingredients fed and (2) allow use with widely varying objectives.

One of the primary purposes of developing and applying models such as the model presented in this revision of *Nutrient Requirements of Beef Cattle* is to improve nutrient management through refined animal feeding. Predicting nutrient requirements as accurately as possible for animals in a given production setting results in minimized overfeeding of nutrients, increased efficiency of nutrient utilization, maximized performance, and reduced excess nutrient excretion. Agricultural animal excretion of nitrogen, phosphorus, copper, and other minerals poses a risk for groundwater and soil contamination in areas of intensified animal production (U.S. Environmental Protection Agency, 1992). With the use of modeling techniques, however, to more accurately predict requirements and match them with dietary nutrients, producers have made significant strides to optimize performance while addressing environmental impacts. The application of a nutrition model to formulate dairy cattle diets in an area of central New York state resulted in a 25 percent decrease in nitrogen excretion and a substantial reduction in feed costs (Fox et al., 1995). Food-producing animals are also often targeted as a source of atmospheric methane, which contributes to global warming. Cattle typically lose 6 percent of ingested energy as eructated methane, which is equivalent to approximately 300 L methane/day for an average steer (Johnson and John-son, 1995). Development of management strategies, including modeling to predict nutrient requirements more precisely, can mitigate methane emissions from cattle by enhancing nutrient utilization and feed efficiency. Application of models in agricultural animal production thus has the potential to significantly reduce nutrient loading of the environment while providing economic benefits and tangible returns to those who implement these systems for improved animal feeding. Both levels of the model introduced in this revision use the same cattle requirement equations presented in this publication, which the subcommittee believes can be used to compute requirements over wide variations in body sizes and cattle types, milk production levels, and environmental conditions.

Level 2 was designed to obtain additional information about ruminal carbohydrate and protein utilization and amino acid supply and requirements. To achieve these objectives, more mechanistic submodels (Fox et al., 1992; Russell et al., 1992; Sniffen et al., 1992; O'Connor et al., 1993; Fox et al., 1995; Pitt et al., 1996) were included to predict microbial growth from feed carbohydrate and protein fractions and their digestion and passage rates. These submodels provide variable ME, MP, and amino acid supplies from feeds, based on variations in DMI, feed composition, and feed fiber characteristics. In considering the level 2 model for use in this publication, other published models were reviewed (Institut National de la Recherche Agronomique, 1989; Commonwealth Scientific and Industrial Research Organization, 1990; Dikstra et al., 1992; Agricultural and Food Research Council, 1993; Baldwin, 1995). Major limitations of the more mechanistic models (Dikstra et al., 1992; Baldwin, 1995) were a lack of field available inputs to drive them, including feed libraries, and no improvement in predictability than the level 2 model chosen (Kohn et al, 1994; Tylutki et al., 1994; Pitt et al., 1996). Major limitations of the other more highly aggregated models (Institut National de la Recherche

Agronomique, 1989; Commonwealth Scientific Industrial Research Organization, 1990; Agricultural and Food Research Council, 1993) were inability to use inputs available in a specific production setting in North America to mechanistically predict feed net energy values and supply of amino acids.

Level 1 should be used when limited information on feed composition is available and the user is not familiar with how to use, interpret, and apply the inputs and results from level 2. Potential uses of level 2 are (Fox et al., 1995)

• as a teaching tool to improve skills in evaluating the interactions of feed composition, feeding management and animal requirements in varying farm conditions;
• to develop tables of feed net energy and metabolizable protein values and adjustment factors that can extend and refine the use of conventional diet formulation programs;
• as a structure to estimate feed utilization for which no values have been determined and on which to design experiments to quantify those values;
• to predict requirements and balances for nutrients for which more detailed systems of accounting are needed, such as peptides, total rumen nitrogen, and amino acid balances;
• as a tool for extending research results to varying farm conditions; and
• as a diagnostic tool to evaluate feeding programs and to account for more of the variation in performance in a specific production setting.

The equations for each level are presented in "pseudo code" form for convenience of programming them into any language. The data on which the equations are based are discussed in the appropriate sections of the text.

In this revision, much more emphasis is placed on predicting the supply of nutrients because animal requirements and diet are interactive, including calculating feed digestibility under specific conditions, heat increment to compute lower critical temperature, calculation of efficiency of ME use for maintenance, growth and lactation, and adjusting microbial protein production for diet effective NDF content. Therefore, accuracy of prediction of nutrient requirements and performance under specific conditions depends on accuracy of description of feedstuff composition and DMI.

In developing more mechanistic models for determining the nutrient requirements of beef cattle, the subcommittee considered recent models that describe some or all aspects of postabsorptive metabolism (Oltjen et al., 1986; France et al., 1987). The France model is mechanistic in its approach to metabolism but has received no, or limited, validation with field data. The Oltjen model was considered by the subcommittee and compared with predictions of the proposed models with respect to growth (see Chapter 3). For further presentation on alternative techniques to

modeling responses to nutrients in farm animals, the reader is referred to the report of the Agricultural and Food Research Council (AFRC) Technical Subcommittee on Responses to Nutrients (Agricultural and Food Research Council, 1991).

REQUIREMENTS FOR BOTH LEVELS

The requirement section is subdivided into four main sections: maintenance, growth, lactation, and pregnancy.

Maintenance

Maintenance requirements are computed by adjusting the base NE_m requirement for breed, physiological state, activity, and heat loss vs heat production, which is computed as ME intake minus retained energy. Heat loss is affected by animal insulation factors and environmental conditions.

ENERGY

$$a_1 = 0.077$$

Adjustment for previous temperature:

$$a_2 = 0.0007(20 - T_p)$$

Adjustment for breed, lactation, sex, and previous plane of nutrition:

See "Errata" from Cover

$$NE_m = [a_1 \, SBW^{0.75} \, (BE) \, (L) \, (SEX) \, (COMP)] + a_2$$
$$COMP = 0.8 + ((CS - 1) * 0.05)$$

Adjustment for activity if on pasture:

$$NE_{mact} = ((0.006 * pI * (0.9 - (TDN_p/100))) + (0.05 * TERRAIN/(pAVAIL + 3))) * BW/4.184$$

otherwise

$$NE_{mact} = 0$$
$$I_m = (NE_m + NE_{mact})/(NE_ma * ADTV)$$

for growing cattle (used to compute heat increment);

$$RE = (DMI - I_m) * NE_{ga}$$
$$YE_n = 0$$
$$LE = 0$$

for lactating cattle (used to compute heat increment);

$$(RE + YE_n + NE_{preg}) = (DMI - I_m) * NE_{ma};$$

adjustment for cold stress:

$$SA = 0.09 * SBW^{0.67}$$
$$HE = (MEI - (RE + YE_n + NE_{preg}))/SA$$
$$EI = (7.36 - 0.296 \, WIND + 2.55 \, HAIR) * MUD2 * HIDE$$

[MUD2 code 1 factor = 1; MUD2 code factor 2 = 0.8; MUD2 code factor 3 = 0.5; MUD2 code 4 factor = 0.2]

[HIDE code 1 factor = 0.8; HIDE code 2 factor = 1; HIDE code 3 factor = 1.2]

TI = 2.5 for newborn calf; 6.5 for 1-month-old calf; 5.1875 + (0.3125 * CS) for yearlings; 5.25 + (0.75 * CS) for adult cattle

IN = TI + EI

LCT = 39 − (IN * (HE) * 0.85)

ME_{cs} = SA (LCT − Tc)/IN

NE_{mcs} = k_m * ME_{cs}

NE_mtotal = (NE_m + NE_{mact} + NE_{mcs}),

or if heat stressed (panting),

NE_mtotal = (NE_m * NE_{mhs}) + NEm_{act}

$I_m$total = NE_mtotal / NE_{ma}

where

a_1 = thermal neutral maintenance requirement (Mcal/day/$SBW^{0.75}$);

a_2 = maintenance adjustment for previous ambient temperature, (Mcal/day/$BW^{0.75}$);

T_p = previous average monthly temperature, °C;

NE_m = net energy required for maintenance adjusted for acclimatization,

BE = breed effect on NE_m requirement (Table 10-1);

L = lactation effect on NE_m requirement (1 if dry or 1.2 if lactating);

SEX = 1.15 if bulls, otherwise 1;

CS = condition score, 1-9 scale;

COMP = effect of previous plane of nutrition on NE_m requirement;

NE_{mact} = activity effect on NE_m requirement;

DMI = dry matter intake, kg/day;

pI = pasture dry matter intake, kg/day;

TDN_p = total digestible nutrient content of the diet, %;

TERRAIN = Terrain factor [1 = level land, 2 = hilly];

pAVAIL = pasture mass available for grazing, T/ha;

I_m = I for maintenance (no stress), kg DM/day;

$I_m$total = I for maintenance with stress, kg/DM/day;

NE_{ma} = net energy value of diet for maintenance, Mcal/kg;

ADTV = 1.12 for diets containing ionophores, otherwise, 1.0;

NE_{ga} = net energy value of diet for gain, Mcal/kg;

YE_n = net energy milk;

NE_{preg} = net energy retained as gravid uterus;

MEC = metabolizable energy content of diet, Mcal/kg;

SA = surface area, m²;

HE = heat production, Mcal/day;

MEI = metabolizable energy intake, Mcal/day;

LCT = animal's lower critical temperature, °C;

TABLE 10-1 Breed Maintenance Requirement Multipliers, Birth Weights, and Peak Milk Production[a]

Breed	Code	NE_m (BE)	Birth wt. kg (CBW)	Peak Milk Yield, kg/day (PKYD)
Angus	1	1.00	31	8.0
Braford	2	0.95	36	7.0
Brahman	3	0.90	31	8.0
Brangus	4	0.95	33	8.0
Braunvieh	5	1.20	39	12.0
Charolais	6	1.00	39	9.0
Chianina	7	1.00	41	6.0
Devon	8	1.00	32	8.0
Galloway	9	1.00	36	8.0
Gelbvieh	10	1.10 1.00	39	11.5
Hereford	11	1.00	36	7.0
Holstein	12	1.20	43	15.0
Jersey	13	1.20	31	12.0
Limousin	14	1.00	37	9.0
Longhorn	15	1.00	33	5.0
Maine Anjou	16	1.00	40	9.0
Nellore	17	0.90	40	7.0
Piedmontese	18	1.00	38	7.0
Pinzgauer	19	1.00	40	11.0
Polled Hereford	20	1.00	33	7.0
Red Poll	21	1.00	36	10.0
Sahiwal	22	0.90	38	8.0
Salers	23	1.00	35	9.0
S.Gertudis	24	0.90	33	8.0
Shorthorn	25	1.00	37	8.5
Simmental	26	1.20	39	12.0
South Devon	27	1.00	33	8.0
Tarentaise	28	1.00	33	9.0

[a]Variable names (BE, CBW, PKYD) are used in various equations to predict cow requirements.

See "Errata" Front cover

IN = insulation value, °C/Mcal/m²/day;

TI = tissue (internal) insulation value, °C/Mcal/m²/day;

EI = external insulation value, °C/Mcal/m²/day;

WIND = wind speed, kph;

HAIR = effective hair depth, cm;

MUD2 = mud adjustment factor for external insulation [1 = dry and clean, 2 = some mud on lower body, 3 = wet and matted, 4 = covered with wet snow or mud];

HIDE = hide adjustment factor for external insulation [1 = thin, 2 = average, 3 = thick];

Tc = current temperature, °C;

MEcs = metabolizable energy required due to cold stress, Mcal/day;

k_m = diet NE_m/diet ME; (Assume 0.576 in derivation)

NE_{mcs} = net energy required due to cold stress, Mcal/day;

NE_{mhs} = 1.07 for rapid shallow panting, and 1.18 for open mouth panting if temperature is ≥30° C;

NE_mtotal = net energy for maintenance required adjusted for breed, lactation, sex, grazing, acclimatization and stress effects, Mcal/day; and

FFMtotal = feed for maintenance (adjusted for stress), kg DM/day.

MAINTENANCE PROTEIN REQUIREMENT

$$MP_{maint} = 3.8 * SBW^{0.75}$$

where

MP$_{maint}$ = metabolizable protein requirement for maintenance, g/day
SBW = body weight.

Growth

Requirements for growth are calculated using shrunk body weight, shrunk weight gain, body composition, and relative body size.

ENERGY AND PROTEIN REQUIREMENTS

EBW = 0.891 SBW
EBG = 0.956 SWG
SRW = 478 kg for animals finishing at small marbling (28% body fat) and replacement heifers,
    = 462 kg for animals finishing at slight marbling (26.8% body fat),
    = 435 kg for animals finishing at trace marbling (25.2% body fat).
EQSBW = SBW * (SRW)/(FSBW)
EQEBW = 0.891 EQSBW
RE = 0.0635 EQEBW$^{0.75}$ EBG$^{1.097}$
NP$_g$ = SWG (268 − (29.4 (RE/SWG))).

If EQEBW ≤300 kg,

MP$_g$ = NP$_g$/(0.83 − (EQEBW * 0.00114))

otherwise,

MP$_g$ = NP$_g$/0.492

where

EQSBW = equivalent shrunk body weight, kg;
EBW = empty body weight, kg;
SBW = shrunk body weight, kg (typically 0.96 * full weight);
EBG = empty body gain, kg;
SWG = shrunk weight gain, kg;
RE = retained energy, Mcal/day;
EQEBW = equivalent empty body weight, kg;
FSBW = actual final shrunk body weight at maturity for breeding heifers or at the body fat end point selected for feedlot steers and heifers;
NP$_g$ = net protein requirement for growth, g/day;
MP$_g$ = metabolizable protein requirement for growth, g/day.

Prediction of average daily gain (ADG) when net energy available for gain (RE) is known:

EBG = 12.341 EQEBW$^{-0.6837}$RE$^{0.9116}$.
SWG = 13.91 RE$^{0.9116}$ EQSBW$^{-0.6837}$.

Growth Requirements of Replacement Heifers

Coefficients for computing target breeding weights at puberty are based on the summary in Chapter 3. Coefficients for computing target breeding weights after first calving are based on USMARC data summarized by Gregory et al. (1992).

PREDICTING TARGET WEIGHTS AND RATES OF GAIN

TPW = MW * (0.55 for dual purpose and dairy, 0.60 for *Bos taurus*, and 0.65 for *Bos indicus*)
TPA = TCA − 280
BPADG = (TPW − SBW)/(TPA − T$_{age}$)
TCW1 = MW * 0.80
TCW2 = MW * 0.92
TCW3 = MW * 0.96
TCW4 = MW * 1
APADG = (TCW1 − TPW)/(280)
ACADG = (TCWxx − TCWx)/CI

where

MW is mature weight, kg;
SBW is shrunk body weight, kg;
TPW is target puberty weight, kg;
TCW1 is target first calf calving weight, kg;
TCW2 is target second calf calving weight, kg;
TCW3 is target third calf calving weight, kg;
TCW4 is target fourth calving weight,
TCWx is current target calving weight, kg;
TCWxx is next target calving weight, kg;
TCA is target calving age in days
TPA is target puberty age in days
BPADG = prepubertal target ADG, kg/day;
APADG = postpubertal target ADG, kg/day;
ACADG = after calving target ADG, kg/day;
T$_{age}$ is heifer age, days;
CI is calving interval, days.

The equations in the growth section are used to compute requirements for the target ADG. For pregnant animals, gain due to gravid uterus growth should be added to predicted daily gain (SWG), as follows:

ADG$_{preg}$ = CBW * (0.3656 − 0.000523t) *
    $e^{((0.0200 * t) - (0.000143 * t^2))}$;

For pregnant heifers, weight of fetal and associated uterine tissue is deducted from EQEBW to compute growth requirements. The conceptus weight (CW) can be calculated as follows:

$$CW = (CBW * 0.01828) * e^{((0.0200 * t) - (0.000143 * t^2))};$$

where

CBW = calf birth weight, kg;
CW = conceptus weight, g;
t = days pregnant.

Lactation

Lactation requirements are calculated using age of cow, time of lactation peak, peak milk yield, day of lactation, duration of lactation, milk fat composition, milk solids not fat composition, and milk protein composition:

$$k = 1/T$$
$$a = 1/(PKYD * k * e)$$
$$Yn = n/(a * e^{(kn)})$$
$$TotalY = -7/[(a * k) * (D * e^{(-kD)}) + ((1/k) * e^{(-kD)}) - (1/k)];$$

if age = 2,

Yn = 0.74 Yn
TotalY = 0.74 TotalY;

if age = 3,

Yn = 0.88 Yn
TotalY = 0.88 TotalY.

$$E = 0.092 * MF + 0.049 * SNF - 0.0569$$
$$YEn = E * Yn$$
$$YFatn = MF/100 * Yn$$
$$YProtn = Prot/100 * Yn$$
$$TotalE = E * TotalY$$
$$TotalFat = MF/100 * TotalY$$
$$Total\ Prot = Prot/100 * TotalY$$
$$MP_{lact} = (YProtn/0.65) * 1000$$

where

age = age of cow, years;
W = current week of lactation;
PKYD = peak milk yield, kg/day (Table 10-1);
T = week of peak lactation;
D = duration of lactation, weeks;
MF = milk fat composition, %;
SNF = milk solids not fat composition, %;
Prot = milk protein composition, %
k = intermediate rate constant;
a = intermediate rate constant;
Yn = daily milk yield at current week of lactation, kg/day;
TotalY = total milk yield for lactation, kg;
E = energy content of milk, Mcal (NE$_m$)/kg;

YE_n = daily energy secretion in milk at current stage of lactation, Mcal (NE$_m$)/day;
YFatn = daily milk fat yield at current stage of lactation, kg/day;
YProtn = daily milk protein yield at current stage of lactation, kg/day;
TotalE = total energy yield for lactation, Mcal;
TotalFat = total fat yield for lactation, kg;
Total Prot = total protein yield for lactation, kg;
MP_{lact} = metabolizable protein requirement for lactation, g/day.

Pregnancy

Calf birthweight and day of gestation are used to calculate pregnancy requirements.

$$NE_m\ req, Mcal = CBW * (k_m/0.13) * (0.4504 - 0.0000996t) * e^{((0.03233 - 0.0000275t)t)}$$
$$Ypn, g/day = CBW * (0.001669 - 0.00000211t) * e^{(0.0278 - 0.0000176t)t} * 6.25$$
$$MP_{preg}, g/day = Ypn/0.65$$

where

CBW = calf birth weight, kg;
t = day of pregnancy;
MP_{preg} = MP for pregnancy, g/day.

ENERGY AND PROTEIN RESERVES

Body condition score, body weight, and body composition are used to calculate energy and protein reserves. The equations were developed from data on chemical body composition and visual appraisal of condition scores on 106 mature cows of diverse breed types and body sizes and were validated on an independent data set of 65 mature cows (data from C. L. Ferrell, USMARC, personal communication, 1995).

(1) Body composition is computed for the current CS:
$$AF = 0.037683\ CS$$
$$AP = 0.200886 - 0.0066762\ CS$$
$$AW = 0.766637 - 0.034506\ CS$$
$$AA = 0.078982 - 0.00438\ CS$$
$$EBW = 0.851\ SBW$$
$$TA = AA * EBW$$

where

AF = proportion of empty body fat;
AP = proportion of empty body protein;
AW = proportion of empty body water;
AA = proportion of empty body ash;
SBW = shrunk body weight, kg;
EBW = empty body weight, kg;
TA = Total Ash, kg.

(2) For CS = 1 ash, fat, and protein composition are as follows:

$$AA1 = 0.074602$$
$$AF1 = 0.037683$$
$$AP1 = 0.194208$$

where

AA1 = proportion of empty body ash @ CS = 1
AF1 = proportion of empty body fat @ CS = 1
AP1 = proportion of empty body protein @ CS = 1

(3) Assuming that ash mass does not vary with condition score, EBW and component body mass at condition score 1 is calculated:

$$EBW1 = TA/AA1$$
$$TF = AF * EBW$$
$$TP = AP * EBW$$
$$TF1 = EBW1 * AF1$$
$$TP1 = EBW1 * AP1$$

where

EBW1 = calculated empty body weight at CS = 1, kg;
TF = total body fat, kg;
TP = total body protein, kg;
TF1 = total body fat @ CS = 1, kg;
TP1 = total body protein @ CS = 1, kg.

(4) Mobilizable energy and protein are computed:

$$FM = (TF - TF1)$$

$$PM = (TP - TP1)$$
$$ER = 9.4 \, FM + 5.7 \, PM$$

where

FM = mobilizable fat, kg;
PM = mobilizable protein, kg;
ER = energy reserves, Mcal.

(5) EBW, AF, and AP are computed for the next CS to compute energy and protein gain or loss to reach the next CS:

$$EBWN = TA/AAN$$

where

EBWN is EBW at the next score;
TA is total kg ash at the current score;
AAN is proportion of ash at the next score.

AF, AP, TF, and TP are computed as in steps 1 and 3 for the next CS and FM, PM, and ER are computed as the difference between the next and current scores.

During mobilization, 1 Mcal of RE will substitute for 0.80 mcal of diet NE_m; during repletion, 1 mcal diet NE_m will provide 1 mcal of RE.

MINERAL AND VITAMIN REQUIREMENTS

Mineral and vitamin requirements are summarized in Tables 10-2 and 10-3. Requirements are identified for maintenance, growth, lactation, and pregnancy.

TABLE 10-2   Calcium and Phosphorus Requirements

Mineral	Requirements, g/day				Maximum Tolerable
	Maintenance	Growth	Lactation	Pregnancy (last 90 d)	
Ca, g/day	0.0154 * W/0.5	RPN * 0.071/0.5	Milk * 1.23/0.5	CBW * (13.7/90)/0.5	0.2 * DMI
P, g/day	0.016 * W/0.68	RPN * 0.045/0.68	Milk * 0.95/0.68	CBW * (7.6/90)/0.68	0.1 * DMI

NOTE: BW, body weight, kg; DMI, dry matter intake, kg; RPN is retained protein, g; Milk, milk production, kg; CBW, calf birth weight, kg.

TABLE 10-3   Other Mineral Requirements and Maximum Tolerable Concentrations and Vitamin Requirements

Mineral/Vitamin	Unit	Growing and Finishing[a]	Cows		Maximum Tolerable Level
			Gestation	Early Lactation	
Magnesium	%	0.10	0.12	0.20	0.40
Potassium	%	0.60	0.60	0.70	3.00
Sodium	%	0.06–0.08	0.06–0.08	0.10	—
Sulfur	%	0.15	0.15	0.15	0.40
Cobalt	mg/kg	0.10	0.10	0.10	10.00
Copper	mg/kg	10.00	10.00	10.00	100.00
Iodine	mg/kg	0.50	0.50	0.50	50.00
Iron	mg/kg	50.00	50.00	50.00	1000.00
Manganese	mg/kg	20.00	40.00	40.00	1000
Selenium	mg/kg	0.10	0.10	0.10	2
Zinc	mg/kg	30.00	30.00	30.00	500
Vitamins					
A	IU/kg	2200	2800	3900	
D	IU/kg	275	275	275	

[a]Also for breeding bulls.

PREDICTING DRY MATTER INTAKE

The equations below are used to predict intake for various cattle types; adjustments for various factors are given in Table 10-4 and can be used with these or other intake estimates.

For growing calves:

$$DMI = ((SBW^{0.75} * (0.2435 \, NE_{ma} - 0.0466 \, NE_{ma}^2 - 0.1128))/NE_{ma})((BFAF)(BI)(ADTV)(TEMP1)(MUD1)).$$

For growing yearlings:

$$DMI = (SBW^{0.75} * (0.2435 \, NE_{ma} - 0.0466 \, NE_{ma}^2 - 0.0869))/NE_{ma})((BFAF)(BI)(ADTV)(TEMP1)(MUD1)).$$

For nonpregnant beef cows:

$$DMI = ((SBW^{0.75} * (0.04997 \, NE_{ma}^2 + 0.03840)/NE_{ma})(TEMP1)(MUD1) + 0.2 \, MM)$$

for diets with an $NE_{ma} < 1.0$ Mcal/kg; $NE_{ma} = 0.95$.

For pregnant cows (last two-thirds of pregnancy):

$$DMI = ((SBW^{0.75} * (0.04997 \, NE_{ma}^2 + 0.04361)/NE_{ma})(TEMP1)(MUD1) + 0.2 \, Yn)$$

for diets with an $NE_{ma} < 1.0$ Mcal/kg; $NE_{ma} = 0.95$

TABLE 10-4 Adjustment Factors for Dry Matter Intake for Cattle[a]

Adjustment Factor	Multiplier
Breed (BI)	
Holstein	1.08
Holstein × Beef	1.04
Empty body fat effect (BFAF)	
21.3 (to 350 kg EQW)	1.00
23.8 (400 kg EQW)	0.97
26.5 (450 kg EQW)	0.90
29.0 (500 kg EQW)	0.82
31.5 (550 kg EQW)	0.73
Anabolic implant (ADTV)	1.00
No anabolic stimulant	0.94
Temperature, °C (TEMP1)	
>35, no night cooling	0.65
>35, with night cooling	0.90
25 to 35	0.90
15 to 25	1.00
5 to 15	1.03
−5 to 5	1.05
−15 to −5	1.16
Mud (MUD1)	
Mild (10–20 cm)	0.85
Severe (30–60 cm)	0.70

[a]National Research Council, 1987.

where

DMI is dry matter intake, kg/day,
SBW is shrunk live body weight, kg,
NE_{ma} is net energy value of diet for maintenance, Mcal/kg,
Yn is milk production, kg/day,
BI is breed adjustment factor for DMI (Table 10-1),
BFAF is body fat adjustment factor (Table 10-4),
ADTV is feed additive adjustment factor for DMI (Table 10-4),
TEMP1 is temperature adjustment factor for DMI (Table 10-4), and
MUD1 is mud adjustment factor for DMI (Table 10-4).

The same environmental adjustments are used to adjust intake for all cattle types.

Adjustment of dry matter intake relative to forage allowance for animals grazing:

$$pI = GRAZE * DMI;$$
$$FA = 1000 * GU * IPM/(SBW * N * DOP)$$

If FA > (DMI * 4) or IPM > 1150 kg/ha,
    GRAZE = 1.0

otherwise,

$$GRAZE = (0.17 \, IPM - 0.000074 \, IPM^2 + 2.4)/100$$

See "Errata" Front Cover

DMI = kg predicted dry matter intake using previous equations;
pI = kg predicted dry matter intake adjusted for grazing situations;
FA = daily forage allowance, kg/day/head;
GRAZE = forage availability factor if grazing, %;
IPM = initial pasture mass (kg DM/ha);
GU = grazing unit size (ha);
SBW = shrunk body weight;
N = number of animals; and
DOP = days on pasture.

SUPPLY OF NUTRIENTS

Amounts are computed from actual DMI when available or from predicted intake provided at the end of this section. Risk of use increases when predicted intakes are used vs actual DMI.

Level One

ENERGY

Ration energy values are computed by summing the energy contribution of each feed to arrive at a total energy content of the ration, using tabular energy values. Tabular energy values used include % TDN, ME (Mcal/kg), NE_{ma} (Mcal/kg), and NE_{ga} (Mcal/kg).

PROTEIN

Supply of metabolizable protein (MP) is the sum of digested ruminally undegraded feed protein and digested microbial protein. Feed composition parameters used include percentage CP, percentage UIP, and percentage DIP.

Undegraded available feed protein is assumed to be 80 percent digestible. Hence,

$$MP_{feed} = UIP_{intake} * 0.8.$$

The contribution of microbial protein to the MP supply is estimated from the microbial crude protein yield.

$$MCP = 0.13 * TDN * eNDF_{adj}$$

where

MCP = microbial crude protein, g/day;
$eNDF_{adj}$ = 1.00 if the effective NDF (eNDF) of the ration is >20%
= $1.0 - ((20 - eNDF) * 0.025$ when eNDF ≤20%;
TDN = total digestible nutrients, g/day;

MCP is assumed to be 80 percent true protein and 80 percent digestible, hence,

$$MP_{bact} = MCP * 0.64$$
$$MP_{tot} = MP_{bact} + MP_{feed}.$$

Level Two

Level 2 computes amino acid requirements and predicts energy and protein supply from feed physical and chemical properties. (All energy and protein requirements are the same as level 1.)

AMINO ACID REQUIREMENTS FOR MAINTENANCE

$$MPAA_i = AATISS_i * 0.01 * MP_{maint}$$

where

MP_{maint} = metabolizable protein required for maintenance, g/day;
$MPAA_i$ = metabolizable requirement for the ith absorbed amino acid, g/day;
$AATISS_i$ = amino acid composition of tissue (Table 10-5).

AMINO ACID REQUIREMENTS FOR GROWTH

$$RPN = PB * 0.01 * EBG$$
$$RPAA_i = AATISS_i * RPN/EAAG_i$$

where

PB = protein content of empty body gain, g/100 g,
EBG = empty body gain, g/day,

TABLE 10-5   Amino Acid Composition of Tissue and Milk Protein (g/100 g protein)

Amino Acid	Tissue[a]	Milk[b]
Methionine	2.0	2.71
Lysine	6.4	7.62
Histidine	2.5	2.74
Phenylalanine	3.5	4.75
Tryptophan	0.6	1.51
Threonine	3.9	3.72
Leucine	6.7	9.18
Isoleucine	2.8	5.79
Valine	4.0	5.89
Arginine[c]	3.3	3.40

[a]Average of three studies summarized by whole empty body values of Ainslie et al., 1993.
[b]Waghorn and Baldwin, 1984.
[c]Based on hindlimb uptake studies (Robinson et al., 1995).

RPN = net protein required for growth, g/day,
$EAAG_i$ = efficiency of use of the ith amino acid for growth (Table 10-6), g/g, and
$RPAA_i$ = metabolizable requirement for growth for the ith absorbed amino acid, g/day.

AMINO ACID REQUIREMENTS FOR LACTATION

$$LPAA_i = AALACT_i * 0.01 * YProtn/EAAL_i$$

where

$AALACT_i$ = i^{th} amino acid content of milk true protein, g/100 g (Table 10-5);
$EAAL_i$ = efficiency of use of the ith amino acid for milk protein formation, g/g (Table 10-6); and
$LPAA_i$ = metabolizable requirement for lactation for the ith absorbed amino acid, g/day.

AMINO ACID PREGNANCY REQUIREMENTS

$$MPAA_i = AATISS_i * YPN / EAAP_i$$

where

YPN = net protein required for gestation, g/day;

TABLE 10-6   Utilization of Individual Absorbed Amino Acids for Physiological Functions (g/g)[a]

Amino acid	Gestation	Lactation
Methionine	0.85	0.98
Lysine	0.85	0.88
Histidine	0.85	0.90
Phenylalanine	0.85	1.00
Tryptophan	0.85	0.85
Threonine	0.85	0.83
Leucine	0.66	0.72
Isoleucine	0.66	0.62
Valine	0.66	0.72
Arginine	0.66	0.85

[a]Requirement for growth varies with stage of growth as determined by Ainslie et al. (1993): EQ if SBW < 300 kg, EAAG = 0.83 − (0.00114 EBW), otherwise 0.492; EAAG is efficiency factor and EBW is equivalent body weight as described by Fox et al. (1992). Other values are from Evans and Patterson (1985).

$EAAP_i$ = efficiency of use of the ith amino acid for gestation, g/g (Table 10-6); and

$MPAA_i$ = metabolizable requirement for gestation for the ith absorbed amino acid, g/day (Table 10-6).

SUPPLY OF ENERGY, PROTEIN, AND AMINO ACIDS

Predicting the energy content of the ration is accomplished by estimating apparent TDN of each feed and for the total ration and utilizing equations and conversion factors to estimate ME, NE_m, NE_g, and NE_l values. To calculate apparent TDN, apparent digestibilities for carbohydrates, proteins, and fats are estimated. These apparent digestibilities are determined by simulating the degradation, passage, and digestion of feedstuffs in the rumen and small intestine. Also, microbial yields and fecal composition are estimated. Feed composition values used include: NDF, lignin, CP, fat, ash, NDIP, as a percent of the diet DM and starch and sugar expressed as a percentage of noncarbohydrates.

INTAKE CARBOHYDRATE

Based on chemical analyses (Appendix Table 1), equations used to calculate carbohydrate composition of the jth feedstuff are listed below:

CHO_j = 100 − CP_j(%DM) − FAT_j(%DM) − ASH_j(%DM)

CC_j = NDF_j(%DM) * 0.01 * $LIGNIN_j$(%NDF) * 2.4

$CB2_j$ = NDF_j(%DM) − $NDIP_j$(%CP) * 0.01 * CP_j(%DM)

NFC_j = CHO − $CB2_j$ − CC_j

$CB1_j$ = $STARCH_j$ (%NFC) * (NFC_j)/100

CA_j = (NFC_j − CBI_j)

where

CP_j (%DM) = percentage of crude protein of the jth feedstuff,

CHO_j (%DM) = percentage of carbohydrate of the jth feedstuff,

FAT_j (%DM) = percentage of fat of the jth feedstuff,

ASH_j (%DM) = percentage of ash of the jth feedstuff,

NDF_j (%DM) = percentage of the jth feedstuff that is NDF,

$NDIP_j$ (%DM) = percentage of neutral detergent insoluble protein of the jth feedstuff,

$LIGNIN_j$ (%NDF) = percentage of lignin of the jth feedstuff's NDF,

$STARCH_j$ (%NFC) = percentage of starch in the nonfiber carbohydrate of the jth feedstuff,

CA_j (%DM) = percentage of DM of the jth feedstuff that is sugar,

$CB1_j$ (%DM) = percentage of DM of the jth feedstuff that is starch,

$CB2_j$ (%DM) = percentage of DM of the jth feedstuff that is available fiber,

CC_j (%DM) = percentage of DM in the jth feedstuff that is unavailable fiber,

NFC_j = percentage of DM in the jth feedstuff that is nonfiber carbohydrates.

INTAKE PROTEIN

The *Ruminant Nitrogen Usage* (National Research Council, 1985) equation is used to predict recycled nitrogen:

$$U = 121.7 − 12.01X + 0.3235 X^2$$

where

U = urea N recycled (percent of N intake), and

X = diet CP, as a percent of diet dry matter.

The following equations are used to calculate the five protein fractions contained in the jth feedstuff from percent of crude protein, percent of protein solubility, percent of NDIP, and percent of ADIP:

PA_j (%DM) = NPN_j * 0.0001 * $SOLP_j$ * CP;

$PB1_j$ (%DM) = $SOLP_j$ * CP * 0.01 − A_j;

PC_j (%DM) = $ADIP_j$ * CP * 0.01;

$PB3_j$ (%DM) = ($NDIP_j$ − $ADIP_j$) * CP * 0.01;

$PB2_j$ (%DM) = CP * 0.01; − A_j − $B1_j$ − $B3_j$ − C_j

where

CP_j (%DM) = percentage of crude protein of the jth feedstuff,

NPN_j (%CP) = percentage of crude protein of the jth feedstuff that is nonprotein nitrogen times 6.25,

$SOLP_j$ (%CP) = percentage of the crude protein of the jth feedstuff that is soluble protein,

$NDIP_j$ (%DM) = percentage of the jth feedstuff that is neutral detergent insoluble protein,

$ADIP_j$ (%CP) = percentage of the jth crude protein that is acid detergent insoluble protein,

PA_j (%CP) = percentage of DM in the jth feedstuff that is nonprotein nitrogen,

$PB1_j$ (%CP) = percentage of DM in the jth feedstuff that is rapidly degraded protein,

$PB2_j$ (%CP) = percentage of DM in the jth feedstuff that is intermediately degraded protein,

$PB3_j$ (%CP) = percentage of DM in the jth feedstuff that is slowly degraded protein, and

PC_j (%CP) = percentage of DM in the jth feedstuff that is bound protein.

Adjusting Degradation Rates of Available Fiber for the Effect of pH

(1) Predict rumen pH (Pitt et al., 1996) if
eNDF < 0.245: pH = 5.425 + 0.04229 eNDF;

otherwise,

pH = 6.46.

(2) Compute original yield for each feed:
Y = 1/((0.05/(Kd − 0.02)) + (2.5)).

(3) Compute relative yield adjustment:
relY = (1 − exp(−5.624 * ((pH − 5.7) 0.909)))/0.9968.

(4) Compute new yield for each feed:
Y′ = relY * Y.

(5) Compute new Kd for each feed:
if pH <5.7, Kd′ = 0;

otherwise,

A = (−0.01490722 + (0.012024 * pH) − (0.0010152 * pH^2))
Kd′ = A * ((Y′/((−0.1058 + (0.0752 * pH)) − Y′)) + 1)

where

eNDF = % effective NDF in ration (decimal form)
Kd = feed specific degradation rate of available fiber fraction (decimal form) must be ≥0.02 h^{-1}
Kd′ = pH adjusted feed specific degradation rate of available fiber fraction (decimal form).

Computing Ruminal Escape of Carbohydrate and Protein

Ruminal degradation and escape of carbohydrate and protein fractions are determined by the following formulas, using digestion rates for each carbohydrate and protein fraction, and the passage rate equation which uses % forage and percent of effective NDF:

RD = Kd/(Kd + Kp)
RESC = Kp/(Kd + Kp)

where

RD = proportion of component of a feedstuff degraded in the rumen
RESC = proportion of component of feedstuff escaping ruminal degradation
Kd = degradation rate of feedstuff component
Kp = passage rate of feedstuff

PASSAGE RATE EQUATION

Kp[forages] = 0.388 + (22.0 DMI/SBW$^{0.75}$) + (0.0002 FORAGE^2)
Kp[conc] = −0.424 + (1.45 Kp[forages])

where

DMI = dry matter intake, kg/day;
SBW = shrunk body weight, kg/day;
FORAGE = forage concentration of the diet, %
Kp is adjusted for individual feeds using a multiplicative adjustment factor (Af) for particle size using diet effective NDF (eNDF):
Af[forages] = 100/(eNDF + 70)
Af[conc] = 100/(eNDF + 90)

where

eNDF = effective NDF concentration of individual feedstuff, percent (decimal form).

The following equations calculate the amounts of protein fractions that are ruminally degraded.

$RDPA_j = I_j * PA_j$
$RDPB1_j = I_j * PB1_j * (Kd_{1j}/(Kd_{1j} + Kp_j))$
$RDPB2_j = I_j * PB2_j * (Kd_{2j}/(Kd_{2j} + Kp_j))$
$RDPB3_j = I_j * PB3_j * (Kd_{3j}/(Kd_{3j} + Kp_j))$
$RDPEP_j = RDPB1_j + RDPB2_j + RDPB3_j$

where

I_j = intake of the jth feedstuff, g/day
Kd_{1j} = rumen rate of digestion of the rapidly degraded protein fraction of the jth feedstuff, h^{-1},
Kd_{2j} = rumen rate of digestion of the intermediately degraded protein fraction of the jth feedstuff, h^{-1},
Kd_{3j} = rumen rate of digestion of the slowly degraded protein fraction of the jth feedstuff, h^{-1},
Kp_j = rate of passage from the rumen of the jth feedstuff, h^{-1},
$RDPA_j$ = amount of ruminally degraded NPN in the jth feedstuff, g/day,
$RDPB1_j$ = amount of ruminally degraded B1 true protein, in the jth feedstuff, g/day,
$RDPB2_j$ = amount of ruminally degraded B2 true protein, in the jth feedstuff, g/day,
$RDPB3_j$ = amount of ruminally degraded B3 true protein in the jth feedstuff, g/day,
$RDPEP_j$ = amount of rumen degraded peptides from the jth feedstuff, g/day.

The undegraded protein is passed to the small intestine and the following equations calculate the amount of each protein fraction that escapes rumen degradation:

$REPB1_j = I_j * PB1_j * (Kp_j/(Kd1_j + Kp_j))$
$REPB2_j = I_j * PB2_j * (Kp_j/(Kd_{2j} + Kp_j))$

$REPB3_j = I_j * PB3_j * (Kp_j/(Kd_{3j} + Kp_j))$

$REPC_j = I_j * PC_j$

where

$REPB1_j$ = amount of ruminally escaped B1 true protein, in the jth feedstuff, g/day,

$REPB2_j$ = amount of ruminally escaped B2 true protein, in the jth feedstuff, g/day,

$REPB3_j$ = amount of ruminally escaped B3 true protein in the jth feedstuff, g/day, and

$REPC_j$ = amount of rumen escaped bound C protein from the jth feedstuff, g/day.

The following equations are used to calculate the amounts of each of the carbohydrate fractions of the jth feedstuff that are ruminally digested:

$RDCA_j = I_j * CA_j * (Kd_{4j}/(Kd_{4j} + Kp_j))$

$RDCB1_j = I_j * CB1_j * (Kd_{5j}/(Kd_{5j} + Kp_j))$

$RDCB2_j = I_j * CB2_j * (Kd_{6j}/(Kd_{6j} + Kp_j))$

where

Kd_{4j} = rumen rate of sugar digestion of the jth feedstuff, h^{-1},

Kd_{5j} = rumen rate of starch digestion of the jth feedstuff, h^{-1},

Kd_{6j} = rumen rate of available fiber digestion of the jth feedstuff, h^{-1},

$RDCA_j$ = amount of ruminally degraded sugar from the jth feedstuff, g/day,

$RDCB1_j$ = amount of ruminally degraded starch from the jth feedstuff, g/day,

$RDCB2_j$ = amount of ruminally degraded available fiber from the jth feedstuff, g/day.

The following equations are used to calculate the amounts of each of the carbohydrate fractions of the jth feedstuff that escape the rumen:

$RECA_j = I_j * CA_j * (Kp_j/(Kd_{4j} + Kp_j))$

$RECB1_j = I_j * CB1_j * (Kp_j/(Kd_{5j} + Kp_j))$

$RECB2_j = I_j * CB2_j * (Kp_j/(Kd_{6j} + Kp_j))$

$RECC_j = I_j * CC_j$

where

$RECA_j$ = amount of ruminally escaped sugar from the jth feedstuff, g/day,

$RECB1_j$ = amount of ruminally escaped starch from the jth feedstuff, g/day,

$RECB2_j$ = amount of ruminally escaped available fiber from the jth feedstuff, g/day, and

$RECC_j$ = amount of ruminally escaped unavailable fiber from the jth feedstuff, g/day.

Calculation of microbial yield

Bacterial yields for structural and nonfiber carbohydrate fermenting bacteria are given by the following:

if eNDF < 20, then $YG_1 = YG_1 * (1 - ((20 - eNDF) * 0.025))$

if eNDF < 20, then $YG_2 = YG_2 * (1 - ((20 - eNDF) * 0.025))$

$1/Y_{1j} = (KM_1/Kd_{6j}) + (1/YG_1)$

$1/Y_{2j} = (KM_2/Kd_{4j}) + (1/YG_2)$

$1/Y_{3j} = (KM_2/Kd_{5j}) + (1/YG_2)$

$RATIO_j = RDPEP_j/(RDCA_j + RDCB1_j + RDPEP_j)$

If RATIO >0.18 RATIO = 0.18

$IMP_j = EXP(0.404 * LOG(RATIO_j * 100) + 1.942)$

$FCBACT_j = Y_{1j} * RDCB2_j$

$Y_{2j} = Y_{2j} * (1 + IMP_j * 0.01)$

$Y_{3j} = Y_{3j} * (1 + IMP_j * 0.01)$

$NFCBACT_j = (Y_{2j} * RDCA_j) + (Y_{3j} * RDCB1_j)$

$BACT_j = NFCBACT_j + FCBACT_j$

$BACTN_j = 0.1 * BACT_j$

$NFCBACTN_j = 0.10 * NFCBACT_j$

$FCBACTN_j = 0.10 * FCBACT_j$

$PEPUP_j = RDPEP_j$

$PEPUPN_j = PEPUP_j/6.25$

$EN = PEPUPN + RDPA/6.25 + ((MP_a - MP_r)/6.25) - BACTN$

$PEPBAL = (PEPUP/6.25) - 2/3\ NFCBACTN$

$BACTNBAL = (((PEPUP + RDPA)/6.25) + U) - BACTN$

where

Y_{1j} = yield efficiency of FC bacteria from the available fiber fraction of the jth feedstuff, g FC bacteria/g FC digested,

Y_{2j} = yield efficiency of NFC bacteria from the sugar fraction of the jth feedstuff, g NFC bacteria/g NFC digested,

Y_{3j} = yield efficiency of NFC bacteria from the starch fraction of the jth feedstuff, g NFC bacteria/g NFC digested,

KM_1 = maintenance rate of the fiber carbohydrate bacteria, 0.05 g FC/g bacteria/h,

KM_2 = maintenance rate of the nonfiber carbohydrate bacteria, 0.15 g NFC/g bacteria/h,

YG_1 = theoretical maximum yield of the fiber carbohydrate bacteria, 0.4 g bacteria/g FC,

YG_2 = theoretical maximum yield of the nonfiber carbohydrate bacteria, 0.4 g bacteria/g NFC,

$RATIO_j$ = ratio of peptides to peptide plus NFC in the jth feestuff,

$RDPEP_j$ = peptides in the jth feedstuff,

$RDCA_j$ = g NFC in the A (sugar) fraction of the jth feedstuff ruminally degraded,

$RDCB1_j$ = g NFC in the B1 (starch and pectins) fraction of the jth feedstuff ruminally degraded,

$RDCB2_j$ = g FC in the B2 (available fiber) fraction in the jth feedstuff ruminally degraded,

Kd_{4j} = growth rate of the sugar fermenting carbohydrate bacteria, h^{-1},

Kd_{5j} = growth rate of the starch fermenting carbohydrate bacteria, h^{-1},

Kd_{6j} = growth rate of the fiber carbohydrate bacteria, h^{-1},

IMP_j = percent improvement in bacterial yield, %, due to the ratio of peptides to peptides plus nonfiber CHO in j^{th} feedstuff,

$FCBACT_j$ = yield of fiber carbohydrate bacteria from the jth feedstuff, g/day,

$NFCBACT_j$ = yield of nonfiber carbohydrate bacteria from the jth feedstuff, g/day,

$BACT_j$ = yield of bacteria from the jth feedstuff, g/day,

$BACTN_j$ = bacterial nitrogen, g/day,

$FCBACTN_j$ = fiber carbohydrate bacterial nitrogen, g/day,

$NFCBACTN_j$ = nonfiber carbohydrate bacterial nitrogen, g/day,

$PEPUP_j$ = bacterial peptide from the jth feedstuff, g/day,

$PEPUPN_j$ = bacterial peptide nitrogen from the jth feedstuff, g/day,

MP_a = metabolizable protein supplied, g/day,

MP_{req} = metabolizable protein required, g/day,

EN = nitrogen in excess of rumen bacterial nitrogen and tissue needs, g/day,

$PEPBAL$ = peptide balance, g nitrogen/day,

$BACTNBAL$ = bacterial nitrogen balance, g/day,

U = recycled nitrogen, g/day.

Microbial composition

Bacterial fractions escaping the rumen are

$REBTP_j = 0.60 * 0.625 * BACT_j$
$REBCW_j = 0.25 * 0.625 * BACT_j$
$REBNA_j = 0.15 * 0.625 * BACT_j$
$REBCHO_j = 0.21 * BACT_j$
$REBFAT_j = 0.12 * BACT_j$
$REBASH_j = 0.044 * BACT_j$

where

$REBTP_j$ = amount of bacterial true protein passed to the intestine by the jth feedstuff, g/day,

$REBCW_j$ = amount of bacterial cell wall protein passed to the intestine by the jth feedstuff, g/day,

$REBNA_j$ = amount of bacterial nucleic acids passed to the intestine by the jth feedstuff, g/day,

$REBCHO_j$ = amount of bacterial carbohydrate passed to the intestine by the jth feedstuff, g/day,

$REBFAT_j$ = amount of bacterial fat passed to the intestine by the jth feedstuff, g/day,

$REBASH_j$ = amount of bacterial ash passed to the intestine by the jth feedstuff, g/day.

Intesinal digestibilities and absorption

Equations for calculating digested protein from feed and bacterial sources are listed below:

$DIGPB1_j = REPB1_j$
$DIGPB2_j = REPB2_j$
$DIGPB3_j = 0.80 * REPB3_j$
$DIGFP_j = DIGPB1_j + DIGPB2_j + DIGPB3_j$
$DIGBTP_j = REBTP_j$
$DIGBNA_j = REBNA_j$
$DIGP_j = DIGFP_j + DIGBTP_j + DIGBNA_j$

where

$DIGPB1_j$ = digestible B1 protein from the jth feedstuff, g/day,

$DIGPB2_j$ = digestible B2 protein from the jth feedstuff, g/day,

$DIGPB3_j$ = digestible B3 protein from the jth feedstuff, g/day,

$DIGFP_j$ = digestible feed protein from the jth feedstuff, g/day,

$DIGBTP_j$ = digestible bacterial true protein produced from the jth feedstuff, g/day,

$DIGBNA_j$ = digestible bacterial nucleic acids produced from the jth feedstuff, g/day,

$DIGP_j$ = digestible protein from the jth feedstuff, g/day.

The equations for calculating digested carbohydrate due to the jth feedstuff are listed below:

$DIGFC_j = RECA_j + stdig * RECB1_j + 0.20 * RECB2_j$
$DIGBC_j = 0.95 * REBCHO_j$
$DIGC_j = DIGFC_j + DIGBC_j$

where

$stdig$ = postruminal starch digestibility, g/g,

$DIGFC_j$ = intestinally digested feed carbohydrate from the jth feedstuff, g/day,

$DIGBC_j$ = digested bacterial carbohydrate produced from the jth feedstuff, g/day,

$DIGC_j$ = digestible carbohydrate from the jth feedstuff, g/day.

The following equation is used to calculate ruminally escaped fat from the jth feedstuff:

$REFAT_j = I_j * FAT_j$

where

$REFAT_j$ = amount of ruminally escaped fat from the jth feedstuff, g/day.

FAT_j = fat composition of the jth feedstuff, g/day.

Equations for calculating digestible fat from feed and bacterial sources are listed below:

$$DIGFF_j = 0.95 * REFAT_j$$
$$DIGBF_j = 0.95 * REBFAT_j$$
$$DIGF_j = DIGFF_j + DIGBF_j$$

where

$DIGFF_j$ = digestible feed fat from the jth feedstuff, g/day,

$DIGBF_j$ = digestible bacterial fat from the jth feedstuff, g/day, and

$DIGF_j$ = digestible fat from the jth feedstuff, g/day.

Fecal Output

The following equations calculate undigested feed residues appearing in the feces from NDIP, ADIP, starch, fiber, fat, and ash fractions, based on data summarized by Van Soest (1994):

$$FEPB3_j = (1 - 0.80) * REPB3_j$$
$$FEPC_j = REPC_j$$
$$FEFP_j = FEPB3_j + FEPC_j$$
$$FECB1_j = (1 - stdig) * RECB1_j$$
$$FECB2_j = (1 - 0.20) * RECB2_j$$
$$FECC_j = RECC_j$$
$$FEFC_j = FECB1_j + FECB2_j + FECC_j$$
$$FEFA_j = I_j * ASH_j * 0.5$$

where

$FEPB3_j$ = amount of feed B3 protein fraction in feces from the jth feedstuff, g/day,

$FEPC_j$ = amount of feed C protein fraction in feces from the jth feedstuff, g/day,

$FEFP_j$ = amount of feed protein in feces from the jth feedstuff, g/day,

$FECB1_j$ = amount of feed starch in feces from the jth feedstuff, g/day,

$FECB2_j$ = amount of feed available fiber in feces from the jth feedstuff, g/day,

$FECC_j$ = amount of feed unavailable fiber in feces from the jth feedstuff, g/day,

$FEFC_j$ = amount of feed carbohydrate in feces from the jth feedstuff, g/day,

$FEFA_j$ = amount of undigested feed ash in feces from the jth feedstuff, g/day.

ASH_j = ash composition of the jth feedstuff, g/day.

Microbial matter in the feces is composed of indigestible bacterial cell walls, bacterial carbohydrate, fat, and ash (Van Soest, 1994):

$$FEBCW_j = REBCW_j$$
$$FEBCP_j = FEBCW_j$$
$$FEBC_j = (1 - 0.95) * REBCHO_j$$

$$FEBF_j = (1 - 0.95) * REBFAT_j$$
$$FEBASH_j = REBASH_j$$
$$FEBACT_j = FEBCP_j + FEBC_j + FEBF_j + FEBASH_j$$

where

$FEBCW_j$ = amount of fecal bacterial cell wall protein from the jth feedstuff, g/day,

$FEBCP_j$ = amount of fecal bacterial protein from the jth feedstuff, g/day,

$FEBC_j$ = amount of bacterial carbohydrate in feces from the jth feedstuff, g/day,

$FEBF_j$ = amount of bacterial fat in feces from the jth feedstuff, g/day,

$FEBASH_j$ = amount of bacterial ash in feces from the jth feedstuff, g/day,

$FEBACT_j$ = amount of bacteria in feces from the jth feedstuff, g/day.

Endogenous protein, carbohydrate, and ash are

$FEENGP_j = 0.09 * IDM_j$ (National Research Council, 1989)

$FEENGF_j = 0.017 * DMI_j$ (Lucas et al., 1961)

$FEENGA_j = 0.0119 * DMI_j$ (Lucas et al., 1961)

where

DMI_j = feed DM consumed, g/day,

$FEENGP_j$ = amount of endogenous protein in feces from the jth feedstuff, g/day,

$FEENGF_j$ = amount of endogenous fat in feces from the jth feedstuff, g/day,

$FEENGA_j$ = amount of endogenous ash in feces from the jth feedstuff, g/day.

IDM_j = indigestible dry matter, g/day.

Total fecal DM is calculated by summing protein, carbohydrate, fat, and ash DM contributions from undigested feed residues, microbial matter, and endogenous matter:

$$FEPROT_j = FEFP_j + FEBCP_j + FEENGP_j$$
$$FECHO_j = FEFC_j + FEBC_j$$
$$FEFAT_j = FEBF_j + FEENGF_j$$
$$FEASH_j = FEFA_j + FEBASH_j + FEENGA_j$$
$$IDM_j = (FEFP_j + FEBCP_j + FECHO_j + FEFAT_j + FEASH_j)/0.91$$

where

$FEPROT_j$ = amount of fecal protein from the jth feedstuff, g/day,

$FECHO_j$ = amount of carbohydrate in feces from the jth feedstuff, g/day,

$FEFAT_j$ = amount of fat in feces from the jth feedstuff, g/day,

$FEASH_j$ = amount of ash in feces from the jth feedstuff, g/day, and

IDM_j = amount of DM in feces from the jth feedstuff, g/day.

Total digestible nutrients and energy values of feedstuffs

Apparent TDN is potentially digestible nutrient intake minus indigestible bacterial and feed components appearing in the feces:

$$TDNAPP_j = (DIET\ PROT_j - FEPROT_j) + (DIET\ CHO_j - FECHO_j) + (2.25 * (DIET\ FAT_j - FEFAT_j))$$

where

$TDNAPP_j$ = apparent TDN from the jth feedstuff, g/day.

The ME values for each feed are based on assuming 1 kg of TDN is equal to 4.409 Mcal of DE and 1 Mcal of DE is equal to 0.82 Mcal of ME (National Research Council, 1976):

$$ME_{aj} = 0.001 * TDNAPP_j * 4.409 * 0.82$$
$$MEC_j = ME_{aj}/(I_j * 0.001)$$
$$MEI = \sum_{j=1}^{n} ME_{aj}$$
$$MEC = MEI/DMI$$

where

ME_{aj} = metabolizable energy available from the jth feedstuff, Mcal/day,

MEC_j = metabolizable energy concentration of the jth feedstuff, Mcal/kg,

MEI = metabolizable energy supplied by the diet, Mcal/day,

MEC = metabolizable energy concentration of the diet, Mcal/kg.

CALCULATION OF NET ENERGY VALUES

$$NEga_j = (1.42 * MEC_j - 0.174 * MEC_j^2 + 0.0122 * MEC_j^3 - 1.65)\ (National\ Research\ Council, 1984);$$
$$NEma_j = (1.37 * MEC_j - 0.138 * MEC_j^2 + 0.0105 * MEC_j^3 - 1.12)\ (National\ Research\ Council, 1984)$$

where

$NEga_j$ = net energy for gain content of the jth feedstuff, Mcal/kg,

$NEma_j$ = net energy for maintenance content of the jth feedstuff, Mcal/kg.

METABOLIZABLE PROTEIN

Total feed MP is the sum of each feed MP:
$$MP_{aj} = DIGP_j - DIGBNA_j$$
$$MP_a = \sum_{j=1}^{n} MP_{aj}$$

where

MP_{aj} = metabolizable protein from the jth feedstuff, g/day,

MP_a = metabolizable protein available in the diet, g/day.

AMINO ACID SUPPLY

Essential amino acid composition of the undegradable protein of each feedstuff is used to calculate supply of amino acids from the feeds. Microbial composition of essential amino acids are used to calculate the supply of amino acids from bacteria.

Bacterial amino acid supply to the duodenum

$$REBAA_i = \sum_{j=1}^{n} (AABCW_i * 0.01 * REBCW_j) + (AABNCW_i * 0.01 * REBTP_j)$$

where

$AABCW_i$ = i^{th} amino acid composition of rumen bacteria cell wall protein, g/100 g (Table 10-7),

$AABNCW_i$ = i^{th} amino acid composition of rumen bacteria noncell wall protein, g/100 g (Table 10-7),

$REBCW_j$ = bacterial cell wall protein appearing at the duodenum as a result of fermentation of the jth feedstuff, g/day,

$REBTP_j$ = bacterial noncell wall protein appearing at the duodenum as a result of fermentation of the jth feedstuff, g/day, and

$REBAA_i$ = amount of the ith bacterial amino acid appearing at the duodenum, g/day.

TABLE 10-7 Amino Acid Composition of Rumen Microbial Cell Wall and Noncell Wall Protein (g/100 g protein)

Amino acid	Cell Wall	Noncell Wall	Ruminal Bacteria[a] Mean	SD
Methionine	2.40	2.68	2.60	0.7
Lysine	5.60	8.20	7.90	0.9
Histidine	1.74	2.69	2.00	0.4
Phenylalanine	4.20	5.16	5.10	0.3
Tryptophan	1.63[b]	1.63	—	—
Threonine	3.30	5.59	5.80	0.5
Leucine	5.90	7.51	8.10	0.8
Isoleucine	4.00	5.88	5.70	0.4
Valine	4.70	6.16	6.20	0.6
Arginine	3.82	6.96	5.10	0.7

[a] Average composition and SD of 441 bacterial samples from animals fed 61 dietary treatments in 35 experiments (Clark et al., 1992). Included for comparison to the cell wall and noncell wall values used in this model.

[b] Data were not available, therefore, content of cell wall protein was assumed to be same as noncell wall protein (O'Connor et al., 1993).

Bacterial amino acid digestion

$$DIGBAA_i = \sum_{j=1}^{n} AABNCW_i * 0.01 * REBTP_j$$

where

$DIGBAA_i$ = amount of the ith absorbed bacterial amino acid

$$REFAA_i = \sum_{j=1}^{n} AAINSP_{ij} * 0.01 * (REPB1_j + REPB2_j + REPB3_j + REPC_j)$$

where

$AAINSP_{ij}$ = i^{th} amino acid content of the insoluble protein for the jth feedstuff, g/100 g,

$REPB1_j$ = rumen escaped B1 protein from the jth feedstuff, g/day,

$REPB2_j$ = rumen escaped B2 protein from the jth feedstuff, g/day,

$REPB3_j$ = rumen escaped B3 protein from the jth feedstuff, g/day,

$REPC_j$ = rumen escaped C protein from the jth feedstuff, g/day, and

$REFAA_i$ = amount of ith dietary amino acid appearing at the duodenum, g/day.

Total duodenal amino acid supply

$$REAA_i = REBAA_i + REFAA_i$$

where

$REAA_i$ = total amount of the ith amino acid appearing at the duodenum, g/day.

Feed amino acid digestion

$$DIGFAA_i = \sum_{j=1}^{n} AAINSP_{ij} * 0.01 * (REPB1_j + REPB2_j + 0.8 * REPB3_j)$$

where

$DIGFAA_i$ = amount of the ith absorbed amino acid from dietary protein escaping rumen degradation, g/day.

Total metabolizable amino acid supply

$$AAA_{si} = DIGBAA_i + DIGFAA_i$$

where

AAA_{si} = total amount of the ith absorbed amino acid supplied by dietary and bacterial sources, g/day.

FEED COMPOSITION VALUES FOR USE OF THE NRC MODEL

A feed library developed for use with the computer model (Appendix Table 1) contains feed composition values needed to predict the supply of nutrients available to meet animal requirements. In this library, feeds are described by their chemical, physical, and biological characteristics. Level 1 uses the tabular net energy and protein values, which are consistent where possible with those in Chapter 11. Level 2 uses the feed carbohydrate and protein fractions and their digestion and passage rates to predict net energy and metabolizable protein values for each feed based on the interaction of these variables. For ease of use, the model feed composition table (Appendix Table 1) is organized to make it easy to find and compare feeds of the same type and to find all values for a feed in the same column. It is arranged with feed names listed alphabetically within feed classes of forages-legumes, forages-grasses, forages-cereal grains, high-energy concentrates, high-protein plant concentrates, plant by-products, and animal byproducts. All the chemical, physical, and biological values for each feed are in the column below the feed name. The international feed number (IFN) is given for each feed, where appropriate, for comparison with previous feed composition tables.

Chemical composition of feeds is described by feed carbohydrate and protein fractions that are used to predict microbial protein production, ruminal degradation, and escape of carbohydrates and proteins and ME and MP in level 2. Feed library values for carbohydrate and protein fractions are based on Sniffen et al. (1992) and Van Soest (1994).

Feedstuffs are composed of chemically measurable carbohydrate, protein, fat, ash, and water. The Weende system for proximate analysis has been used for more than 150 years to measure these components as crude fiber, ether extract, dry matter, and total nitrogen, with nitrogen-free extract (NFE) being calculated by difference. However, this system cannot be used to mechanistically predict microbial growth because crude fiber does not represent all of the fiber, NFE does not accurately represent the nonfiber carbohydrates, and protein must be described by fractions related to its ruminal degradation characteristics.

The level 2 model was developed to mechanistically predict microbial growth and ruminal degradation and escape of carbohydrate and protein to more dynamically predict ME and MP feed values. To accomplish this objective, the detergent fiber system of feed analysis is used to compute carbohydrate (fiber carbohydrates, CHO FC, and nonfiber carbohydrates, CHO NFC) and protein fractions according to their fermentation characteristics (A = fast, B = intermediate and slow, and C = not fermented and unavailable to the animal), as described by Sniffen et al. (1992).

Validations of the system implemented in level 2 for predicting feed biological values from feed analysis of carbohydrate and protein fractions have been published (Ainslie et al., 1993; O'Connor et al., 1993; Fox et al., 1995). However, the subcommittee recognizes that considerable research is needed to refine this structure. The decision to implement the second level was based on the need to identify a system that will allow for implementing accumulated knowledge that can lead to accounting for more of the variation in performance. It is then assumed that further research between this revision and the next one will result in refinement of sensitive coefficients to improve the accuracy of its use under specific conditions.

The procedures used to determine each fraction are described as follows (Sniffen et al., 1992); the methods of crude protein fractionation have been recently standardized (Licitra et al., 1996).

1. Residual from neutral detergent fiber (NDF) procedure is total insoluble matrix fiber (cellulose, hemicellulose, and lignin) (Van Soest et al., 1991).

2. Lignin procedure is an indicator of indigestible fiber (Van Soest et al., 1991). Then the unavailable fiber is estimated as lignin $* 2.4$. The factor 2.4 is not constant across feeds. It may overestimate the CHO C fraction in feeds that are of low lignification. However, it appears to be of sufficient accuracy for the current state of the model.

3. Available fiber (CHO fraction B2) is NDF $-$ (NDFN $* 6.25$) $-$ CHO fraction C, and is used to predict ruminal fiber digestion and microbial protein production on fiber. Intestinal digestibility of the B2 fraction that escapes the rumen is assumed to be 20%.

4. Total nitrogen is measured by Kjeldahl (Association of Official Analytical Chemists, 1980).

5. Soluble nitrogen (NPN + soluble true protein) is measured to identify total N rapidly degraded in the rumen (Krishnamoorthy et al., 1983).

6. True protein is precipitated from the soluble fraction to separate the NPN (protein fraction A) from true rapidly degraded protein (protein fraction B1). Protein fraction B1 typically contains albumin and globulin proteins and provides peptides for meeting NFC microbial requirements for maximum efficiency of growth. A small amount of this fraction escapes ruminal degradation and 100 percent is assumed to be digested intestinally. Protein fraction A provides ammonia for both FC and NFC growth.

7. The detergent analysis system (Van Soest et al., 1991) was designed to analyze for carbohydrate and protein fractions in forages. It has limitations in the analysis of other feedstuffs, particularly in the case of animal byproducts and treated plant protein sources. Nitrogen that is insoluble in neutral detergent (without sodium sulfite) and acid detergent (Van Soest et al., 1991) measures slowly degraded plus unavailable protein. Animal proteins do not contain fiber. However, because of filtering problems, analysis with this procedure will yield unrealistic values for ADF and NDF pools. To correct for this problem, all animal proteins have been assigned ADIP values that reflect average unavailable protein due to heat damage and keratins. The residual protein fraction (B2) has been assigned rates reflecting their relatively slower rates.

8. Acid detergent insoluble protein (ADIP) (Van Soest et al., 1991) is used to identify unavailable protein (protein fraction C) and is assumed to have 0 ruminal and intestinal digestibility, realizing some studies have shown digestive disappearance of ADIP. The levels of ADIP can be adjusted where appropriate.

9. NDIP $-$ ADIP identifies slowly degraded available protein (protein fraction B3). This fraction typically contains prolamin and extensin type proteins and nearly all escapes degradation in the rumen and is assumed to have an intestinal digestibility of 80%.

10. (Total nitrogen $* 6.25$) $-$ A $-$ B1 $-$ B3 $-$ C = protein intermediate in degradation rate (protein fraction B2), except for animal protein as described above. This fraction typically contains glutelin protein and extent of ruminal degradation and escape is variable, depending on individual feed characteristics and level of intake. The ruminally escaped B2 is assumed to have an intestinal digestibility of 100 percent.

11. Ash (Association of Official Analytical Chemists, 1980).

12. Solvent-soluble fat (Association of Official Analytical Chemists, 1980). All of this fraction is assumed to escape ruminal degradation and is assumed to have an intestinal digestibility of 95 percent. Only the glycerol and galactolipid are fermented and the fatty acids escape rumen digestion.

13. Nonfiber carbohydrates (sugar, starch; NFC) are computed as 100 $-$ CP $-$ ([NDF $-$ NDF protein] $-$ fat $-$ ash). Pectins are included in this fraction. Pectins are more rapidly degraded than starches but do not give rise to lactic acid.

14. CHO fraction A is nonfiber CHO $-$ starch. It is assumed that these nonstarch polysaccharides are more rapidly degradable than most starches. Nearly all of this fraction is degraded in the rumen, but the small amount that escapes is assumed to have an intestinal digestibility of 100 percent.

15. CHO fraction B1 is nonfiber CHO $-$ sugar. This fraction has a variable ruminal degradability, depending on level of intake, type of grain, degree of hydration and type of processing. Microbial protein

production is most sensitive to ruminal starch degradation in model level 2. The B1 fraction that escapes is assumed to have a variable digestibility, depending on type of grain and type of processing.

Feed physical characteristics are described as effective NDF (eNDF) by Sniffen et al. (1992). The basic eNDF is described as the percent of NDF remaining on a 1.18 mm screen after dry sieving (Smith and Waldo, 1969; Mertens, 1985). This value was then adjusted for density, hydration and degree of lignification of the NDF within classes of feeds (Appendix Table 1). The eNDF was found to be an accurate predictor of rumen pH (Pitt et al., 1996);

$$\text{Rumen pH} = 5.425 + 0.04229 * \%\text{eNDF for }\%\text{eNDF} < 35\% \text{ in DM; } (R^2 = 0.52).$$

The rumen pH is directly related to microbial protein yield (Russell et al., 1992) and FC microbial growth (Pitt et al., 1996). In level 1, the microbial yield multiplier = 1 if eNDF >20 percent and is reduced 2.5 percent for each percentage unit reduction in eNDF <20 percent. Level 2 adjusts microbial protein yield for rumen pH using this same approach but with a more mechanistic adjustment based on predicted microbial growth rates. Adjustment to FC digestion rate is made in level 2, based on the predicted rumen pH.

"Effective NDF" is the percentage of the NDF effective in stimulating chewing and salivation, rumination, and rumen motility. The data of Russell et al. (1992) and Pitt et al. (1996) show that rumen pH < 6.2 results in linear reductions in microbial protein production and FC digestion. Using data in the literature, Pitt et al. (1966) evaluated several approaches for predicting rumen pH; diet content of forage, diet NDF, a mechanistic model of rumen fermentation, or the effective NDF values published by Sniffen et al. (1992). Effective NDF gave predictions of rumen pH similar to the mechanistic model and has the advantage of simplicity and flexibility in application. The tabular values for eNDF can be used as a guide, with adjustments based on field observations and experience. The importance of stimulating salivary flow in buffering the rumen is well documented (Beauchemin, 1991). Additional factors not accounted for in the eNDF system that can influence rumen pH are total grain intake and its digestion rate, and form of grain (whole corn will stimulate rumination but processed corn may not; a higher proportion of the starch in whole corn will escape ruminal fermentation compared to processed corn and other grains). Therefore adjustments or functional equivalents of eNDF must be assigned to feeds in these cases to make the system reflect these conditions. Ionophores will inhibit the growth of *Streptococcis bovis* (*S. bovis*), which produces lactic acid, which is 10 times stronger than the normal volatile fatty acids produced in the rumen. Highly digestible feeds that are high in pectins (soybean hulls, beet pulp, etc.) will not produce the drop in pH as grains do.

Estimated eNDF requirements are provided in Table 10-8 and are based on the data of Pitt et al. (1996).

Feed Biological Values

Level 1 uses tabular energy and protein values in traditional approaches to ration formulation; level 2 permits the user to integrate intake, digestion, and passage rates of carbohydrate and protein fractions to predict metabolizable energy and protein values of feeds for each unique situation.

The tabular TDN values are from summaries of digestion trial data (National Research Council, 1989; Van Soest, 1994), experimental data of subcommittee members, and represent 1 times maintenance, which is appropriate for gestating beef cows. Level 2 computes a TDN value that reflects the integration of level of intake and ruminal digestion and passage rates. Tabular net energy values are computed from TDN with the NRC (1984) equations. Tabular DIP/UIP values are based on Van Soest (1994), NRC (1989), data in the literature, experimental data of subcommittee members, or generated from the model level 2.

TABULAR NET ENERGY VALUES

The net energy system implemented by the 1976 Subcommittee on Beef Cattle Nutrition (National Research Council, 1976) for growing cattle has been successfully used since then to adjust for methane, urinary, and heat increment losses in meeting net energy requirements for maintenance and tissue deposition. This system accounts for differences in usefulness of absorbed energy depending on source of energy and physiological function (National Research Council, 1984). However, these values are not directly measurable in feeds and do not account for the variation in ME and MP derived from feeds with varying levels of intake and extent of ruminal and intestinal digestion. Level 2 allows the prediction of NE values with these variables accounted for.

TABLE 10-8   Estimated eNDF Requirements

Diet Type	Minimum eNDF Required, % of DM
High concentrate to maximize gain/feed fed mixed diet, good bunk mgt, and ionophores	5 to 8[a]
Fed mixed diet, variable bunk mgt, or no ionophore fed	20
High concentrate to maximize NFC use and microbial protein yield	20[b]

[a]To keep rumen pH more than 5.6 to 5.7, the threshold below which cattle stop eating, based on the data of Britton (1989).

[b]To keep rumen pH more than 6.2 to maximum cell wall digestion and/or microbial protein yield.

Both use the 1984 NRC equations to predict NE_m and NE_g values as shown in the equations section. These equations are mechanistic in predicting NE values from the standpoint of reducing the efficiency of use of ME for maintenance and growth (with a relatively greater effect on NE_g) as ME value of the feed declines (National Research Council, 1984). Diet NE_m and NE_g values determined in the body composition data base described by Fox et al. (1992) were regressed against NE_m and NE_g predicted with the 1984 NRC equations. Diet NE_g concentrations varied from approximately 0.90 to 1.50 Mcal/kg. There was no bias in either NE_m or NE_g predicted values, and the R^2 was 0.89 and 0.58, respectively. The lower R^2 for NE_g prediction is the result of feed for gain reflecting all cumulative errors in predicting requirements in this system, because NE_m requirement and feed for maintenance is computed using a fixed 0.077 Mcal/$BW^{0.75}$. Thus, it is likely that this is a "worst-case" scenario for predicted feed NE_g because maintenance requirement can be highly variable (Fox et al., 1992).

TABULAR UIP/DIP VALUES

The system of UIP/DIP values was introduced in *Ruminant Nitrogen Usage* (National Research Council, 1985) and was implemented in the dairy cattle revision (National Research Council, 1989) to more accurately predict protein available to meet rumen microbial requirements and to supplement microbial protein in meeting animal requirements. Level 2 allows the determination of these values mechanistically, based on the integration of feed carbohydrate and protein fractions and microbial growth. The tabular values for use in level 1 are from various sources and represent determinations by various methods. Analytically, DIP and UIP tabular values are determined by either in vitro or in situ methods, which have limitations in predicting ruminal degradation and escape of protein because of the limitations of the procedures and no accounting for variation in effects of digestion and passage rates.

MODEL PREDICTED NET ENERGY AND METABOLIZABLE PROTEIN VALUES

Level 2 permits the user to integrate intake, digestion, and passage rates of carbohydrate and protein fractions to predict metabolizable energy and protein values of feeds for each unique situation. Digestion rates have been assigned to each feed as described by Sniffen et al. (1992). The equations describe how these are used to predict metabolizable energy and protein values. Essential amino acid values have been assigned to feeds to represent their concentration in the undegraded protein fraction, based on O'Connor et al. (1993).

REFERENCES

Agricultural and Food Research Council. 1991. AFRC Technical Committee on Responses to Nutrients Report Number 7. Theory of Response to Nutrients by Farm Animals. Nutr. Abstr. Rev. Ser. B. 61:683–722.

Agricultural and Food Research Council. 1993. Energy and Protein Requirements of Ruminants. Wallingford, U.K.: CAB International.

Ainslie, S. J., D. G. Fox, T. C. Perry, D. J. Ketchen, and M. C. Barry. 1993. Predicting metabolizable protein and amino acid adequacy of diets fed to lightweight Holstein steers. J. Anim. Sci. 71:1312.

Association of Official Analytical Chemists. 1980. Official Methods of Analysis, 13th Ed. Washington, D.C.: Association of Official Analytical Chemists.

Baldwin, R. L., 1995. Modeling Ruminant Digestion and Metabolism. New York: Chapman and Hall.

Baldwin, R. L., J. H. M. Thornley, and D. E. Beever. 1987. Metabolism of the lactating cow. II. Digestive elements of a mechanistic model. J. Dairy Res. 54:107.

Beauchemin, K. A. 1991. Ingestion and mastication of feed by dairy cattle. Vet. Clin. N. Am. Food Anim. Prac. 7(2).

Britton, R. A. 1989. Acidosis: A continual problem in cattle fed high grain diets. Proceedings of the Cornell Nutrition Conference 8–15.

Clark, J. H., T. H. Klusmeyer, and M. R. Cameron. 1992. Microbial protein synthesis and flows of nitrogen fractions to the duodenum of dairy cows. J. Dairy Sci. 75:2304.

Commonwealth Scientific and Industrial Research Organization. 1990. Feeding standards for Australian livestock. East Melbourne: CSIRO Publications.

Dikstra, J. H., D. C. Neal, D. E. Beever, and J. France. 1992. Simulation of nutrient digestion, absorption and outflow in the rumen, model description. J. Nutr. 122:2239.

Evans, E. H., and R. J. Patterson. 1985. Use of dynamic modelling seen as good way to formulate crude protein, amino acid requirements for cattle diets. Feedstuffs 57(42):24.

Fox, D. G., C. J. Sniffen, J. D. O'Connor, P. J. Van Soest, and J. B. Russell. 1992. A net carbohydrate and protein system for evaluating cattle diets. III. Cattle requirements and diet adequacy. J. Anim. Sci. 70:3578.

Fox, D. G., M. C. Barry, R. E. Pitt, D. K. Roseler, and W. C. Stone. 1995. Application of the Cornell net carbohydrate and protein model for cattle consuming forages. J. Anim. Sci. 73:267.

France, J., M. Gill, J. H. M. Thornley, and P. England. 1987. A model of nutrient utilization and body composition in beef cattle. Anim. Prod. 44:371–385.

Gregory, K. E., L. V. Cundiff, and R. M. Koch. 1992. Composite breeds to use in heterosis and breed differences to improve efficiency of beef production. USDA-ARS Misc. Pub. Washington, D.C.: U.S. Department of Agriculture.

Institut National de la Recherche Agronomique. 1989. Ruminant Nutrition. Montrouge, France: John Libbey Eurotext.

Johnson, K. A., and D. E. Johnson. 1995. Methane emmissions from cattle. J. Anim. Sci. 73:2483.

Kohn, A., R. C. Boston, J. D. Ferguson, and W. Chalupa. 1994. The integration and comparison of dairy cow models. Proceedings of the Fourth International Workshop on Modelling Nutrient Utilisation in Farm Animals. Denmark. October, 1994.

Krishnamoorthy, U. C., C. J. Sniffen, M. D. Stern, and P. J. Van Soest. 1983. Evaluation of a mathematical model of digesta and in-vitro simulation of rumen proteolysis to estimate the rumen undegraded nitrogen content of feedstuffs. Br. J. Nutr. 50:555.

Licitra, G., T. M. Hernandez, and P. J. Van Soest. 1996. Standardization of procedures for nitrogen fractionation of ruminant feeds. Anim. Sci. Feed Technol. 57.347.

Lucas, H. L., Jr., W. W. G. Smart, Jr., M. A. Cipolloni, and H. D. Gross. 1961. Relations Between Digestibility and Composition of Feeds and Feeds, S-45 Report. Raleigh: North Carlina State College.

Mertens, D. R. 1985. Effect of fiber on feed quality for dairy cows. P. 209 in Proceedings of the 46th Minnesota Nutrition Conference. St. Paul, Minn.

National Research Council. 1976. Nutrient Requirements of Beef Cattle, Fifth Revised Ed. Washington, D.C.: National Academy Press.

National Research Council. 1984. Nutrient Requirements of Beef Cattle, Sixth Rev. Ed. Washington, D.C.: National Academy Press.

National Research Council. 1985. Ruminant Nitrogen Usage. Washington, D.C.: National Academy Press.

National Research Council. 1987. Predicting Feed Intake of Food-Producing Animals. Washington, D.C.: National Academy Press.

National Research Council. 1989. Nutrient Requirements of Dairy Cattle, Sixth Revised Ed. Update. Washington, D.C.: National Academy Press.

O'Connor, J. D., C. J. Sniffen, D. G. Fox, and W. Chalupa. 1993. A net carbohydrate and protein system for evaluating cattle diets. IV. Predicting amino acid adequacy. J. Anim. Sci. 71:1298.

Oltjen, J. W., A. C. Bywater, R. L. Baldwin, and W. N. Garrett. 1986. Development of a dynamic model of beef cattle growth and composition. J. Anim. Sci. 62:86–97.

Pitt, R. E., J. S. Van Kessel, D. G. Fox, M. C. Barry, and P. J. Van Soest. 1996. Prediction of ruminal volatile fatty acids and pH within the net carbohydrate and protein system. J. Anim Sci. 74:226.

Robinson, T. F., D. H. Beermann, T. M. Byrem, D. E. Ross, and D. G. Fox. 1995. Effects of abomasal casein infusion on mesenteric drained viscera amino acid absorption, hindlimb amino acid net flux and whole body nitrogen balance in Holstein steers. J. Anim. Sci. 73(Suppl. 1):140.

Russell, J. B., J. D. O'Connor, D. G. Fox, P. J. Van Soest, and C. J. Sniffen. 1992. A net carbohydrate and protein system for evaluating cattle diets. I. Ruminal fermentation. J. Anim. Sci. 70:3551.

Smith, L. W., and D. R. Waldo. 1969. Method for sizing forage cell wall particles. J. Dairy Sci. 52:2051.

Sniffen, C. J., J. D. O'Connor, P. J. Van Soest, D. G. Fox, and J. B. Russell. 1992. A net carbohydrate and protein system for evaluating cattle diets. II. Carbohydrate and protein availability. J. Anim. Sci. 70:3562.

Tylutki, T. P., D. G. Fox, and R. G. Anrique. 1994. Predicting net energy and protein requirements for growth of implanted and nonimplanted heifers and steers and nonimplanted bulls varying in body size. J. Anim. Sci. 72:1806-1813.

Van Soest, P. J. 1994. Nutritional Ecology of the Ruminant, 2nd Ed. Ithaca, N.Y.: Cornell University Press.

Van Soest, P. J., J. B. Robertson, and B. A. Lewis. 1991. Methods for dietary fiber, neutral detergent fiber, and nonstarch polysaccharides in relation to animal nutrition. J. Dairy Sci. 74:3583.

Waghorn, G. D., and R. L. Baldwin. 1984. Model of metabolite flux with mammary gland of the lactating cow. J. Dairy Sci. 67:531.

Composition of Selected Feeds

see "Errata" - front cover

Table 11-1 contains nutrient composition data for commonly used beef cattle feeds from, primarily, nine commercial laboratories in the United States and Canada. Entries for feed energy values and other entries for which no number of samples (N) or standard deviation (SD) is given are from *Nutrient Requirements of Dairy Cattle, Sixth Revised Edition* (National Research Council, 1989). Wet-chemistry techniques were used to determine nutrient concentrations. International feed numbers have been included; however, they have not been included for data sets from the commercial laboratories that combine feeds with more than one international feed number. For example, most laboratories only described the feed as, for example, "alfalfa hay" without giving the maturity.

Feeds in Appendix Table 1A that have the same International Feed Number as feeds in Table 11-1 were made to match those in Table 11-1 as nearly as possible. The majority of the nutrient analyses given in Table 11-1 were conducted after 1988 and thus reflect the values obtained with recent production and manufacturing processes, and analytical techniques. The table shows the feed name, mean concentration of nutrients, number of samples analyzed, and standard deviation (SD). Because crop varieties, weather, soil fertility and type, processing method, storage conditions, and sampling technique all influence nutrient concentrations, an average value without an estimate of the normal variation is of limited value. An estimate of the variation associated with the nutrient concentration of a given feed can also be used in stoichastic programming to reduce ration costs (D'Alfonso et al., 1992).

Data from this table is intended to help producers evaluate whether data they receive on their own feedstuffs are within normal ranges. In comparing table values with an individual sample, keep in mind that the larger the number of samples analyzed, the more reliable the table value. The SD is an estimate of the variation existing among samples of the same feed. For example, 5,883 samples of alfalfa hay had a mean protein concentration of 18.61 percent and an SD of 2.84. This means that 66.6 percent of the alfalfa samples analyzed had a crude protein concentration between 15.77 and 21.45 percent (mean ± 1 SD) and 95 percent of the samples were between 12.93 and 24.29 percent (mean ± 2 SD). Nutrient concentration varies for many feedstuffs, but if the SD value for an individual sample is greater than 2 SD from the mean, verification of that value is recommended.

Estimates of the ruminal undegradability of crude protein are included in Table 11-1. The mean values given in the table are probably lower than what would be observed with cattle allowed to consume feed ad libitum, because the experimental techniques used in measuring protein degradability often require restricted intakes. Although the use of undegradable protein in diet formulation is not an exact science, ignoring the differences in degradability among feedstuffs is no longer practical, and many factors affect the amount of dietary protein escaping ruminal degradation (National Research Council, 1985). In addition, monensin slows protein degradation (Poos et al., 1979; Isichei and Bergen, 1980; Whetsone et al., 1981), however, monensin also inhibits bacterial protein synthesis (Poos et al., 1979; Chalupa, 1980), so total protein supply to the intestine may not be increased. Also proteins such as soybean meal with an isoelectric point within the range of the normal rumen pH (5.5 to 7.0) may have higher undegradabilities when included in high concentrate diets that decrease rumen pH (Loerch et al., 1983; Zinn and Owens, 1983). Consequently, the subcommittee recommends increasing the undegradability value of the more degradable protein sources by 1 SD when used in higher energy diets with access ad libitum.

EFFECTS OF PROCESSING TREATMENT

Many treatments are used to improve the nutritive value of feedstuffs for beef cattle. The treatments as such are

TABLE 11-1   Means and Standard Deviations for the Composition Data of Feeds Commonly Used in Beef Cattle Diets

Entry No.	Feed Name/ Description	International Feed No.	TDN (%)	DE (Mcal/ kg)	ME (Mcal/ kg)	NE_m	NE_g	Dry Matter (%)	Crude Protein (%)	Ruminal Unde-grad-ability (%)	Ether Extract (%)	Fiber (%)	NDF (%)	ADF (%)
			Value as Determined at Maintenance Intake			Net Energy Values for Growing-Cattle Mcal/kg								
01	ALFALFA (*Medicago sativa*) Fresh		62	2.73	2.24	1.38	0.80	23.40	18.90	22	3.15	26.50	47.10	36.80
	N		—	—	—	—	—	22	3146	—	9	10	2092	3126
	SD		—	—	—	—	—	3.66	3.00	—	0.65	2.28	7.02	5.11
02	Fresh, late vegetative	2-00-181	66	2.91	2.39	1.51	0.92	23.20	22.20	22	2.90	24.20	30.90	24.00
	N		—	—	—	—	—	14	17	—	4	14	12	6
	SD		—	—	—	—	—	3.39	2.00	—	0.95	2.29	4.79	3.66
03	Fresh, full bloom	2-00-188	50	2.22	1.81	0.97	0.42	23.80	19.3	22	2.6	30.4	38.6	35.9
	N		—	—	—	—	—	8	8	—	2	2	12	2
	SD		—	—	—	—	—	3.88	3.70	—	0.57	1.83	6.14	2.82
04	Hay		60	2.65	2.17	1.31	0.74	90.60	18.6	28	2.39	26.1	43.9	33.8
	N		—	—	—	—	—	5,895	5883	12	169	122	4675	5764
	SD		—	—	—	—	—	1.76	2.84	7	1.16	4.54	6.44	4.67
05	Hay, sun-cured, early bloom	1-00-059	60	2.65	2.17	1.31	0.74	90.50	19.90	22	2.9	28.5	39.3	31.9
	N		—	—	—	—	—	43	63.00	—	28	29	14	15
	SD		—	—	—	—	—	1.92	2.25	—	1.35	3.98	3.58	2.40
06	Hay, sun-cured, mid-bloom	1-00-063	58	2.56	2.10	1.24	0.68	91.00	18.70	—	2.6	28.0	47.1	36.7
	N		—	—	—	—	—	60	56.00	—	23	22	22	26
	SD		—	—	—	—	—	1.88	2.93	—	1.82	4.25	6.53	2.58
07	Hay, sun-cured, full bloom	1-00-068	55	2.43	1.99	1.14	0.58	90.90	17.0	22	3.4	30.1	48.8	38.7
	N		—	—	—	—	—	210	20.00	—	12	14	10	9
	SD		—	—	—	—	—	2.06	2.50	—	1.73	4.27	3.49	2.42
08	Meal		62	2.73	2.24	1.38	0.80	91.70	18.9	59	2.70	26.5	42.0	33.2
	N		—	—	—	—	—	145	97.00	10	60	73	11	26
	SD		—	—	—	—	—	1.93	2.01	17	0.48	2.48	7.7	4.7
09	Meal, dehydrated, 15% protein	1-00-022	59	2.60	2.13	1.27	0.70	90.40	17.30	59	2.4	29.0	55.4	37.5
	N		—	—	—	—	—	23	21	—	13	18	1	2
	SD		—	—	—	—	—	2.18	1.75	—	0.44	3.17	—	1.47
10	Meal, dehydrated, 17% protein	1-00-023	61	2.69	2.21	1.34	0.77	91.80	18.90	59	3.00	26.2	45.0	34.3
	N		—	—	—	—	—	72	50	—	37	46	1	2
	SD		—	—	—	—	—	1.50	0.68	—	0.49	2.25	—	0.95
11	Silage	3-00-216	63	2.78	2.28	1.41	0.83	44.10	19.5	23	3.70	25.4	47.5	37.5
	N		—	—	—	—	—	8289	8315	6	84	38	6842	8295
	SD		—	—	—	—	—	11.6	2.93	8	0.92	2.9	6.6	4.9
12	BARLEY (*Hordeum vulgare*) Grain	4-00-549	88	3.84	3.03	2.06	1.40	88.1	13.20	27	2.2	3.37	18.1	5.77
	N		—	—	—	—	—	1743	1884	16	8	6	1216	1399
	SD		—	—	—	—	—	0.86	1.50	10	0.44	1.6	4.8	2.2
13	Silage		60	2.65	2.17	1.31	0.74	37.10	11.90	23	2.92	—	56.8	33.9
	N		—	—	—	—	—	188	186	—	5	—	44	185
	SD		—	—	—	—	—	9.30	2.70	—	0.61	—	5.7	4.2
14	Straw	1-00-498	40	1.76	1.45	0.60	0.08	91.20	4.40	25	1.90	41.5	72.5	48.8
	N		—	—	—	—	—	29	35	—	7	26	2	3
	SD		—	—	—	—	—	3.31	0.91	—	0.27	4.03	1.83	4.65
15	BEET SUGAR (*Beta vulgaris altissima*) Pulp, dehydrated	4-00-669	74	3.26	2.68	1.76	1.14	91.00	9.8	45	0.6	20.0	44.6	27.5
	N		—	—	—	—	—	47	31	4	25	29	2	5
	SD		—	—	—	—	—	1.37	1.04	14	0.15	2.40	20.4	6.79
16	BERMUDAGRASS, COASTAL (*Cynodon dactylon*) Fresh	2-00-719	64	2.82	2.31	1.44	0.86	30.30	12.6	20	3.7	28.4	73.3	36.8
	N		—	—	—	—	—	15	48	—	10	11	41	41
	SD		—	—	—	—	—	6.91	2.88	—	0.95	1.77	5.10	4.64
17	Hay, sun-cured, 43-56 days growth	1-09-210	49	2.16	1.77	0.93	0.39	93.0	7.8	23	2.7	32.6	—	—
	N		—	—	—	—	—	1	4	—	2	2	3	3
	SD		—	—	—	—	—	—	1.19	—	1.83	4.73	2.45	4.18
18	BLUEGRASS, KENTUCKY (*Poa pratensis*) Fresh, early vegetative	2-00-777	72	3.17	2.60	1.70	1.08	30.80	17.4	20	3.5	25.2	55	29
	N		—	—	—	—	—	4	2	—	2	2	1	1
	SD		—	—	—	—	—	0.69	0.14	—	0.07	0.21	—	—
19	BLOOD Meal	5-00-380	66	2.91	2.49	1.51	0.92	90.50	93.8	75	1.69	1.35	41.6	2.81
	N		—	—	—	—	—	52	40	7	19	2	28	37
	SD		—	—	—	—	—	5.9	12.1	12	3.4	14	20.2	2.60

Ash (%)	Calcium (%)	Phosphorus (%)	Magnesium(%)	Potassium (%)	Sodium (%)	Sulfur (%)	Copper (mg/kg)	Iodine (mg/kg)	Iron (mg/kg)	Manganese (mg/kg)	Selenium (mg/kg)	Zinc (mg/kg)	Cobalt (mg/kg)	Molybdenum (mg/kg)
10.50	1.29	0.26	0.26	2.78	0.01	0.27	4.47	—	191	26.3	—	15.2	0.44	0.94
41	3079	3079	3079	3079	2750	401	2748	—	2749	2750	—	2748	6	2742
0.75	0.30	0.08	0.08	0.59	0.03	0.05	4.82	—	350	29.60	—	29.7	0.05	1.00
10.20	1.71	0.30	0.36	2.27	0.21	0.36	10.7	—	111	41	—	—	0.17	—
10	10	10	10	10	2	9	1	—	1	2	—	—	1	—
0.83	0.48	0.04	0.10	0.50	0.01	0.09	—	—	—	18	—	—	—	—
10.9	1.19	0.26	0.40	3.62	0.16	0.31	14.9	—	293	41	—	32	—	0.49
8	6	6	6	6	6	1	5	—	6	6	—	6	—	5
2.35	0.24	0.04	0.10	0.89	0.07	—	2.33	—	232	35.2	—	16.2	—	0.06
8.57	1.40	0.28	0.28	2.43	0.05	0.28	7.3	—	198	30.3	0.41	18.8	0.65	0.93
378	5771	5769	5319	5324	2813	654	2896	—	2904	2895	158	2904	38	1,354
0.92	0.32	0.05	0.07	0.53	0.06	0.07	6.5	—	319	27	0.31	12	0.34	1.30
9.2	1.63	0.21	0.34	2.56	0.15	0.30	12.7	0.17	227	36	0.55	30	—	0.29
36	98	91	93	96	7	1	93	1	97	95	86	97	—	9
1.61	0.39	0.05	0.10	0.61	0.13	—	3.0	—	137	25.5	0.39	7.6	—	0.24
8.5	1.37	0.22	0.35	1.56	0.12	0.28	17.7	0.16	225	28	—	31	—	0.39
41	9	13	7	8	5	3	3	1	4	4	—	3	—	2
1.48	0.28	0.05	0.11	0.51	0.05	0.03	5.64	—	182	7.7	—	14.1	—	0.05
7.8	1.19	0.24	0.27	1.56	0.07	0.27	9.9	0.13	155	42	—	26	—	0.23
16	6	7	6	7	3	1	6	1	8	6	—	4	—	4
1.07	0.14	0.08	0.11	0.75	0.07	—	4.2	—	28.1	8.6	—	2.8	—	0.28
10.3	1.53	0.27	0.29	2.48	0.09	0.25	11.4	—	396	39.4	0.33	35.8	0.31	3.0
41	53	56	31	34	31	14	25	—	27	27	4	24	5	17
0.75	0.25	0.03	0.04	0.19	0.05	0.02	3.1	—	66	5.0	0.35	9.3	0.04	0.70
9.9	1.38	0.25	0.29	2.46	0.08	0.21	10.4	0.13	309	30.7	0.31	21.4	—	0.19
12	5	5	5	6	4	4	2	1	3	2	2	2	—	1
0.93	0.07	0.03	0.04	0.14	0.01	0.02	1.7	—	54.7	2.5	0.32	1.4	—	—
10.6	1.51	0.25	0.32	2.61	0.11	0.24	9.3	0.16	441	34	0.36	21	—	0.33
21	25	28	12	11	10	8	6	1	7	7	2	5	—	3
0.61	0.13	0.02	0.04	0.29	0.05	0.03	1.74	3.74	0.40	7.5	0.04	—	—	—
9.5	1.32	0.31	0.26	2.85	0.02	0.28	12.1	—	252	32.4	0.18	19.5	0.65	1.27
26	8190	8190	8164	8164	4307	1251	4307	—	4307	4307	7	4307	2	4307
1.4	0.27	0.05	0.06	0.55	0.03	0.08	23.7	—	407	29.2	0.07	24.8	0.15	0.97
2.4	0.05	0.35	0.12	0.57	0.01	0.15	5.3	—	59.5	18.3	—	13.0	0.35	1.16
1153	1395	1906	1409	257	1408	63	1408	—	1408	1408	—	1408	16	196
0.18	0.03	0.05	0.02	0.18	0.01	0.02	2.8	—	56.3	8.5	—	5.03	0.28	0.55
8.3	0.52	0.29	0.19	2.57	0.12	0.24	7.7	—	375	44.8	0.15	24.5	0.72	1.56
2	187	187	82	82	82	32	82	—	82	82	32	82	6	82
0.32	0.16	0.07	0.05	0.83	0.32	0.07	2.9	—	602	28	0.12	13.7	0.41	0.94
7.5	0.30	0.07	0.23	2.36	0.14	0.17	5.40	—	200.	16	—	7	—	0.07
8	34	40	22	22	5	5	18	—	20	4	17	1	—	—
1.40	0.09	0.03	0.05	0.48	0.01	0.01	1.33	—	72.0	0.73	0.58	—	—	—
5.3	0.68	0.10	0.28	0.22	0.20	0.22	13.8	—	293	37.6	0.12	1.0	—	0.08
22	18	23	21	12	8	9	5	—	13	10	1	3	—	3
1.29	0.07	0.01	0.05	0.07	0.07	0.01	0.07	—	62.8	1.3	—	0.03	—	0.04
8.1	0.49	0.27	0.17	1.70	0.06	—	6.0	—	2.44	—	—	—	—	—
34	8	8	1	1	1	—	1	—	1	—	—	—	—	—
1.86	0.07	0.03												
76.6	38.3	8.0	0.26	0.18	0.13	1.30	0.08	0.21	9	—	290	—	—	.12
2	1	1	1	1	1	1	1	—	1	—	—	—	1	—
1.34	—	—	—	—	—	—	—	—	—	—	—	—	—	—
9.4	0.50	0.44	0.18	2.27	0.14	0.17	—	—	300	—	—	—	—	—
1	2	2	1	1	1	1	—	—	1	—	—	—	—	—
	0.09	0.04												
2.62	0.40	0.32	0.04	0.31	0.40	0.80	13.9	—	2281	11.7	—	33.0	—	0.53
15	39	39	39	39	39	27	39	—	39	39	—	39	—	39
2.4	0.74	0.37	0.06	0.22	0.26	0.39	6.4	—	469	6.4	—	13.9	—	1.03

TABLE 11-1   Means and Standard Deviations for the Composition Data of Feeds Commonly Used in Beef Cattle Diets —*Continued*

Entry No.	Feed Name/ Description	International Feed No.	Value as Determined at Maintenance Intake			Net Energy Values for Growing-Cattle Mcal/kg		Dry Matter (%)	Crude Protein (%)	Ruminal Undegradability (%)	Ether Extract (%)	Fiber (%)	NDF (%)	ADF (%)
			TDN (%)	DE (Mcal/kg)	ME (Mcal/kg)	NE_m	NE_g							
	BREWER'S GRAINS													
20	Dehydrated	5-02-141	66	2.39	2.39	1.51	0.91	90.20	29.2	50	10.8	7.8	48.7	31.2
		N	—	—	—	—	—	581	571	10	10	40	133	320
		SD	—	—	—	—	1.51	3.70	13	3.25	1.47	10.2	4.4	0.34
	BROOME, SMOOTH (*Bromus inermis*)													
21	Fresh, early vegetative	2-00-956	74	3.26	2.68	1.76	1.14	26.1	21.3	23	4.0	23.0	47.9	31.0
		N	—	—	—	—	—	8	6	—	3	3	4	5
		SD	—	—	—	—	—	6.39	2.47	—	0.35	0.53	3.63	3.16
22	Hay, sun-cured, mid-bloom	1-05-633	56	2.47	2.03	1.18	0.61	87.6	14.4	23	2.2	31.9	57.7	36.8
		N	—	—	—	—	—	2	4	—	3	3	1	3
		SD	—	—	—	—	—	—	3.22	—	0.16	3.21	—	4.58
23	Hay, sun-cured, mature	1-00-944	53	2.34	1.92	1.07	0.52	92.6	6.0	23	2.0	32.2	70.5	44.8
		N	—	—	—	—	—	6	2	—	1	2	1	1
		SD	—	—	—	—	—	0.54	0.28	—	—	2.82	—	—
	CANARY GRASS, REED (*Phalaris arundianacea*)													
24	Fresh	2-01-113	60	2.65	2.17	1.31	0.74	22.8	17.0	19	4.1	24.4	46.4	28.3
		N	—	—	—	—	—	4	3	—	2	2	1	1
		SD	—	—	—	—	—	4.89	3.65	—	0.49	3.39	—	—
25	Hay, sun-cured	1-01-104	55	2.43	1.99	1.14	0.58	89.3	10.2	22	3.0	33.9	70.5	36.6
		N	—	—	—	—	—	10	14	—	10	10	6	6
		SD	—	—	—	—	—	2.08	2.06	—	0.64	3.80	1.14	0.78
	CANOLA (*Brassica dapus*)													
26	Grain		70	3.09	2.53	1.63	1.03	92.2	30.7	20	7.4	12.5	55.4	22.1
		N	—	—	—	—	—	39	346	—	7	6	66	150
	Canola	SD	—	—	—	—	—	1.55	4.32	—	0.71	1.82	10.4	3.89
27	Meal, sun-cured	5-03-871	69	3.04	2.49	1.60	1.0	82.0	40.9	28	3.47	13.3	27.2	17.0
		N	—	—	—	—	—	154	129	10	105	120	24	19
	CITRUS (*Citrus* spp)	SD - See "errata"													
28	Pomace without fines, dehydrated	4-01-237	82	3.62	2.96	2.00	1.35	91.1	6.7	30	3.7	12.8	23.0	23.0
		N	—	—	—	—	—	275	365	—	260	314	1	1
		SD	—	—	—	—	—	1.52	0.40	—	0.86	1.19	—	—
	CLOVER, LADINO (*Trifolium pratense*)													
29	Fresh, early vegetative	2-01-380	68	3.00	2.46	1.57	0.97	19.3	25.8	20	4.6	13.9	35	33
		N	—	—	—	—	—	4	3	—	3	3	1	1
		SD	—	—	—	—	—	1.44	1.21	—	1.87	0.40	—	—
30	Hay, sun-cured	1-01-378	60	2.65	2.17	1.31	0.74	89.1	22.4	22	2.7	20.8	36.0	32.0
		N	—	—	—	—	—	5	4	—	3	3	1	1
		SD	—	—	—	—	—	2.71	1.18	—	0.750	2.90	—	—
	CLOVER, RED (*Trifolium pratense*)													
31	Fresh, early bloom	2-01-428	69	3.04	2.49	1.6	1.00	19.6	20.8	20	5.0	23.2	40.0	31.0
		N	—	—	—	—	—	5	3	—	2	3	1	1
		SD	—	—	—	—	—	0.46	3.06	—	0.07	4.25	—	—
32	Fresh, full bloom	2-01-429	64	2.82	2.31	1.44	0.86	26.2	14.6	22	2.9	26.1	43.0	35.0
		N	—	—	—	—	—	4	3	—	2	2	1	1
		SD	—	—	—	—	—	3.00	0.46	—	1.55	5.02	—	—
33	Hay, sun-cured	1-01-415	55	2.43	1.99	1.14	0.58	88.4	15.0	24	2.8	30.7	46.9	36.0
		N	—	—	—	—	—	21	13	—	11	11	2	2
		SD	—	—	—	—	—	1.91	1.91	—	0.32	3.96	12.9	9.19
	CORN, DENT YELLOW (*Zea mays indentata*)													
34	Cobs, ground	1-28-234	50	2.21	1.81	0.97	0.42	90.1	2.8	50	0.6	35.4	87.0	39.5
		N	—	—	—	—	—	3	3	—	3	3	2	2
		SD	—	—	—	—	—	0.25	0.28	—	0.148	0.40	2.82	6.36
35	Distiller's grains with solubles dehydrated	5-28-236	90	3.88	3.18	2.18	1.50	90.3	30.4	52	10.7	6.9	46.0	21.3
		N	—	—	—	—	—	450	439	6	166	76	158	370
		SD	—	—	—	—	—	2.19	3.55	20	3.12	1.33	8.71	4.82
36	Gluten feed	5-28-243	80	3.53	2.89	1.94	1.30	90.0	23.8	22	3.91	7.5	36.2	12.7
		N	—	—	—	—	—	33	57	2	10	6	25	48
		SD	—	—	—	—	—	1.69	3.59	11	1.04	2.41	6.8	2.62
37	Gluten meal	5-28-242	89	3.92	3.22	2.20	1.52	88.2	66.3	59	2.56	5.5	8.9	7.9
		N	—	—	—	—	—	20	29	8	12	1	12	25
		SD	—	—	—	—	—	2.10	2.97	12	0.30	—	2.86	4.1
38	Grain, cracked	4-20-698	90	3.92	3.25	2.24	1.55	90.0	9.8	55	4.06	2.29	10.8	3.3
		N	—	—	—	—	—	3708	3579	14	134	127	2488	3481
		SD	—	—	—	—	—	0.88	1.06	19	0.64	0.90	3.57	1.83
39	Silage, well-eared	3-28-250	72	3.17	2.60	1.69	1.08	34.6	8.65	30	3.09	19.5	46.0	26.6
		N	—	—	—	—	—	32231	32364	4	314	54	27777	32315
		SD	—	—	—	—	—	7.25	1.28	6	0.81	4.44	6.50	4.19

Ash (%)	Calcium (%)	Phosphorus (%)	Magnesium(%)	Potassium (%)	Sodium (%)	Sulfur (%)	Copper (mg/kg)	Iodine (mg/kg)	Iron (mg/kg)	Manganese (mg/kg)	Selenium (mg/kg)	Zinc (mg/kg)	Cobalt (mg/kg)	Molybdenum (mg/kg)
4.18	0.29	0.70	0.27	0.58	0.15	0.40	11.3	—	221	44	—	82.0	—	3.16
100	267	267	267	267	267	90	267	—	267	267	—	267	—	267
0.18	0.10	0.05	0.18	0.23	0.08	6.4	—	104	12.7	—	13.7	—	0.74	—
10.4	0.55	0.45	0.32	3.16	—	0.20	—	—	—	—	—	21	—	—
6	2	2	—	1	—	1	—	—	—	—	—	1	—	—
0.45	0.10	0.18	—	—	—	—	—	—	—	—	—	—	—	—
10.9	0.29	0.28	0.10	1.99	0.01	—	25.0	—	91	40	—	30	—	0.58
3	1	1	1	1	1	—	1	—	1	1	—	1	—	1
1.75	—	—	—	—	—	—	—	—	—	—	—	—	—	—
7.2	0.26	0.22	0.12	1.85	0.01	—	10.4	—	80	73	—	24	—	0.19
2	3	2	3	3	2	—	2	—	2	2	—	1	—	2
1.41	0.15	0.01	0.07	0.80	—	—	5.1	—	28.2	45.8	—	—	—	0.06
10.2	0.36	0.33	—	3.64	—	—	—	—	—	—	—	—	—	—
3	2	2	—	1	—	—	—	—	—	—	—	—	—	—
1.85	0.06	0.04	—	—	—	—	—	—	—	—	—	—	—	—
8.1	0.36	0.24	0.22	2.91	0.02	0.14	11.9	—	150	92	—	18	—	—
10	12	12	8	8	2	1	1	—	1	1	—	1	—	—
0.80	0.09	0.04	0.06	0.47	0.01	—	—	—	—	—	—	—	—	—
4.0	0.30	0.59	0.21	0.16	0.03	0.42	12.4	—	253	47.7	—	88.3	—	4.2
11	126	126	126	126	126	17	126	—	126	126	—	126	—	126
0.03	0.12	0.09	0.04	0.17	0.10	0.06	5.2	—	370	9.8	—	16.8	—	0.85
7.10	0.70	1.20	0.57	1.37	0.03	1.17	7.95	—	211	55.8	—	71.5	—	1.79
31	102	133	27	38	25	14	14	—	25	27	—	27	—	22
6.6	1.88	0.13	0.17	0.77	0.08	0.08	6.14	—	360	7	—	15	—	0.19
335	20	16	9	14	5	6	6	—	11	8	—	6	—	3
0.80	0.42	0.02	0.02	0.17	0.02	0.04	0.42	—	335	0.7	—	2.6	—	0.10
11.9	1.27	0.35	0.42	2.40	0.12	0.16	—	—	—	—	—	20	—	—
3	1	1	1	1	1	1	—	—	—	—	—	1	—	—
1.38	—	—	—	—	—	—	—	—	—	—	—	—	—	—
9.4	1.45	0.33	0.47	2.44	0.13	0.21	9.41	0.30	470	123	—	17	—	0.16
2	3	3	3	3	1	3	3	1	4	3	—	1	—	1
0.16	0.22	0.06	0.07	0.27	—	0.01	1.2	—	211	60.9	—	—	—	—
10.2	2.26	0.38	0.51	2.49	0.20	0.17	9.0	0.25	300	50	—	19	—	0.16
2	1	1	1	1	1	1	1	1	1	1	—	1	—	1
0.567	1	—	—	—	—	—	—	—	—	—	—	—	—	—
7.8	1.01	0.27	0.51	1.96	0.20	0.17	10.0	0.25	300	47	—	16	—	0.12
2	1	1	1	1	1	1	1	1	1	1	—	1	—	1
0.70	—	—	—	—	—	—	—	—	—	—	—	—	—	—
7.5	1.38	0.24	0.38	1.81	0.18	0.16	11.0	0.25	238	108	—	17	—	0.16
9	11	11	7	11	2	2	4	1	8	4	—	3	—	1
0.88	0.22	0.06	0.13	0.58	0.04	0.01	12.6	—	121	46.5	—	17.1	—	—
1.8	0.12	0.04	0.07	0.89	0.08	0.47	7.00	—	230	6	0.08	5	—	0.13
1	2	2	2	2	1	2	1	—	1	1	1	1	—	1
—	0.01	0.01	0.01	0.02	—	0.01	—	—	—	—	—	—	—	—
4.60	0.26	0.83	0.33	1.08	0.30	0.44	10.6	—	358	27.6	—	67.8	—	1.80
18	384	384	383	383	382	113	383	—	383	383	—	383	—	291
0.86	0.23	0.15	0.08	0.27	0.26	0.12	7.81	—	858	11.7	—	23.9	—	0.45
6.9	0.07	0.95	0.40	1.40	0.26	0.47	6.98	—	226	22.1	—	73.3	—	1.80
8	61	61	61	61	61	20	61	—	61	61	—	61	—	49
1.74	0.05	0.29	0.10	0.34	0.20	0.09	2.55	—	127	7.28	—	19.4	—	0.49
2.86	0.07	0.61	0.15	0.48	0.06	0.90	4.76	—	159	20.6	—	61.4	—	0.93
7	33	33	33	33	33	8	33	—	33	33	—	33	—	33
0.52	0.09	0.29	0.16	0.06	0.13	0.16	6.5	—	86.9	38.1	—	86.6	—	0.63
1.46	0.03	0.32	0.12	0.44	0.01	0.11	2.51	—	54.5	7.89	0.14	24.2	—	0.60
87	3516	3515	3437	3437	1749	382	1743	—	1738	1741	17	1743	—	1691
0.33	0.07	0.04	0.03	0.06	0.05	0.02	1.98	—	43.2	7.1	0.12	11.1	—	0.31
3.59	0.25	0.22	0.18	1.14	0.01	0.12	4.18	—	131	23.5	—	17.7	—	0.53
56	32195	32195	32125	32127	13313	3335	13316	—	13323	13316	—	13323	—	10815
0.78	0.09	0.04	0.03	0.26	0.03	0.03	5.14	—	340	25.1	—	16.1	—	0.58

TABLE 11-1   Means and Standard Deviations for the Composition Data of Feeds Commonly Used in Beef Cattle Diets —*Continued*

Entry No.	Feed Name/ Description	International Feed No.	Value as Determined at Maintenance Intake TDN (%)	DE (Mcal/kg)	ME (Mcal/kg)	Net Energy Values for Growing-Cattle Mcal/kg NEm	NEg	Dry Matter (%)	Crude Protein (%)	Ruminal Undegradability (%)	Ether Extract (%)	Fiber (%)	NDF (%)	ADF (%)
	COTTON (*Gossypium* spp.)													
40	Hulls	1-01-599	42	1.85	1.52	0.68	0.15	90.4	4.2	50	1.7	47.8	88.3	65.3
	N		—	—	—	—	—	22	28	—	26	27	2	4
	SD		—	—	—	—	—	1.34	0.74		1.19	3.07	2.41	4.31
41	Seed	5-01-614	90	3.97	3.25	2.24	1.55	89.4	24.4	27	17.5	25.6	51.6	41.8
	N		—	—	—	—	—	241	476	—	167	62	260	418
	SD		—	—	—	—	—	2.51	3.16	—	2.99	3.91	6.04	4.78
42	Seed, meal solv-extd	5-07-873	75	3.31	2.71	1.79	1.16	90.2	46.1	43	3.15	13.2	28.9	17.9
	N		—	—	—	—	—	138	117	21	91	53	25	35
	SD		—	—	—	—	—	1.57	3.17	11	1.72	1.64	7.05	3.27
	FATS													
43	Fat, animal, hydrolyzed	4-00-376	177	7.30	7.30	6.00	4.50	99.2	—	—	99.2	—	—	—
	N		—	—	—	—	—	5	—	—	3	—	—	—
44	Oil, vegetable	4-05-077	177	7.80	6.40	4.75	3.51	99.8	—	—	99.9	—	—	—
	SD		—	—	—	—	—	0.28			1.04			
	N		—	—	—	—	—	5	—	—	6	—	—	—
	SD		—	—	—	—	—	0.29	—	—	0.11	—	—	—
	FEATHERMEAL													
45	Poultry	5-03-795	68	3.00	2.46	1.57	0.97	93.3	85.8	76	7.21	0.9	54.9	18.3
	N		—	—	—	—	—	19	20	2	9	1	11	20
	SD		—	—	—	—	—	2.16	7.41	6	2.28	—	7.56	9.29
	FESCUE, KENTUCKY 31 (*Festuca arundinacea*)													
46	Fresh	2-01-902	61	2.69	2.21	1.34	0.77	31.3	15.0	2.0	5.5	24.6	62.2	34.4
	N		—	—	—	—	—	5	51	—	18	18	8	8
	SD		—	—	—	—	—	3.76	2.02	—	0.75	2.39	8.36	4.39
47	Hay, sun-cured, mature	1-09-189	44	1.94	1.59	0.75	0.22	90.0	10.8	25	4.7	31.2	70.0	39.0
	N		—	—	—	—	—	1	13	—	13	10	1	1
	SD		—	—	—	—	—	—	3.58	—	0.84	2.36	—	—
	FISH, ANCHOVY (*Engraulis ringen*)													
48	Meal, mechanical extracted	5-01-985	79	3.48	2.86	1.91	1.27	92.0	71.2	60	4.6	1.1	—	—
	N		—	—	—	—	—	67	58	26	36	9	—	—
	SD		—	—	—	—	—	1.19	2.24	16	1.62	0.01	—	—
	FISH, MENHADEN (*Brevoortia tyrannus*)													
49	Meal, mechanical extracted	5-02-009	73	3.22	2.64	1.73	1.11	91.7	67.9	60	10.7	0.8	—	—
	N		—	—	—	—	—	79	91	26	96	38	—	—
	SD		—	—	—	—	—	1.18	2.65	16	1.84	0.20	—	—
	MEAT													
50	Meal, rendered	5-00-385	71	3.13	2.57	1.66	1.05	93.8	58.2	56	11.0	2.01	48.2	6.35
	N		—	—	—	—	—	65	53	7	20	9	22	43
	SD		—	—	—	—	—	4.38	7.94	21	2.15	0.92	11.8	3.39
	MOLASSES AND SYRUP													
51	Beet sugar molasses, >48% invert sugar, >79.5 degrees brix	4-00-668	75	3.31	2.71	1.79	1.16	77.9	8.5	20	0.2	0.0	0.0	0.0
	N		—	—	—	—	—	21	12	—	3	—	—	—
	SD		—	—	—	—	—	1.71	1.11	—	0.105	—	—	—
52	Sugarcane, molasses, >46% invert sugar, >79.5 degrees brix (black-strap)	4-04-696	72	3.17	2.60	1.70	1.08	74.3	5.8	20	0.2	0.5	—	0.4
	N		—	—	—	—	—	84	64	—	6	1	—	1
	SD		—	—	—	—	—	3.27	2.03	—	0.240	—	—	—
	OATS (*Avena sativa*)													
53	Grain	4-03-309	77	3.40	2.78	1.85	1.22	89.2	13.6	17	5.2	12.0	29.3	14.0
	N		—	—	—	—	—	97	229	4	125	108	54	111
	SD		—	—	—	—	—	1.80	1.59	3	0.97	1.40	7.03	4.45
54	Hay, sun-cured	1-03-280	53	2.34	1.91	1.08	0.52	90.7	9.5	20	2.4	32.0	63.0	38.4
	N		—	—	—	—	—	27	32	—	13	17	1	1
	SD		—	—	—	—	—	2.55	2.26	—	0.88	3.57	—	—
55	Hulls	1-03-281	35	1.54	1.27	0.41	0.00	92.4	4.1	25	1.5	33.2	72.2	39.6
	N		—	—	—	—	—	26	17	—	15	15	4	4
	SD		—	—	—	—	—	1.14	1.33	—	0.81	3.44	5.72	2.06
56	Silage	3-03-296	59	2.6	2.13	1.27	0.70	36.4	12.7	23	3.12	31.8	58.1	38.6
	N		—	—	—	—	—	635	639	—	5	2	143	631
	SD		—	—	—	—	—	10.8	3.04	—	0.32	4.62	6.71	4.55
57	Straw	1-03-283	50	2.21	1.81	0.97	0.42	92.2	4.4	30	2.2	40.4	74.4	47.9
	N		—	—	—	—	—	71	74	—	16	64	4	5
	SD		—	—	—	—	—	2.10	1.09	—	0.42	2.98	2.70	2.48

Ash (%)	Calcium (%)	Phosphorus (%)	Magnesium (%)	Potassium (%)	Sodium (%)	Sulfur (%)	Copper (mg/kg)	Iodine (mg/kg)	Iron (mg/kg)	Manganese (mg/kg)	Selenium (mg/kg)	Zinc (mg/kg)	Cobalt (mg/kg)	Molybdenum (mg/kg)
2.9	0.15	0.09	0.14	0.88	0.02	0.08	13.3	—	131	119	0.09	22	—	0.02
20	16	16	10	11	7	6	4	—	5	3	1	3	—	3
0.48	0.02	0.02	0.01	0.05	0.01	0.06	4.0	—	49.7	2.2	—	0.1	—	0.01
4.16	0.17	0.62	0.384	1.24	0.01	0.27	7.9	—	107	131	—	37.7	—	1.16
16	383	383	383	383	383	121	383	—	383	383	—	383	—	374
0.29	0.10	0.10	0.05	0.07	0.01	0.05	2.7	—	190	210	—	8.1	—	0.50
7.0	0.20	1.16	0.65	1.65	0.07	0.42	16.5	—	162	26.9	—	73.5	—	25.0
34	164	167	47	167	79	21	41	—	42	43	—	37	—	33
0.47	0.13	0.08	0.09	0.08	0.05	0.12	2.8	—	71	13.2	—	15.3	—	0.87
—	—	—	—	—	—	—	—	—	—	—	—	—	—	—
3.50	1.19	0.68	0.06	0.20	0.24	1.85	14.2	—	702	12.0	—	105	—	0.56
5	18	18	18	18	18	15	18	—	18	18	—	18	—	18
0.40	1.69	0.84	0.04	0.09	0.13	0.45	5.24	—	422	45	—	9.0	—	0.29
7.2	0.51	0.37	0.27	2.30	—	0.18	—	—	—	—	—	22	—	—
2	25	27	24	24	—	24	—	—	—	—	—	1	—	—
3.60	0.10	0.08	0.05	0.48	—	0.03	—	—	—	—	—	—	—	—
6.8	0.41	0.30	0.16	1.96	0.02	—	22.0	—	132	97	—	35	—	—
13	2	2	2	2	1	—	2	—	2	2	—	2	—	—
0.92	0.13	0.07	0.02	0.19	—	—	12.7	—	9.2	22.6	—	1.4	—	—
16.0	4.06	2.69	0.27	0.79	0.96	0.78	9.9	3.41	234	12	1.47	114	—	0.19
47	51	52	32	35	32	4	27	2	28	31	27	31	—	1
1.54	0.54	0.45	0.05	0.27	0.33	0.23	1.80	3.49	63.2	5.9	0.25	16.7	—	—
20.6	5.46	3.14	0.16	0.77	0.44	0.58	11.3	1.19	594	40	2.34	157	—	0.17
87	68	67	19	21	22	4	20	2	21	21	16	18	—	2
2.12	0.800	0.31	0.03	0.16	0.13	0.26	3.5	1.41	271	17.7	0.69	19.0	—	0.07
21.3	9.13	4.34	0.27	0.49	0.80	0.51	21.4	—	758	174	—	265	—	2.3
7	52	52	52	52	52	25	52	—	52	52	—	52	—	52
5.67	2.75	1.21	0.30	0.16	0.33	0.14	68.3	—	609	990	—	995	—	1.8
11.4	0.15	0.03	0.29	6.06	1.48	0.60	21.6	—	87	6	—	18	—	0.46
9	13	11	10	10	8	9	7	—	8	7	1	5	—	—
1.34	0.054	0.01	0.01	0.29	0.08	0.05	1.3	—	25.2	0.3	—	0.032	—	—
13.3	1.00	0.10	0.42	4.01	0.22	0.47	65.7	2.10	263	59	—	21	—	1.59
52	32	31	12	16	9	9	8	1	11	11	—	5	—	4
2.34	0.182	0.02	0.10	0.88	0.02	0.02	26.0	—	34.4	6.4	—	6.0	—	0.75
3.3	0.01	0.41	0.16	0.51	0.02	0.21	8.6	—	94.1	40.3	0.24	40.8	0.06	1.70
94	168	175	152	151	49	22	131	—	132	141	32	144	8	104
0.50	0.03	0.05	0.02	0.09	0.02	0.02	4.1	—	50.0	15.1	0.15	9.5	0.02	0.76
7.9	0.32	0.25	0.29	1.49	0.18	0.23	4.8	—	406	99	—	45	—	0.07
11	7	26	23	11	16	3	4	—	5	4	—	1	—	3
0.85	0.09	0.06	0.27	0.65	0.06	0.06	1.5	—	160	48.2	—	—	—	0.01
6.6	0.16	0.15	0.13	0.59	0.07	0.10	7.1	—	138	27	0.43	29	—	—
12	9	9	6	8	6	2	4	—	3	5	1	3	—	—
0.69	0.04	0.05	0.03	0.05	0.08	0.06	3.2	—	48.4	9.68	—	8.0	—	—
10.1	0.58	0.31	0.21	2.88	0.09	0.24	8.0	—	367	66.3	0.07	29.8	—	1.89
2	627	627	562	562	562	67	562	—	562	562	19	562	—	469
1.20	0.21	0.07	0.06	0.85	0.13	0.06	4.5	—	388	33.5	0.06	8.9	—	0.94
7.8	0.23	0.06	0.17	2.53	0.42	0.22	10.3	—	164	31	—	6	—	—
14	68	66	18	16	5	6	4	—	15	5	—	11	—	—
1.85	0.09	0.04	0.04	0.25	0.07	0.01	0.54	—	47.1	11.8	—	1.1	—	—

TABLE 11-1 Means and Standard Deviations for the Composition Data of Feeds Commonly Used in Beef Cattle Diets —*Continued*

Entry No.	Feed Name/ Description	International Feed No.	TDN (%)	DE (Mcal/ kg)	ME (Mcal/ kg)	NE$_m$	NE$_g$	Dry Matter (%)	Crude Protein (%)	Ruminal Unde- grad- ability (%)	Ether Extract (%)	Fiber (%)	NDF (%)	ADF (%)
			Value as Determined at Maintenance Intake			Net Energy Values for Growing-Cattle Mcal/kg								
	ORCHARD GRASS (*Dactylis glomerata*)													
58	Fresh, early bloom	2-03-442	68	3.00	2.46	1.57	0.97	23.5	12.8	20	3.70	32.00	58.1	30.70
	N		—	—	—	—	—	8	7	—	5	5	3	2
	SD		—	—	—	—	—	3.87	2.37	—	0.80	2.93	8.31	1.98
59	Fresh, mid-bloom	2-03-443	57	2.51	2.06	1.21	0.64	27.4	10.1	22	3.5	33.5	57.6	35.6
	N		—	—	—	—	—	3	4	—	2	2	1	1
	SD		—	—	—	—	—	5.36	3.89	—	0.36	2.25	—	—
60	Hay, sun-cured, early bloom	1-03-425	65	2.87	2.35	1.47	0.88	89.1	12.8	24	2.9	33.9	59.6	33.8
	N		—	—	—	—	—	7	9	—	6	5	4	4
	SD		—	—	—	—	—	3.30	3.51	—	0.82	1.72	5.28	1.25
61	Hay, sun-cured, late bloom	1-03-428	54	2.38	1.95	1.11	0.55	90.6	8.4	24	3.4	37.1	65.0	37.8
	N		—	—	—	—	—	7	1	—	1	1	3	3
	SD		—	—	—	—	—	1.51	—	—	—	—	2.77	0.20
	PEANUT (*Arachis hypogaea*)													
62	Seeds without coats, meal solvent extracted	5-03-650	77	3.40	2.78	1.85	1.22	92.4	52.9	30	2.30	8.40	—	—
	N		—	—	—	—	—	16	12	2	10	10	—	—
	SD		—	—	—	—	—	1.82	3.93	0.06	1.00	1.19	—	—
	PRAIRIE PLANTS, MIDWEST													
63	Hay, sun-cured	1-03-191	51	2.25	1.84	1.00	0.45	91.0	6.4	25	2.3	33.7	62.3	41.7
	N		—	—	—	—	—	8	5	—	5	5	1	1
	SD		—	—	—	—	—	1.42	1.63	—	0.65	1.94	—	—
	RICE (*Oryza sativa*)													
64	Bran with germs	4-03-928	70	3.09	2.53	1.63	1.03	90.5	14.4	25	15.0	12.9	33.00	20.0
	N		—	—	—	—	—	37	34	—	29	25	8	1
	SD		—	—	—	—	—	0.74	1.42	—	2.14	1.46	6.57	—
65	Hulls	1-08-075	12	0.53	0.43	0.00	0.00	91.9	3.1	35	1.1	42.7	82.40	68.7
	N		—	—	—	—	—	21	22	—	18	18	3	2
	SD		—	—	—	—	—	1.45	1.10	—	1.07	3.59	4.95	1.54
	RYE GRASS, ITALIAN (*Lolium multiforum*)													
66	Fresh	2-04-073	84	3.70	3.04	2.06	1.40	22.6	17.9	20	4.1	20.9	61.00	38.0
	N		—	—	—	—	—	5	2	—	2	2	1	1
	SD		—	—	—	—	—	2.35	2.26	—	0.141	1.27	—	—
	SORGHUM (*Sorghum bicolor*)													
67	Grain	4-04-383	82	3.62	2.96	2.00	1.35	90.0	12.6	57	3.03	2.76	16.10	6.38
	N		—	—	—	—	—	226	230	8	68	45	7	10
	SD		—	—	—	—	—	2.29	1.99	8	0.66	0.95	3.36	0.56
68	Silage	3-04-323	60	2.65	2.17	1.31	0.74	30	9.39	29	2.64	26.90	60.80	38.8
	N		—	—	—	—	—	588	584	—	32	16	282	581
	SD		—	—	—	—	—	13.5	2.83	—	0.34	3.74	7.59	5.65
	SOYBEAN (*Glycine max*)													
69	Seed coats	1-04-560	77	3.40	2.98	1.86	1.22	90.3	12.2	25	2.10	39.9	66.3	49.0
	N		—	—	—	—	—	28	27	—	17	23	6	6
	SD		—	—	—	—	—	3.43	2.51	—	0.56	4.79	2.03	2.85
70	Meal	—	84	3.7	3.04	2.06	1.4	90.9	51.8	34	1.67	5.37	10.3	7.0
	N		—	—	—	—	—	807	786	45	204	192	150	283
	SD		—	—	—	—	—	1.88	3.45	12	0.97	0.90	5.80	3.33
71	Seeds, meal solvent extracted, 44% protein	5-20-637	84	3.70	3.04	2.06	1.40	89.1	49.90	34	1.6	7.0	14.9	10.0
	N		—	—	—	—	—	119	111	—	87	92	2	3
	SD		—	—	—	—	—	1.22	1.25	—	0.67	0.95	1.27	0.057
72	Seeds without hulls, meal solvent extd	5-04-612	87	3.84	3.15	2.15	1.48	89.9	54.00	34	1.1	3.8	7.79	6.10
	N		—	—	—	—	—	78	75	—	41	55	1	3
	SD		—	—	—	—	—	1.72	1.72	—	0.38	0.55	—	0.75
73	Seed whole	5-04-610	94	4.14	3.40	2.35	1.64	86.4	40.3	25	18.2	10.1	14.9	11.1
	N		—	—	—	—	—	5	241	—	50	35	55	179
	SD		—	—	—	—	—	2.07	3.84	—	2.64	4.32	6.22	5.71
	SUNFLOWER, COMMON (*Helianthus annuus*)													
74	Seeds without hulls, meal solvent extd	5-04-739	65	2.87	2.35	1.47	0.88	92.5	26	26	2.9	12.7	40.0	30.0
	N		—	—	—	—	—	21	22	9	19	20	1	1
	SD		—	—	—	—	—	1.73	3.96	5	0.63	2.18	—	—
	TIMOTHY (*Phleum pratense*)													
75	Fresh, late vegetative	2-04-903	66	2.91	2.39	1.51	0.91	26.7	12.2	20	3.8	32.1	55.7	29.0
	N		—	—	—	—	—	5	8	—	2	2	6	1
	SD		—	—	—	—	—	1.86	3.87	—	0.25	1.93	3.65	
76	Hay, sun-cured, early bloom	1-04-882	59	2.6	2.13	1.28	0.71	89.1	10.8	22	2.8	33.6	61.4	35.2
	N		—	—	—	—	—	13	12	—	10	8	5	5
	SD		—	—	—	—	—	1.72	3.35	—	0.54	1.36	1.22	2.38

Ash (%)	Calcium (%)	Phosphorus (%)	Magnesium (%)	Potassium (%)	Sodium (%)	Sulfur (%)	Copper (mg/kg)	Iodine (mg/kg)	Iron (mg/kg)	Manganese (mg/kg)	Selenium (mg/kg)	Zinc (mg/kg)	Cobalt (mg/kg)	Molybdenum (mg/kg)
8.1	0.25	0.39	0.31	3.38	0.04	0.26	33.1	—	785	104	—	—	—	—
6	1	1	1	1	1	1	1	—	2	1	—	—	—	—
1.68	—	—	—	—	—	—	—	—	21.2	—	—	—	—	—
7.5	0.23	0.17	0.33	2.09	0.26	—	50.1	—	68	136	—	25	—	0.10
4	1	2	1	1	1	—	1	—	1	1	—	1	—	1
0.53	—	0.08	—	—	—	—	—	—	—	—	—	—	—	—
8.5	0.27	0.34	0.11	2.91	0.01	0.26	19.0	—	93	157	—	40	—	0.43
6	1	1	1	1	1	1	1	—	1	1	—	1	—	1
1.60	—	—	—	—	—	—	—	—	—	—	—	—	—	—
10.1	0.26	0.30	0.11	2.67	0.01	—	20.0	20.0	84	167	0.03	38	—	0.30
3	1	1	1	1	1	—	1	—	1	1	1	1	—	1
3.10	—	—	—	—	—	—	—	—	—	—	—	—	—	—
6.3	0.32	0.66	0.17	1.28	0.03	0.33	16.0	0.07	155	29	—	36	—	0.12
7	2	3	1	2	1	2	1	1	1	1	—	1	—	1
1.02	0.247	0.05	—	0.03	—	0.01	—	—	—	—	—	—	—	—
8.0	0.35	0.14	0.26	1.0	—	—	—	—	88	—	—	34	—	—
4	3	3	2	1	—	—	—	—	1	—	—	1	—	—
1.07	0.01	0.06	0.02	—	—	—	—	—	—	—	—	—	—	—
11.5	0.10	1.73	0.97	1.89	0.03	0.20	12.2	—	229	396	0.44	33	—	1.53
27	21	21	13	18	6	9	6	—	9	8	1	7	—	2
2.16	0.06	0.40	0.24	0.22	0.03	0.01	3.80	—	80.6	125	—	23.8	—	0.25
20.6	0.12	0.07	0.37	0.65	0.02	0.08	3.4	—	99	320	0.15	24	—	—
12	15	14	3	8	1	5	1	—	1	4	1	1	—	—
1.51	0.06	0.02	0.40	0.62	—	0.03	—	—	—	27.1	—	—	—	—
17.4	0.65	0.41	0.35	2.00	0.01	0.10	—	—	1000	—	—	—	—	—
2	2	2	—	1	1	1	—	—	1	—	—	—	—	—
2.33	0.01	0.01	—	—	—	—	—	—	—	—	—	—	—	—
1.87	0.04	0.34	0.17	0.44	0.01	0.14	4.7	—	80.8	15.4	0.46	0.99	—	—
62	40	39	37	28	27	4	26	—	36	34	3	13	—	—
0.43	0.04	0.07	0.04	0.11	0.01	0.03	1.9	—	45.1	4.6	0.58	0.64	—	—
5.9	0.49	0.22	0.28	1.72	0.01	0.12	9.2	—	383	68.5	0.03	1.31	—	—
1	572	572	567	573	567	85	567	—	567	567	2	567	—	—
—	0.26	0.07	0.10	0.65	0.02	0.03	5.7	—	88.4	60.0	0.01	0.75	—	—
4.9	0.53	0.18	0.22	129	0.03	0.11	17.8	—	409	10	0.14	48	0.12	—
10	10	8	2	5	4	2	1	—	2	3	1	2	1	—
0.48	0.134	0.07	0.07	0.26	0.02	0.03	—	—	120	5.0	—	34	—	—
6.9	0.46	0.73	0.32	2.42	0.07	0.46	19.1	—	277	48.3	0.46	67.9	—	6.67
121	348	352	276	281	268	99	271	—	267	270	12	270	—	250
0.58	0.80	0.20	0.06	0.20	0.31	0.06	17.8	—	159	48.6	0.25	57.3	—	2.85
7.2	0.40	0.71	0.31	2.22	0.04	0.46	22.4	—	185	35	0.51	57	—	0.12
66	26	29	19	21	12	6	15	—	15	15	10	13	—	1
0.58	0.11	0.04	0.03	0.24	0.03	0.04	7.9	—	39.0	3.5	0.28	7.5	—	—
6.7	0.29	0.71	0.33	2.36	0.01	0.48	22.5	0.12	145	41	0.22	63	—	0.12
34	19	19	6	9	4	2	6	1	2	5	2	7	—	1
0.68	0.05	0.05	0.02	0.15	0.01	0.01	5.0	—	35.3	8.66	0.14	7.7	—	—
4.56	0.27	0.65	0.27	2.01	0.04	0.35	14.6	—	182	345	—	59.0	—	3.98
1	156	156	156	156	156	17	156	—	156	156	—	156	—	156
—	0.20	0.08	0.03	0.12	0.31	0.04	4.2	—	197	15.6	—	34.3	—	3.42
8.1	0.45	1.02	0.70	1.27	0.03	0.33	4.0	—	33	20	2.30	105	—	—
14	11	11	7	7	2	2	1	—	1	2	1	1	—	—
0.34	0.08	0.25	0.12	0.33	0.02	0.14	—	—	—	6.0	—	—	—	—
7.5	0.40	0.26	0.16	2.73	0.11	0.13	8.9	—	132	127	—	36	—	0.15
8	4	4	4	4	4	2	2	—	4	2	—	2	—	2
0.97	0.12	0.08	0.04	0.40	0.09	—	2.5	—	78.2	32.7	—	7.1	—	0.082
5.7	0.51	0.29	0.13	2.41	0.01	0.13	11	—	203	103	—	62	—	—
9	3	3	2	2	1	1	1	—	2	1	—	1	—	—
0.92	0.08	0.07	0.21	2.10	—	—	—	—	4.2	—	—	—	—	—

TABLE 11-1   Means and Standard Deviations for the Composition Data of Feeds Commonly Used in Beef Cattle Diets —*Continued*

Entry No.	Feed Name/ Description	International Feed No.	TDN (%)	DE (Mcal/kg)	ME (Mcal/kg)	NE$_m$	NE$_g$	Dry Matter (%)	Crude Protein (%)	Ruminal Undegradability (%)	Ether Extract (%)	Fiber (%)	NDF (%)	ADF (%)
77	Hay, sun-cured, full bloom	1-04-884	56	2.47	2.03	1.18	0.61	89.4	8.1	25	2.9	35.2	64.2	37.5
		N	—	—	—	—	—	8	15	—	7	7	8	8
		SD	—	—	—	—	—	2.43	1.03	—	0.73	1.20	2.19	2.27
	TREFOIL, BIRDSFOOT (*Lotus corniculatus*)													
78	Fresh	2-20-786	66	2.91	2.39	1.51	0.91	19.3	20.6	20	4.0	21.2	46.7	—
		N	—	—	—	—	—	9	12	—	3	3	11	—
		SD	—	—	—	—	—	4.28	3.97	—	1.30	7.74	11.7	—
79	Hay, sun-cured	1-05-044	59	2.60	2.13	1.28	0.71	90.6	15.9	23	2.1	32.3	47.5	36.0
		N	—	—	—	—	—	9	8	—	7	7	1	1
		SD	—	—	—	—	—	1.46	2.31	—	0.52	5.32	—	—
	WHEAT (*Triticum aestivum*)													
80	Bran	4-05-190	70.0	3.09	2.53	1.63	1.03	89.0	17.4	20	4.3	11.3	42.8	14.0
		N	—	—	—	—	—	86	64	4	56	54	6	6
		SD	—	—	—	—	—	1.23	1.13	10	0.80	1.28	8.68	1.46
81	Flour by-product, less than 9.5% fiber	4-05-205	69	3.04	2.50	1.6	1.00	89.3	18.7	21	4.7	8.5	35.9	11.7
		N	—	—	—	—	—	96	59	3	94	66	26	38
		SD	—	—	—	—	—	1.49	1.15	2	0.85	1.00	6.81	0.93
82	Fresh, early vegetative	2-05-176	73	3.22	2.64	1.73	1.11	22.2	27.4	20	4.4	17.4	46.2	28.4
		N	—	—	—	—	—	2	2	—	1	1	1	1
		SD	—	—	—	—	—	0.99	1.62	—	—	—	—	—
83	Grain	4-05-211	88	3.88	3.18	2.18	1.5	90.2	14.2	23	2.34	3.66	11.8	4.17
		N	—	—	—	—	—	136	100	5	34	25	14	43
		SD	—	—	—	—	—	1.97	1.96	6	1.21	1.14	2.02	3.58
84	Hay, sun-cured	1-05-172	58	2.56	2.10	1.24	0.68	88.7	8.7	23	2.2	29.0	68.0	41.0
		N	—	—	—	—	—	12	8	—	6	9	1	1
		SD	—	—	—	—	—	3.09	2.22	—	0.90	2.01	—	—
85	Silage	3-05-184	57	2.51	2.06	1.21	0.64	34.2	12.5	20	6.09	26.8	60.7	39.2
		N	—	—	—	—	—	181	181	—	2	3	82	181
		SD	—	—	—	—	—	11.1	2.96	—	2.1	3.80	7.62	5.28
86	Straw	1-05-175	41	1.81	1.48	0.64	0.11	91.3	3.5	40	2.0	41.7	78.9	55.0
		N	—	—	—	—	—	37	68	—	15	25	14	16
		SD	—	—	—	—	—	3.12	1.29	—	1.10	5.81	4.82	4.95

NOTE: Undegradability values that do not have N (number) or SD (standard deviation) entries are based on in situ data and are estimates only. The energy values (TDN, DE, etc.) are based on book values and were not adjusted for the mean composition data. The energy values can be influenced by all the factors that affect the other nutrients as well as amount of intake, processing technique, grain:forage ratio, and thermal stress. For most feeds there is no data base providing means and SD for the energy values. Some trace minerals and the fat-soluble vitamins are not listed in the table because their values were not routinely determined by the laboratories contributing data to this summary. Int. Ref. #, international reference number.

not reviewed in this section, but the effects of the most commonly used treatments affecting nutritive value are discussed. However, many of the references useful in providing further insight on methods and details of methods are available in other reviews (e.g., Beeson and Perry, 1982; Berger et al., 1994). Although, processing is used across a wide array of feedstuffs, it is not an issue with many for which uniform methodology applies. This presentation is confined to roughages and grains; methods applied to roughages and grains often vary and/or unprocessed feed is an alternative.

Roughages

The nutritive value of roughages is often improved through the use of physical and, occasionally, chemical or biological treatment methods. Responses to physical processing such as steaming, chopping, wafering, and grinding (with or without pelleting) are usually in inverse proportion to the quality of the starting forage (Minson, 1963). Coarse chopping, with or without wafering, usually has only a slight influence on nutritive value, although intake might be enhanced through indirect effects such as ease of handling and presentation to the animals. Alternatively, fine grinding, with or without pelleting, can have a major influence, particularly on intake but also on available energy. Potential benefit depends on appropriate supplementation, especially with protein (Campling and Freer, 1966; Weston, 1967). Increased intake usually is observed when mean particle size is reduced to 5 mm, and intake is increased in proportion to further reduction in size with maximal intake achieved when mean particle size is 1 mm

Ash (%)	Calcium (%)	Phosphorus (%)	Magnesium(%)	Potassium (%)	Sodium (%)	Sulfur (%)	Copper (mg/kg)	Iodine (mg/kg)	Iron (mg/kg)	Manganese (mg/kg)	Selenium (mg/kg)	Zinc (mg/kg)	Cobalt (mg/kg)	Molybdenum (mg/kg)
5.2	0.43	0.20	0.09	1.99	0.07	0.14	29.0	—	140	93	—	54	—	—
8	3	4	3	4	3	3	2	—	2	2	—	1	—	—
0.813	0.09	0.01	0.04	0.51	0.09	0.01	33.9	—	24.9	16.9	—	—	—	—
11.2	1.74	0.26	0.40	3.26	0.11	0.25	12.8	—	176	83	—	31	—	0.49
7	8	8	6	8	6	1	5	—	5	5	—	5	—	6
3.25	0.40	0.05	0.12	1.66	0.05	—	3.4	—	125	13.6	—	7	—	0.21
7.4	1.70	0.23	0.51	1.92	0.07	0.25	9.26	—	227	29	—	77	—	0.11
5	3	3	3	4	1	1	1	—	3	1	—	1	—	1
0.79	0.09	0.01	0.20	0.25	—	—	—	—	149	—	—	—	—	—
6.6	0.14	1.27	0.63	1.37	0.06	0.24	14.2	—	163	134	0.57	110	108	—
37	30	29	17	17	13	8	8	—	10	8	5	6	3	—
0.60	0.03	0.21	0.07	0.10	0.02	0.02	1.8	—	56	14	0.25	36	0.03	—
5.0	0.17	1.01	0.40	1.81	0.02	0.19	12.6	—	170	124	—	102	—	2.1
30	69	70	55	56	44	18	50	—	51	49	—	45	—	39
0.99	0.15	0.13	0.09	0.14	0.06	0.04	3.13	—	118	23	—	35	—	37
13.3	0.42	0.40	0.21	3.50	0.18	0.22	—	—	100	—	—	—	—	—
1	1	1	1	1	2	2	—	—	1	—	—	—	—	—
—	—	—	—	—	0.14	0.03	—	—	—	—	—	—	—	—
2.01	0.05	0.44	0.13	0.40	0.01	0.14	6.48	—	45.1	36.6	0.05	38.1	—	0.12
25	90	91	16	16	2	15	16	—	16	16	1	15	—	1
0.26	0.03	0.14	0.01	0.02	0.01	0.01	1.3	—	5.6	2.4	—	2.8	—	—
7.9	0.15	0.20	0.12	0.99	0.21	0.22	—	—	200	—	—	—	—	—
4	8	8	1	5	2	2	—	—	1	—	—	—	—	—
2.05	0.02	0.08	—	0.44	0.1	0.03	—	—	—	—	—	—	—	—
7.5	0.44	0.29	0.17	2.24	0.04	0.21	9.0	—	386	79.5	—	28.0	—	1.61
1	177	177	169	169	168	36	159	—	169	169	—	169	—	169
—	0.32	0.09	0.15	0.73	0.10	0.06	6.0	—	322	47	—	11.0	—	1.06
7.7	0.17	0.05	0.12	1.40	0.14	0.19	3.6	—	157	41	—	6	—	0.05
46	51	48	37	39	5	5	34	—	35	34	—	30	—	2
2.61	0.07	0.02	0.02	0.70	0.01	0.01	1.2	—	39.5	13.7	—	0.77	—	0.01

or less. Pelleting is an improvement over grinding because it produces less dust. The average effect of pelleting and grinding was an 11 percent increase in intake for cattle, with a greater response from young compared to mature animals (Greenhalgh and Reid, 1973). In a summary of research with bulls, Sundstol (1991) reported that grinding by itself and grinding with pelleting enhanced intake of straw by 7 and 37 percent, respectively. The above summary applies mostly to hays and straws. Silages are rarely processed as finely as dry forages although the amount of chopping and particle size reduction that occurs during harvesting can vary significantly. From a summary of available literature on corn (Wilkinson, 1978) and grass silage (McDonald et al., 1991) and within the range of particle lengths commonly observed for silage (mean length, 5 to 15 mm), there is a negative relationship of length to intake;

however, the intake decrease is generally less than 10 percent.

Digestibility of roughages is decreased by grinding, with or without pelleting, and the decrease is usually in proportion to the intake increase (Blaxter et al., 1956). For 21 studies, Minson (1963) found an average 3.3 percent decrease in dry matter digestibility. Thomson and Beever (1980) reported greater decreases for ground grasses (0 to 15 percent) than for ground legumes (3 to 6 percent). Digestibility decreases are usually attributed to a faster rate of passage of food, with more digestion occurring in the hindgut. In contrast, pelleting and grinding roughages results in lowering heat increment so that the net dietary energy from these roughages is often higher than for the parent product (Osbourne et al., 1976).

Chemical alkali is used to upgrade roughages; it hydrolyzes chemical bonds between fibrous components in the cell wall. Sodium hydroxide is more effective than ammonia or urea, but it is more expensive and has greater environmental consequences, so ammonia or urea are more widely used. Berger et al. (1994) concluded, from 21 studies on crop residues and 6 on grasses, that ammoniation improved dry matter intake by 22 and 14 percent, respectively. With regard to digestibility, 32 studies on ammoniated crop residues and 10 on grasses demonstrated a 15 and 16 percent improvement, respectively. Urea enhanced intake by 13 percent and digestibility by 23 percent. Oxidation is an alternative chemical procedure that has been used to upgrade roughages and microbial and enzymatic methods have been developed and tested as well. Steam treatment is an additional physical process that has been developed. However, none of these latter processes are widely used in North America at present. For details, the reader is referred to Berger et al. (1994).

Grains

GENERAL

Processing can significantly improve the nutritive value of cereal grains for beef cattle. The most common physical processes used are rolling or grinding the grain, with or without additional moisture; and this is done chiefly to rupture the pericarp and expose starch granules to aid digestion (Beauchemin et al., 1994). In a few cases (see below), processing of whole grain for beef cattle is not beneficial; but this is the exception rather than the rule. When processing is used, results are often variable and unpredictable. Furthermore, processing can affect nutrient requirements in a subtle fashion. To rationalize these effects, significant principles about grain processing will be discussed first.

PRINCIPLES OF GRAIN PROCESSING

Cattle are less able than other ruminants in the ability to masticate whole grain (Theurer, 1986). Sorghum presents the greatest difficulty followed by wheat, barley, corn, and oats. Morgan and Campling (1978) found that younger cattle can digest whole grain better than older cattle; however, Campling (1991) concluded that further studies on a possible relationship between cattle age or weight digestion of grain are necessary. The ability of rumen microbes to digest grain depends on particle size (Galyean et al., 1981; Beauchemin et al., 1994)—fine particles are digested more rapidly than coarse particles. Microbial digestion proceeds from the inside to the outside of the kernel, and the protein matrix, which surrounds starch granules in the endosperm, is a barrier to the effective digestion of starch (McAllister et al., 1990a). For this and related reasons,

there are major differences between the rates at which grains are digested; for example, barley is digested more rapidly than corn (McAllister et al., 1990b). Rapid acid production from the fermentation of starch in the rumen is undesirable; thus, starch bypassing digestion in the rumen altogether can be beneficial, hence processes that inhibit digestion of grain protein will decrease starch digestion in the rumen (Fluharty and Loerch, 1989). Because heat has a major influence on protein digestion, any process using heat treatment is likely to influence grain nutritive value. Unfortunately, in the heat treatment of grain, the relationships of time, temperature, and moisture to protein digestibility are ill-defined; therefore, effects of heat treatments on grain nutritive value would be difficult to interpret. This is further complicated because heat gelatinizes starch, which facilitates microbial digestion (Theurer, 1986) and could therefore offset some or all of the effects of heat. Enhanced microbial protein synthesis and decreased grain protein degradability were associated with steam processing and rolling of sorghum to produce a lighter flake (Xiong et al., 1991). Zinn (1990a) found that the longer the corn was steamed, the faster nonammonia nitrogen was processed in the duodenum of cattle. Roughage source and amount influence dynamics of rumen liquid and particulate flow and may, therefore, influence grain digestion in the rumen (Goetsch et al., 1987).

Instrinsic characteristics of grains affect the rate or extent of starch digestion and can reduce benefits from processing. One factor is the form of starch and the other is the presence of tannins. Amylopectin is more digestible than amylose; hence, waxy grains are more digestible than other grains (Sherrod et al., 1969). Tannins present in bird-resistant grains, for example, sorghums, reduce digestibility (Maxson et al., 1973). Within varieties of the same grain, total digestible nutrients (TDN) varied as much as 7 percent (Parrot et al., 1969). Grain quality for beef cattle is positively associated with grain density or fiber content, as shown for barley by Mathison et al. (1991a) and Engstrom et al. (1992).

Grain that is fermented less rapidly and extensively in the rumen can escape microbial digestion and may be digested enzymatically in the small intestine. In a review of many trials, Owens et al. (1986) estimated that cattle are 42 percent more efficient in utilizing starch when it is digested in the abomasum and small intestine compared to the forestomach. Thus, processes that cause starch to escape rumen digestion could be beneficial, provided it is effectively digested in the intestine and not passed further to the caecum, where fermentation can resume and significant depletion of nitrogen from the animal may result (Owens et al., 1986). The concept of limited starch digestion in the small intestine does not seem plausible. Furthermore, digestion in the hindgut does not usually compensate for reduced digestion in the rumen (Goetsch et al., 1987).

For these reasons, processed grain that escapes rumen fermentation may not enhance provision of net energy or improve nitrogen utilization in the animal.

There are two important points to consider that will affect digestible energy derived by the animal and could further modify the benefits of processing. First, positive effects on digestion can result by combining grains and different forms of grain, as reported between ground, high-moisture corn and dry-rolled sorghum (Stock et al., 1991); between dry corns of different particle size (Turgeon et al., 1983); between dry and high-moisture corn (Stock et al., 1987); between wheat and high-moisture corn (Bock et al., 1991); and between high-moisture sorghum grain and dry-rolled corn (Streeter et al., 1989). Positive associative effects are not consistent (Mader et al., 1991) and not completely understood. The second consideration is level of feeding. Moe and Tyrrell (1979) reported that the metabolizable energy of corn grain for dairy cows was reduced from 3.58 Mcal/kg at maintenance to 2.92 Mcal/kg at 2.5 times maintenance. More recently, Bines et al. (1988) reported that intake effects on digestibility of mixed diets containing processed grain may be significant in young cattle but not in lactating cows. Although interest exists in restricted feeding of feedlot steers and heifers, effects on digestibility attributable to intake levels used in practice are small.

CORN

In diets containing less than 20 percent roughage, differences in DE and NE for corn—whole or rolled, or ground coarse or fine—are usually fairly small (Goodrich and Meiske, 1966; Vance et al., 1970, 1972; Preston, 1975). Differences in the DE and NE values of these forms of corn in low-roughage diets may be greater for the high-moisture grain (>20 percent water); diets containing unprocessed grain had superior feeding value to diets containing rolled grain, and diets containing rolled grain had superior feeding value to diets containing the ground form (Mader et al., 1991). Relative to whole dry corn, steam processing and flaking improved NE by at least 10 percent when inert roughage was included in the diet but had no effect in an all-concentrate diet (Vance et al., 1970). From studies on diets containing 50 percent corn and 20 percent whole cottonseed, Zinn (1987) concluded that steam flaked corn contained 13.4 and 14.2 percent more NE_m and NE_g, respectively, than dry-rolled corn. Zinn (1990b) reported that decreasing flake density of steam-processed corn from 0.42 to 0.30 kg/L enhanced starch digestion and improved diet nitrogen utilization. However, effect of flake density on corn NE was small and tended to favor flakes of intermediate density (Zinn, 1990b). Duration of steaming prior to flaking was associated with improved flow of nonammonia nitrogen to the duodenum (Zinn, 1990a). Although an intermediate steaming time of 47 min reduced digestibility of the starch, effect on diet DE was very slight (<2 percent; Zinn, 1990b). Intake of high- or all-concentrate corn-based diets is usually greatest when the corn is whole or is steam processed and flaked.

In diets containing intermediate or higher concentrations of roughage (>25 percent), corn is usually ground, adversely affecting digestibility (Moe and Tyrrell, 1977, 1979); fine-ground corn can be detrimental to utilization of the roughage (Moe et al., 1973; Orskov, 1976, 1979).

In many areas of North America, corn is preserved wet as a high-moisture grain. Digestible dry matter and energy of diets containing high-moisture corn are at least equal and may be as much as 5 percent higher than the same diet containing dry corn (McCaffree and Merrill, 1968; McKnight et al., 1973; Tonroy et al., 1974; Galyean et al., 1976; MacLeod et al., 1976). These results are also evident in dry corn reconstituted with moisture and stored for a short period of time prior to feeding (Tonroy et al., 1974). Corn containing 25 to 30 percent moisture has greater value than corn that is either drier or wetter than this (Mader et al., 1991) but this may be the result if intake rather than utilization (Clark, 1975). A minor concern about high-moisture grain and corn in particular is that most if not all of the vitamin E may be lost during storage (Young et al., 1975).

SORGHUM

Whole sorghum is not digested easily by cattle; dry grinding or steam processing and rolling significantly improves the digestibility of sorghum starch and energy. In low-roughage diets and relative to dry grinding, steam processing and flaking increased starch digestibility from 3 to 5 percent (McNeill et al., 1971; Hinman and Johnson, 1974) and DE by 5 to 10 percent (Buchanan-Smith et al., 1968; Husted et al., 1968). In contrast to the above, the NE value was equal in steam-processed and flaked sorghum and ground dry sorghum (Garrett, 1968). This may be explained by the fact that fine grinding enhanced NE by 8 percent, relative to the coarse rolled product (Brethour, 1980). Effectiveness of steam processing and rolling of sorghum may depend on the density of flake produced. Xiong et al. (1991) found dry matter intake and feed efficiency tended to be higher for diets containing sorghum grain with a density of 283 as opposed to 437 g/L. These researchers estimated the lighter grain contained 2.34 Mcal/kg NE_m and 1.63 Mcal/kg NE_g, as opposed to 2.21 and 1.52, respectively, for the heavier product. Ground, reconstituted sorghum had equivalent DE to the steam-processed and rolled product (Buchanan-Smith et al., 1968; McNeill et al., 1971; Kiesling et al., 1973); however, the latter process may enhance intake (Franks et al., 1972). Dry-heat treatments—for example, micronizing, popping, exploding, and roasting—may improve sorghum nutritive value as much as steam processing and rolling (Beeson and Perry, 1982). Starch digestibility was enhanced as much by micronizing and popping as it was by steam processing and rolling (Riggs et al., 1970; Hinman and Johnson, 1974; Croka and Wagner, 1975). Again, dry-heat treatments may not be as effective as steam processing to promote intake.

In intermediate- and high-roughage diets, dry-rolled sorghum is better utilized than in low-roughage diets (Keating et al., 1965). Thus, provided the whole grain is rolled, this process is likely to have a much smaller influence in these types of diets compared to those containing less roughage.

BARLEY

Although cattle ate more feed when they were given diets containing whole, as opposed to rolled barley, efficiency of utilization was greater for the rolled barley diets (Mathison et al., 1991b). Yaramecio et al. (1991) reported NE_g values of 1.15 and 1.80 Mcal/kg for diets containing whole or rolled barley and most of this difference appeared to be due to improved digestibility. There is greater controversy about the value of steam-processed and rolled barley compared to dry-rolled barley. Zinn (1993) found steam-processed barley contained 2.24 Mcal/kg NE_m and 1.56 Mcal/kg NE_g, respectively, vs 2.14 and 1.47 for the dry-rolled grain. In the same experiment, benefits of a thin flake (0.19 kg/L) as opposed to a thick flake (0.39 kg/L) were evident. By contrast, steam processing of barley failed to improve the feeding value of a barley diet in two Canadian studies (Mathison et al., 1991a; Engstrom et al., 1992). Parrot et al. (1969) reported that steam processing and rolling did not improve digestibility of barley compared to dry rolling except when the initial DE value of the barley was low. Steam processing prior to rolling may be useful to maximize intake of barley diets, particularly in dry areas where dry-rolled or ground barley becomes too dusty. When barley is rolled or ground, fines should also be avoided to minimize digestive disturbances such as bloat (Hironaka et al., 1979). High-moisture barley has a feeding value equal to dry barley (Kennelly et al., 1988) and is superior in the rolled as opposed to whole form (Rode et al., 1986).

In medium- to high-roughage diets, dry-rolled barley was equivalent to the ammoniated high-moisture whole grain (Mandell et al., 1988) and steam-rolled dry barley was superior to the whole dry grain (Morgan et al., 1991).

OATS

Starch digestibility of a high-grain whole oat diet was 0.61 which contrasts to 0.69 when the oats were dry-rolled (Orskov et al., 1980). In mixed diets, whole oat grains seem to be well digested by cattle and there is little benefit in further processing (Campling, 1991).

WHEAT

Starch digestibility of a high-grain whole wheat diet was 0.83 and this was increased to 0.99 when the wheat wheat was rolled (Orskov et al., 1980). In contrast to oats, digestibility of starch in mixed diets containing whole wheat was only 0.60, as opposed to 0.86 for the same diet when the wheat was rolled and crushed (Toland, 1978). Steam-processed and rolled wheat, with a thick flake, has the same value as coarse ground or dry-rolled wheat (Brethour, 1970). Finely ground wheat should be avoided in beef cattle diets to maximize intake and prevent acidosis.

REFERENCES

Beauchemin, K. A., T. A. McAllister, Y. Dong, B. I. Farr, and K. J. Cheng. 1994. Effects of mastication on digestion of whole cereal grains by cattle. J. Anim. Sci. 72:236–246.

Beeson, W. M., and T. W. Perry. 1982. Effect of processing on nutritive value of feeds: Cereal grains. Pp. 193–212 in Handbook of Nutritive Value of Processed Food, M. Rechcigl, ed. Boca Raton, Fla.: CRC Press.

Berger, L. L., G. C. Fahey, L. D. Bourquin, and E. C. Titgemeyer. 1994. Modification of forage quality after harvest. Pp. 922–966 in Forage Quality, Evaluation, and Utilization, G. C. Fahey, M. Collins, D. R. Mertens, and L. E. Moser, eds. Madison, Wisc.: American Society of Agronomy, Crop Science Society, Soil Science Society.

Bines, J. A., W. H. Broster, J. D. Stutton, V. J. Broster, D. J. Napper, T. Smith, and J. W. Siviter. 1988. Effect of amount consumed and diet composition on the apparent digestibility in cattle and sheep. J. Agric. Sci. Camb. 110:249–259.

Blaxter, K. L., N. McC. Graham, and F. W. Wainman. 1956. Some observations on the digestibility of food by sheep, and on related problems. Br. J. Nutr. 10:69–91.

Bock, B. J., R. T. Brandt, D. L. Harmon, S. J. Anderson, J. K. Elliott, and T. B. Avery. 1991. Mixtures of wheat and high-moisture corn in finishing diets: Feedlot performance and in situ rate of starch digestion in steers. J. Anim. Sci. 69:2703–2710.

Brethour, J. R. 1970. The use and value of wheat in beef cattle feeding. Pp. 177–190 in Wheat in Livestock and Poultry Feeds: Proceedings of an International Symposium at Oklahoma State University. June 18–19.

Brethour, J. R. 1980. Nutritional value of milo for cattle. Report of Progress 384, Roundup 67, pp. 5–8. Fort Hays Branch, Kansas State University.

Buchanan-Smith, J. G., R. Totusek, and A. D. Tillman. 1968. Effect of methods of processing on digestibility and utilization of grain sorghum by cattle and sheep. J. Anim. Sci. 27:525–530.

Campling, R. C. 1991. Processing cereal grains for cattle—A review. Livestock Prod. Sci. 28:223–234.

Campling, R. C., and M. Freer. 1966. Factors affecting the voluntary intake of food by cows. 8. Experiments with ground, pelleted roughages. Br. J. Nutr. 20:229–244.

Chalupa, W. 1980. Chemical control of rumen microbial metabolism. P. 325 in Digestive Physiology and Metabolism in Ruminants: Proceedings of the 5th International Symposium on Ruminant Physiology, Y. Ruckebusch and P. Thivend, eds. Lancaster, England: MTP Press.

Clark, J. H. 1975. Utilization of high-moisture grains by dairy and beef cattle. Pp. 205–238 in Proceedings of the 2nd International Silage Research Conference. Cedar Falls, Iowa: National Silo Association, Inc.

Croka, D. C., and D. G. Wagner. 1975. Micronized sorghum grain. III. Energetic efficiency for feedlot cattle. J. Anim. Sci. 40:936–939.

D'Alfonso, T. H., W. B. Roush, and J. A. Ventura. 1992. Least-cost poultry rations with nutrient variability: A comparison of linear programming with a margin of safety and stochastic programming models. Poult. Sci. 71:255–262.

Engstrom, D. F., G. W. Mathison, and L. A. Goonewardene. 1992. Effect of beta-glucan, starch, and steam vs dry rolling of barley grain on its degradability and utilization by steers. Anim. Feed Sci. Tech. 37:33–46.

Fluharty, F. L., and S. C. Loerch. 1989. Chemical treatment of ground corn to limit starch digestion. Can. J. Anim. Sci. 69:173–180.

Franks, L. G., J. R. Newsom, R. E. Renbarger, and R. Totusek. 1972. Relationship of rumen volatile fatty acids to type of grains, sorghum grain processing method and feedlot performance. J. Anim. Sci. 35:404–409.

Galyean, M. L., D. G. Wagner, and F. N. Owens. 1976. Site and extent of starch digestion in steers fed processed corn rations. J. Anim. Sci. 43:1088–1094.

Galyean, M. L., D. G. Wagner, and F. N. Owens. 1981. Dry matter and starch disappearance of corn and sorghum as influenced by particle size and processing. J. Dairy Sci. 64:1804–1812.

Garrett, W. N. 1968. Influence of method of processing on the feeding value of milo and wheat. Eighth Annu. Calif. Feeders Day Rept.

Goetsch, A. L., F. N. Owens, M. A. Funk, and B. E. Doran. 1987. Effects of whole or ground corn with different forms of hay in 85% concentrate diets on digestion and passage rates in beef heifers. Anim. Feed Sci. Tech. 18:151–164.

Goodrich, R. D., and J. C. Meiske. 1966. Whole corn grain vs. ground corn grain, long hay vs. ground hay and 10 lb. vs 15 lb. corn silage for finishing cattle. Minn. Beef Cattle Feeders Day Res. Rept. B-76:61–67.

Greenhalgh, J. F. D., and G. W. Reid. 1973. Long- and short-term effects of pelleting a roughage for sheep. Anim. Prod. 19:77–86.

Hinman, D. D., and R. R. Johnson. 1974. Influence of processing methods on digestion of sorghum starch in high-concentrate beef cattle rations. J. Anim. Sci. 39:417–422.

Hironaka, R., N. Kimura, and G. C. Kozub. 1979. Influence of feed particle size on rate and efficiency of gain, characteristics of rumen fluid and rumen epithelium, and numbers of rumen protozoa. Can. J. Anim. Sci. 59:395–402.

Husted, W. T., S. Mehen, W. H. Hale, M. Little, and B. Theurer. 1968. Digestibility of mil processed by different methods. J. Anim. Sci. 27:531–534.

Isichei, C. O., and W. G. Bergen. 1980. The effect of monensin on the composition of abomasal nitrogen flow in steers fed grain and silage rations. J. Anim. Sci. 51(Suppl. 1)371.

Keating, E. R., W. J. Saba, W. H. Hale, and B. Taylor. 1965. Further observations on the digestion of mil and barley by steers and lambs. J. Anim. Sci. 24:1080–1085.

Kennelly, J. J., G. W. Mathison, and G. de Boer. 1988. Influence of high-moisture barley on the performance and carcass characteristics of feedlot cattle. Can. J. Anim. Sci. 68:811–820.

Kiesling, H. E., J. E. McCroskey, and D. G. Wagner. 1973. A comparison of energetic efficiency of dry-rolled and reconstituted rolled sorghum grain by steers using indirect calorimetry and the comparative slaughter technique. J. Anim. Sci. 37:790–795.

Loerch, S. C., L. L. Berger, D. Gianola, and G. C. Fahey, Jr. 1983. Effect of dietary protein source and energy level on in situ nitrogen disappearance of various protein sources. J. Anim. Sci. 56:206–216.

MacLeod, G. K., D. N. Mowat, and R. A. Curtis. 1976. Feeding value for finishing steers and Holstein male calves of whole dried corn and of whole and rolled high moisture acid-treated corn. Can. J. Anim. Sci. 56:43–49.

Mader, T. L., J. M. Dahlquist, R. A. Britton, and V. E. Krause. 1991. Type and mixtures of high-moisture corn in beef cattle finishing diets. J. Anim. Sci. 69:3480–3486.

Mandell, I. B., H. H. Nicholson, and G. I. Christison. 1988. The effects of barley processing on nutrient digestion within the gastro-intestinal tract of beef cattle fed mixed diets. Can. J. Anim. Sci. 68:191–198.

Mathison, G. W., R. Hironaka, B. K. Kerrigan, I. Vlach, L. P. Milligan, and R. D. Wesenburger. 1991a. Rate of starch degradation, apparent digestibility and rate and efficiency of steer gain as influenced by grain volume-weight and processing method. Can. J. Anim. Sci. 71:867–878.

Mathison, G. W., D. F. Engstrom, and D. D. MacLeod. 1991b. Effect of feeding whole and rolled barley to steers in the morning or afternoon in diets containing differing proportions of hay and grain. Anim. Prod. 53:321–330.

Maxson, W. E., R. L. Shirley, J. E. Bertrand, and A. Z. Palmer. 1973. Energy values of corn, bird-resistant and non bird-resistant sorghum grain in rations fed to steers. J. Anim. Sci. 37:1451–1457.

McAllister, T. M., K.-J. Cheng, L. M. Rode, and J. G. Buchanan-Smith. 1990a. Use of formaldehyde to regulate digestion of barley starch. Can. J. Anim. Sci. 70:581–589.

McAllister, T. M., L. M. Rode, D. J. Major, K. -J. Cheng, and J. G. Buchanan-Smith. 1990b. Effect of ruminal microbial colonization on cereal grain digestion. Can. J. Anim. Sci. 70:571–579.

McCaffree, J. D., and W. G. Merrill. 1968. High moisture corn for dairy cows in early lactation. J. Dairy Sci. 51:553–560.

McDonald, P., A. R. Henderson, and S. J. E. Heron. 1991. The Biochemistry of Silage, 2nd Ed. Marlow, Bucks., U.K.: Chalcombe Publications.

McKnight, D. R., G. K. MacLeod, J. G. Buchanan-Smith, and D. N. Mowat. 1973. Utilization of ensiled or acid-treated high-moisture shelled corn by cattle. Can. J. Anim. Sci. 53:491–496.

McNeill, J. W., G. D. Potter, and J. K. Riggs. 1971. Ruminal and postruminal carbohydrate utilization in steers fed processed sorghum grain. J. Anim. Sci. 33:1371–1374.

Minson, D. J. 1963. The effect of pelleting and wafering on the feeding value of roughage—A review. J. Br. Grassland Soc. 18:39–44.

Moe, P. W., H. F. Tyrrell, and N. W. Hoover. 1973. Physical form and energy value of corn in timothy hay diets for lactating cows. J. Dairy Sci. 60:752–758.

Moe, P. W., and H. F. Tyrrell. 1977. Effects of feed intake and physical form on energy value of corn in timothy hay diets for lactating cows. J. Dairy Sci. 60:752–758.

Moe, P. W., and H. F. Tyrrell. 1979. Effect of endosperm type on incremental energy value of corn grain for dairy cows. J. Dairy Sci. 62:447–454.

Morgan, C. A., and R. C. Campling. 1978. Digestibility of whole barley and oat grains by cattle of different ages. Anim. Prod. 27:323–329.

Morgan, E. K., M. L. Gibson, M. L. Nelson, and J. R. Males. 1991. Utilization of whole or steamrolled barley fed with forages to wethers and cattle. Anim. Feed Sci. Tech. 33:59–78.

National Research Council. 1985. Ruminant Nitrogen Usage. Washington, D.C.: National Academy Press.

National Research Council. 1989. Nutrient Requirements of Horses, Fifth Revised Ed. Washington, D.C.: National Academy Press.

Orskov, E. R. 1976. The effect of processing on digestion and utilization of cereals by ruminants. Proc. Nutr. Soc. 35:245–252.

Orskov, E. R. 1979. Recent information on processing of grain for ruminants. Livestock Prod. Sci. 6:335–347.

Orskov, E. R., R. J. Barnes, and B. A. Lukins. 1980. A note on the effect of different amounts of NaOH application on digestibility by cattle of barley, oats, wheat and maize. J. Agric. Sci. Camb. 94:271–273.

Osbourne, D. F., D. E. Beever, and D. J. Thomson. 1976. The influence of physical processing on the intake, digestion and utilization of dried herbage. Proc. Nutr. Soc. 35:191–200.

Owens, F. N., R. A. Zinn, and Y. K. Kim. 1986. Limitations to starch digestion in the ruminant small intestine. J. Anim. Sci. 63:1634–1648.

Parrot, J. C., S. Mehen, W. H. Hale, M. Little, and B. Theurer. 1969. Digestibility of dry rolled and steam processed flaked barley. J. Anim. Sci. 28:425–428.

Poos, M. I., T. L. Hanson, and T. J. Klopfenstein. 1979. Monensin effect on diet digestibility, ruminal protein bypass and microbial protein synthesis. J. Anim. Sci. 48:1516.

Preston, R. L. 1975. Net energy evaluation of cattle finishing rations containing varying proportions of corn grain and corn silage. J. Anim. Sci. 41:622–624.

Riggs, J. K., J. W. Sorenson, J. L. Adame, and L. M. Schake. 1970. Popped sorghum grain for finishing beef cattle. J. Anim. Sci. 30:634–638.

Rode, L. M., K. J. Cheng, and J. W. Costerton. 1986. Digestion by cattle of urea-treated, ammonia-treated or rolled high moisture barley. Can. J. Anim. Sci. 66:711–721.

Sherrod, L. B., R. C. Albin, and R. D. Furr. 1969. Net energy of regular and waxy sorghum grains for finishing steers. J. Anim. Sci. 29:997–1000.

Stock, R. A., D. R. Brink, R. T. Brandt, J. K. Merrill, and K. K. Smith. 1987. Feeding combinations of high moisture corn and dry corn to finishing cattle. J. Anim. Sci. 65:282–289.

Stock, R. A., M. H. Sindt, R. M. Cleale, and R. A. Britton. 1991. High-moisture corn utilization in finishing cattle. J. Anim. Sci. 69:1645–1656.

Streeter, M. N., D. G. Wagner, F. N. Owens, and C. A. Hibberd. 1989. Combinations of high-moisture harvested sorghum grain and dry-rolled corn: Effects on site and extent of digestion in beef heifers. J. Anim. Sci. 67:1623–1633.

Sundstol, F. 1991. Large scale utilization of straw for ruminant production systems. Pp. 55–60 in Recent Advances on the Nutrition of Herbivores, Y. W. Ho, H. K. Wong, N. Abdullah, Z. A. Tajuddin, eds. Kuala Lumpur: Malaysian Society of Animal Production.

Theurer, C. B. 1986. Grain processing effects on starch utilization by ruminants. J. Anim. Sci. 63:1649–1662.

Thomson, D. J., and D. E. Beever. 1980. The effect of conservation on the digestion of forages by ruminants. Pp. 291–308 in Digestive Physiology and Metabolism in Ruminants: Proceedings of the 15th International Symposium on Ruminant Physiology, Y. Ruckebusch and P. Thivend, eds. Lancaster, U.K.: MTP Press Ltd.

Toland, P. C. 1978. Influence of some digestive processes on the digestion by cattle of cereal grains fed whole. Aust. J. Exp. Agric. Anim. Husb. 18:29–33.

Tonroy, B. R., T. W. Perry, and W. M. Beeson. 1974. Dry, ensiled high-moisture, ensiled reconstituted high moisture and volatile fatty acid treated high moisture corn for growing-finishing beef cattle. J. Anim. Sci. 39:931–936.

Turgeon, O. A., D. R. Brink, and R. A. Britton. 1983. Corn particle size mixtures, roughage level and starch utilization in finishing steer diets. J. Anim. Sci. 57:739–749.

Vance, R. D., R. R. Johnson, E. W. Klosterman, B. W. Dehority, and R. L. Preston. 1970. All-concentrate rations for growing-finishing cattle. Ohio Agricultural Research and Development Center Research Summary 49.

Vance, R. D., R. L. Preston, E. W. Klosterman, and V. R. Cahill. 1972. Utilization of whole shelled and crimped corn grain with varying proportions of corn silage by growing-finishing steers. J. Anim. Sci. 35:598–605.

Weston, R. H. 1967. Factors limiting the intake of feed by sheep. II. Studies with wheaten hay. Aust. J. Agric. Res. 18:983–1002.

Whetstone, H. D., C. L. Davis, and M. P. Bryant. 1981. Effect of monensin on breakdown of protein by ruminal microorganisms in vitro. J. Anim. Sci. 53:803–809.

Wilkinson, J. M. 1978. The ensiling of forage maize: Effects on composition and nutritive value. Pp. 201–237 in Forage Maize, E. S. Bunting, B. F. Pain, R. H. Phipps, J. M. Wilkinson, amd R. E. Gunn, eds. London: Agricultural Research Council.

Xiong, Y., S. J. Bartle, and R. L. Preston. 1991. Density of steam-flaked sorghum grain, roughage level, and feeding regimen for feedlot steers. J. Anim. Sci. 69:1707–1718.

Yaramecio, B. J., G. W. Mathison, D. F. Engstrom, L. A. Roth, and W. R. Caine. 1991. Effect of ammoniation on the preservation and feeding value of barley grain for growing-finishing cattle. Can. J. Anim. Sci. 71:439–455.

Young, L. G., A. Lun, J. Pos, R. P. Forshaw, and D. Edmeades. 1975. Vitamin E stability in corn and mixed feed. J. Anim. Sci. 40:495–499.

Zinn, R. A. 1987. Influence of lasalocid and monensin plus tylosin on comparative feeding value of steam-flaked versus dry-rolled corn in diets for feedlot cattle. J. Anim. Sci. 65:256–266.

Zinn, R. A. 1990a. Influence of steaming time on site of digestion of flaked corn in steers. J. Anim. Sci. 68:776–781.

Zinn, R. A. 1990b. Influence of flake density on the comparative feeding value of steam-flaked corn for feedlot cattle. J. Anim. Sci. 68:767–778.

Zinn, R. A. 1993. Influence of processing on the comparative feeding value of barley for feedlot cattle. J. Anim. Sci. 71:3–10.

Zinn, R. A., and F. N. Owens. 1983. Site of protein digestion in steers: predictability. J. Anim. Sci. 56:707.

Appendix

Nutrient Requirements of Beef Cattle

Seventh Revised Edition, 1996

A User's Guide for NRC Model Application

National Research Council
Board on Agriculture
Committee on Animal Nutrition
Subcommittee on Beef Cattle Nutrition

SUBCOMMITTEE ON BEEF CATTLE NUTRITION

JOCK G. BUCHANAN-SMITH, *Chair*, University of Guelph, Canada
LARRY L. BERGER, University of Illinois
CALVIN FERRELL, U.S. Department of Agriculture, Agricultural Research Service, Clay Center, Nebraska
DANNY G. FOX, Cornell University
MICHAEL GALYEAN, Clayton Livestock Research Center, Clayton, New Mexico
DAVID P. HUTCHESON, Animal Agricultural Consulting, Inc., Amarillo, Texas
TERRY J. KLOPFENSTEIN, University of Nebraska
JERRY W. SPEARS, North Carolina State University

This guide was prepared by Danny G. Fox,
with the assistance of Michael C. Barry.

Table of Contents

1 Introduction

A computer disk containing two stand-alone programs is provided as a companion to the National Research Council's (NRC's) *Nutrient Requirements of Beef Cattle, Seventh Revised Edition 1996* to demonstrate how to use the NRC model Levels 1 and 2. The two computer programs include (1) a table generator program and (2) the NRC model program containing two levels of equations. These programs allow the user to apply the equations summarized in Chapter 10 of the report. (See the report's Glossary for definitions of acronyms used.) An understanding of ruminant nutrition and knowledge of the underlying biological concepts presented in this report are essential for use of the models.

The programs predict requirements and energy and protein allowable production from the dietary ingredients fed. All programs use the same cattle requirement equations, which can be used to compute requirements over wide variations in body sizes and cattle types, milk production levels, and environmental conditions. Rate of gain or energy reserves balance are predicted based on ME available for productive purposes after maintenance, growth, gestation, and milk production requirements have been satisfied.

We have attempted to make the software accurate and user friendly. The programs were developed as a Lotus 1-2-3® spreadsheet. Baler® was used to protect the spreadsheet and develop the user interface. Program help screens provide guidelines for choosing inputs and in interpreting and applying outputs. Pop-up evaluator screens in the NRC model program interpret output and provide application recommendations.

TUTORIALS

The focus of this user's guide is to demonstrate how to apply the model Levels 1 and 2. Tutorials provide a quick overview of the program applications. Examples are provided that allow the user to input data from an actual feedlot and cow-calf ranch, analyze the diets, and evaluate the results. The user is referred to the following chapters for detailed information on biological bases for equations and assumptions used in the software:

- maintenance, Chapter 1;
- growth and energy reserves, Chapter 3;
- pregnancy and lactation, Chapter 4;
- rumen fermentation and protein metabolism, Chapters 2 and 10;
- minerals, Chapter 5;
- dry matter intake, Chapter 7;
- feed analysis and feed library, Chapter 10;
- analysis of common feeds by commercial laboratories, Chapter 11; and
- a list of all equations, Chapter 10.

COMPUTER PROGRAMS

Tables This program allows the user to compute tables of nutrient requirements and diet nutrient density required over a feeding period indicated. It also allows a rapid determination of how well a diet meets the requirements of the group of cattle being fed that diet and whether modifications are needed.

NRC Model This program contains two levels of solution for predicting energy and protein supply from actual rations, using a feed library (Appendix Table 1). Level 1 uses tabular NE_m, NE_g, and DIP values to compute energy and protein supply, microbial growth, and nitrogen requirements for fermentation. Level 2 predicts feed carbohydrate and protein ruminal degradation, microbial growth, and fermentation nitrogen requirements, and escape of carbohydrate and protein to dynamically predict ME and MP derived from each feed fed, and amino acid balances.

Feed Library (Appendix Table 1) A critical component of the NRC model program is the feed library developed from research data and the values in Table 11-1 of the report; Table 11-1 lists some of the same feeds and International Feed Numbers found in Appendix Table 1, and values correspond wherever possible. The feed library, Appendix Table 1, contains feed composition values needed to predict the supply of nutrients available to meet animal requirements in both model Levels 1 and 2. A detailed description of the feed library can be found in Chapter 10 of the report.

Feeds can be added to the feed library, and any of the library composition values can be changed. The user should use actual values whenever possible. Appendix Table 1 differs from Table 11-1 of the report because of the additional carbohydrate and protein fractions needed for Level 2. When feeds are added to the library on the disk, use Appendix Tables 6 through 9 to assign digestion rates and effective NDF values.

Because of the many variables involved and judgments that must be made in choosing inputs and interpreting outputs, the NRC makes no claim for the accuracy of this software and the user is solely responsible for risk of use.

HARDWARE AND SOFTWARE REQUIREMENTS AND INSTALLATION

This software is designed to operate on microcomputers that run MS-DOS. The NRC model requires the following hardware:

1. an IBM personal computer or "compatible" running MS-DOS or PC-DOS Version 3.0 or later,
2. at least one floppy-disk drive,
3. at least one hard drive, and
4. 640 KB random access memory (RAM).

Additional memory (2MB), a hard disk, math co-processor, and printer are optional, but highly recommended.

The NRC model requires the following software:

1. PC-DOS or MS-DOS Version 3.0 or later,
2. NRC disk.

To install this software:

1. Make a back-up copy of the original disk for safety and archival purposes, then use the back-up and store the original disk.

2. Create a subdirectory on your hard drive to store the program files.
   For example, at the **C:/** prompt, *type* **MD NRC**.
3. Copy all the files from the backup copy of the distribution diskettes to that subdirectory.
   For example, at the **C:/** prompt, type **CD NRC,** *then type* **copy a:*.*** ⟨**Enter**⟩
4. *Type* **INSTALL**.

PROGRAM OPERATION AND USE

1. Select the directory on your computer that contains the NRC files. If you installed the software on your C drive, you should be at the **C:/NRC** directory prompt.
2. At the directory prompt, you may choose one of the three following options:

- To start the table generator program, type **TABLES**
- To start the NRC model program, type **NRC**
- To open the feed library, type **FEEDS**

A "**Welcome to the Software**" screen will appear. Press any key to continue. To go from one program to the other, you must return to the NRC directory.

After the program is loaded and the "**Welcome to the Software**" screen appears, press any key to continue. The main menu screen will appear. The program returns to this screen whenever the ⟨**ESC**⟩ key is pressed. This program contains a context-sensitive help system that is accessed by pressing the ⟨**F1**⟩ key when the cursor is on the input or output cell in question. Other "hot" keys have been defined and are shown below. Cell locations are shown above and to the left of each screen for reference.

Key	Description
⟨**F1**⟩	Access on-line help system
⟨**F6**⟩	Go to feed import screen
⟨**F7**⟩	Go to feed energy and protein values screen
⟨**F10**⟩	Go to feed amounts screen
⟨**F11**⟩	Go to detailed diet evaluation screen
⟨**ESC**⟩	Go to main menu

MAIN MENU

"MAIN MENU" SCREEN

```
1996 Nutrient Requirements of Beef Cattle

Describe Units and Levels   Print Results
Describe Animal
Describe Management        View Feed Digestion
Describe Environment       View Requirements
Describe Feed              View Amino Acid Balances
View Balance Screen        View Mineral Balances

                           Save Inputs
Quit                       Retrieve Inputs

Press ⟨F1⟩ at any time for context sensitive help
Press ⟨ESC⟩ at any time to escape to this screen
```

Position the cursor over the appropriate option and press ⟨**ENTER**⟩ to select that option. Help is available for each option by pressing ⟨**F1**⟩ when the cursor is positioned on that option. The options are described below.

Describe Units and Levels is used to name the diet, choose the grading system, solution level (Level 1 or Level 2), units (English or metric), and diet basis (dry matter or as fed).

Describe Animal is used to describe the animal (type, age, sex, body weight, condition score, mature weight), and reproductive cycle (days pregnant, days in milk, lactation #, peak milk production, time of lactation peak, duration of lactation, milk composition, age @ puberty, calving interval, expected calf birth weight).

Describe Management is used to describe feed additives used, grazing conditions, and to make adjustments to efficiency of use of ME and microbial yield.

Describe Environment is used to describe environmental conditions (wind, temperature, hair coat condition).

Describe Feed is used to bring in feeds from the feed library, view and change composition of feeds chosen from the feed library, and change amounts (actual consumption of each feed in the diet).

View Balance Screen is used to view the supply-requirements balances of energy and protein for the animal, management, environment, and feed inputs, predicted performance, diet net energy, and protein concentrations.

Quit is used to exit the program.

Print Results is used to obtain a printout of this evaluation.

View Feed Digestion is used to view each feed calculation from the rumen simulation in Level 2 (*degradation and passage rates, carbohydrate and protein fraction amounts ruminally degraded and escaped, bacterial growth and nitrogen (N) balance, intestinal digestion, fecal output, predicted feed NE and MP values*).

View Requirements is used with both levels to view calculations of animal requirements by physiological function (maintenance, growth, lactation, pregnancy).

View Amino Acid Balances is used to view each essential amino acid requirement, supply, balance (supply-requirement), and percent of requirement met.

View Mineral Balances is used to view each mineral requirement, amounts supplied from the diet, balances, and percent in the diet.

Save Inputs is used to save the inputs for this evaluation.

Retrieve Inputs is used to retrieve inputs for previous evaluations saved (⟨/⟩, ⟨*file get*⟩) so they can be updated.

TUTORIAL LESSON 1: FEEDLOT CASE STUDY

Begin the tutorial by opening the NRC model program (at the NRC directory prompt, *type* **NRC**); select the **Describe Units and Levels** option on the main screen. Press ⟨**Enter**⟩

This case study is a 20,000 head capacity western Canada feedlot. Cattle are fed in open dirt lots surrounded by windbreaks. Typical pens contain 250 head. The basal ration is dry rolled barley and barley silage. The questions are as follows.

1. Should the roughage level in the ration be lowered to increase energy intake?
2. Should the barley silage be chopped finer, and is the barley grain rolled fine enough?
3. Are feed "bypass" protein or protected amino acids needed?
4. How can I adjust gain predictions for cattle type and weather conditions?

Data from closeouts will be used to adjust the model so it predicts accurately for that feedlot, and then inputs will be changed to answer the questions. The data base is 1969 Hereford × Charolais crossbred steers fed in 8 pens in the fall with an initial weight of 837 lb and final weight of 1,284 lb with an average grade of Canadian AA. The cattle received an estrogenic implant and were fed an ionophore. The average weight during the feeding period was 1,060 lb, with an ADG of 3.48 and conversion of 6.98 lb DM/lb gain. The average diet DMI was 5 lb coarse chopped barley silage, 19 lb coarse rolled barley grain, and 0.3 lb minerals. Feed analysis available indicated the barley silage was 48.7% NDF with 65% estimated to be eNDF, 10.4% CP, 3% fat, and 8% ash; and barley grain was 19% NDF with 34% estimated to be eNDF, 13% CP, 2.1% fat, and 3% ash. Environmental conditions were 5 mph average wind on the cattle in the pens; the previous month's average temperature was 40° F and average temperature during the feeding period was 30° F. Other inputs were average hide thickness, hair depth of 0.2 inch (typical of early summer-fall; 0.5 inch is typical of winter), and average hair coat condition is clean and dry.

Describe Units and Levels

"DESCRIBE UNITS AND LEVELS" SCREEN

1996 NRC Nutrient Requirements of Beef Cattle Describe Units and Levels		
Diet	NRC Feedlot1	Grading System 2
Level	1 Tabular System	
Units	1 English	
Feed H2O	0 Dry Matter	
	Main Menu	

Press ⟨F1⟩ at any time for context sensitive help
Press ⟨ESC⟩ at any time to escape to the main menu

Diet: Enter an identifying name for the particular diet being evaluated in cell C1024.
Entry for the example is NRC feedlot1. ⟨Enter⟩

Grading System: In cell H1024 enter the grading system. Choices are 1 (USDA Standard or Canadian A, which are related to 25.2% body fat), 2 (USDA Select or Canadian AA, which are related to 26.8% body fat), and 3 (USDA Choice or Canadian AAA, which are related to 27.8% body fat). The program uses this grade to identify the standard reference weight. The standard reference weight is divided by the finished weight, and this result is multiplied by the actual weight. These calculations provide the weight used

in the equations that compute net energy and protein in the gain. (See Chapter 3 for the biological basis and validation of this method.)
Entry for the example is 2. ⟨**Enter**⟩

Level: In cell C1026 enter either a 1 (uses tabular feed net energy and protein degradability values) or 2 (feed energy and absorbed protein values based on feed carbohydrate and protein fractions and their digestion rates). It is often practical to adjust the diet until balanced with Level 1, then evaluate it with Level 2 to get predicted feed net energy values and amino acid balances, based on actual feed analysis for carbohydrate and protein fractions.
Entry for the example is 1 (will later be changed to 2 for further evaluation). ⟨**Enter**⟩

Units: In cell C1028 enter either a 0 for metric or 1 for English. Be sure all data are entered in the same units as chosen here.
Value for the example is 1 (English). ⟨**Enter**⟩

Feed H₂0: In cell C1030 enter 0 (dry matter) or 1 (as fed). This is used to determine DMI from the feed amounts fed that is entered later.
Value for the example is 0 (Dry Matter). ⟨**Enter**⟩

Context sensitive help (⟨**F1**⟩) is available to guide the user in selecting appropriate values to enter in these cells. After you are satisfied with the inputs for this section, press ⟨**Enter**⟩ to return to the main menu. Then select **Describe Animal** ⟨**Enter**⟩.

Describe Animal

"DESCRIBE ANIMAL" SCREEN

Describe Animal		
Animal Type	1	Growing/Finishing
Age	14	Months
Sex	2	Steer
Body Weight	1060	lb
Condition Score	5	1 = emaciated 9 = very fat
Mature Weight	1284	lb at 27% fat (slight marbling)
Breeding System	2	2-way cross
	1	
Dam's Breed	11	Hereford
Sire's Breed	6	Charolais
	1	
	1	
Next		

Press ⟨F1⟩ at any time for context sensitive help
Press ⟨ESC⟩ at any time to escape to the main menu

When entering values, press ⟨**Enter**⟩ twice to move to the next input cell and to cause chosen category to be displayed.

Animal Type: In cell D1043 enter the correct code for the class of cattle. Choices are 1 (growing and finishing), 2 (lactating cow), 3 (dry cow), 4 (herd replacement heifer), 5 (breeding bull). This invokes the inputs and equations needed to compute requirements, predict DMI, and evaluate the diet for that class.
The entry for this example is 1. ⟨**Enter**⟩

Animal Age: In cell D1044 enter the average age in months. This value influences expected DMI and tissue insulation.
The entry for this example is 14. ⟨**Enter**⟩

Sex: In cell D1045 enter the code for the sex of the animal. Choices are 1 for a bull, 2 for a steer, 3 for a heifer, and 4 for a cow. A heifer is entered as a cow after calving the first time.
The entry for this example is 2. ⟨**Enter**⟩

Body Weight: In cell D1046 enter the shrunk body weight that best represents the group being fed together. Body weight is a major determinant of DMI, maintenance, and growth requirements.
The entry for this example is 1060. ⟨**Enter**⟩

Condition Score: In cell D1047 enter the average condition score of the cattle in the group (Appendix Table 2). See Chapters 1 and 3 for a detailed discussion of the 1 to 9 condition scoring system used and its biological basis. The choices are 1 through 9 (1 = emaciated, 5 = moderate, 9 = very fat). Condition score is used to describe tissue insulation, the potential for compensatory growth in growing cattle, and energy reserves in cows. Appendix Table 3 gives estimates of the relationship between previous nutrition and body condition score in growing cattle.
Entry for the example is 5. ⟨**Enter**⟩

Mature Weight: In cell D1048 enter the expected average weight at the grade chosen in the Units and Levels screen. For cows, replacement heifers, or breeding bulls, enter the expected mature weight at condition score 5. The weight that best corresponds to the cattle in question based on the user's experience for the type of growing animal, implant strategy, and ration should be entered. A general guide is that the finishing weight should be reduced 50–75 lb if rations that contain more than 70% grain are fed continuously after weaning or if anabolic steroids are not used. Finishing weight should be increased 50–75 lb if animals are grown slowly or if they are implanted with estrogen in combination with trenbolone acetate.
Entry for the example is 1284. ⟨**Enter**⟩

Breeding System: In cell E1050 enter the code for the breeding system. Choices are 1 (straightbred), 2 (2-way crossbred), and 3 (3-way crossbred). E1051 is used for animal breed if straightbred, E1052 and E1053 are used if describing 2-way crossbred, and E1053 to E1055 are used when describing 3-way crossbred. When cells are not in use, the previous number (or NA) will appear. Breeding system for growing cattle influences maintenance energy requirement and predicted DMI. No adjustments are made for special breed effects other than dairy or *Bos indicus* types, as the data to date indicate most identifiable breed effects are due to differences in mature size, fat distribution, and hair and hide factors which are considered independently. The equations and biological basis for these effects are discussed in Chapters 1, 3, and 4.
Entry for this example is 2. ⟨**Enter**⟩

Breed Codes: In cells E1051 to E1055 enter breed codes for the parent breeds in the breeding system specified. If breeding system 1 is chosen, then animal breed appears in C1051 and the breed code is entered in E1051. Ignore all other cells in this section. If breeding system 2 is chosen, then the dam's breed will appear in B1052 (enter breed code in E1052) and the sire's breed will appear in B1053 (enter breed code in E1053). Valid breed codes are shown in the help system and in Appendix Table 4. Stored breed values are used to determine maintenance energy requirements, and defaults for calf birth weight and peak milk production. See Chapters 1 and 4 for the biological basis for these breed adjustments.
Entries for this example are 11 for dam's breed (Hereford) and 6 for sire's breed (Charolais). ⟨**Enter**⟩

Press ⟨**F1**⟩ to display chosen breeds.
Place cursor on **Next** ⟨**Enter**⟩
Select **Describe Management** ⟨**Enter**⟩

Describe Management

"DESCRIBE MANAGEMENT" SCREEN

Describe Management	
Additive	4 implant + ionophore
On Pasture?	0 no
	30
	1,500
	3
	45
	1
Diet NEm Adjuster	100% (Level 1 only)
Diet NEg Adjuster	100% (Level 1 only)
Diet Microbial Yield	13.0% TDN (Level 1 only)
Main Menu	
Press ⟨F1⟩ at any time for context sensitive help	
Press ⟨ESC⟩ at any time to escape to the main menu	

Additives: In cell E1084 enter the code that describes additives used. The choices and their effects are shown in Appendix Table 4. The biological basis for these adjustments are discussed in Chapters 3 and 5.
The entry for this example is 4 (implant plus ionophore). ⟨**Enter**⟩

On Pasture: In cell E1085 enter 0 if the animals are not grazing; enter 1 if they are. If 1 is chosen, other inputs must be chosen to compute maintenance requirements and predict DMI.
The entry for this example is 0. ⟨**Enter**⟩ However, the following are requested when 1 is chosen.

Grazing Unit Size: In cell E1086 enter the number of hectares (metric) or acres (English) per head grazed in the pasture. If the distance traveled is minimal, enter 0. This input is used to adjust energy maintenance requirements for walking activity. ⟨**Enter**⟩

Initial Pasture Mass: In cell E1087 enter the kg DM/hectare (metric) or lb DM/acre (English) when the cattle are turned into the pasture. This can be estimated from hay harvesting experience, clippings, or calibrated measuring devices such as height and/or density estimates, Plexiglas weight plates, or electronic pasture probes. ⟨**Enter**⟩

Days on Pasture and Number of Animals: In cell E1088 enter the number of days on the pasture and in E1089 the number of animals. Initial pasture mass, number of days on pasture, and number of animals are used to predict pasture DMI. ⟨**Enter**⟩

Terrain: In cell E1090 enter 1 (relatively level) or 2 (rolling). This value is used to adjust maintenance requirement. ⟨**Enter**⟩

Diet NE_m and NE_g Adjusters (Level 1 only): *Leave these at 1 (100%) unless you are certain you want to adjust the diet NE values.* In cells E1092 or E1093 enter a value between 0.8 and 1.2 if you wish to change the diet NE_{ma} or NE_{ga}. The appropriate way to use this is to move it up or down until predicted and actual ADG agree after all other inputs are carefully checked. Unrealistically high ADG and feed efficiency may be predicted for calves consuming high-energy rations; unrealistically low ADG and feed efficiency may be predicted for these same calves when approaching the fatness of choice grade. ***These entries are left at 1 (100%) for this example.*** ⟨**Enter**⟩

Microbial Yield (Level 1 only): *Leave the entry in cell E1094 at 13% unless you have information that indicates you should lower microbial yield in cattle fed low-quality forage diets.* In Level 1, microbial yield is a constant 13% of TDN as discussed in Chapter 2, except it is reduced on high-concentrate rations based on the eNDF level. However, there is no adjustment in the model for diets with low energy contents or low intakes. In either case, if rate of passage is reduced, then microbial turnover is increased and efficiency of microbial protein synthesis is reduced. Literature values for microbial yield for cattle fed low-quality forages average 7.8% of TDN; the DIP requirement was determined to be 7.1% of DM for cows grazing dormant forage. Therefore, it is recommended that microbial yield be reduced to 7.5–10% of TDN for cows or calves consuming low-quality diets. ***The entry is left at 13 (13%) for this example.*** ⟨**Enter**⟩

Place cursor on **Main Menu** ⟨**Enter**⟩
Select **Describe Environment** ⟨**Enter**⟩

Describe Environment

"DESCRIBE ENVIRONMENT" SCREEN

Describe Environment		
Wind Speed	5	mph
Previous Temp.	40	Degrees F
Current Temp.	30	Degrees F
Night Cooling	2	yes
Hair Depth	0.2	in
Hide	2	average
Hair Coat	1	clean and dry
Heat Stress	1	none

Main Menu
Press ⟨F1⟩ at any time for context sensitive help
Press ⟨ESC⟩ at any time to escape to the main menu

The equations driven by the inputs in the environmental description section are used to compute the lower critical temperature of the animal and to adjust predicted DMI for environmental effects. Cattle usually compensate for short-term environmental effects, so the inputs chosen should generally reflect average environmental conditions for at least 2 weeks. Predicted maintenance requirements are very sensitive to these effects after the animal reaches its lower critical temperature, so these inputs should be chosen carefully.

Wind Speed: In cell D1104 enter the average wind speed the cattle are exposed to. Wind speed influences maintenance requirements by reducing the external insulation of the animal. Increasing wind speed decreases the external insulation value of the animal and thus results in increased energy maintenance requirements. *The model is very sensitive to this input after the lower critical temperature is reached, so choose carefully.* ***Entry for this example is 5 mph because of the windbreaks (wind speed in open areas outside the pens is 15 mph).*** ⟨**Enter**⟩

Previous Temperature: In cell D1105 enter the average temperature for the previous month. This value is used to increase NE_m requirement as it gets colder or reduce it as it gets warmer. ***Entry for this example is 40° F.*** ⟨**Enter**⟩

Current Temperature: In cell D1106 enter the average temperature the cattle are exposed to. In most situations, the current average daily temperature is the most practical to use. This value is used to adjust predicted DMI for temperature effects and is also used in the calculations for the effects of cold stress on energy maintenance require-

ments. The model is very sensitive to this input after the lower critical temperature is reached.
Entry for this example is 30° F. ⟨Enter⟩

Night Cooling: In cell D1108 enter either 1 (no night cooling) or 2 (cools off at night). If 1 is chosen, predicted DMI is reduced as described in Chapter 7 with hot daytime temperatures. If 2 is chosen, it is assumed that cattle can dissipate heat at night and DMI is not affected.
Entry for this example is 2 (nights cool off). ⟨Enter⟩

Hair Depth: In cell D1109 enter the average hair depth. This input is used to compute the external insulation of the animal. Enter the effective hair coat depth of the animal, in increments of 0.1. As hair length increases, so does the external insulation value provided by the animal. A general guide to use is an effective coat depth of 0.2 inches (0.6 cm) during the summer and 0.5 inches (1.3 cm) during the winter. *The model is very sensitive to this input after the lower critical temperature is reached, and this entry should be chosen carefully.*
Entry for this example is 0.2 inches. ⟨Enter⟩

Hide: In cell D1110 enter either 1 (thin hide—i.e., dairy or *Bos indicus* types); 2 (average—i.e., most European breeds); or 3 (thick—i.e., Hereford or similar breeds). This value influences the external insulation value of the animal. Increased hide thickness implies increased external insulation. *The model is very sensitive to this value below the animal's lower critical temperature, and this entry should be chosen carefully.*
Entry for this example is 2 (average). ⟨Enter⟩

Hair Coat: In cell D1111 enter either 1 (clean and dry), 2 (some mud on lower body), 3 (some mud on lower body and sides), or 4 (heavily covered with mud). This value is used to adjust external insulation. *The model is very sensitive to this value below the animal's lower critical temperature, and this entry should be chosen carefully.*
Entry for this example is 1 (clean and dry). ⟨Enter⟩

Heat Stress: In cell D1112 enter either 1 (no panting; not heat stressed), 2 (rapid shallow panting, or 3 (open mouth panting). This value is used to adjust maintenance energy requirements for the energy cost of dissipating heat.
Entry for this example is 1 (no heat stress). ⟨Enter⟩

Place cursor on **Main Menu** ⟨Enter⟩
Select **Describe Feed** ⟨Enter⟩

Describe Feed

"DESCRIBE FEED" SCREEN

Describe Feed
Feed Composition
Feed Amounts
New Feeds
Main Menu
Press ⟨F1⟩ at any time for context sensitive help
Press ⟨ESC⟩ at any time to escape to the main menu

These choices are used to change the composition of feeds in the current ration and to add new feeds to the existing ration or develop a new ration. For this example, we will start with developing a new ration.

Select **New Feeds** (cell B1127) ⟨**Enter**⟩.
*Note: This option can be accessed from any point in the program by pressing ⟨**F6**⟩.*

The screen will change (see below) after the feeds for this ration are retrieved from the feed library. The feed library contains average compositional values for net energy, protein, carbohydrate, and protein fractions and their digestion rates, and minerals and vitamins. It is critical to choose feeds that most accurately describe the actual feeds in the ration. To aid in making these choices, the default feed library is printed in its entirety in Appendix Table 1. It can be accessed on disk by entering **FEEDS** after returning to the NRC directory. Composition values can be modified and new feeds added.

"NEW FEEDS" SCREEN

New Feeds	
Code # for feed to be imported:	Look up feed codes
	999 Minerals
	Main Menu
	Import
Current Feeds:	
Barley Grain heavy	blank
Barley Silage	blank
Minerals	blank
blank	blank
blank	blank
blank	blank
Press ⟨F1⟩ at any time for context sensitive help	
Press ⟨ESC⟩ at any time to escape to the main menu	

Look Up Feed Codes: Takes you to a listing of all available feeds in the main feed library. The listing is organized in alphabetical order by grass forages, legume forages, grain crop forages, energy concentrates, plant protein concentrates, food processing byproducts, animal processing byproducts, and minerals. Blanks follow each category to allow the users to add their own feeds. Feed numbers 101–129 are grass forages, 130–134 are blank, 135–139 are grass pastures, 140–148 are range forages, 201–223 are legume forages, 224–229 are blank, 230–231 are legume pastures, 232–250 are blank, 301–323 are grain crop forages, 324–350 are blank, 401–435 are energy concentrates (note: all cotton products including whole cotton and cottonseed meal are in this category), 436–450 are blank, 501–522 are protein concentrates, 523–550 are blank, 601–607 are food processing byproducts, 608–620 are blank, 701–707 are animal byproducts, 801–834 are mineral feeds, 900–910 are blank.

Write down the code numbers of the feeds you want to import and then press ⟨F6⟩ to return to the New Feeds screen.

Code for Feed to Be Imported: Position the cursor on cell F1223 and enter the code number corresponding to the feed you want to import. Press ⟨**Enter**⟩ until the cursor moves down one cell. If the code number is entered correctly, the name of the feed should appear to the right of the code number. If this name is correct, position the cursor on **Import** and press ⟨**Enter**⟩.

A new screen will appear. Place the cursor in the row where you want the new feed; you can begin with line one for the first feed, line two for the second, etc. When the cursor is in the right place, press ⟨**Enter**⟩ and the new feed will be retrieved from the feed library. Repeat this process until all feeds desired are obtained.

For the example, bring in feed #s
301 (barley silage)
402 (barley grain-heavy)
999 (minerals)

Up to 14 feeds can be imported. Blank code 130 can be imported into the remaining 11 lines so that the only feeds showing are those in this ration. When all feeds are entered, return to the **Main Menu ⟨ESC⟩**
Select **Describe Feed ⟨Enter⟩**
Select **Feed Composition ⟨Enter⟩**

Feed Composition: Press the right arrow key or tab key to scroll across table values to be modified. **Enter desired value. ⟨Enter⟩**
For the example, modify feed analytical values as follows:

Feed	Cost $/Ton	NDF, % DM	Effective NDF % NDF	CP	Fat	Ash
Barley Silage	25	48.7	65	10.4	3.0	8.0
Barley Grain	120	19.0	34	13.0	2.1	3.0
Minerals	200					

After desired feed composition values are entered, press ⟨F10⟩ to get **Feed Amounts and Performance Summary** screen.

"FEED AMOUNTS AND PERFORMANCE SUMMARY" SCREEN

Feed Amounts and Performance Summary	
5.00 Barley Silage	0.00 Blank
19.00 Barley Grain Heavy	0.00 Blank
0.30 Mineral	0.00 Blank
0.00 Blank	0.00 Blank
0.00 Blank	0.00 Blank
0.00 Blank	0.00 Blank
0.00 Blank	0.00 Blank

Pred DMI	24.0 lbs	
Act. DMI	24.3 lbs	Cost $1.49/day
ADG	3.67 lb	Intake Scalar 100.0%
MP Balance	111 g/d	Basis: Dry Matter
DIP Balance	42 g/d	Units: Pounds

Press ⟨F1⟩ at any time for context sensitive help
Press ⟨ESC⟩ at any time to escape to the main menu

Feed Amounts: Place cursor next to feed name and enter desired value. Enter the amount (use the same moisture basis as indicated on general screen) for each feed listed. *Be sure to enter 0 for all other cells not in use.* ⟨**Enter**⟩

Intake Scalar: *This input is only used to change each feed amount fed, by the same proportion; enter the proportional change in total DMI desired.* The scalar can be used to evaluate this diet formula for other conditions where the intake is predicted to change. For example, if the body weight is changed to 600 lb, DMI is predicted to be 16.7 lb, which is 68.7% of the current actual DMI. Entering 0.687 as the intake scalar reduces the actual DMI to 16.7 without having to change the feed amounts. Also a dry matter formula can be entered as lb/10 lb, then the scalar adjusted until the actual DMI is correct. Then the formula can be used to adjust to any DMI expected for various conditions. For example, this diet is 20.58% barley silage, 78.19% barley grain, and 1.23% minerals. Dividing each by 10 and entering as feed amounts, and entering 2.43 (24.3/10) as the scalar gives the correct DMI. *Entry for this example is 1 (decimal for 100%).* ⟨**Enter**⟩

Performance Summary: Press ⟨F9⟩ to calculate. Actual DMI is close to predicted (24.3 vs 24.0). If actual and predicted DMI differ by more than 5–10%, carefully check all inputs that influence DMI (breed, body weight, mature size, temperature, mud and storm exposure, diet energy density, ionophores, implant). The diet is evaluated with actual DMI. Predicted ADG exceeds actual ADG by 5.5%. Rumen microbial nitrogen requirements are being met (DIP balance is +42 g/day). Animal MP balance (111 g/day) is adequate for 3.67 lb ADG. The ration cost is $1.49/day. The lower portion of the **Feed Amounts and Performance Summary** screen provides a quick and simple evaluation of expected performance once the inputs are entered.

Press ⟨**F11**⟩

NRC MODEL DIET EVALUATION
Execute a Diet Evaluation with NRC Model Level 1

Evaluate: Place cursor on **Evaluate** and press ⟨**Enter**⟩ to start through "pop-up" screens of prioritized evaluations of the results; continue to press ⟨**Enter**⟩ to continue through the evaluation. Pop-up screens are described below.

"LEVEL 1 DIET EVALUATION" SCREEN

Level 1 Diet Evaluation					
Diet			NRC Feedlot1		Evaluate
NE Diet Mcal/d	NE Reqd Mcal/d	Differ Mcal/d	MP Diet g/d	MP Reqd g/d	Differ g/d
Totals			906	794	111
Maint	23.2	9.0	14.2	906	390	516
Preg	14.2	0.0	14.2	516	0	516
Lact	14.2	0.0	14.2	516	0	516
Gain	8.4	8.4	0.0	516	404	111
Reserves	0.0			111		

DMI predicted	23.99 lb/d		DIP required	911
DMI actual	24.30 lb/d		DIP Supplied	953
ME Allowed ADG	3.67 lb/d		DIP Balance	42.2 g/d
			0	

eNDF required	1.94 lb/d		MP from Bacteria	583 g/d
eNDF supplied	2.81 lb/d		MP from UIP	323 g/d
NDF in Ration	25%DM		Diet CP	12.3%DM
Diet TDN	78%DM		DIP	70.2%CP
Diet ME	1.28 Mcal/lb		Total ration NSC	56.3%DM
Diet NEm	0.96 Mcal/lb		Cost/d	$1.49/d
Diet NEg	0.57 Mcal/lb			0
DMI/Maint DMI	2.57		MP allowed ADG	4.68 lb/d
Est. Ruminal pH	5.91			

DMI Predicted and Actual (IC21 and IC22):   Predicted DMI can be used as a guide, particularly to evaluate the effects of different input variables on DMI. If actual DMI are not available, predicted DMI can be used to compute "actual" DMI.

Diet TDN, NE$_m$, and NE$_g$: Using the tabular feed composition values in the feed library, this diet is computed to contain 78% TDN, 1.28 ME Mcal/lb diet DM, 0.96 NE$_m$ Mcal/lb diet DM, and 0.57 NE$_g$ Mcal/lb diet DM, respectively.

NE and MP Available vs Required: Balances are shown after requirements are met for each physiological

function. Energy balances are reflected in ME allowed ADG (cell IC23; 3.59 lb) and MP allowed ADG (cell IG33; 4.76 lb). Energy is first limiting in this example.

Effective Fiber Level (IC26):   Check the assignment of eNDF; it is used in computing rumen pH and passage rate. Rumen pH is predicted from eNDF, which is used to adjust fiber digestion rate and microbial yield. The adjustment is based on a linear decrease in pH, microbial growth, and fiber digestion rate less than pH 6.2 (20% eNDF in the diet DM). This diet contains 11.6% eNDF (cell IC27/cell IC22). The eNDF required (cell IC26) in high-energy diets is 8%, which is considered to be the concentration necessary to keep rumen pH above 5.7, below which cattle have been shown to dramatically reduce DMI. Under these low pH conditions (pH < 6), microbial yield will be reduced at least one-third and very little energy will be derived from the fiber in forages consumed. As much as 25% eNDF may be required to maintain an adequate pH for maximum forage digestion and microbial growth, depending on feeding management. If eNDF is too low, passage rate may be high, reducing predicted NE value. If the effective fiber is too low, it can be increased by coarse chopping or adding sources higher in effective fiber. Appendix Table 9 gives guidelines for adjusting stored values in the feed library and guidelines for estimating eNDF values for forages.

DIP Balance (IG23): This value should be positive to ensure that rumen microbial nitrogen needs are met. If DIP is deficient, add urea or other highly degradable nitrogen sources. If MP supply exceeds requirements, replace UIP with DIP. In this example, the DIP balance is 42 g, which exceeds requirements by 4.6%, which is a reasonable safety factor.

MP from Bacteria and Feed: The MP from bacteria (cell IG26; 583 g) provide all but 211 g of the required MP (cell IF13; 794 g), but the natural feeds supply 323 g (cell IG27), leaving an excess of 111 g (cell IG13). The microbial yield has been adjusted for the effect of pH (IC34; 5.91). Rumen pH is predicted from eNDF. The adjustment is based on a linear decrease in pH and microbial growth below pH 6.2 (20% eNDF in the diet DM). This diet contains 11.6% eNDF (cell IC27/cell IC22).

Diet CP: The total diet CP is 12.3% of the DM (cell IG28), with 70.2% of the protein degradable (IG29). This CP concentration provides approximately the right amount of DIP and provides more UIP than needed.

Execute a Diet Evaluation with NRC Model Level 2

Press ⟨ESC⟩ to go to the **Main Menu**; select **Describe Units and Levels** ⟨Enter⟩; select **Level 2.** ⟨Enter⟩; press ⟨F11⟩

"LEVEL 2 DIET EVALUATION" SCREEN

Level 2 Diet Evaluation						
Diet	NRC Feedlot1				Evaluate	
	NE Diet Mcal/d	NE Reqd Mcal/d	Differ Mcal/d	MP Diet g/d	MP Reqd g/d	Differ g/d
Totals				1060	775	285
Maint	22.5	9.4	13.2	1060	390	670
Preg	13.2	0.0	13.2	670	0	670
Lact	13.2	0.0	13.2	670	0	670
Gain	7.9	7.9	0.0	670	384	285
Reserves	0.0			285		

				% of requirement
DMI predicted	24.39 lb/d	Bact N Bal	−3 g/d	−1.2%
DMI actual	24.30 lb/d	Peptide Bal	8 g/d	6.3%
ME allowed ADG	3.46 lb/d	Urea Cost	0.3 Mcal/d	

eNDF required	1.94 lb/d	MP from Bacteria	819 g/d
eNDF supplied	2.81 lb/d	MP from UIP	241 g/d
NDF in Ration	25% DM	Diet CP	12.3% DM
Diet TDN	76% DM	DIP	76.5% CP
Diet ME	1.25 Mcal/lb	Total NSC in ration	56.3% DM
Diet NEm	0.93 Mcal/lb	Cost/day	$1.49/d
Diet NEg	0.54 Mcal/lb	Total N Balance	46 g/d
DMI/Maint DMI	2.49	MP allowed ADG	6.03 lb/d
Est. Ruminal pH	5.91	EAA Allowed ADG	6.39 lb/d
		Most Limit AA HIS	141.9%

AA	Requirement	Amino acids, G/day Supply	% of Requirement	Input summary Growing/finishing steer
MET	15	24	155	BW = 1060 lb;MW = 1284 lb
LYS	50	75	151	CS = 5.
ARG	26	68	267	
THR	30	53	177	
LEU	52	78	150	
ILE	22	58	265	
VAL	31	62	201	
HIS	19	27	142	
PHE	27	56	205	
TRYP	5	16	353	

To save your results (check to make sure they agree with values presented here), press ⟨**ESC**⟩, select **Save Inputs**, a prompt will appear to [Enter filename (maximum eight characters) to save], type **Feedlot1**, press ⟨**Enter**⟩.

Differences between NRC Model Levels 1 and 2

The differences between Level 2 (the evaluation above) and the Level 1 diet evaluation are described below.

1. *ME allowed ADG* is computed from NE values predicted from tabular values in Level 1. Level 2 predicts energy and protein availability based on simulations of ruminal fermentation and intestinal digestion. The simulations account for the effects of (1) rates of digestion and passage of feed ingredients, (2) effect of rumen pH on fiber digestibility, (3) intestinal digestion of starch and fiber, and (4) the energy cost of excreting excess N (urea cost, IG23 is added to the NE_m requirement). In this example, Level 2 ADG is lower (3.46 vs 3.67) because the cost of excreting excess N is added to the NE_m requirement, and diet NE values are lower because the rumen pH of 5.91

reduced fiber digestion rate. Appendix Table 11 demonstrates the sensitivity of the Level 2 model to these variables and is discussed below in the paragraph headed "Evaluate." In this evaluation, predicted and observed ADG are nearly identical (3.46 vs 3.48).

2. *Microbial protein yield* in Level 1 is fixed at 13% TDN, which is not sensitive to extent of ruminal digestion. MP from bacteria is computed in Level 2 from bacterial growth on fiber and nonfiber carbohydrates, which are sensitive to amounts of dietary fiber and nonfiber carbohydrates and their digestion rates, and rumen pH. MP from feeds are computed from feed protein escaping digestion in the rumen, which is sensitive to feed amounts of protein fractions with medium and slow digestion rates. In this example, the MP balance is higher in Level 2 than in Level 1 because of a higher microbial protein production (819 g vs 583 g) and a lower predicted ADG. A major factor in this diet is the high nonfiber digestion rate in the barley grain resulting in a high extent of ruminal degradation (90% of starch digested in the rumen). As a result, the MP allowable ADG is higher than in Level 1 (6.03 vs 4.68).

3. *Rumen nitrogen balances* are given as total bacterial N balance (IG21) and peptide balance (IG22). The total ruminal N balance is lower than in Level 1 (-3 g N vs 42 g DIP) because of a higher predicted microbial yield. This difference would be greater except recycled N is included in Level 2. In model Level 2, peptides stimulate growth of bacteria that grow on nonfiber carbohydrates. Therefore microbial yield of nonfiber carbohydrate bacteria will be increased when the peptide balance is increased from negative to 0. This is accomplished by adding natural protein sources of protein such as soybean meal that have rapid or medium digestion rates. *Supplementation with peptide sources to get peptides balanced should be considered only when MP or essential amino acids are deficient.*

4. *Essential amino acid balances* (⟨**Page Down**⟩ to lines 38–51) and first limiting amino acid allowable ADG (IG34). The requirements are computed as described in Chapters 3 and 10 and the supplies are computed from MP from bacteria and MP from the essential amino acids in the undegraded feed protein. The balances should be not less than 5% of requirements. The importance of ratios between essential amino acids are discussed in Chapter 3, but no attempt is made in this revision to make specific amino acid ratio recommendations. In this example, energy is first limiting because the ME allowable ADG is 3.46 vs 6.39 for essential amino acid allowable ADG.

Evaluate: Place cursor on **Evaluate** and press ⟨**Enter**⟩ to start through "pop-up" screens of prioritized evaluations

of the results; continue to press ⟨**Enter**⟩ to continue through the evaluation. The guidelines below are provided to assist in interpeting results and making changes for fine-tuning the diet. The following can be used as a diagnostic tool or to make actual and observed performance agree, to ensure that the model is accurately describing the cattle so evaluation of alternatives will be accurate.

Dry Matter Intake: Compare total feed dry matter entered vs model predicted DMI (IC21 vs IC22). If more than 5 to 10% different, check input variables that influence predicted DMI (ration DM and quality control, accuracy of weights, body weight, current temperature, ionophore and implant, diet energy density, feed processing). The actual DMI must be accurately determined, taking into account bunk clean out, moisture content of feeds, and scale accuracy. The accuracy of any model prediction is highly dependent on the DMI used.

Diet Energy and Protein (IC29 to IC32): These values (IC29 to IC32) are computed from feed carbohydrate fractions and their digestion and passage rate adjusted for rumen pH. Appendix Table 11 shows the sensitivity of feed biological values to level of intake and rumen pH, using several common feeds. The efficiency of ME use for NE_g ranges from 27% for the brome hay to 47% for corn. The negative NE_g value for brome hay at a low pH shows the effect of extrapolating equations beyond the range of the data. Shown next are biological values generated by the Level 2 model for 2, 4, 6, and 8% passage rate/hr, the range in passage rates typical for the feeds at $1\times$ to $4\times$ maintenance level of intake. Passage rate would be 2 to 4%/hr at $1\times$ level of intake, which is typical for dry cows, and would be about double that at $3\times$ to $4\times$ level of intake, which can occur with thin, compensating, yearling feedlot cattle. The passage rate is also sensitive to feed eNDF value.

Within each of these categories, feed TDN, NE_g, and MP from microbial true protein (MTP) are predicted, and at 8%/hr are predicted for both the high (6.5) and low (5.7) ruminal pH that can occur. The percent of protein escaping ruminal fermentation varies considerably depending on passage rate. This is especially true in feeds high in B2 protein, such as soybean meal. Passage rate has little effect on escape protein in feeds (such as corn silage) with a high proportion of B1 and B3 protein. The adequacy of tabular values for DIP and UIP depends on the level of intake. Passage rate had the greatest effect on feed energy values for forages because of their lower intestinal digestibility. Rumen pH has a dramatic effect on both forage energy value and MTP. These values reflect a 0% digestion rate for the available NDF at the low pH and approximately 40% less MTP yield from A and B1 carbohydrates.

ME Allowed ADG: If predicted ADG (IC23) is not as expected for the conditions described (cattle type, diet type, environment, and management conditions), first carefully check all inputs. ***Input errors are the greatest source of prediction errors.*** Mistakes or incorrect judgements about inputs such as body size, milk production and its composition, environmental conditions, or feed additives are often made.

Adjust feed carbohydrate fractions and their digestion rates as necessary. If inputs are correct and performance is still not as expected, predicted diet energy values are likely the cause. First, see if predicted total diet net energy values (IC31 and IC32) and for each feed are near those expected. *Predicted energy values for individual feeds can be accessed by pressing* ⟨**F7**⟩; *use the tab key to find the NE and DIP values in metric or English units.* Feed factors may be influencing energy derived from the diet as the result of feed compositional changes and possible effects on digestion and passage rates. The NE derived from forages are most sensitive to NDF amount and percent of the NDF that is lignin, available NDF digestion rate (CHO B2), and eNDF value. For example, if the NDF% of a feed is increased, the starch and sugar fractions in the feed will be decreased automatically by the model because more feed dry matter will escape digestion and the feed will have a lower net energy value. Dry matter digestibility can be further decreased by lowering the NDF digestion rate; after making sure the feed composition values are appropriate, the digestion rate is considered. Adjustments are made using the ranges and descriptions in Appendix Tables 6 through 8.

The major factors influencing energy derived from feeds high in nonfiber carbohydrates are ruminal and intestinal starch digestion rate (CHO B1). This is mainly a concern when feeding corn grain, corn silage, sorghum grain, or sorghum silage.

Check postruminal starch digestibility to make sure that it is appropriate for the starch source being fed. Intestinal digestibilities can be modified by choosing feed digestion from the main menu and selecting Intestinal Digestibilities. The model assumes an average starch digestibility (CHO B1) of 75%, however, this may not be appropriate for all starch sources. Appendix Table 10 can be used to adjust the starch digestibility for effects of processing.

Effective Fiber Level (IC26):   Check the assignment of eNDF; it is used in computing rumen pH and passage rate. Rumen pH is predicted from eNDF, which is used to adjust fiber digestion rate and microbial yield. The adjustment is based on a linear decrease in pH, microbial growth, and fiber digestion rate less than pH 6.2 (20% eNDF in the diet DM). This diet contains 11.6% eNDF (cell IC27/cell IC22). The eNDF required (cell IC26) in

high-energy diets is 8%, which is considered to be the concentration necessary to keep rumen pH > 5.7. Below this level, cattle may dramatically reduce DMI. Under these low pH conditions (pH < 6), microbial yield will be reduced at least one-third and very little energy will be derived from the fiber in forages consumed. As much as 25% eNDF may be required to maintain an adequate pH for maximum forage digestion and microbial growth, depending on feeding management. If eNDF is too low, passage rate will be high, reducing predicted NE value. If the effective fiber is too low, it can be increased by coarse chopping or adding sources higher in effective fiber. Appendix Table 9 gives guidelines for adjusting stored values in the feed library and guidelines for estimating eNDF values for feeds.

Rumen Nitrogen (N) Balance (IG21 and IG22):   If the peptide balance is negative and MP is deficient, add feeds such as soybean meal that are high in degradable true protein until ruminal peptide balance is ≥0 g to increase microbial yield from NFC. If MP is adequate, it is not necessary to balance ruminal peptides. Adjust remaining ruminal N requirements with feeds high in NPN or soluble protein until total ruminal N is balanced. Because of the number of assumptions required to adequately predict total N balance, it may be desirable under some conditions to have supply exceed requirements by about 5% (105% of requirement) to allow for prediction errors.

Metabolizable Protein (IG13):   This component represents an aggregate of nonessential and essential amino acids. The MP requirement is determined by the animal type and the energy allowable ADG. The adequacy of the diet to meet these requirements depends on microbial protein produced from fiber and nonfiber carbohydrate fermentation and feed protein escaping fermentation. If MP balance appears to be unreasonable, check first the starch (B1 carbohydrate) digestion rates, using the ranges and descriptions in Appendix Tables 6 through 8 for carbohydrate B1. Altering the amount of degradable starch will also alter the peptide and total ruminal N balance because of altered microbial growth. Often the most economical way to increase MP supply is to increase microbial protein production by adding highly degradable sources of starch, such as processed grains. Further adjustments are made with feeds high in slowly degraded or rumen escape (bypass) protein (low B2 protein digestion rates; see Appendix Tables 6 through 8).

Check total ration protein degradability (IG29) and individual values for the feeds (press ⟨**F7**⟩ *to obtain the predicted Biological Values to compare with the tabular values.* If considerably different, you may have an entry error or need to adjust the protein fraction digestion rates. First, check protein fractions entered. Next, check their digestion rates. In altering degradability, the most sensitive fraction is the medium or B2 fraction since most of the fast or B1 fraction will be degraded in the rumen and most of the slow or B3 fraction will escape. Thus the easiest way to alter amounts—degraded vs escaping—is to change the amount of soluble protein and/or the digestion rate of the B2 fraction.

Essential Amino Acids (IA39 to ID51): The amino acid with the lowest supply as a percent of requirement is assumed to be the most limiting for the specified performance. Because of the number of assumptions required to adequately predict amino acid adequacy, it may be desirable under some conditions to have first limiting amino acids exceed requirements by about 5% (105% of requirement) to allow for prediction errors.

The adjustment for amino acids is done last because the amino acid balance is affected by the preceding steps. Essential amino acid balances can be estimated with Level 2 because the effects of the interactions of intake, digestion, and passage rates on microbial yield, available undegraded feed protein, and estimates of their amino acid composition can be predicted along with microbial, body tissue, and milk amino acid composition. However, the development of more accurate feed composition and digestion rates, and more mechanistic approaches to predict utilization of absorbed amino acids, will result in improved predictability of diet amino acid adequacy for cattle. To improve the amino acid profile of a ration, use feeds high in the first limiting amino acids.

Diet CP (IG28): After all of the above factors are correctly evaluated, the diet CP content will be the CP requirement. The CP requirement represents the amount of DIP needed in the rumen and the amount of UIP needed to supplement the microbial protein to meet the MP requirement.

PREDICTING RESPONSES TO ALTERNATIVE FEEDLOT CONDITIONS

After adjustments are made in the NRC model Level 2 to account for the factors influencing ADG, so that predicted and observed values agree, the effect of other conditions can be predicted. Twenty evaluations were made to predict responses to other variables of interest. All evaluations began with the inputs described previously, then one variable was changed at a time to evaluate its effect. Each variable was changed back to the original value before changing the next variable to be considered, unless indicated otherwise. A summary of evaluations made to predict these effects is shown in Table 2a (animal and environmen-

tal factors), 2b (effective fiber and rumen pH), and 2c (body weight and protein requirements).

Sensitivity to Animal and Environmental Factors: The results of these evaluations are shown in Table 2a. The first line shows the actual performance, and the second line shows that Level 2 predicted the actual performance. The third line shows that a 10% decrease in DMI below actual will reduce ADG 14% and feed efficiency 5.1%. The fourth line shows that cattle with a finished weight of 1500 lb would be predicted to gain 11% faster at the same case study mean weight (1060 lb). However, since they must be fed to a heavier weight to be finished, their overall feed efficiency would be similar (data not shown). The fifth, sixth, and seventh data lines show that CS 1 cattle would be expected to make compensatory growth, while CS 9 cattle would be expected to gain more slowly. The next section shows the effects of winter feeding conditions on performance at an average winter temperature of 10° F (previous temperature, 10° F; hair depth, 0.5 inches). The eighth data line shows that ADG decreases in the winter at the same DMI as a result of an increase in NE_m requirement. Typically DMI does not increase in the winter in commercial feedlots in the Plains states, so DMI was not changed for these evaluations. The ninth line shows that if the cattle were exposed to wind of 15 mph instead of the current 5 mph, ADG would be reduced dramatically. The next four lines show that if the insulation is reduced by matted hair, thin hide, or short hair, performance can be reduced. The last two lines show that the potential for compensatory growth with CS 1 or depression in performance with CS 9 depends on cold stress. In this case, the CS 9 steer would outperform the CS 1 steer because of

TABLE 2a Effect of Animal and Environmental Factors on Performance

Factor	Daily Gain, lb	DMI/ADG
Actual performance	3.48	6.98
Model Level 2 predicted performance	3.46	7.02
Predicted effect of 10% decreased DMI	2.97	7.36
Predicted effect of larger mature size: 1,500 lb at Canadian AA	3.85	6.31
Predicted effect of body condition		
CS 1 @ same DMI	3.54	6.86
CS 1 @ 10% increased DMI	4.22	6.33
CS 9	3.09	7.86
Predicted effect of cold stress		
Winter, same DMI	3.14	7.74
Winter, wind at 15 mph	1.59	15.3
Winter, with matted hair	2.41	10.1
Winter, with thin hide	2.89	8.41
Winter, with short hair	3.07	7.92
Winter, with CS 1	2.58	9.42
Winter, with CS 9	2.81	8.65

the insulation benefits of body fat when the effective environmental temperature is below the animal's lower critical temperature.

Sensitivity to Feed Effective NDF and Rumen pH:
Table 2b shows the effect of fine processing the silage or grain, or both. The effect of fine chopping the barley silage was simulated by reducing the barley silage eNDF to 30%. The effect of fine rolling the barley grain was simulated by reducing the barley grain eNDF to 17%. The ruminal pH is predicted to drop, reducing cell wall digestion and therefore net energy derived from the fiber. Also microbial protein production (MCP) declined at a lower pH.

TABLE 2b   Influence of Effectiveness of Feed Fiber in Controlling Rumen pH

	ADG, lb/day	Ruminal pH	Silage NE_g	Grain NE_g	MCP[a] g/day
Current predicted	3.46	5.9	0.47	0.58	819
Silage processed fine	3.24	5.8	0.37	0.57	698
Silage and grain both processed fine	2.46	5.7	0.12	0.51	576

[a]MCP is microbial crude protein produced, g/day.

Sensitivity of Protein Requirements to Stage of Growth: Table 2c shows the weight at which the dietary undegraded and microbial protein will not meet the energy allowable ADG requirement for protein or essential amino acids. The ruminal requirement for degradable protein is essentially met (-1.2%) and does not change with cattle weight because the requirement for degradable protein is proportional to the fermentable carbohydrates in the diet. In this study, the breakpoint for cattle size is 600 lb. Below this weight, supplemental UIP to provide amino acids will be needed. The last line shows that the cattle being evaluated would have a first limiting amino acid allowable ADG 1.17 lb/day below the energy allowable ADG.

TABLE 2c   Effect of Body Weight on Protein Requirements

Average SBW	ME allowed ADG	Rumen balance	Protein allowed ADG MP	Amino acids
at 1,060 lb	3.46	-1.2%	6.03	5.98
at 800 lb	3.64	-1.2%	4.34	4.17
at 600 lb	3.64	-1.2%	3.50	3.37
at 400 lb	3.63	-1.2%	2.56	2.46

EVALUATION OF THE FEEDLOT CASE STUDY USING THE TABLE GENERATOR

If in the NRC model, choose ⟨**Quit**⟩.
At the **C:/NRC** prompt, *type* **TABLES**.

The same case study data used in Chapter 3 of the report is used here to demonstrate how to use the table generators. *(See "Introduction" for a comparison of the table generator with the NRC model.)*   Initial data were entered in the table generator as shown below.

OPENING SCREEN OF THE TABLE GENERATOR

1996 NRC Beef Cattle Requirements
Table Generator Menu
Growing & Finishing Requirements
Growing & Finishing Diet Evaluations
Replacement Heifer Requirements Diet Evaluations
Beef Cow Requirements & Diet Evaluations
Breeding Bull Requirements
Breeding Bull Diet Evaluations
Exit Program
Units   1   0 = Metric; 1 = English
For Backgrounding, Stocker, & Feedlot Systems:
Grading System   2   1 = Trace; 2 = Slight; 3 = Small
Note: User entry cells are highlighted
Press ⟨ESC⟩ at any time to escape to this screen

Follow the following steps, in order, to begin.

Units: Select the system of units of measure—metric (0) or English (1).
The value for this example is 1. ⟨**Enter**⟩

Grading System: Enter the code for the grading system—trace (1), slight (2), or small (3).
The value for this example is 2 (slight). ⟨**Enter**⟩
Press ⟨**ESC**⟩ to select the class of cattle.

Type of Evaluation: Move the cursor over the first word in the line that describes the class of cattle and type of evaluation (requirements or diet evaluation) you want, and press enter.
The choice for this evaluation is Growing and Finishing Requirements. ⟨**Enter**⟩

Input the following information as it appears in the table below:

Nutrient Requirements for Growing and Finishing Cattle		
Wt @ slight Marbling	1284	Lbs
Weight range	600 1100	Lbs
ADG range	1 4	Lbs
Breed code	6	Charolais

A table containing daily requirements for net energy, MP, Ca, and P for cattle of this body size over the range specified will be calculated. (See Chapter 9 for discussion.)

Press ⟨**ESC**⟩ to return to the **Main Menu**
Select **Growing and Finishing Diet Evaluations**

The following information from the Level 1 evaluation was entered for diet D in the screen that appears. Input the following: 79% TDN, 12.3% CP, and 70.3% DIP. A table containing predicted DMI, ADG, DIP balance (g/day), UIP balance (g/day), MP balance (g/day), and Ca and P requirements (% in the DM) will be computed for six weights. The minimum weight is 55% of the finished weight entered (rounded to the nearest 25 kg) and the heaviest weight is 80% of the weight entered, with four equal increments in between. See Chapter 9 of the report for a discussion of this approach. This table functions independently from the requirements table because it computes daily requirements for a specified weight range and ADG.

ADJUSTERS

The DMI adjuster for diet D nearest the mean weight during the feeding period (1,025 lb; actual mean was 1,060 lb) was changed (percents entered as decimals) to make predicted and observed DMI agree (1.02 entered resulted in 24.3).

The predicted ADG was 4.10 lb/day compared to the actual of 3.48 lb/day. This ADG prediction included no adjustments for environmental conditions and the Level 1 tabular TDN value, which is not sensitive to the actual composition of the feed or pH conditions. According to Appendix Table 14, NE_m is 19% above thermoneutral conditions (maintenance multiplier is 1.19 for clean and dry @ 30° F @ 5 mph). This same effect can be accounted for by reducing both NE_m and NE_g available by 10% (entered as 0.9) in the NE adjusters. *Note: This is adjusted by body weight category.* The predicted ADG for diet D at 1,025 lb will now be 3.87 lb/day.

The diet TDN predicted by Level 2, which is adjusted for actual feed composition and effects of ruminal pH on fiber digestion, was entered for diet D in place of the tabular TDN (76 vs 79% TDN). After incorporating this change, the DMI adjuster must be changed to 100% to obtain actual intake because predicted DMI increases. The predicted ADG will now be 3.46 lb/day, compared to the actual of 3.48 lb/day.

The diet can now be evaluated across the expected mean weights for this group of cattle during the feeding period. The DIP balance is adequate at all weights, and the UIP is adequate above 900 lb.

3

Cow-Calf Ranch Case Study

TUTORIAL LESSON 2: COW-CALF RANCH CASE STUDY

Begin the tutorial by opening the NRC model program (at the NRC directory prompt, *type* **NRC**) and select the **Describe Units and Levels** option on the main screen. Press ⟨**Enter**⟩

The ranch used in this case study is in the northern plains and carries approximately 600 beef cows and 100 replacement heifers. Cows are predominantly Simmental sired females from Angus × Hereford cows with a mature size of approximately 1,300 lb at condition score 5. Calving season for mature cows is March and April, and calf birth weight averages 80 lb. Calves are weaned approximately October 15. Average steer calf weaning weight at 200 days is 575 lb, and average heifer calf weaning weight at 200 days is 525 lb. Replacement heifers wean at 45% of mature weight the middle of October, conceive at 60% of mature weight during the first week of May, and are 85% of mature weight at calving.

Body condition scores average 3 to 4 at weaning. The goal is to have them back to CS 5 by December 1 to provide insulation for winter, and maintain them at CS 5 until calving. They will lose a score by pasture turnout (approximately May 1); but the goal is to have them gain one score by start of breeding May 15. Over the 12-month reproductive cycle the energy balance should average near 0.

The winter feed resources available include two qualities of hay. Corn and range cake are fed as needed to supplement the hay.

This information will be used to demonstrate how to use the model Levels 1 and 2 to evaluate the feeding program for this herd, beginning with an evaluation of the winter feeding program.

Describe Units and Levels

"DESCRIBE UNITS AND LEVELS" SCREEN

1996 NRC Nutrient Requirements of Beef Cattle Describe Units and Levels Screen		
Diet	NRC Dry Cow 1	Grading System 3
Level	1 Tabular System	
Units	1 English	
Feed H₂0	0 Dry Matter	
Press ⟨F1⟩ at any time for context sensitive help Press ⟨ESC⟩ at any time to escape to the main menu		

Diet: Enter an identifying name for the particular diet being evaluated in cell C1024.
Entry for the example is NRC Dry Cow 1. ⟨**Enter**⟩

Grading System: This section is for growing-finishing cattle. In cell H1024 enter the grading system. Choices are 1 (USDA Standard or Canadian A, which are related to 25.2% body fat); 2 (USDA Select or Canadian AA, which are related to 26.8% body fat); and 3 (USDA Choice or Canadian AAA, which are related to 27.8% body fat). The program uses this to identify the standard reference weight that is divided by the finished weight, with the result multiplied times the actual weight to get the weight to use in the equation that computes net energy and protein in the gain. (See report Chapter 3 for the biological basis and validation of this method.)
Entry for the example is 3. ⟨**Enter**⟩

Level: In cell C1026 enter either a 1 (uses tabular feed net energy and protein degradability values) or 2 (feed energy and absorbed protein values based on feed carbohydrate and protein fractions and their digestion rates). It is

172

often practical to adjust the diet until balanced with Level 1, then evaluate it with Level 2 to get predicted feed net energy values and amino acid balances, based on actual feed analysis for carbohydrate and protein fractions. *Entry for the example is 1, then will be changed later to 2 for further evaluation.* ⟨Enter⟩

Units: In cell C1028 enter either a 0 for metric or 1 for English. *Be sure all data is entered in the same units entered here.*
Value for the example is 1 (English). ⟨Enter⟩

Feed H₂0: In cell C1030 enter 0 (dry matter) or 1 (as fed). This is used to determine DMI from the feed amounts fed that is entered later.
Value for the example is 0 (Dry Matter). ⟨Enter⟩

Context sensitive help (⟨F1⟩) is available to guide the user in selecting appropriate values to enter in these cells. After you are satisfied with the inputs for this section, press ⟨ESC⟩ to return to the **Main Menu**.

Describe Animal

"DESCRIBE ANIMAL" SCREEN 1

Describe Animal		
Animal Type	3	Dry Cow
Age	60	Months
Sex	4	Cow
Body Weight	1300	lb
Condition Score	5	1 = v.thin—9 = v.fleshy
Mature Weight	1300	lb @ maturity
Breeding System	3	3-way cross
	1	
	1	
Sire's Breed	26	Simmental
Maternal Grandsire	1	Angus
Maternal Granddam	11	Hereford
Next		
Press ⟨F1⟩ at any time for context sensitive help		
Press ⟨ESC⟩ at any time to escape to the main menu		

When entering values, press ⟨**Enter**⟩ twice to move to the next input cell and to cause chosen category to be displayed.

Animal Type: In cell D1043 enter the correct code for the class of cattle. Choices are 1 (growing and finishing), 2 (lactating cow), 3 (dry cow), 4 (herd replacement heifer), 5 (breeding bulls). This invokes the inputs and equations needed to compute requirements, predict DMI, and evaluate the diet for that class.
The entry for this example is 3 (dry cow). ⟨Enter⟩

Age: In cell D1044 enter the average age in months. This value influences expected DMI, growth requirements, and tissue insulation.
The entry for this example is 60. ⟨Enter⟩

Sex: In cell D1045 enter the code for the sex of the animal. Choices are 1 for a bull, 2 for a steer, 3 for a heifer, and 4 for a cow. A heifer is entered as a cow after calving the first time.
The entry for this example is 4. ⟨Enter⟩

Body Weight: In cell D1046 enter the shrunk body weight that best represents the group being fed together. Body weight is a major determinant of DMI, maintenance, and growth requirements.
The entry for this example is 1300. ⟨Enter⟩

Condition Score: In cell D1047 enter the average condition score of the cattle in the group (Appendix Table 2). (See report Chapter 3 for a detailed discussion of the 1 to 9 condition scoring system used and its biological basis.) The choices are 1 through 9, with 1 indicating very thin, 5 indicating average, and 9 indicating very fat. Condition is used to describe tissue insulation, the potential for compensatory growth in growing cattle, and energy reserves in cows.
Entry for the example is 5. ⟨Enter⟩

Mature Weight: In cell D1048 enter the expected average weight at the grade chosen in the Units and Levels screen. If cows, replacement heifers, or breeding bulls, enter the expected mature weight at CS 5.
Entry for the example is 1300. ⟨Enter⟩

Breeding System: In cell E1050 enter the code for the breeding system. Choices are 1 (straightbred), 2 (2-way crossbred), and 3 (3-way crossbred). Breeding system for growing cattle influences maintenance energy requirement and predicted DMI. No adjustments are made for special breed effects other than dairy or *Bos indicus* types, as the data to date indicate most identifiable breed effects are the result of differences in mature size, fat distribution, and hair and hide factors, which are considered independently. (The equations and biological basis for these effects are discussed in report Chapters 1, 3, and 4.)
Entry for this example is 3. ⟨Enter⟩

Breed Codes: In cells E1051 to E1055 enter breed codes for the parent breeds in the breeding system specified. Valid breed codes are shown in the help system and in Appendix Table 4, along with stored breed values used to determine maintenance energy requirements, and defaults for calf birth weight and peak milk production. (See report

Chapters 1 and 4 for the biological basis for these breed adjustments.)

Entries for this example are 26 for sire's breed (Simmental), 1 for maternal dam's breed (Angus), and 11 for maternal sire's breed (Hereford). ⟨Enter⟩

Press ⟨F9⟩ to display chosen breeds.

Place cursor on **NEXT**

Press ⟨**Enter**⟩

The second screen appears for describing reproductive cycle parameters.

"DESCRIBE ANIMAL" SCREEN 2

Days Pregnant	190	Days	
Days in Milk	0	Days	
Lactation Number	0	0 = dry or heifer	
Peak milk production	0	lb	21.5
Time of Lactation Peak	0	Weeks	8.5
Duration of Lactation	0	Weeks	30
Milk Fat	0	%	4
Milk Protein	0	%	3.4
Milk SNF	0	%	8.3
Age @ 1st Conception	15	Months	
Calving Interval	12	Months	
Expected Calf Birth Weight	80	lb	79.9

Main Menu

Press ⟨F1⟩ at any time for context sensitive help
Press ⟨ESC⟩ at any time to escape to the main menu

Note: Stored default values appear in column H. These will be used when the value in column F is 0.

Days Pregnant: In cell F1063 enter the days the cow is pregnant. This is used along with expected birth weight to compute pregnancy requirements, conceptus weight, and ADG as described in report Chapter 4.
Entry for this example is 190. ⟨Enter⟩

Days in Milk: In cell F1064 enter the number of days since calving. This is used along with peak milk and lactation number to predict milk production for the day entered.
Entry for this example is 0 (dry). ⟨Enter⟩

Lactation Number: In cell F1065 enter the lactation number.
If evaluating the lactating cows, the value of 3 would be entered. ⟨Enter⟩

Peak Milk Production: In cell F1066 enter either the default value or a value estimated from Appendix Table

12 (predicted weaning weights for different mature sizes and milk production levels). In this example, the default value is 21.5. Appendix Table 12 indicates a male calf weaning weight of approximately 587 lb at 7 months for a 1,300 lb cow at 21.5 lb peak milk compared to the actual steer 200 day weaning weight of 575 lb. Thus the default milk production is acceptable. The peak milk along with time of peak and duration of lactation is used to develop a lactation curve for predicting milk production for the day entered, as described in Chapter 4.
If evaluating the lactating cows, the default value of 21.5 is used for this case study. ⟨Enter⟩

Time of Peak: Enter the default value to the right unless other information is available. This is used in computing a lactation curve.
If evaluating the lactating cows, the value of 8.5 is used for this case study. ⟨Enter⟩

Duration of Lactation: In cell F1068 enter the length of lactation. This is used in computing a lactation curve.
If evaluating the lactating cows, the value used for this case study is 30 weeks. ⟨Enter⟩

Milk Fat, Protein, and SNF (solids not fat):   In cells F1069, F1070, and F1071 enter the default values displayed to the right, unless values are available. Both quantity and composition are used to predict lactation requirements.
If evaluating the lactating cows, the default values of 4, 3.4, and 8.3, respectively, are used for this case study. ⟨Enter⟩

Age at First Conception and Calving Interval: In cells F1072 and F1073 enter these values. They are used to predict growth requirements as described in report Chapter 3.
The values for this example are 15 months and 12 months, respectively. ⟨Enter⟩

Expected Calf Birth Weight: In cell F1074 enter the default value to the right or enter your own. This is used along with days pregnant to compute pregnancy requirements, conceptus weight, and ADG, as described in report Chapter 4.
The value for this example is 80 lb. ⟨Enter⟩

Place cursor on **Main Menu** ⟨Enter⟩
Select **Describe Management** ⟨Enter⟩

Describe Management

"DESCRIBE MANAGEMENT" SCREEN

Describe Management		
Additive	1	none
On Pasture?	0	no
	30	
	1500	
	3	
	45	
	1	
Diet NEm Adjuster	100%	(Level 1 only)
Diet NEg Adjuster	100%	(Level 1 only)
Diet Microbial Yield 13.0%	TDN	(Level 1 only)

Main Menu

Press ⟨F1⟩ at any time for context sensitive help
Press ⟨ESC⟩ at any time to escape to the main menu

Additive: In cell E1084 enter the code that describes additives used. The choices and their effects are shown in Appendix Table 5. (The biological basis for these adjustments are discussed in report Chapters 3 and 5.)
Entry for this example is 1 (no implant). ⟨Enter⟩

On Pasture: In cell E1085 enter 0 if the animals are not grazing and 1 if they are. If 1 is chosen, other inputs must be chosen to compute maintenance requirements and to predict DMI.
Entry for this example is 0. ⟨Enter⟩ However, the other inputs needed for grazing will be discussed when 1 is chosen.

Grazing Unit Size: In cell E1086 enter the number of hectares (metric) or acres (English) per head grazed in the pasture. If the distance traveled is minimal, enter 0. This input is used to adjust energy maintenance requirements for forage availability. ⟨Enter⟩

Initial Pasture Mass: In cell E1087 enter the kg DM/ hectare (metric) or lb DM/acre (English) when the cattle are turned into the pasture. This can be estimated from hay harvesting experience, clippings, or calibrated measuring devices such as height and/or density estimates, Plexiglas weight plates, or electronic pasture probes. ⟨Enter⟩

Days on Pasture and Number of Animals: In cell E1088 enter the days on the pasture and in E1089 the number of animals. Initial pasture mass, days on pasture, and number of animals are used to predict pasture DMI. ⟨Enter⟩

Terrain: In cell E1090 enter a 1 (relatively level) or 2 (rolling) in units of 0.1. This value is used to adjust maintenance requirement. ⟨Enter⟩

Diet NE$_m$ and NE$_g$ Adjusters (Level 1 only): *Leave these at 1 (100%) unless you are certain you want to adjust the diet NE values.* In cells E1092 or E1093 enter a value between 0.8 and 1.2 if you wish to change the diet NE$_{ma}$ or NE$_{ga}$. The appropriate way to use this is to move it up or down until predicted and actual ADG agree after all other inputs are carefully checked. Unrealistically high ADG and feed efficiency may be predicted for calves consuming high-energy rations; unrealistically low ADG and feed efficiency may be predicted for these same calves when approaching the fatness of choice grade.
These entries are left at 1 (100%) for this example. ⟨Enter⟩

Microbial Yield (Level 1 only): *Leave the entry in cell E1094 at 13% unless you have information that indicates you should lower microbial yield in cattle fed low-quality forage diets.* In Level 1, microbial yield is a constant 13% of TDN as discussed in Chapter 2, except it is reduced on high-concentrate rations based on the eNDF level. However, there is no adjustment in the model for diets with low energy contents or low intakes. In either case, if rate of passage is reduced, then microbial turnover is increased and efficiency of microbial protein synthesis is reduced. Literature values for microbial yield for cattle fed low-quality forages average 7.8% of TDN; the DIP requirement was determined to be 7.1% of DM for cows grazing dormant forage. Therefore, it is recommended that microbial yield be reduced to 7.5–10% of TDN for cows or calves consuming low-quality diets.
The entry is left at 13 (13%) for this example. ⟨Enter⟩

Place cursor on **Main Menu** ⟨Enter⟩
Select **Describe Environment** ⟨Enter⟩

Describe Environment

"DESCRIBE ENVIRONMENT" SCREEN

Describe Environment		
Wind Speed	5	mph
Previous Temp.	40	Degrees F
Current Temp.	30	Degrees F
Night Cooling	2	yes
Hair Depth	0.5	in
Hide	2	average
Hair Coat	1	clean & dry
Heat Stress	1	none
Main Menu		

Press ⟨F1⟩ at any time for context sensitive help
Press ⟨ESC⟩ at any time to escape to the main menu

The equations driven by the inputs in the environmental description section are used to compute lower critical temperature of the animal and to adjust predicted DMI for the effects of environment. Cattle usually compensate for short-term environmental effects, so the inputs chosen should generally reflect average environmental conditions for at least 2 weeks. Predicted maintenance requirements are very sensitive to these effects after the animal reaches its lower critical temperature, so these inputs should be chosen carefully.

Wind Speed: In cell D1104 enter the average wind speed the cattle are exposed to. Wind speed influences maintenance requirements. Increasing wind speed decreases the external insulation value of the animal and thus results in increased energy maintenance requirements. The model is very sensitive to this input after the lower critical temperature is reached.
Entry for this example is 5 mph. ⟨Enter⟩

Previous Temperature: In cell D1105 enter the average temperature for the previous month. This value is used to increase NE_m requirement as it gets colder or reduces it as it gets warmer.
Entry for this example is 40° F. ⟨Enter⟩

Current Temperature: In cell D1106 enter the average current temperature the cattle are exposed to. In most situations, the average daily temperature is the most practical to use. This value is used to adjust predicted DMI for temperature effects and is also used in the calculations for the effects of cold stress on energy maintenance requirements. The model is very sensitive to this input after the lower critical temperature is reached.
Entry for this example is 30° F. ⟨Enter⟩

Night Cooling: In cell D1108 enter either 1 (no night cooling) or 2 (cools off at night). If 1 is chosen, predicted DMI is reduced as described in report Chapter 7 with hot daytime temperatures. If 2 is chosen, it is assumed that cattle can dissipate heat at night and DMI is not affected.
Entry for this example is 2 (nights cool off). ⟨Enter⟩

Hair Depth: In cell D1109 enter the average hair depth. This value is used to increase the external insulation of the animal. Enter the effective hair coat depth of the animal, in increments of 0.1. As hair length increases, so does the external insulation value provided by the animal. A general guide to use is an effective coat depth of 0.25 inches (0.6 cm) during the summer and 0.5 inches (1.3 cm) during the winter. The model is very sensitive to this value below the animal's lower critical temperature.
Entry for this example is 0.5 inches. ⟨Enter⟩

Hide: In cell D1110 enter either 1 (thin hide; dairy or *Bos indicus* types); 2 (average; most European breeds); or 3 (Hereford or similar breeds with thick hides). This value influences the external insulation value of the animal. Increased hide thickness implies increased external insulation. *The model is very sensitive to this value below the animal's lower critical temperature.*
Entry for this example is 2 (average). ⟨Enter⟩

Hair Coat: In cell D1111 enter either 1 (clean and dry), 2 (some mud on lower body), 3 (some mud on lower body and sides), or 4 (heavily covered with mud). This value is used to adjust external insulation. *The model is very sensitive to this value below the animal's lower critical temperature; this entry should be chosen carefully.*
Entry for this example is 1 (clean and dry). ⟨Enter⟩

Heat Stress: In cell D1112 enter either 1 (no panting; not heat stressed), 2 (rapid shallow panting), or 3 (open mouth panting). This value is used to adjust maintenance energy requirements for the energy cost of dissipating heat.
Entry for this example is 1 (no heat stress). ⟨Enter⟩

Place cursor on **Main Menu** ⟨**Enter**⟩.
Select **Describe Feed** ⟨**Enter**⟩.

Describe Feed

"DESCRIBE FEED" SCREEN

Describe Feed
Feed Composition
Feed Amounts
New Feeds
Main Menu

Press ⟨F1⟩ at any time for context sensitive help
Press ⟨ESC⟩ at any time to escape to the main menu

These choices are used to change the composition and of feeds in the current ration and to add new feeds to the existing ration or develop a new ration. For this example, we will start with developing a new ration.

Select **New Feeds** (cell B1127) and press ⟨**Enter**⟩.

Note: This option can be accessed from any point in the program by pressing F6.

The screen will look as below after the feeds for this ration are retrieved from the feed library. The feed library contains average compositional values for net energy, protein, carbohydrate, and protein fractions and their digestion rates. It is critical to choose feeds that most accurately describe the actual feeds in the ration. To aid in making these choices, the default feed library is printed in its entirety in Appendix Table 1. It can be accessed by typing **FEEDS** at the **C:\NRC** prompt. Feed composition values can be modified and new feeds added.

"NEW FEEDS" SCREEN

New Feeds	
Code # for feed to be imported:	Look up feed codes
	999 Minerals
	Main Menu
	Import
Current Feeds:	
Brome hay mid bloom	blank
Brome hay mature	blank
minerals	blank
blank	blank
blank	blank
Press ⟨F1⟩ at any time for context sensitive help	
Press ⟨ESC⟩ at any time to escape to the main menu	

Look Up Feed Codes: Takes you to a listing of all available feeds in the main feed library. The listing is organized in alphabetical order by grass forages, legume forages, grain crop forages, energy concentrates, plant protein concentrates, food processing byproducts, and animal processing byproducts; blanks follow each category to allow the users to add their own feeds. Feed numbers 101–129 are grass forages, 130–134 are blank, 135–139 are grass pastures, 140–148 are range forages, 201–223 are legume forages, 224–229 are blank, 230–231 are legume pastures, 232–250 are blank, 301–323 are grain crop forages, 324–350 are blank, 401–435 are energy concentrates (note: all cotton products including whole cotton and cottonseed meal are in this category), 436–450 are blank, 501–522 are protein concentrates, 523–550 are blank, 601–607 are food processing byproducts, 608–620 are blank, 701–707

are animal byproducts, 801–834 are mineral feeds, 900–910 are blank.

Write down the code numbers of the feeds you want to import and then press ⟨F6⟩ to return to the New Feeds screen.

Code for Feed to Be Imported: Position the cursor on cell F1223 and enter the code number corresponding to the feed you want to import. Press ⟨Enter⟩ until the cursor moves down one cell. If the code number is entered correctly, the name of the feed should appear to the right of the code number. If this name is correct, position the cursor on **Import** and press ⟨Enter⟩.

A new screen will appear. Place the cursor on the row where you want the new feed. When the cursor is in the right place, press ⟨Enter⟩ and the new feed will be brought in from the feed library. Repeat this process until all feeds desired are obtained.

For the example, bring in feed #s
 105 (brome hay mid bloom)
 107 (brome hay mature)
 999 (Minerals)

Up to 14 feeds can be imported. Blank code 130 can be imported into the remaining 9 lines so that the only feeds showing are those in this ration. When all feeds are entered, return to the **Main Menu** by pressing ⟨ESC⟩.

Select **Describe Feed** ⟨**Enter**⟩
Select **Feed Composition**

Feed Composition: Press the right arrow or tab keys to scroll across table values to be modified. **Enter the desired value. ⟨Enter⟩**

For the example, modify feed analytical values as follows:

Feed	Cost $/Ton	NDF % DM	CP% DM	DIP	TDN
105	70	56	12	77	57
107	50	65	7	75	50
999	200				

After desired feed composition values are entered, press ⟨F10⟩ to get **Feed Amounts and Performance Summary** screen.

"FEED AMOUNTS AND PERFORMANCE SUMMARY" SCREEN

Feed Amounts and Performance Summary		
0.00 Brome hay mid bloom		0.00 Blank
26.4 Brome hay mature		0.00 Blank
0.00 grass pasture spring		0.00 Blank
0.00 grass pasture summer		0.00 Blank
0.30 minerals		0.00 Blank
0.00 Blank		0.00 Blank
0.00 Blank		0.00 Blank

Pred DMI	26.6 lbs	
Act. DMI	26.7 lbs	Cost $0.75 per day
NEm Balance	−0.10 Mcal	Intake Scalar 100.0%
MP Balance	165 g/d	Basis: Dry Matter
DIP Balance	−150 g/d	Units: Pounds
Days to lose	1 condition	score: 3199

Press ⟨F1⟩ at any time for context sensitive help
Press ⟨ESC⟩ at any time to escape to the main menu

Feed Amounts: Place cursor next to feed name and enter desired value. Enter amount (use same moisture basis as indicated on general screen) for each feed listed. *Be sure to enter 0 for all other cells not in use.* ⟨**Enter**⟩

Intake Scalar: *This input is only used to change each feed amount fed, by the same proportion; enter the proportional change in total DMI wanted.* The scalar will increase or decrease each amount entered by the same percentage by changing the scalar to evaluate this diet formula for other conditions where the intake is predicted to change. For example, if the body weight is changed to 1100 lb,

DMI is predicted to be 23.5 lb, which is 88.0% of the current actual DMI. Entering 0.88 as the intake scalar reduces the actual DMI to 23.6 without having to change the feed amounts. Also a dry matter formula can be entered as lb/10 lb, then the scalar adjusted until the actual DMI is correct. Then the formula can be used to adjust to any DMI expected for various conditions. *Entry for this example is 1 (decimal for 100%).* ⟨**Enter**⟩

Performance Summary: Press ⟨F9⟩ to calculate. Predicted DM is used for actual because actual is not available. If the difference between actual and predicted DMI exceeds 5 to 10%, check all inputs that influence DMI (breed, body weight, mature size, temperature, mud and storm exposure, diet energy density). The diet is evaluated with actual DMI. Predicted NE balance is near 0. Rumen microbial N requirements are not being met (DIP balance is −150 g/day). Animal MP supply exceeds requirements by 165 g. The ration cost is $.75/day.

Re-evaluation: Next, lower the microbial growth to 10% of TDN (management screen cell E1094). This results in a DIP balance of +30 g/day. In Level 1 microbial yield is a constant 13% of TDN as discussed in Chapter 2, except it is reduced on high concentrate rations based on the eNDF level. However, there is no adjustment in the model for diets with low energy contents or low intakes. In either case if rate of passage is reduced, then microbial turnover is reduced and efficiency of microbial protein synthesis is reduced. Literature values for cattle fed low quality forages average 7.8% of TDN, and the DIP requirement was determined to be 7.1% of DM for cows grazing dormant forage. Therefore, it is recommended that microbial yield be reduced to 7.5–10% of TDN for cows or calves consuming low quality diets.

NRC MODEL DIET EVALUATION

Execute a Diet Evaluation with NRC Model Level 1

Press ⟨F11⟩

"LEVEL 1 DIET EVALUATION" SCREEN

Level 1 Diet Evaluation

Diet	NRC Example Dry Cow			Evaluate		
	NEm Diet Mcal/d	NEm Reqd Mcal/d	Differ Mcal/d	MP Diet g/d	MP Reqd g/d	Differ g/d
Totals	11.6	11.6	−0.1	551	501	50
Maint	11.6	10.6	1.0	551	455	96
Preg	1.0	1.1	−0.1	96	46	50
Lact	−0.1	0.0	−0.1	50	0	50
Gain	−0.1	0.0	−0.1	50	0	50
Reserves	−0.1			50		

DMI predicted	26.65 lb/d	DIP Required	599
DMI actual	26.70 lb/d	DIP Supplied	629
		DIP Bal	30 g/d
Days to lose 1 cond. score:		3199	

Effective NDF required	5.34 lb/d	MP from Bacteria	383 g/d
Effective NDF supplied	16.82 lb/d	MP from UIP	168 g/d
NDF in ration	64%DM	Diet CP	6.9%DM
Diet TDN	49%DM	DIP	75%CP
Diet ME	0.81 Mcal/lb	Total NSC in ration	20.7%DM
Diet NEm	0.43 Mcal/lb	Cost/d	$0.75/d
Diet NEg	0.19 Mcal/lb		
DMI/Maint DMI	1.09		
Est. Ruminal pH	6.46		

Evaluate: Place cursor on **Evaluate** and press ⟨**Enter**⟩ to start through "pop-up" screens of prioritized evaluations of the results; continue to press ⟨**Enter**⟩ to continue through the evaluation. The pop-up screens are described next.

DMI Predicted and Actual (IC21 and IC22):   Predicted DMI can be used as a guide, particularly to evaluate the effects of different input variables on DMI. Predicted DMI is used to compute actual DMI because actual DMI is not available.

Diet TDN, NE$_m$, and NE$_g$:   Using the tabular feed composition values in the feed library, this diet is computed to contain 49% TDN, and ME, NE$_m$, and NE$_g$ concentrations of 0.81, 0.43, and 0.19 Mcal/lb of diet DM, respectively.

NE and MP Available vs Required: Balances are shown after requirements are met for each physiological

function. Energy balances are reflected in days to change a condition score (IE24).

Effective Fiber: Not applicable for beef cows fed typical high-forage diets. See the feedlot case study for application of effective fiber.

DIP Balance (IG23):   This value should be positive to be sure rumen microbial N needs are met. If deficient, add urea or other highly degradable N sources, or replace UIP with DIP if MP supply exceeds requirements. In this example, the DIP balance is 30 g because the microbial yield was reduced to 10% of TDN.

MP from Bacteria and Feed: The MP from bacteria (cell IG21; 383 g) provide all but 118 g of the required 501 g, but the natural feeds supply 168 g (cell IG27), leaving an excess of 50 g (cell IG13).

Diet CP: Although diet CP does not appear as a pop-up screen, the total diet CP is 6.9% of the DM (cell IG26), with 75% of the protein degradable (IG29). This CP level provides about the right amount of DIP and provides more UIP than needed.

Execute a Diet Evaluation with NRC Model Level 2

Press ⟨**ESC**⟩ to go to the **Main Menu**; select **Describe Units and Levels** ⟨**Enter**⟩; select **Solution 2.** ⟨**Enter**⟩; press ⟨**F11**⟩

"LEVEL 2 DIET EVALUATION" SCREEN

			Level 2 Diet Evaluation			
Diet	NRC Dry Cow 1			Evaluate		
	NEm Avail Mcal/d	NEm Reqd Mcal/d	Differ Mcal/d	MP Avail g/d	MP Reqd g/d	Differ g/d
Totals	10.5	12.3	−1.8	908	501	407
Maint	10.5	11.2	−0.7	908	455	453
Preg	−0.7	1.1	−1.8	453	46	407
Lact	−1.8	0.0	−1.8	407	0	407
Gain	−1.8	0.0	−1.8	407	0	407
Reserves	−1.8		−1.8	407		

DMI predicted	26.65 lb/d	Bact N Bal	−24 g/d	13.7%
DMI actual	26.70 lb/d	Peptide Bal	−27 g/d	36.6%
		Urea Cost	0.5 Mcal/d	

Days to lose 1 cond. score: 170

Effective NDF required	5.34 lb/d	MP from Bacteria	660 g/d
Effective NDF supplied	16.82 lb/d	MP from UIP	248 g/d
NDF in ration	64%DM	Diet CP	6.9%DM
Diet TDN	47%DM	DIP	59.2%CP
Diet ME	0.77 Mcal/lb	Total NSC in ration	20.7%DM
Diet NEm	0.39 Mcal/lb	Cost/d	$0.75 /d
Diet NEg	0.15 Mcal/lb	Total N Balance	65 g/d
DMI/Maint DMI	0.98		
Est. Ruminal pH	6.46		
		Most Limit AA HIS	165.1%

Amino Acids, G/d

AA	Requirement	Supply	% of Requirement
MET	10	19	197
LYS	31	61	195
ARG	17	53	321
THR	19	44	230
LEU	34	63	188
ILE	14	46	327
VAL	20	50	251
HIS	12	20	165
PHE	17	43	249
TRP	3	22	745

To save your results (check to make sure they agree with values presented here), press ⟨**ESC**⟩, select **Save Inputs**, a prompt will appear to [Enter filename (maximum eight characters) to save], *type in a file name* **DRYCOW1**, press ⟨**Enter**⟩.

Differences between NRC Model Levels 1 and 2

The differences between Level 2 (the evaluation above) and Level 1 diet evaluation are described below.

1. *ME allowed condition score change* is computed in Level 2 from energy availability based on simulations of ruminal fermentation and intestinal digestion. The simulations account for the effects of (1) individual feed content of carbohydrate and protein fractions, (2) ruminal rates of digestion and passage, (3) effect of rumen pH on fiber digestibility, (4) intestinal starch and fiber digestibility, and (5) the energy cost of excreting excess N (urea cost, IG 23 is added to the NE_m requirement). In this example, Level 2 NE balance is lower (-1.8 vs -0.1) because the cost of excreting excess N is added to the NE_m requirement and diet NE values are lower, because it was predicted from actual NDF values, where in Level 1 it was predicted from tabular NE values. *This evaluation indicates that supplementation with the range cake is required, which is in agreement with field observations. The next step would be to improve energy and total N balance by adding the range cake to the diet.*

2. *MP from bacteria* is computed from bacterial growth on fiber and nonfiber carbohydrates, which are sensitive to feed amounts of fiber and nonfiber carbohydrates and their digestion rates, and rumen pH. MP from feeds are computed from feed protein escaping digestion in the rumen, which is sensitive to feed amounts of protein fractions with medium and slow digestion rates. In this example, the MP balance is higher in Level 2 than in Level 1 evaluation with microbial yield at 1390 of TDN because of a higher microbial protein production (660 g vs 571 g).

3. *Rumen N balances* are given as total bacterial N balance (IG21) and peptide balance (IG22). The total N balance is lower than in Level 1 (-24 vs 30 g) because of a higher predicted microbial yield. This difference would be greater except recycled N is included in Level 2. Peptides stimulate growth in bacteria that grow on nonfiber carbohydrates. Therefore microbial yield of nonfiber carbohydrate bacteria will be increased when the peptide balance is increased from negative to 0. This is accomplished by adding natural protein sources of protein such as soybean meal that have rapid or medium digestion rates. *If peptide balance is less than 0, supplementation with peptide sources should be considered only when MP or essential amino acids are deficient. In this case, MP and all essential amino acids are in excess, so the peptide balance should be ignored.*

4. *Essential amino acid balances* (⟨**Page Down**⟩ to lines 38–51). The requirements are computed as described in report Chapters 3 and 10, and the supplies are computed from MP from bacteria and MP from the essential amino acids in the undegraded feed protein. The balances should be not less than 5% of requirements. The importance of ratios between essential amino acids are discussed in report Chapter 3 but no attempt was made to make specific amino acid recommendations. In this example, energy is first limiting because energy balance is negative and MP and amino acid balances are positive.

Evaluate: Place cursor on **Evaluate** and press ⟨**Enter**⟩ to start through a prioritized evaluation of the results; continue to press ⟨**Enter**⟩ to continue through the evaluation. The following guidelines are given in part for interpreting results and making changes for fine-tuning the diet. The guides in the "evaluate" section can be used as a diagnostic tool or to identify why actual and observed performance agreed.

Dry Matter Intake: Compare total feed dry matter entered vs model predicted DMI (IC21 vs IC22). If there is a more than 5 to 10% difference, check input variables that influence predicted DMI (rations DM and quality control; accuracy of weights; body weight; current temperature; ionophore use; diet energy density; feed processing). The actual DMI must be accurately determined, taking into account feed wasted, moisture content of feeds, and scale accuracy. The accuracy of any model prediction is highly dependent on the DMI used. Intake of each feed must be as uniform as possible over the day because as far as we know all field application models assume a total mixed ration with steady state conditions.

ME Allowed Condition Score Change (IE24):   If predicted days for condition score change is not as expected for the conditions described (cattle type, diet type, environment, and management conditions), first carefully check all inputs. ***Input errors are the greatest source of prediction errors.*** Mistakes or incorrect judgements about inputs such as body size, milk production and its composition, environmental conditions, or feed additives are often made.

Adjust feed carbohydrate fractions and their digestion rates as necessary. If inputs are correct and performance is still not as expected, predicted diet energy values are likely the cause. First, see if predicted total diet net energy values (IC31 and IC32) and for each feed are near those expected. *Predicted energy values for individual feeds can be accessed by pressing* ⟨**F7**⟩;

use the tab key to find the NE and DIP values in metric or English units. Feed factors may be influencing energy derived from the diet as the result of feed compositional changes and, possibly, effects on digestion and passage rates. The NE derived from forages are most sensitive to NDF amount and percent of the NDF that is lignin, available NDF digestion rate (CHO B2), and eNDF value. For example, if the NDF% of a feed is increased, the starch and sugar fractions in the feed will be decreased automatically by the model, more feed dry matter will escape digestion and the feed will have a lower net energy value. Dry matter digestibility can be further decreased by lowering the NDF digestion rate. After making sure the feed composition values are appropriate, the digestion rate is considered. Adjustments are made, using the ranges and descriptions in Appendix Tables 6 through 8.

The major factors influencing energy derived from feeds high in nonfiber carbohydrates are ruminal and intestinal starch digestion rate (CHO B1). This is mainly a concern when feeding corn grain, corn silage, sorghum grain, or sorghum silage.

Check postruminal starch digestibility to make sure that it is appropriate for the starch source being fed. Intestinal digestibilities can be modified by choosing feed digestion from the main menu and selecting Intestinal Digestibilities. The model assumes an average starch digestibility (CHO B1) of 75 percent, however, this may not be appropriate for all starch sources. Appendix Table 10 can be used to adjust the starch digestibility for effects of processing on corn sources. For example, if cows were supplemented with whole corn, the intestinal starch digestibility should be lowered.

Effective Fiber Level (IC26):   Generally not a problem with high-forage based beef cow diets. See the feedlot case study for application of effective fiber.

Rumen N Balance (IG21 and IG22):   If peptide balance is less than 0 and MP balance is negative, feeds such as soybean meal that are high in degradable true protein can be added until ruminal peptide balance is ≥0 to increase microbial yield. Then adjust remaining ruminal N requirements with feeds high in NPN or soluble protein until total rumen N is balanced. Because of the number of assumptions required to adequately predict total N balance, it may be desirable under some conditions to have supply exceed requirements by about 5% (105% of requirement) to allow for prediction errors. This ration needs to contain slightly more degradable protein to overcome the

−24 g deficiency (cell IG21), which is 13.7% below requirements (cell II21).

Metabolizable Protein (IG13):   This component represents an aggregate of nonessential amino acids and essential amino acids. The MP requirement is determined in cows by the body weight and growth requirement, conceptus growth rate, and milk amounts and composition. The adequacy of the diet to meet these requirements will depend on microbial protein produced from fiber and nonfiber carbohydrate fermentation and feed protein escaping fermentation. If MP balance appears to be unreasonable, check first the starch (Carbohydrate B1) digestion rates, using the ranges and descriptions in Appendix Tables 6 through 8 for Carbohydrate B1. Altering the amount of degradable starch will also alter the peptide and total rumen N balance because of altered microbial growth. Often the most economical way to increase MP supply is to increase microbial protein production by adding highly degradable sources of starch, such as processed grains. Further adjustments are made with feeds high in slowly degraded or rumen escape (bypass) protein (low Protein B2 digestion rates; see Appendix Tables 6 through 8).

Check total ration protein degradability (IG29) and individual values for the feeds (press ⟨F7⟩ to obtain the predicted biological values) to compare with the tabular values. If considerably different, you may have an entry error or need to adjust the protein fraction digestion rates. First, check protein fractions entered. Next, check their digestion rates. In altering degradability the most sensitive fraction is the medium or B2 fraction since most of the fast or B1 fraction will be degraded in the rumen and most of the slow or B3 fraction will escape. Thus the easiest way to alter amounts degraded vs escaping is to change the amount of soluble protein and/or the digestion rate of the B2 fraction.

Essential Amino Acids (IA39 to ID51):   The amino acid with the lowest supply as a percent of requirement is assumed to be the most limiting for the specified performance. Because of the number of assumptions required to predict amino acid adequacy, it may be desirable under some conditions to have first limiting amino acids exceed requirements by about 5% (105% of requirement) to allow for prediction errors.

The adjustment for amino acids is done last because the amino acid balance is affected by the preceding steps. Essential amino acid balances can be estimated with Level 2 because the effects of the interactions of intake, digestion, and passage rates on microbial yield and available undegraded feed protein, and estimates of their amino acid composition can be predicted along with microbial, body tissue, and milk amino acid composition. However, the development of more accurate feed composition and digestion rates, and more mechanistic approaches to predict utilization of absorbed amino acids, will result in improved predictability of diet amino acid adequacy for cattle. To improve the amino acid profile of a ration, use feeds high in the first limiting amino acids.

Diet CP (IG28): After all of the above are correctly evaluated, the diet CP content will be the CP requirement. The CP requirement represents the amount of DIP needed in the rumen and the amount of UIP needed to supplement the microbial protein to meet the MP requirement.

EVALUATING COW HERD REQUIREMENTS OVER THE REPRODUCTIVE CYCLE USING THE TABLE GENERATORS

Application of the Table Generators for the Cow Herd

The table generators were designed to compute nutrient requirements and to evaluate diets for beef cows and bred heifers of a specific mature size, expected calf birth weight, and milk production level for each of the 12 months of the reproductive cycle. Month 1 of the reproductive cycle is the first month of pregnancy for bred heifers and is the first month after calving for cows. Requirements for replacement heifers between weaning and breeding are determined as described for the feedlot case study. The table generators contain the same requirement equations as both model levels except for the environmental effects. To account for environmental effects on maintenance requirements, the first line under each month is for entering NE_m multipliers. Appendix Table 14 gives suggested adjustment factors for this purpose. The goal of the diet evaluation section is to find the best match of available forages (up to three) and requirements for each month of the reproductive cycle and to identify needs for supplementation. DMI adjusters are provided for each diet to allow adjustment for intakes other than those predicted. For example, if pasture availability allows only 75% of expected DMI for part of the year, enter **.75** for the DMI adjuster for diet C to get balances for those months. Model Levels 1 or 2 can be used to compute amounts of supplement to feed where needed.

Bred Heifers Tutorial

At the opening menu of the Table Generator, choose English units (1 = English). Then choose **Replacement Heifer Requirements and Diet Evaluation** and press ⟨**Enter**⟩.

Animal Descriptions: Enter animal descriptions (1300 lb mature weight, 80 lb birth weight, 15 month age at breeding,). Only one breed code can be chosen; in this case, Angus (code 1), one of the dam parent breeds, is entered, which results in no change in the maintenance requirement due to breed. The effect of breed on maintenance requirement (Appendix Table 4) can be accounted for in crossbreds by averaging the adjustments for the parent breeds. This NE_m multiplier is entered in the NE_m requirement factor line.

NE_m Adjusters for Environmental Conditions: The reproductive cycle for heifers begins at breeding (month 1) and ends at calving. Month 9 (January in this case study) is the last full month. Based on expected mean monthly temperatures and environmental conditions in this case study, the appropriate NE_m multiplier to use (from Appendix Table 14) is 1 for all months, except 1.19 for month 8 (December) and 1.29 for month 9 (January).

Nutrient Requirements of Pregnant Replacement Heifers

Mature Weight	1300 lb
Calf Birth Weight	80 lb
Age @ Breeding	15 months
Breed Code	1 Angus

	Months since Conception								
	1	2	3	4	5	6	7	8	9
NE_m Req. Factor	100%	100%	100%	100%	100%	100%	100%	119%	129%
NE_m required, Mcal/day									
Maintenance	6.46	6.63	6.80	6.97	7.14	7.30	7.47	9.08	10.06
Growth	2.56	2.63	2.70	2.77	2.83	2.90	2.97	3.03	3.10
Pregnancy	0.03	0.06	0.14	0.29	0.58	1.07	1.88	3.12	4.88
Total	9.05	9.33	9.64	10.03	10.55	11.28	12.32	15.24	18.03
MP required, g/day									
Maintenance	319	327	336	344	352	360	369	377	385
Growth	130	130	131	131	131	128	126	123	121
Pregnancy	2	3	7	13	24	45	80	137	227
Total	450	461	473	488	507	534	574	637	733
Minerals									
Calcium required, g/day									
Maintenance	11	12	12	13	13	13	14	14	15
Growth	10	10	9	9	9	9	9	9	8
Pregnancy	0	0	0	0	0	0	11	11	11
Total	21	21	22	22	22	22	34	34	34
Phosphorus required, g/day									
Maintenance	9	9	9	10	10	10	11	11	11
Growth	4	4	4	4	4	4	4	3	3
Pregnancy	0	0	0	0	0	0	6	6	6
Total	13	13	13	13	14	14	20	21	21
ADG, lb/day									
Growth	0.95	0.95	0.95	0.95	0.95	0.95	0.95	0.95	0.95
Pregnancy	0.06	0.09	0.15	0.24	0.38	0.56	0.81	1.14	1.54
Total	1.01	1.05	1.10	1.19	1.33	1.51	1.76	2.09	2.49
Body weight, lb									
Shrunk Body	809	838	867	896	924	953	982	1011	1040
Gravid Uterus Mass	3	5	9	15	24	38	59	88	129
Total	812	843	876	911	949	992	1041	1099	1169

Table of Nutrient Requirements: Nutrient requirements, target ADG, and body weights are shown above for each month from breeding to calving. The nutrient requirements for maintenance are increasing monthly due to increasing body weight and environmental conditions in months 8 and 9. The target ADG for growth is computed as target calving weight minus target breeding weight/280 days of gestation. Target calving weight is 80% of mature weight and target puberty weight is 60% of mature weight (65% for *Bos indicus*). The requirements for the fixed ADG for growth (0.95 lb/day) is increasing because body weight is increasing, which increases the energy content of the ADG. The total ADG includes gravid uterine weight, so total ADG (lb/day) starts at 1.01, increases to 1.5 by month 6, and reaches 2.49 the last month of pregnancy. The body weight changes reflect similar changes, beginning with 812 lb at breeding, increases to 992 lb by month 6, and reaches 1,169 by month 9.

⟨Page Down⟩ three times (to line 258) to compute and view the diet evaluations. The requirements and table generator tables are linked, so only the following will need to be entered.

Describe Diets: Enter in the diet evaluation section the three primary forages available (mature and midbloom forage and pasture) as Diet A, B, C. Forage TDN (% of DM), CP (% of DM), and DIP (% of CP) are entered for the diet evaluation.

Diet	TDN % of DM	CP % of DM	DIP % of CP
A (mature forage)	50	7	75
B (midbloom forage)	57	12	77
C (pasture)	70	15	80

Table of Diet Evaluations: The diet evaluation below shows nutrient balances for each month of pregnancy, using the nutrient requirements from the first table, the predicted DMI, and the TDN, CP, and DIP values entered for the three forages available on the ranch. The calcium and phosphorus percent of DM are diet density requirements based on predicted DMI.

Diet Evaluation for Pregnant Replacement Heifers

Mature Weight	1300 lb	Breed Code	1 Angus			
Calf Birth Wt.	80 lb					
Age @ Breeding	15 months					

Ration	TDN % DM	NE$_m$ Mcal/lb	NE$_g$ Mcal/lb	CP % DM	DIP % DM	DMI Factor
A	50	0.45	0.20	7	75	100%
B	57	0.56	0.31	12	77	100%
C	70	0.76	0.48	15	80	100%

Months since Conception

		1	2	3	4	5	6	7	8	9
	NE$_m$ Req. Factor	100%	100%	100%	100%	100%	100%	100%	119%	129%
A	DM, lb	20.3	20.9	21.4	21.9	22.5	23.0	23.5	24.0	24.5
	NE allowed ADG, lb	0.82	0.81	0.79	0.75	0.68	0.56	0.37	0.00	0.00
	DIP Balance, g/day	−116	−119	−122	−125	−128	−131	−134	−137	−140
	UIP Balance, g/day	93	98	103	107	111	114	112	116	9
	MP Balance, g/day	74	78	82	85	89	91	89	92	7
	Ca% DM	0.22%	0.22%	0.21%	0.21%	0.20%	0.18%	0.27%	0.23%	0.23%
	P% DM	0.17%	0.17%	0.17%	0.16%	0.16%	0.14%	0.19%	0.16%	0.16%
B	DM, lb	21.4	21.9	22.5	23.1	23.6	24.2	24.7	25.3	25.8
	NE allowed ADG, lb	1.86	1.85	1.83	1.79	1.71	1.58	1.37	0.64	0.00
	DIP Balance, g/day	177	182	186	191	196	200	205	209	214
	UIP Balance, g/day	121	132	142	151	162	173	179	239	253
	MP Balance, g/day	97	105	113	121	129	139	143	191	202
	Ca% DM	0.32%	0.31%	0.30%	0.29%	0.28%	0.26%	0.34%	0.28%	0.22%
	P% DM	0.24%	0.24%	0.23%	0.23%	0.22%	0.21%	0.25%	0.20%	0.15%
C	DM, lb	21.0	21.6	22.1	22.7	23.2	23.7	24.3	24.8	25.3
	NE allowed ADG, lb	3.46	3.45	3.43	3.38	3.30	3.15	2.92	2.15	1.43
	DIP Balance, g/day	276	283	290	298	305	312	319	326	33
	UIP Balance, g/day	5	19	33	47	63	85	100	165	181
	MP Balance, g/day	4	15	27	38	50	68	80	132	145
	Ca% DM	0.48%	0.47%	0.26%	0.45%	0.41%	0.39%	0.46%	0.40%	0.34%
	P% DM	0.37%	0.36%	0.21%	0.35%	0.33%	0.32%	0.35%	0.29%	0.25%

Note: Energy balance based on target weight and rate of gain. Requirements are for target weight and diet NE allowed ADG.

The energy and UIP balances indicate the deficiency or excess over that needed to meet target ADG and weights. The NE allowed ADG is also shown, which should be compared with the target growth ADG in the nutrient requirements table. The NE allowed ADG (for heifer growth) for forage A indicates it is not adequate alone for any month without supplementation, suggesting that the bred heifers need to be fed separately from the mature cows. Forage B is adequate for all months except for the last two, when it would require supplementation. Forage C exceeds requirements for all months.

Press ⟨ESC⟩ to return to the Main Menu; then choose **Beef Cow Requirements and Diet Evaluations** ⟨Enter⟩

Mature Cow Tutorial

Animal Descriptions: Enter the animal descriptions (1,300 lb mature weight, 80 lb birth weight, 21.5 lb

peak milk, and 29 weeks of age at weaning). Only one breed code can be chosen; in this case study, Angus (code 1), one of the dam parent breeds, is entered, which results in no changes in the maintenance requirement due to breed. The effect of breed on maintenance requirement (Appendix Table 4) can be accounted for in crossbreds by averaging the adjustments for the parent breeds. This NE$_m$ multiplier is entered in the NE$_m$ requirement factor line.

NE$_m$ Adjusters for Environmental Conditions: The reproductive cycle for cows begins at calving (month 1). Based on expected mean monthly temperatures and environmental condition in this case study, the appropriate NE$_m$ multipliers from Appendix Table 14 are: months 1 (March), 2 (April), and 10 (December), 1.19; and months 11 and 12 (January and February), 1.29. All other months are entered as 1.

Nutrient Requirements of Beef Cows

Mature Weight	1300 Lbs	Milk Fat	4.0 %
Calf Birth Wt.	80 Lbs	Milk Protein	3.4 %
Age @ Calving	60 Months	Calving Interval	12 Months
Age @ Weaning	29 Weeks	Time Peak	8.5 Weeks
Peak Milk	21.5 Lbs	Lact. Duration	29 Weeks
Breed Code	1 Angus	Milk SNF	8.3 %

Months since Calving

	1	2	3	4	5	6	7	8	9	10	11	12
NE_m Req. Factor	119%	119%	100%	100%	100%	100%	100%	100%	100%	119%	129%	129%
NE_m required, Mcal/day												
Maintenance	13.16	13.16	11.06	11.06	11.06	11.06	9.22	9.22	9.22	10.97	11.89	11.89
Growth	0.00	0.00	0.00	0.00	0.00	0.00	0.00	0.00	0.00	0.00	0.00	0.00
Lactation	5.83	7.00	6.30	5.04	3.78	2.72	0.00	0.00	0.00	0.00	0.00	0.00
Pregnancy	0.00	0.00	0.01	0.03	0.06	0.14	0.29	0.58	1.07	1.88	3.12	4.88
Total	18.99	20.16	17.37	16.13	14.90	13.92	9.51	9.79	10.29	12.85	15.01	16.76
MP required, g/day												
Maintenance	455	455	455	455	455	455	455	455	455	455	455	455
Growth	0	0	0	0	0	0	0	0	0	0	0	0
Lactation	425	510	459	367	275	198	0	0	0	0	0	0
Pregnancy	0	0	1	2	3	7	13	24	45	80	137	227
Total	880	965	915	824	733	660	468	479	500	535	592	682
Calcium required, g/day												
Maintenance	18	18	18	18	18	18	18	18	18	18	18	18
Growth	0	0	0	0	0	0	0	0	0	0	0	0
Lactation	20	24	22	17	13	9	0	0	0	0	0	0
Pregnancy	0	0	0	0	0	0	0	0	0	11	11	11
Total	38	42	40	35	31	27	18	18	18	29	29	29
Phosphorus required, g/day												
Maintenance	14	14	14	14	14	14	14	14	14	14	14	14
Growth	0	0	0	0	0	0	0	0	0	0	0	0
Lactation	12	14	12	10	7	5	0	0	0	0	0	0
Pregnancy	0	0	0	0	0	0	0	0	0	5	5	5
Total	26	28	27	24	22	19	14	14	14	19	19	19
ADG, lb/day												
Growth	0.00	0.00	0.00	0.00	0.00	0.00	0.00	0.00	0.00	0.00	0.00	0.00
Pregnancy	0.00	0.00	0.03	0.06	0.09	0.15	0.24	0.38	0.56	0.81	1.14	1.54
Total	0.00	0.00	0.03	0.06	0.09	0.15	0.24	0.38	0.56	0.81	1.14	1.54
Milk lb/day	17.9	21.5	19.3	15.5	11.6	8.4	0.0	0.0	0.0	0.0	0.0	0.0
Body weight, lb												
Shrunk Body	1300	1300	1300	1300	1300	1300	1300	1300	1300	1300	1300	1300
Conceptus	0	0	2	3	5	9	15	24	38	59	88	129
Total	1300	1300	1302	1303	1305	1309	1315	1324	1338	1359	1388	1429

Table of Nutrient Requirements: Maintenance requirements change with month due to the effect of lactation (increased 20 percent) and environment. No requirements are computed for growth because the cows are mature. Lactation requirements change with amount of milk for each month. Pregnancy requirements and ADG due to pregnancy change do not increase much until the last 90 days. The shrunk body weight assumes a constant condition score of 5, and the conceptus weight is added to obtain the monthly total. See Appendix Table 13 for factors to compute body weight changes due to condition score changes.

⟨**Page Down**⟩ three times (to line 453) to compute and view the diet evaluations. The requirements and diet evalu-ation tables are linked, so only the following will need to be entered.

Describe Diets: Enter in the diet evaluation section the three primary forages available (mature and midbloom forage and pasture) as Diet A, B, and C. Forage TDN (% of DM), CP (% of DM) and DIP (% of CP) are entered for the diet evaluation. All composition values are on a DM basis.

Diet	TDN % of DM	CP % of DM	DIP % of CP
A (mature forage)	50	7	75
B (midbloom forage)	57	12	77
C (pasture)	70	15	80

Diet Evaluation for Beef Cows

Mature Weight	1300 Lbs	Milk Fat	4.0%	
Calf Birth Wt.	80 Lbs	Milk Protein	3.4%	
Age @ Calving	60 Months	Calving Interval	12 Months	
Age @ Weaning	29 Weeks	Time Peak	8.5 Weeks	
Peak Milk	21.5 Lbs	Milk SNF	8.3%	
Breed Code	1· Angus			

Ration	TDN % DM	ME Mcal/lb	NE$_m$ Mcal/lb	CP % DM	DIP % DM	DMI Factor
A	50	0.84	0.45	7.0	75.0	100%
B	57	0.95	0.56	12.0	77.0	100%
C	70	1.17	0.76	15.0	80.0	100%

Months since Calving

	1	2	3	4	5	6	7	8	9	10	11	12
NE$_m$ Req. Factor	119%	119%	100%	100%	100%	100%	100%	100%	100%	119%	129%	129%
Milk lb/day	17.9	21.5	19.3	15.5	11.6	8.4	0.00	0.00	0.00	0.00	0.00	0.00
A DM, lb	26.89	27.61	29.27	28.49	27.72	27.07	25.40	25.40	25.40	25.40	25.40	25.40
Energy Balance	−6.79	−7.64	−4.09	−3.20	−2.33	−1.64	2.01	1.73	1.23	−1.33	−3.49	−5.24
DIP Balance, g/day	−153	−157	−167	−162	−158	−154	−145	−145	−145	−145	−145	−145
UIP Balance, g/day	−252	−335	−220	−131	−42	29	202	202	176	133	61	−52
MP Balance, g/day	74	−201	−268	−176	−104	−34	24	173	162	141	106	49
Ca% DM	0.31%	0.34%	0.30%	0.27%	0.25%	0.22%	0.16%	0.16%	0.16%	0.25%	0.25%	0.25%
P% DM	0.21%	0.22%	0.20%	0.19%	0.17%	0.16%	0.12%	0.12%	0.12%	0.16%	0.16%	0.16%
Reserves Flux/mo	−258	−290	−156	−122	−88	−62	61	53	37	−51	−133	−199
B DM, lb	28.13	28.84	30.09	29.31	28.54	27.89	26.22	26.22	26.22	26.22	26.22	26.22
Energy Balance	−3.11	−3.88	−0.38	0.42	1.21	1.83	5.29	5.01	4.51	1.95	−0.21	−1.96
DIP Balance, g/day	233	239	249	243	236	231	217	217	217	217	217	217
UIP Balance, g/day	9	−69	43	127	209	275	328	328	328	328	294	181
MP Balance, g/day	97	7	−55	35	101	167	220	360	348	327	292	235
Ca % DM	0.30%	0.32%	0.29%	0.27%	0.24%	0.22%	0.15%	0.15%	0.15%	0.25%	0.25%	0.25%
P% DM	0.20%	0.21%	0.19%	0.18%	0.17%	0.15%	0.12%	0.12%	0.12%	0.16%	0.16%	0.16%
Reserves Flux/mo	−118	−147	−15	13	37	55	161	152	137	59	−8	−74
C DM, lb	31.70	32.42	33.23	32.46	31.68	31.03	29.36	29.36	29.36	29.36	29.36	29.36
Energy Balance	5.08	4.46	7.87	8.53	9.16	9.65	12.79	12.51	12.01	9.45	7.29	5.54
DIP Balance, g/day	416	426	436	426	416	408	386	386	386	386	386	386
UIP Balance, g/day	379	306	407	442	431	422	400	400	400	400	400	400
MP Balance, g/day	303	245	326	388	449	499	628	617	596	561	504	414
Ca% DM	0.27%	0.29%	0.26%	0.24%	0.22%	0.20%	0.14%	0.14%	0.14%	0.22%	0.22%	0.22%
P% DM	0.18%	0.19%	0.18%	0.16%	0.15%	0.14%	0.11%	0.11%	0.11%	0.14%	0.14%	0.14%
Reserves Flux/mo	154	136	239	259	279	293	389	380	365	287	222	168

Table of Diet Evaluations: The diet evaluation table shows nutrient balances for each month of pregnancy, using the nutrient requirements from the first table, the predicted DMI, and the TDN, CP, and DIP values entered for the three forages available on the ranch. The calcium and phosphorus percent of DM are diet density requirements.

DIP is adequate for all diets except Diet A. This deficiency should be evaluated with the model Levels 1 and 2 as described previously.

Diet A (mature forage) meets energy requirements for the first 90 days after weaning, then it becomes deficient in month 10 (−1.33 Mcal diet NE$_m$/day), causing 51 Mcal to be mobilized that month. If continued in months 11 and 12 until calving, the deficiency totals 383 Mcal, which represents a condition score loss of approximately 1.5 (383/

245; Appendix Table 13). If Diet A is fed for months 7, 8, and 9, and diet B is fed for months 10, 11, and 12, energy balance would be 128 Mcal. However, if Diet A is grazed mature range at DMI at 75% of predicted for months 7, 8, and 9, in the previous scenario, then energy balance is −163 Mcal, a loss of 2/3 of a condition score. If Diet B is fed for months 1 and 2 between calving and pasture turnout, 265 Mcal will be mobilized, a loss of over 1 condition score. If followed by the pasture at the predicted voluntary DMI, this loss will be nearly replenished in month 3. However, if Diet C (pasture) DMI is changed to 75% of predicted the first month of grazing (month 3) followed by 100% of predicted DMI in month 4 (second month of grazing), the condition score loss will not be replenished until month 4.

4 Guideline Diet Nutrient Density Requirement Tables

As described in the previous chapters, the NRC model allows the user to predict for beef cattle, nutrient requirements and performance under specific animal, environmental, and dietary conditions. Many variables (i.e., maintenance, growth, milk, microbial growth) are continuous and interact with feed composition. Simple tables of dietary requirements cannot do as good a job of accounting for animal, feed, and environmental variation as the NRC model. However, in many situations, the simple tables (Appendix Tables 15 through 23) with dietary nutrient requirements are sufficient. These tables were computed with modifications of the table generators and were designed to give guidelines for simple diagnostic or teaching purposes for the most common classes of beef cattle. The information in these tables is similar to that in Appendix Tables 10 and 11 of the 1984 edition of *Nutrient Requirements of Beef Cattle*. Information on quantities of required nutrients rather than density is available using the table generator. The model must be used for all other situations.

In most beef production situations, groups of cattle are fed to appetite either high forage (stocker, backgrounding, cow-calf) or high-grain diets (growing and finishing cattle) and are supplemented to support the energy allowable production, based on group averages. The tables were designed with that in mind. Requirements are not given for specific rates of ADG for growing and finishing cattle, but are for the energy allowable ADG when cattle are fed a particular diet and consume the predicted DMI. The five tables for growing cattle (Appendix Tables 15–19) cover final weight ranges of 1,000 to 1,400 lb in 100 lb increments. This range was selected based on the demand in the United States for carcass weights of 600 to 900 lb. Following the procedure used by the table generator, the ranges included are 55 to 80% of final weight. As a result, the weights and ADG within the tables differ from table to table, but all are at a similar stage of growth across

the tables for growing cattle. The table for bred heifers (Appendix Table 20) contains diet density requirements for mature sizes in 100 lb increments from 1,000 to 1,400 lb. The three tables for mature cows (Appendix Tables 21–23) include the requirements for animals with mature weights of 1,000, 1,200, and 1,400 lb and three levels of peak milk production during a 29-week lactation for each weight class. The milk production levels cover the range of expected peak milk given for the 28 breeds in Appendix Table 4. Calves born from 1,200 lb cows were assumed to weigh 80 lb. A similar ratio of calf:cow birth weights was used for the other cow and bred heifer weight classes.

The simplifications of the model and table generator used to make these tables are outlined below. In addition, some of the limitations in using these tables are discussed.

1. For all cattle, all values are driven by tabular diet net energy concentration, which is used to predict DMI with the equations described in chapter 5 of the report. Thus all values are a function of the predicted DMI, and the user cannot adjust the predicted DMI to match observed values. All diet density requirements reflect these predicted DMI and tabular feed NE values, and are not sensitive to local variations due to cattle type, normal intake patterns during the growth period, environmental conditions, and the effects of rumen pH on cell wall digestion, feed net energy values, and microbial yield.

2. For all cattle, the dietary % CP requirement was computed as ((grams MP required/0.67)/grams DMI) * 100. This method assumes that on average, 80% of MP comes from microbial protein and 20% comes from UIP in the typical beef cattle production system. Model Level 1 assumes 64% of MCP and 80% of UIP is absorbed; thus $(0.64 * 0.80) + (0.8 * 0.20) = 0.67$. An evaluation of each table with model Level 1 indicates that this method results in an adequate

pool of DIP + UIP in most situations. This method has two primary limitations:

The MP requirement and % CP needed in the diet depend on the predicted DMI, which is subject to the errors discussed above.

The resulting CP intake does not directly reflect expected microbial growth and may not be adequate to meet the DIP requirement for maximum carbohydrate digestibility. In typical beef production systems, the MP supplied often exceeds MP requirements when DIP is adequate to support maximum carbohydrate digestion. As discussed in Chapter 10, maximizing ruminal digestion of carbohydrates increases absorbed microbial amino acids as well as feed NE value. Model Level 2 allows the user to predict ruminal carbohydrate degradation and corresponding ruminal nitrogen requirements for specific conditions.

3. For growing cattle tables, the predicted ADG is a function of predicted DMI and diet NE values, with no adjustment for environmental conditions. The predicted ADG and body weight are then used to compute MP, Ca, and P requirements. As discussed in Chapter 9 of the report, these DMI were developed on averages during the feeding period and give the same rate of gain for all weights. Thus the density requirements reflect the changes in composition of gain with weight but they do not reflect typical intake and ADG patterns. During the feeding period, cattle consume a higher proportion of body weight early and a lower proportion later. Further, the user cannot adjust DMI and NE efficiency until performance agrees with observed performance. Both the model and table generators allow for adjustment for these conditions. The weights given should be related to group averages fed a specific diet. *For lighter cattle or situations where amino acid deficiencies are most likely to limit growth and protein supplementation is likely to be the most expensive, the user should use model Level 2 to determine dietary requirements.*

4. To develop the bred heifer table, an iterative procedure was used to determine diet NE_m and NE_g (and resulting DMI) needed to support the target ADG for the bred heifers with no environmental stress. If the predicted DMI is overestimated, the diet density requirement is underestimated. The birth weight was fixed as described above. The model and table generator allow for variable DMI, environmental conditions, and birth weights.

5. To develop the tables for the mature cows, an iterative procedure was used to determine diet NE_m concentration (and resulting predicted DMI) needed to meet the cow's requirements with no gain or loss in energy reserves and with no environmental stress. If the predicted DMI is overestimated, the diet density requirement is underestimated. The birth weight and lactation length were fixed as described above. Both the model and table generator allow for a variable DMI, environmental conditions, birth weights, lactation length, and reserves fluxes that occur in most situations.

COMPARISONS WITH THE 1984 NRC TABLES

Growing Cattle

The major differences are as follows:

1. Tables are presented by weight at 28% fat (growing and finishing) or maturity (replacement heifers) compared to 1984 NRC tables, which have requirements for each sex within two frame sizes. The biological basis for this approach is discussed in report Chapter 3. The approach in this revision assumes all cattle have similar requirements at the same stage of growth and accounts for the effects of body size, implant strategy, and feeding program on finished weight.

2. Within each final weight, the weight ranges given are 55-80% of final weight to reflect typical feeding group averages, which is the same procedure used for the table generator. As a result, the weights and ADG within the tables differ from table to table, but all are at a similar stage of growth across the five tables for growing cattle. The 1984 NRC weights started at 300 lb for all types and continued in 100 lb increments to 1,000 or 1,100 lb. The user can compute requirements for a particular weight with the model.

3. Within each table, diet density requirements are given for the diet energy allowable ADG rather than for a specific ADG. It is assumed that cattle are fed a particular diet to ad libitum intake and that only replacement heifers are being fed for a target ADG. The user can estimate the requirement for a particular ADG by finding the nearest value in the table.

4. The ADG for a particular cattle type and energy density is higher than the 1984 NRC because predicted DMI is higher.

Bred Heifers: The requirements are presented by mature size for target ADG and weight each month between conception and calving. In the 1984 NRC report, the requirements for three levels of ADG for each of six body weights with no adjustment for mature size were included. The predicted requirement for dietary energy is lower in this revision because the predicted DMI is higher.

Mature Cows: The 1984 NRC report provided tables for seven weights of dry pregnant cows during the middle and last third of pregnancy and two levels of milk production. In this revision, peak milk is used to compute a lactation curve, and milk production varies with month of lactation. Birth weight is used to compute a conceptus growth curve and pregnancy requirements that vary by month of pregnancy. To reflect the effect of these continuous variables on requirements, diet nutrient density requrements are provided for each month of the reproductive cycle across the range of milk production levels expected for the North American beef cattle population. Included are three categories of cow mature sizes across the range that generally will produce calves with carcass weights that will be within current industry standards without discounts. Dietary nutrient requirement concentrations are lower in many cases than in the 1984 NRC report primarily because of a higher predicted DMI in this revision. The model or table generator should be used to compute concentrations needed for the DMI observed under specific conditions.

Growing Bull Calves: Use expected bull mature weight * 0.60 to choose the table (15 through 19) to use for growing bull calves. A mature weight of 1,723 lb in the model or table generator predicts requirements similar to the 1984 NRC implanted medium-frame steer requirements. A mature weight of 2,083 lb predicts requirements similar to the 1984 NRC medium-frame bull requirements. A mature weight of 2,328 lb predicts requirements similar to the 1984 NRC large-frame bull requirements.

Two-Year Old Heifers and Mature Bulls: Tables are not provided for these classes of cattle. Their requirements can be computed with the model or table generator. The NE_m requirement is 15% higher than for the 1984 NRC bull requirements.

Appendix Tables

APPENDIX TABLE 1A   Feed Library—Energy and Crude Protein Values, Plant Cell Wall Constituents, and Digestibility Rates

Feed No.	Common Name	Int. Ref. No.	Cost $/ton	Conc. AF %DM	Forage %DM	DM %AF	NDF %DM	Lignin %NDF	eNDF %NDF	TDN %DM	ME Mcal/kg	NE$_{ma}$ Mcal/kg	NE$_{ga}$ Mcal/kg
101	Bahiagrass 30% Dry Matter	2-00-464	0.00	0.00	100.00	30.00	68.00	10.29	41.00	54.00	1.95	1.11	0.55
102	Bahiagrass Hay	1-00-462	0.00	0.00	100.00	90.00	72.00	11.11	98.00	51.00	1.84	1.00	0.45
103	Bermudagrass Late Vegetative	1-09-210	0.00	0.00	100.00	91.00	76.60	8.57	98.00	49.00	1.77	0.93	0.39
104	Brome Hay Pre-bloom	1-00-887	0.00	0.00	100.00	88.00	55.00	7.69	98.00	60.00	2.17	1.31	0.74
105	Brome Hay Mid Bloom	1-05-633	0.00	0.00	100.00	88.00	57.70	6.06	98.00	56.00	2.02	1.18	0.61
106	Brome Hay Late bloom	1-00-888	0.00	0.00	100.00	91.00	68.00	11.11	98.00	55.00	1.99	1.14	0.58
107	Brome Hay Mature	1-00-944	0.00	0.00	100.00	92.00	70.50	11.27	98.00	53.00	1.92	1.07	0.52
108	Fescue Meadow Hay	1-01-912	0.00	0.00	100.00	88.00	65.00	10.77	98.00	56.00	2.02	1.18	0.61
109	Fescue, Alta Hay	1-05-684	0.00	0.00	100.00	89.00	70.00	9.29	98.00	55.00	1.99	1.14	0.58
110	Fescue, K31 Hay	1-09-187	0.00	0.00	100.00	91.00	62.20	6.35	98.00	61.00	2.21	1.34	0.77
111	Fescue K31 Hay Full bloom	1-09-188	0.00	0.00	100.00	91.00	67.00	7.46	98.00	58.00	2.10	1.24	0.68
112	Fescue, K31 Mature	1-09-189	0.00	0.00	100.00	91.00	70.00	10.00	98.00	44.00	1.59	0.75	0.22
113	Napiergrass Fresh 30 day DM	2-03-158	0.00	0.00	100.00	20.00	70.00	14.29	41.00	55.00	1.99	1.14	0.58
114	Napiergrass Fresh 60 day DM	2-03-162	0.00	0.00	100.00	23.00	75.00	18.67	41.00	53.00	1.92	1.07	0.52
115	Orchardgrass Hay, Early bloom	1-03-425	0.00	0.00	100.00	89.00	59.60	7.70	98.00	65.00	2.35	1.47	0.88
116	Orchardgrass Hay, Late bloom	1-03-428	0.00	0.00	100.00	90.60	65.00	11.40	98.00	54.00	1.95	1.11	0.55
117	Pangolagrass Fresh	2-03-493	0.00	0.00	100.00	21.00	70.00	11.40	41.00	55.00	1.99	1.14	0.58
118	Red Top Fresh	2-03-897	0.00	0.00	100.00	29.00	64.00	12.50	41.00	63.00	2.28	1.41	0.83
119	Reed Canarygrass Hay	1-00-104	0.00	0.00	100.00	89.00	64.00	6.25	98.00	55.00	1.99	1.14	0.58
120	Ryegrass Hay	1-04-077	0.00	0.00	100.00	88.00	41.00	4.88	98.00	64.00	2.31	1.44	0.86
121	Sorghum Sudan Ha	1-04-480	0.00	0.00	100.00	91.00	66.00	6.06	98.00	56.10	2.03	1.18	0.62
122	Sorghum-Sudan Pasture	2-04-484	0.00	0.00	100.00	18.00	55.00	5.45	41.00	65.00	2.35	1.47	0.88
123	Sorghum-Sudan Silage	3-04-499	0.00	0.00	100.00	28.00	68.00	7.04	41.00	55.00	1.99	1.14	0.58
124	Timothy Hay Late Vegetative	1-04-881	0.00	0.00	100.00	89.00	55.00	5.45	98.00	62.00	2.24	1.38	0.80
125	Timothy Hay Early bloom	1-04-882	0.00	0.00	100.00	89.00	61.40	6.56	98.00	59.00	2.13	1.28	0.71
126	Timothy Hay Mid bloom	1-04-883	0.00	0.00	100.00	89.00	63.70	7.46	98.00	57.00	2.06	1.21	0.64
127	Timothy Hay Full bloom	1-04-884	0.00	0.00	100.00	89.00	64.20	8.82	98.00	56.00	2.02	1.18	0.61
128	Timothy Hay Seed stage	1-04-888	0.00	0.00	100.00	89.00	72.00	12.50	98.00	47.00	1.70	0.86	0.32
129	Wheatgrass crest., hay	1-05-351	0.00	0.00	100.00	92.00	65.00	9.23	98.00	53.00	1.92	1.07	0.52
130	Blank		0.00	100.00	0.00	100.00	0.00	0.00	0.00	0.00	0.00	0.00	0.00
131	Blank		0.00	100.00	0.00	100.00	0.00	0.00	0.00	0.00	0.00	0.00	0.00
132	Blank		0.00	100.00	0.00	100.00	0.00	0.00	0.00	0.00	0.00	0.00	0.00
133	Blank		0.00	100.00	0.00	100.00	0.00	0.00	0.00	0.00	0.00	0.00	0.00
134	Blank		0.00	100.00	0.00	100.00	0.00	0.00	0.00	0.00	0.00	0.00	0.00
135	Grass Pasture Spring	2-00-956	0.00	0.00	100.00	23.00	47.90	6.00	41.00	74.00	2.68	1.76	1.14
136	Grass Pasture Summer		0.00	0.00	100.00	25.00	55.00	7.00	41.00	67.00	2.42	1.54	0.94
137	Grass Pasture Fall		0.00	0.00	100.00	24.00	67.00	6.50	41.00	53.00	1.92	1.07	0.52
138	Mix Pasture Spring		0.00	0.00	100.00	21.00	41.50	7.00	41.00	79.00	2.86	1.91	1.27
139	Mix Pasture Summer		0.00	0.00	100.00	22.00	46.50	7.80	41.00	67.00	2.42	1.54	0.94
140	Range June Diet		0.00	0.00	100.00	20.00	65.60	5.00	41.00	64.90	2.35	1.47	0.88
141	Range July Diet		0.00	0.00	100.00	20.00	67.70	5.50	41.00	62.30	2.25	1.39	0.81
142	Range Aug. Diet		0.00	0.00	100.00	20.00	63.70	8.00	41.00	59.40	2.15	1.29	0.72
143	Range Sep. Diet		0.00	0.00	100.00	20.00	66.60	9.00	41.00	57.30	2.07	1.22	0.65
144	Range Winter		0.00	0.00	100.00	80.00	66.10	11.00	41.00	50.50	1.83	0.99	0.44
145	Meadow Spring		0.00	0.00	100.00	15.00	53.00	8.00	41.00	44.80	1.62	0.78	0.25
146	Meadow Fall		0.00	0.00	100.00	20.00	52.00	8.00	41.00	51.90	1.88	1.03	0.48
147	Meadow Hay		0.00	0.00	100.00	90.00	67.60	5.00	98.00	60.00	2.17	1.31	0.74
148	Prairie Hay	1-03-191	0.00	0.00	100.00	91.00	72.70	6.00	98.00	48.00	1.74	0.90	0.35
149	Blank		0.00	0.00	100.00	100.00	0.00	0.00	0.00	0.00	0.00	0.00	0.00
150	Blank		0.00	0.00	100.00	100.00	0.00	0.00	0.00	0.00	0.00	0.00	0.00
201	Alfalfa Hay Early Vegetative	1-00-54-S	0.00	0.00	100.00	91.00	33.00	18.18	92.00	66.00	2.39	1.51	0.91
202	Alfalfa Hay Early Vegetative	1-00-N	0.00	0.00	100.00	91.00	36.00	14.72	92.00	67.00	2.42	1.54	0.94
203	Alfalfa Hay Late Vegetative	1-00-059-S	0.00	0.00	100.00	91.00	37.00	18.92	92.00	63.00	2.28	1.41	0.83
204	Alfalfa Hay Late Vegetative	1-00-N	0.00	0.00	100.00	91.00	39.00	16.67	92.00	64.00	2.31	1.44	0.86
205	Alfalfa Hay Early Bloom	1-00-059-S	0.00	0.00	100.00	91.00	39.30	20.00	92.00	60.00	2.17	1.31	0.74
206	Alfalfa Hay Early Bloom	1-00-N	0.00	0.00	100.00	91.00	42.00	16.90	92.00	62.00	2.24	1.38	0.80
207	Alfalfa Hay Mid Bloom	1-00-063-S	0.00	0.00	100.00	91.00	47.10	22.73	92.00	58.00	2.10	1.24	0.68
208	Alfalfa Hay Mid Bloom	1-00-N	0.00	0.00	100.00	91.00	49.00	18.91	92.00	60.00	2.17	1.31	0.74
209	Alfalfa Hay Full Bloom	1-00-068-S	0.00	0.00	100.00	91.00	48.80	22.92	92.00	55.00	1.99	1.14	0.58
210	Alfalfa Hay Full Bloom	1-00-N	0.00	0.00	100.00	91.00	51.00	20.39	92.00	56.00	2.02	1.18	0.61
211	Alfalfa Hay Late Bloom	1-00-070-S	0.00	0.00	100.00	91.00	53.00	23.02	92.00	52.00	1.88	1.04	0.49
212	Alfalfa Hay Late Bloom	1-00-N	0.00	0.00	100.00	91.00	55.00	22.18	92.00	53.00	1.92	1.07	0.52
213	Alfalfa Hay Mature	1-00-71-S	0.00	0.00	100.00	91.00	58.00	24.83	92.00	50.00	1.81	0.97	0.42
214	Alfalfa Hay Seeded		0.00	0.00	100.00	91.00	70.00	24.30	92.00	45.00	1.63	0.79	0.25
215	Alfalfa Hay Weathered		0.00	0.00	100.00	89.00	58.00	25.86	92.00	48.00	1.74	0.90	0.35
216	Alfalfa Meal dehydrated 15%CP	1-00-022	0.00	0.00	100.00	90.00	55.40	26.00	6.00	59.00	2.13	1.28	0.71

NOTE: See the glossary for definitions of acronyms and Chapter 10 for a discussion of tabular energy and protein values, feed carbohydrate and protein fractions, and recommended analytical procedures.

[a]Carbohydrate digestive rate.

[b]Protein digestion rates.

CP %DM	DIP %CP	UIP %CP	SolP %CP	NPN %SolP	NDFIP %CP	ADFIP %CP	Starch %NSC	Fat %DM	Ash %DM	A kd[a] %/hr	B1 kd[a] %/hr	B2 kd[a] %/hr	B1 kd[b] %/hr	B2 kd[b] %/hr	B3 kd[b] %/hr
8.90	83.00	17.00	41.00	2.40	14.50	2.00	5.00	2.10	10.00	250.00	30.00	3.00	135.00	11.00	0.09
8.20	63.00	37.00	25.00	96.00	31.00	6.50	6.00	1.60	11.00	250.00	30.00	3.00	135.00	11.00	0.09
7.80	85.00	15.00	25.90	25.40	34.20	8.90	6.00	2.70	8.00	250.00	30.00	3.00	135.00	11.00	0.09
16.00	79.00	21.00	25.00	96.00	31.00	6.50	6.00	2.60	10.00	250.00	30.00	3.00	135.00	11.00	0.09
14.40	79.00	21.00	25.00	96.00	31.00	6.50	6.00	2.20	10.90	250.00	30.00	3.00	135.00	11.00	0.09
10.00	59.00	41.00	25.00	96.00	31.00	6.50	6.00	2.30	9.00	250.00	30.00	3.00	135.00	11.00	0.09
6.00	48.00	52.00	25.00	96.00	31.00	6.50	6.00	2.00	7.20	250.00	30.00	3.00	135.00	11.00	0.09
9.10	67.00	33.00	25.00	96.00	31.00	6.50	6.00	2.40	8.00	250.00	30.00	3.00	135.00	11.00	0.09
10.20	71.00	29.00	25.00	96.00	31.00	6.50	6.00	2.20	10.00	250.00	30.00	3.00	135.00	11.00	0.09
15.00	82.00	18.00	25.90	25.40	34.20	8.90	6.00	5.50	9.00	250.00	30.00	3.00	135.00	11.00	0.09
12.90	77.00	23.00	25.90	25.40	34.20	8.90	6.00	5.30	8.00	250.00	30.00	3.00	135.00	11.00	0.09
10.80	86.00	14.00	25.90	25.40	34.20	8.90	6.00	4.70	6.80	250.00	30.00	3.00	135.00	11.00	0.09
8.70	83.00	17.00	46.00	2.20	10.00	2.20	8.00	3.00	9.00	250.00	30.00	3.00	135.00	11.00	0.09
7.80	81.00	19.00	46.00	2.20	10.00	2.20	8.00	1.00	6.00	250.00	30.00	3.00	135.00	11.00	0.09
12.80	77.00	23.00	25.00	96.00	31.00	5.70	6.00	2.90	8.50	250.00	30.00	3.00	135.00	11.00	0.09
8.40	64.00	36.00	25.00	96.00	31.00	6.10	6.00	3.40	10.10	250.00	30.00	3.00	135.00	11.00	0.09
9.10	84.00	16.00	42.00	4.80	24.00	2.20	5.00	2.30	7.60	250.00	30.00	3.00	135.00	11.00	0.09
11.60	87.00	13.00	42.00	4.80	24.00	2.20	5.00	3.90	8.00	350.00	25.00	9.00	200.00	14.00	2.00
10.30	71.00	29.00	25.00	96.00	31.00	6.10	6.00	3.10	10.00	250.00	30.00	3.00	135.00	11.00	0.09
8.60	65.00	35.00	25.00	96.00	31.00	5.70	6.00	2.20	10.00	250.00	30.00	3.00	135.00	11.00	0.09
11.30	69.00	31.00	20.00	95.00	40.00	11.00	10.00	1.80	9.60	250.00	20.00	3.00	135.00	11.00	0.09
16.80	88.00	12.00	45.00	11.11	30.00	5.00	90.00	3.90	9.00	350.00	25.00	9.00	200.00	14.00	2.00
10.80	72.00	28.00	50.00	90.00	40.00	11.00	100.00	2.80	9.80	250.00	20.00	5.00	175.00	12.00	1.50
14.00	79.00	21.00	25.00	96.00	31.00	5.70	6.00	3.00	8.00	250.00	30.00	3.00	135.00	11.00	0.09
10.80	73.00	27.00	25.00	96.00	31.00	5.70	6.00	2.80	5.70	250.00	30.00	3.00	135.00	11.00	0.09
9.70	69.00	31.00	25.00	96.00	31.00	6.10	6.00	2.70	7.00	250.00	30.00	3.00	135.00	11.00	0.09
8.10	62.00	38.00	25.00	96.00	31.00	6.10	6.00	2.90	5.20	250.00	30.00	3.00	135.00	11.00	0.09
6.00	50.00	50.00	25.00	96.00	31.00	6.50	6.00	2.00	6.00	250.00	30.00	3.00	135.00	11.00	0.09
9.00	67.00	33.00	25.00	96.00	31.00	6.10	6.00	2.30	9.00	250.00	30.00	3.00	135.00	11.00	0.09
0.00	0.00	100.00	0.00	0.00	0.00	0.00	0.00	0.00	100.00	0.00	0.00	0.00	0.00	0.00	0.00
0.00	0.00	100.00	0.00	0.00	0.00	0.00	0.00	0.00	100.00	0.00	0.00	0.00	0.00	0.00	0.00
0.00	0.00	100.00	0.00	0.00	0.00	0.00	0.00	0.00	100.00	0.00	0.00	0.00	0.00	0.00	0.00
0.00	0.00	100.00	0.00	0.00	0.00	0.00	0.00	0.00	100.00	0.00	0.00	0.00	0.00	0.00	0.00
0.00	0.00	100.00	0.00	0.00	0.00	0.00	0.00	0.00	100.00	0.00	0.00	0.00	0.00	0.00	0.00
21.30	94.00	6.00	41.00	2.44	14.50	2.00	5.00	4.00	10.40	350.00	40.00	20.00	200.00	12.00	2.00
15.00	90.00	10.00	42.00	4.76	24.00	2.20	5.00	3.70	9.00	350.00	40.00	18.00	200.00	10.00	2.00
22.00	93.00	7.00	43.00	2.33	16.40	2.00	5.00	3.70	10.00	350.00	45.00	19.00	200.00	12.00	2.00
26.00	94.00	6.00	43.00	2.33	12.40	2.10	8.00	3.20	10.25	350.00	45.00	12.00	200.00	14.00	2.00
19.50	92.00	8.00	44.00	3.41	12.50	2.60	8.00	3.20	9.40	350.00	45.00	10.00	200.00	14.00	2.00
11.00	72.00	28.00	42.00	5.00	24.00	2.00	0.00	3.00	10.00	250.00	30.00	12.00	135.00	12.00	3.00
10.50	70.00	30.00	42.00	5.00	24.00	2.00	0.00	3.00	10.00	250.00	30.00	11.00	135.00	12.00	2.00
9.70	66.00	34.00	42.00	5.00	24.00	2.00	0.00	3.00	10.00	250.00	30.00	10.00	135.00	10.00	0.75
6.90	67.00	33.00	42.00	5.00	24.00	2.00	0.00	3.00	10.00	250.00	30.00	12.00	135.00	12.00	0.75
4.70	63.00	37.00	42.00	5.00	24.00	2.00	0.00	3.00	10.00	250.00	30.00	10.00	135.00	10.00	0.20
20.30	94.00	6.00	60.00	5.00	2.00	1.00	0.00	3.00	10.00	250.00	30.00	1.50	135.00	40.00	6.00
13.40	92.00	8.00	60.00	5.00	2.00	1.00	0.00	3.00	10.00	250.00	30.00	1.50	135.00	40.00	6.00
13.40	77.00	23.00	25.00	5.00	2.00	1.00	0.00	3.00	11.00	250.00	30.00	6.00	135.00	8.00	0.09
5.30	62.00	38.00	25.00	5.00	2.00	1.00	0.00	3.00	8.00	250.00	30.00	3.50	135.00	3.50	0.09
0.00	0.00	100.00	0.00	0.00	0.00	0.00	0.00	0.00	100.00	0.00	0.00	0.00	0.00	0.00	0.00
0.00	0.00	100.00	0.00	0.00	0.00	0.00	0.00	0.00	100.00	0.00	0.00	0.00	0.00	0.00	0.00
30.00	90.00	10.00	30.00	96.00	15.00	10.00	10.00	4.00	10.00	250.00	30.00	4.50	150.00	9.00	1.25
23.40	87.00	13.00	30.00	96.00	15.00	10.00	10.00	3.20	10.00	250.00	30.00	4.50	150.00	9.00	1.25
27.00	89.00	11.00	30.00	93.00	15.00	10.00	10.00	3.80	9.00	250.00	30.00	4.50	150.00	9.00	1.25
21.70	86.00	14.00	30.00	93.00	15.00	10.00	10.00	3.00	10.00	250.00	30.00	4.50	150.00	9.00	1.25
25.00	88.00	12.00	29.00	93.00	18.00	11.00	10.00	2.90	9.20	250.00	30.00	4.50	150.00	9.00	1.25
19.90	84.00	16.00	29.00	93.00	18.00	11.00	10.00	2.90	9.20	250.00	30.00	4.50	150.00	9.00	1.25
22.00	84.00	16.00	28.00	93.00	25.00	14.00	10.00	2.60	8.50	250.00	30.00	4.50	150.00	9.00	1.25
17.00	82.00	18.00	28.00	93.00	25.00	14.00	10.00	2.39	8.57	250.00	30.00	4.50	150.00	9.00	1.25
17.00	82.00	18.00	27.00	93.00	29.00	16.00	10.00	3.40	7.80	250.00	30.00	4.50	150.00	9.00	1.25
13.00	77.00	23.00	27.00	93.00	29.00	16.00	10.00	1.80	9.00	250.00	30.00	4.50	150.00	9.00	1.25
17.00	82.00	18.00	26.00	92.00	33.00	18.00	10.00	1.50	8.00	250.00	30.00	4.50	150.00	9.00	1.25
12.00	75.00	25.00	26.00	92.00	33.00	18.00	10.00	1.60	8.00	250.00	30.00	4.50	150.00	9.00	1.25
14.00	79.00	21.00	25.00	92.00	36.00	20.00	10.00	1.30	7.00	250.00	30.00	4.50	150.00	9.00	1.25
12.00	75.00	25.00	25.00	92.00	36.00	20.00	10.00	1.00	7.00	250.00	30.00	4.50	150.00	9.00	1.25
10.00	70.00	30.00	15.00	100.00	45.00	25.00	10.00	0.00	8.00	250.00	30.00	4.50	150.00	9.00	1.25
17.30	54.00	46.00	28.00	100.00	25.00	17.00	10.00	2.40	9.90	300.00	37.00	9.00	150.00	8.00	0.15

APPENDIX TABLE 1A   Feed Library (*Continued*)

Feed No.	Common Name	Int. Ref. No.	Cost $/ton AF	Conc. %DM	Forage %DM	DM %AF	NDF %DM	Lignin %NDF	eNDF %NDF	TDN %DM	ME Mcal/kg	NE$_{ma}$ Mcal/kg	NE$_{ga}$ Mcal/kg
217	Alfalfa Silage Early Bloom	3-00-216	0.00	0.00	100.00	35.00	43.00	23.26	82.00	63.00	2.28	1.41	0.83
218	Alfalfa Silage Mid Bloom	3-00-217	0.00	0.00	100.00	38.00	47.00	23.40	82.00	58.00	2.10	1.24	0.68
219	Alfalfa Silage Full Bloom	3-00-218	0.00	0.00	100.00	40.00	51.00	23.53	82.00	55.00	1.99	1.14	0.58
220	Birdsfoot Trefoil, hay	1-05-044	0.00	0.00	100.00	91.00	47.50	19.15	92.00	59.00	2.13	1.28	0.71
221	Clover Ladino Hay	1-01-378	0.00	0.00	100.00	89.00	36.00	19.44	92.00	60.00	2.17	1.31	0.74
222	Clover Red Hay	1-01-415	0.00	0.00	100.00	88.00	46.90	17.86	92.00	55.00	1.99	1.14	0.58
223	Vetch Hay	1-05-106	0.00	0.00	100.00	89.00	48.00	16.67	92.00	57.00	2.06	1.21	0.64
224	Blank		0.00	100.00	0.00	100.00	0.00	0.00	0.00	0.00	0.00	0.00	0.00
225	Blank		0.00	100.00	0.00	100.00	0.00	0.00	0.00	0.00	0.00	0.00	0.00
226	Blank		0.00	100.00	0.00	100.00	0.00	0.00	0.00	0.00	0.00	0.00	0.00
227	Blank		0.00	100.00	0.00	100.00	0.00	0.00	0.00	0.00	0.00	0.00	0.00
228	Blank		0.00	100.00	0.00	100.00	0.00	0.00	0.00	0.00	0.00	0.00	0.00
229	Blank		0.00	100.00	0.00	100.00	0.00	0.00	0.00	0.00	0.00	0.00	0.00
230	Leg Pasture Spring		0.00	0.00	100.00	20.00	33.00	8.00	41.00	79.00	2.86	1.91	1.27
231	Leg Pasture Summer	2-00-181	0.00	0.00	100.00	23.20	38.00	8.50	41.00	66.00	2.39	1.51	0.91
232	Blank		0.00	100.00	0.00	100.00	0.00	0.00	0.00	0.00	0.00	0.00	0.00
233	Blank		0.00	100.00	0.00	100.00	0.00	0.00	0.00	0.00	0.00	0.00	0.00
234	Blank		0.00	100.00	0.00	100.00	0.00	0.00	0.00	0.00	0.00	0.00	0.00
235	Blank		0.00	100.00	0.00	100.00	0.00	0.00	0.00	0.00	0.00	0.00	0.00
236	Blank		0.00	100.00	0.00	100.00	0.00	0.00	0.00	0.00	0.00	0.00	0.00
237	Blank		0.00	100.00	0.00	100.00	0.00	0.00	0.00	0.00	0.00	0.00	0.00
238	Blank		0.00	100.00	0.00	100.00	0.00	0.00	0.00	0.00	0.00	0.00	0.00
239	Blank		0.00	100.00	0.00	100.00	0.00	0.00	0.00	0.00	0.00	0.00	0.00
240	Blank		0.00	100.00	0.00	100.00	0.00	0.00	0.00	0.00	0.00	0.00	0.00
241	Blank		0.00	100.00	0.00	100.00	0.00	0.00	0.00	0.00	0.00	0.00	0.00
242	Blank		0.00	100.00	0.00	100.00	0.00	0.00	0.00	0.00	0.00	0.00	0.00
243	Blank		0.00	100.00	0.00	100.00	0.00	0.00	0.00	0.00	0.00	0.00	0.00
244	Blank		0.00	100.00	0.00	100.00	0.00	0.00	0.00	0.00	0.00	0.00	0.00
245	Blank		0.00	100.00	0.00	100.00	0.00	0.00	0.00	0.00	0.00	0.00	0.00
246	Blank		0.00	100.00	0.00	100.00	0.00	0.00	0.00	0.00	0.00	0.00	0.00
247	Blank		0.00	100.00	0.00	100.00	0.00	0.00	0.00	0.00	0.00	0.00	0.00
248	Blank		0.00	100.00	0.00	100.00	0.00	0.00	0.00	0.00	0.00	0.00	0.00
249	Blank		0.00	100.00	0.00	100.00	0.00	0.00	0.00	0.00	0.00	0.00	0.00
250	Blank		0.00	100.00	0.00	100.00	0.00	0.00	0.00	0.00	0.00	0.00	0.00
301	Barley Silage		0.00	25.00	75.00	39.00	56.80	5.44	65.00	60.00	2.17	1.31	0.74
302	Barley Straw	1-00-498	0.00	0.00	100.00	91.00	72.50	13.75	100.00	40.00	1.45	0.60	0.08
303	Corn Cobs Ground	1-28-234	0.00	0.00	100.00	90.00	87.00	7.78	56.00	50.00	1.81	0.97	0.42
304	Corn Silage 25% Grain	3-28-250-N	0.00	25.00	75.00	29.00	52.00	9.62	68.00	68.00	2.46	1.57	0.97
305	Corn Silage 25% Grain	3-28-250-S	0.00	25.00	75.00	29.00	55.00	10.91	81.00	61.00	2.21	1.34	0.77
306	Corn Silage 35% Grain	3-28-250	0.00	35.00	65.00	33.00	46.00	8.70	81.00	69.00	2.49	1.60	1.00
307	Corn Silage 40% Grain	3-28-250	0.00	40.00	60.00	33.00	45.00	8.89	81.00	66.00	2.39	1.51	0.91
308	Corn Silage 40% GR +NPN	3-28-250	0.00	40.00	60.00	33.00	45.00	8.89	81.00	67.00	2.42	1.54	0.94
309	Corn Silage 40% GR +NPN+Ca	3-28-250	0.00	40.00	60.00	33.00	45.00	8.89	81.00	68.00	2.46	1.57	0.97
310	Corn Silage 45% Grain	3-28-250	0.00	45.00	55.00	34.00	43.00	7.32	81.00	72.00	2.60	1.70	1.08
311	Corn Silage 45% GR +NPN		0.00	45.00	60.00	33.00	43.00	7.80	81.00	78.66	2.84	1.90	1.26
312	Corn Silage 45% GR +NPN+Ca		0.00	45.00	60.00	33.00	43.00	7.80	81.00	75.00	2.71	1.79	1.16
313	Corn Silage 50% Grain	3-28-250	0.00	50.00	50.00	35.00	41.00	7.00	71.00	75.00	2.71	1.79	1.16
314	CS50% + NPN+CA		0.00	50.00	50.00	35.00	41.00	7.00	71.00	82.32	2.98	2.01	1.36
315	Corn Silage Immature (no Ears)	3-28-252	0.00	0.00	100.00	25.00	60.00	5.00	81.00	65.00	2.35	1.47	0.88
316	Corn Silage Stalklage	3-28-251	0.00	0.00	100.00	30.00	68.00	10.29	81.00	55.00	1.99	1.14	0.58
317	Corn Stalks Grazing		0.00	0.00	100.00	50.00	65.00	10.00	100.00	65.85	2.38	1.50	0.91
318	Oat Silage Dough	3-03-296	0.00	0.00	100.00	36.40	58.10	16.07	61.00	59.00	2.13	1.28	0.71
319	Oat Straw	1-03-283	0.00	0.00	100.00	92.20	74.40	20.00	98.00	45.00	1.63	0.79	0.25
320	Oat Hay	1-03-280	0.00	0.00	100.00	91.00	63.00	9.09	98.00	53.00	1.92	1.07	0.52
321	Sorghum, Silage	3-04-323	0.00	0.00	100.00	30.00	60.80	9.38	81.00	60.00	2.17	1.31	0.74
322	Wheat Silage dough	3-05-184	0.00	0.00	100.00	35.00	60.70	14.81	61.00	57.00	2.06	1.21	0.64
323	Wheat Straw	1-05-175	0.00	0.00	100.00	89.00	78.90	16.47	98.00	41.00	1.48	0.64	0.11
324	Blank		0.00	100.00	0.00	100.00	0.00	0.00	0.00	0.00	0.00	0.00	0.00
325	Blank		0.00	100.00	0.00	100.00	0.00	0.00	0.00	0.00	0.00	0.00	0.00
326	Blank		0.00	100.00	0.00	100.00	0.00	0.00	0.00	0.00	0.00	0.00	0.00
327	Blank		0.00	100.00	0.00	100.00	0.00	0.00	0.00	0.00	0.00	0.00	0.00
328	Blank		0.00	100.00	0.00	100.00	0.00	0.00	0.00	0.00	0.00	0.00	0.00
329	Blank		0.00	100.00	0.00	100.00	0.00	0.00	0.00	0.00	0.00	0.00	0.00
330	Blank		0.00	100.00	0.00	100.00	0.00	0.00	0.00	0.00	0.00	0.00	0.00

CP %DM	DIP %CP	UIP %CP	SolP %CP	NPN %SolP	NDFIP %CP	ADFIP %CP	Starch %NSC	Fat %DM	Ash %DM	A kd[a] %/hr	B1 kd[a] %/hr	B2 kd[a] %/hr	B1 kd[b] %/hr	B2 kd[b] %/hr	B3 kd[b] %/hr
19.50	92.00	8.00	50.00	100.00	27.00	15.00	10.00	3.70	9.50	250.00	35.00	5.50	150.00	11.00	1.75
17.00	91.00	9.00	45.00	100.00	32.00	18.00	10.00	3.10	9.00	250.00	35.00	5.50	150.00	11.00	1.75
16.00	91.00	9.00	40.00	100.00	37.00	21.00	10.00	2.70	8.00	250.00	35.00	5.50	150.00	11.00	1.75
15.90	82.00	18.00	28.00	96.00	25.20	14.00	10.00	2.10	7.40	250.00	30.00	4.50	150.00	9.00	1.25
22.40	86.00	14.00	30.00	96.00	15.00	10.00	10.00	2.70	9.40	250.00	30.00	4.50	150.00	9.00	1.25
15.00	80.00	20.00	25.00	92.00	35.60	20.00	10.00	2.80	7.50	250.00	30.00	4.50	150.00	9.00	1.25
20.80	86.00	14.00	28.00	96.00	25.20	14.00	10.00	3.00	7.00	250.00	30.00	4.50	150.00	9.00	1.25
0.00	0.00	100.00	0.00	0.00	0.00	0.00	0.00	0.00	100.00	0.00	0.00	0.00	0.00	0.00	0.00
0.00	0.00	100.00	0.00	0.00	0.00	0.00	0.00	0.00	100.00	0.00	0.00	0.00	0.00	0.00	0.00
0.00	0.00	100.00	0.00	0.00	0.00	0.00	0.00	0.00	100.00	0.00	0.00	0.00	0.00	0.00	0.00
0.00	0.00	100.00	0.00	0.00	0.00	0.00	0.00	0.00	100.00	0.00	0.00	0.00	0.00	0.00	0.00
0.00	0.00	100.00	0.00	0.00	0.00	0.00	0.00	0.00	100.00	0.00	0.00	0.00	0.00	0.00	0.00
0.00	0.00	100.00	0.00	0.00	0.00	0.00	0.00	0.00	100.00	0.00	0.00	0.00	0.00	0.00	0.00
28.00	95.00	5.00	46.00	2.17	10.00	2.15	8.00	2.70	10.00	350.00	45.00	10.00	200.00	20.00	2.00
22.20	94.00	6.00	46.00	2.17	12.00	3.00	8.00	2.90	10.20	350.00	45.00	10.00	200.00	18.00	2.00
0.00	0.00	100.00	0.00	0.00	0.00	0.00	0.00	0.00	100.00	0.00	0.00	0.00	0.00	0.00	0.00
0.00	0.00	100.00	0.00	0.00	0.00	0.00	0.00	0.00	100.00	0.00	0.00	0.00	0.00	0.00	0.00
0.00	0.00	100.00	0.00	0.00	0.00	0.00	0.00	0.00	100.00	0.00	0.00	0.00	0.00	0.00	0.00
0.00	0.00	100.00	0.00	0.00	0.00	0.00	0.00	0.00	100.00	0.00	0.00	0.00	0.00	0.00	0.00
0.00	0.00	100.00	0.00	0.00	0.00	0.00	0.00	0.00	100.00	0.00	0.00	0.00	0.00	0.00	0.00
0.00	0.00	100.00	0.00	0.00	0.00	0.00	0.00	0.00	100.00	0.00	0.00	0.00	0.00	0.00	0.00
0.00	0.00	100.00	0.00	0.00	0.00	0.00	0.00	0.00	100.00	0.00	0.00	0.00	0.00	0.00	0.00
0.00	0.00	100.00	0.00	0.00	0.00	0.00	0.00	0.00	100.00	0.00	0.00	0.00	0.00	0.00	0.00
0.00	0.00	100.00	0.00	0.00	0.00	0.00	0.00	0.00	100.00	0.00	0.00	0.00	0.00	0.00	0.00
0.00	0.00	100.00	0.00	0.00	0.00	0.00	0.00	0.00	100.00	0.00	0.00	0.00	0.00	0.00	0.00
0.00	0.00	100.00	0.00	0.00	0.00	0.00	0.00	0.00	100.00	0.00	0.00	0.00	0.00	0.00	0.00
0.00	0.00	100.00	0.00	0.00	0.00	0.00	0.00	0.00	100.00	0.00	0.00	0.00	0.00	0.00	0.00
0.00	0.00	100.00	0.00	0.00	0.00	0.00	0.00	0.00	100.00	0.00	0.00	0.00	0.00	0.00	0.00
0.00	0.00	100.00	0.00	0.00	0.00	0.00	0.00	0.00	100.00	0.00	0.00	0.00	0.00	0.00	0.00
11.90	86.00	14.00	70.00	100.00	7.70	6.10	100.00	2.92	8.30	300.00	50.00	10.00	300.00	10.00	0.50
4.40	30.00	70.00	20.00	95.00	75.00	65.00	100.00	1.90	7.50	250.00	30.00	3.00	135.00	11.00	0.09
2.80	22.00	78.00	25.00	10.00	15.00	10.00	90.00	0.60	1.80	350.00	35.00	4.00	150.00	12.00	0.10
8.30	76.00	24.00	55.00	100.00	16.00	9.00	100.00	2.10	8.00	275.00	25.00	6.00	300.00	10.00	0.20
8.30	76.00	24.00	55.00	100.00	16.00	9.00	100.00	2.10	7.00	275.00	25.00	4.00	300.00	10.00	0.20
8.60	77.00	23.00	50.00	100.00	16.00	9.00	100.00	2.60	7.00	275.00	30.00	4.00	300.00	10.00	0.20
9.20	78.00	22.00	50.00	100.00	16.00	8.00	100.00	3.10	4.00	275.00	30.00	4.00	300.00	10.00	0.20
13.20	85.00	15.00	66.00	100.00	16.00	8.00	100.00	3.10	4.00	275.00	30.00	4.00	300.00	10.00	0.20
13.00	85.00	15.00	66.00	100.00	16.00	8.00	100.00	3.10	6.00	275.00	30.00	4.00	300.00	10.00	0.20
8.65	78.00	22.00	45.00	100.00	16.00	8.00	100.00	3.09	3.59	275.00	30.00	4.00	300.00	10.00	0.20
13.00	85.00	15.00	66.00	100.00	16.00	4.85	100.00	3.05	6.10	275.00	30.00	4.00	300.00	10.00	0.20
13.00	85.00	15.00	66.00	100.00	16.00	4.85	100.00	3.05	4.50	275.00	30.00	4.00	300.00	10.00	0.20
8.00	75.00	25.00	45.00	100.00	16.40	7.88	100.00	3.50	4.20	275.00	30.00	4.00	300.00	10.00	0.20
13.00	85.00	15.00	63.00	100.00	16.00	4.85	100.00	3.50	5.80	275.00	30.00	4.00	300.00	10.00	0.20
9.00	78.00	22.00	45.00	100.00	16.00	4.50	100.00	3.10	11.00	275.00	30.00	4.00	300.00	10.00	0.20
6.30	68.00	32.00	45.00	100.00	16.00	4.50	100.00	2.10	9.00	275.00	30.00	4.00	300.00	10.00	0.20
6.50	69.00	31.00	20.00	95.00	31.43	13.57	10.00	2.10	7.20	250.00	30.00	2.00	135.00	4.00	0.09
12.70	85.00	15.00	50.00	100.00	30.00	10.00	100.00	3.12	10.10	275.00	50.00	5.00	300.00	12.00	0.20
4.40	55.00	45.00	20.00	95.00	75.00	65.00	5.00	2.20	7.80	250.00	30.00	3.00	135.00	11.00	0.09
9.50	68.00	32.00	30.00	93.00	30.00	10.00	90.00	2.40	7.90	250.00	30.00	4.00	135.00	11.00	0.09
9.39	73.00	27.00	45.00	100.00	50.00	5.00	100.00	2.64	5.90	275.00	20.00	5.00	300.00	8.00	0.20
12.50	81.00	19.00	45.00	100.00	27.00	8.00	100.00	2.50	7.50	275.00	50.00	10.00	300.00	10.00	0.20
3.50	31.00	69.00	20.00	95.00	75.00	65.00	100.00	2.00	7.70	250.00	50.00	8.00	135.00	11.00	0.09
0.00	0.00	100.00	0.00	0.00	0.00	0.00	0.00	0.00	100.00	0.00	0.00	0.00	0.00	0.00	0.00
0.00	0.00	100.00	0.00	0.00	0.00	0.00	0.00	0.00	100.00	0.00	0.00	0.00	0.00	0.00	0.00
0.00	0.00	100.00	0.00	0.00	0.00	0.00	0.00	0.00	100.00	0.00	0.00	0.00	0.00	0.00	0.00
0.00	0.00	100.00	0.00	0.00	0.00	0.00	0.00	0.00	100.00	0.00	0.00	0.00	0.00	0.00	0.00
0.00	0.00	100.00	0.00	0.00	0.00	0.00	0.00	0.00	100.00	0.00	0.00	0.00	0.00	0.00	0.00
0.00	0.00	100.00	0.00	0.00	0.00	0.00	0.00	0.00	100.00	0.00	0.00	0.00	0.00	0.00	0.00
0.00	0.00	100.00	0.00	0.00	0.00	0.00	0.00	0.00	100.00	0.00	0.00	0.00	0.00	0.00	0.00

APPENDIX TABLE 1A   Feed Library (*Continued*)

Feed No.	Common Name	Int. Ref. No.	Cost $/ton AF	Conc. %DM	Forage %DM	DM %AF	NDF %DM	Lignin %NDF	eNDF %NDF	TDN %DM	ME Mcal/kg	NE$_{ma}$ Mcal/kg	NE$_{ga}$ Mcal/kg
331	Blank		0.00	100.00	0.00	100.00	0.00	0.00	0.00	0.00	0.00	0.00	0.00
332	Blank		0.00	100.00	0.00	100.00	0.00	0.00	0.00	0.00	0.00	0.00	0.00
333	Blank		0.00	100.00	0.00	100.00	0.00	0.00	0.00	0.00	0.00	0.00	0.00
334	Blank		0.00	100.00	0.00	100.00	0.00	0.00	0.00	0.00	0.00	0.00	0.00
335	Blank		0.00	100.00	0.00	100.00	0.00	0.00	0.00	0.00	0.00	0.00	0.00
336	Blank		0.00	100.00	0.00	100.00	0.00	0.00	0.00	0.00	0.00	0.00	0.00
337	Blank		0.00	100.00	0.00	100.00	0.00	0.00	0.00	0.00	0.00	0.00	0.00
338	Blank		0.00	100.00	0.00	100.00	0.00	0.00	0.00	0.00	0.00	0.00	0.00
339	Blank		0.00	100.00	0.00	100.00	0.00	0.00	0.00	0.00	0.00	0.00	0.00
340	Blank		0.00	100.00	0.00	100.00	0.00	0.00	0.00	0.00	0.00	0.00	0.00
341	Blank		0.00	100.00	0.00	100.00	0.00	0.00	0.00	0.00	0.00	0.00	0.00
342	Blank		0.00	100.00	0.00	100.00	0.00	0.00	0.00	0.00	0.00	0.00	0.00
343	Blank		0.00	100.00	0.00	100.00	0.00	0.00	0.00	0.00	0.00	0.00	0.00
344	Blank		0.00	100.00	0.00	100.00	0.00	0.00	0.00	0.00	0.00	0.00	0.00
345	Blank		0.00	100.00	0.00	100.00	0.00	0.00	0.00	0.00	0.00	0.00	0.00
346	Blank		0.00	100.00	0.00	100.00	0.00	0.00	0.00	0.00	0.00	0.00	0.00
347	Blank		0.00	100.00	0.00	100.00	0.00	0.00	0.00	0.00	0.00	0.00	0.00
348	Blank		0.00	100.00	0.00	100.00	0.00	0.00	0.00	0.00	0.00	0.00	0.00
349	Blank		0.00	100.00	0.00	100.00	0.00	0.00	0.00	0.00	0.00	0.00	0.00
350	Blank		0.00	100.00	0.00	100.00	0.00	0.00	0.00	0.00	0.00	0.00	0.00
401	Barley Malt Sprouts w/hulls	4-00-545	0.00	100.00	0.00	93.00	46.00	6.52	34.00	71.00	2.57	1.66	1.05
402	Barley Grain Heavy	4-00-549	0.00	100.00	0.00	88.00	18.10	10.53	34.00	84.00	3.04	2.06	1.40
403	Barley Grain Light		0.00	100.00	0.00	88.00	28.00	10.36	34.00	77.00	2.78	1.85	1.22
404	Corn Hominy	4-02-887	0.00	100.00	0.00	90.00	23.00	3.64	9.00	91.00	3.29	2.27	1.57
405	Corn Grain Cracked	4-20-698	0.00	100.00	0.00	88.00	10.80	2.22	60.00	90.00	3.25	2.24	1.55
406	Corn Dry Ear 45 lb/bu		0.00	100.00	0.00	86.00	31.00	7.14	56.00	77.00	2.78	1.85	1.22
407	Corn Dry Ear 56 lb/bu	04-28-238	0.00	100.00	0.00	87.00	28.00	7.10	56.00	82.00	2.96	2.00	1.35
408	Corn Dry Grain 45 lb/bu		0.00	100.00	0.00	88.00	10.00	2.22	60.00	88.00	3.18	2.18	1.50
409	Corn Ground Grain 56 lb/bu	04-02-931	0.00	100.00	0.00	88.00	9.00	2.22	0.00	88.00	3.18	2.18	1.50
410	Corn Dry Grain 56 lb/bu	04-02-931	0.00	100.00	0.00	88.00	9.00	2.22	60.00	88.00	3.18	2.18	1.50
411	Corn Grain Flaked	4-20-224	0.00	100.00	0.00	86.00	9.00	2.22	48.00	93.00	3.36	2.33	1.62
412	Corn HM Ear 56 lb/bu		0.00	100.00	0.00	72.00	28.00	7.10	56.00	85.00	3.07	2.09	1.42
413	Corn HM Grain 45 lb/bu		0.00	100.00	0.00	72.00	10.50	2.22	0.00	90.00	3.25	2.24	1.55
414	Corn HM Grain 56 lb/bu	04-20-771	0.00	100.00	0.00	72.00	9.00	2.22	60.00	93.00	3.36	2.33	1.62
415	Cottonseed Black Whole	5-01-614	0.00	100.00	0.00	92.00	40.00	37.50	100.00	95.00	3.43	2.38	1.67
416	Cottonseed High Lint	5-01-614	0.00	100.00	0.00	92.00	51.60	34.04	100.00	90.00	3.25	2.24	1.55
417	Cottonseed Meal—mech	5-01-617	0.00	100.00	0.00	92.00	28.00	21.43	36.00	78.00	2.82	1.88	1.24
418	Cottonseed Meal—Sol-41%CP	5-07-873	0.00	100.00	0.00	92.00	28.90	23.08	36.00	75.00	2.71	1.79	1.16
419	Cottonseed Meal—Sol-43%CP	5-01-630	0.00	100.00	0.00	92.00	28.00	25.00	36.00	75.00	2.71	1.79	1.16
420	Molasses Beet	4-00-668	0.00	100.00	0.00	77.90	0.00	0.00	0.00	75.00	2.71	1.79	1.16
421	Molasses Cane	4-04-696	0.00	100.00	0.00	74.30	0.00	0.00	0.00	72.00	2.60	1.70	1.08
422	Oats 32 lb/bu	4-03-318	0.00	100.00	0.00	91.00	42.00	9.52	34.00	73.00	2.64	1.73	1.11
423	Oats 38 lb/bu	4-03-309	0.00	100.00	0.00	89.00	29.30	9.38	34.00	77.00	2.78	1.85	1.22
424	Rice Bran	4-03-928	0.00	100.00	0.00	90.50	33.00	13.00	0.00	70.00	2.53	1.63	1.03
425	Rice Grain Ground	4-03-938	0.00	100.00	0.00	89.00	16.00	13.00	0.00	79.00	2.86	1.91	1.27
426	Rice Grain Polished	4-03-932	0.00	100.00	0.00	89.00	1.84	0.00	0.00	89.00	3.22	2.21	1.52
427	Rye Grain	4-04-047	0.00	100.00	0.00	88.00	19.00	5.30	34.00	84.00	3.04	2.06	1.40
428	Sorghum, Dry grain	4-04-383	0.00	100.00	0.00	89.00	13.30	6.09	34.00	76.00	2.75	1.82	1.19
429	Sorghum, Rolled grain	4-04-383	0.00	100.00	0.00	90.00	16.10	6.09	34.00	82.00	2.96	2.00	1.35
430	Sorghum, Steam flaked		0.00	100.00	0.00	70.00	23.00	6.09	34.00	88.00	3.18	2.18	1.50
431	Tapioca		0.00	100.00	0.00	89.00	8.00	0.00	0.00	84.00	3.04	2.06	1.40
432	Wheat Ground	4-05-211	0.00	100.00	0.00	89.00	11.80	6.25	0.00	88.00	3.18	2.18	1.50
433	Wheat Middlings	4-05-205	0.00	100.00	0.00	89.00	35.00	5.95	2.00	83.00	3.00	2.03	1.37
434	Wheat Grain Hard red spring	4-05-268	0.00	100.00	0.00	88.00	11.70	6.25	0.00	84.00	3.04	2.06	1.40
435	Wheat Grain Soft white	4-05-337	0.00	100.00	0.00	90.00	9.70	4.29	0.00	85.00	3.07	2.09	1.42
436	Blank		0.00	100.00	0.00	100.00	0.00	0.00	0.00	0.00	0.00	0.00	0.00
437	Blank		0.00	100.00	0.00	100.00	0.00	0.00	0.00	0.00	0.00	0.00	0.00
438	Blank		0.00	100.00	0.00	100.00	0.00	0.00	0.00	0.00	0.00	0.00	0.00
439	Blank		0.00	100.00	0.00	100.00	0.00	0.00	0.00	0.00	0.00	0.00	0.00
440	Blank		0.00	100.00	0.00	100.00	0.00	0.00	0.00	0.00	0.00	0.00	0.00
441	Blank		0.00	100.00	0.00	100.00	0.00	0.00	0.00	0.00	0.00	0.00	0.00
442	Blank		0.00	100.00	0.00	100.00	0.00	0.00	0.00	0.00	0.00	0.00	0.00
443	Blank		0.00	100.00	0.00	100.00	0.00	0.00	0.00	0.00	0.00	0.00	0.00
444	Blank		0.00	100.00	0.00	100.00	0.00	0.00	0.00	0.00	0.00	0.00	0.00
445	Blank		0.00	100.00	0.00	100.00	0.00	0.00	0.00	0.00	0.00	0.00	0.00
446	Blank		0.00	100.00	0.00	100.00	0.00	0.00	0.00	0.00	0.00	0.00	0.00
447	Blank		0.00	100.00	0.00	100.00	0.00	0.00	0.00	0.00	0.00	0.00	0.00

CP %DM	DIP %CP	UIP %CP	SolP %CP	NPN %SolP	NDFIP %CP	ADFIP %CP	Starch %NSC	Fat %DM	Ash %DM	A kd[a] %/hr	B1 kd[a] %/hr	B2 kd[a] %/hr	B1 kd[b] %/hr	B2 kd[b] %/hr	B3 kd[b] %/hr
0.00	0.00	100.00	0.00	0.00	0.00	0.00	0.00	0.00	100.00	0.00	0.00	0.00	0.00	0.00	0.00
0.00	0.00	100.00	0.00	0.00	0.00	0.00	0.00	0.00	100.00	0.00	0.00	0.00	0.00	0.00	0.00
0.00	0.00	100.00	0.00	0.00	0.00	0.00	0.00	0.00	100.00	0.00	0.00	0.00	0.00	0.00	0.00
0.00	0.00	100.00	0.00	0.00	0.00	0.00	0.00	0.00	100.00	0.00	0.00	0.00	0.00	0.00	0.00
0.00	0.00	100.00	0.00	0.00	0.00	0.00	0.00	0.00	100.00	0.00	0.00	0.00	0.00	0.00	0.00
0.00	0.00	100.00	0.00	0.00	0.00	0.00	0.00	0.00	100.00	0.00	0.00	0.00	0.00	0.00	0.00
0.00	0.00	100.00	0.00	0.00	0.00	0.00	0.00	0.00	100.00	0.00	0.00	0.00	0.00	0.00	0.00
0.00	0.00	100.00	0.00	0.00	0.00	0.00	0.00	0.00	100.00	0.00	0.00	0.00	0.00	0.00	0.00
0.00	0.00	100.00	0.00	0.00	0.00	0.00	0.00	0.00	100.00	0.00	0.00	0.00	0.00	0.00	0.00
0.00	0.00	100.00	0.00	0.00	0.00	0.00	0.00	0.00	100.00	0.00	0.00	0.00	0.00	0.00	0.00
0.00	0.00	100.00	0.00	0.00	0.00	0.00	0.00	0.00	100.00	0.00	0.00	0.00	0.00	0.00	0.00
0.00	0.00	100.00	0.00	0.00	0.00	0.00	0.00	0.00	100.00	0.00	0.00	0.00	0.00	0.00	0.00
0.00	0.00	100.00	0.00	0.00	0.00	0.00	0.00	0.00	100.00	0.00	0.00	0.00	0.00	0.00	0.00
0.00	0.00	100.00	0.00	0.00	0.00	0.00	0.00	0.00	100.00	0.00	0.00	0.00	0.00	0.00	0.00
0.00	0.00	100.00	0.00	0.00	0.00	0.00	0.00	0.00	100.00	0.00	0.00	0.00	0.00	0.00	0.00
0.00	0.00	100.00	0.00	0.00	0.00	0.00	0.00	0.00	100.00	0.00	0.00	0.00	0.00	0.00	0.00
0.00	0.00	100.00	0.00	0.00	0.00	0.00	0.00	0.00	100.00	0.00	0.00	0.00	0.00	0.00	0.00
0.00	0.00	100.00	0.00	0.00	0.00	0.00	0.00	0.00	100.00	0.00	0.00	0.00	0.00	0.00	0.00
28.10	64.17	35.83	48.00	83.00	27.00	4.00	85.00	1.40	7.00	250.00	30.00	3.00	135.00	11.00	0.09
13.20	66.93	33.07	17.00	29.00	8.00	5.00	90.00	2.20	2.40	300.00	30.00	5.00	300.00	12.00	0.35
14.00	66.93	33.07	17.00	29.00	8.00	5.00	90.00	2.30	4.00	300.00	30.00	5.00	300.00	12.00	0.35
11.50	47.49	52.51	18.00	78.00	8.00	5.00	90.00	7.30	1.00	150.00	20.00	4.00	150.00	4.00	0.09
9.80	44.67	55.33	11.00	73.00	15.00	5.00	90.00	4.06	1.46	150.00	15.00	5.00	150.00	5.00	0.09
9.00	46.03	53.97	16.00	69.00	18.00	3.00	90.00	3.70	2.00	150.00	18.00	5.00	150.00	5.00	0.09
9.00	46.03	53.97	16.00	69.00	18.00	3.00	90.00	3.70	1.90	150.00	18.00	5.00	150.00	5.00	0.09
9.80	41.22	58.78	12.00	73.00	15.00	5.00	90.00	4.30	1.60	150.00	10.00	5.00	150.00	4.00	0.09
9.80	57.37	42.63	11.00	73.00	15.00	5.00	90.00	4.30	1.60	300.00	35.00	6.00	135.00	10.00	0.15
9.80	44.67	55.33	11.00	73.00	15.00	5.00	90.00	4.30	1.60	150.00	10.00	5.00	150.00	5.00	0.09
9.80	43.04	56.96	8.00	73.00	15.00	5.00	90.00	4.30	2.00	175.00	25.00	7.00	135.00	5.00	0.08
9.00	62.05	37.95	30.00	80.00	18.72	8.28	100.00	3.70	1.90	300.00	35.00	6.00	135.00	10.00	0.15
9.80	67.82	32.18	40.00	100.00	15.90	5.30	100.00	4.30	1.60	300.00	35.00	6.00	135.00	10.00	0.15
9.80	67.82	32.18	40.00	100.00	15.90	5.30	100.00	4.30	1.60	300.00	35.00	6.00	135.00	10.00	0.15
23.00	69.56	30.44	40.00	2.50	6.00	6.00	90.00	17.50	5.00	300.00	25.00	3.00	175.00	8.00	0.25
24.40	69.56	30.44	40.00	2.50	6.00	6.00	90.00	17.50	4.16	300.00	25.00	3.00	175.00	8.00	0.25
44.00	57.00	43.00	20.00	40.00	10.00	8.00	90.00	5.00	7.00	300.00	25.00	3.00	175.00	8.00	0.25
46.10	57.00	43.00	20.00	40.00	10.00	8.00	90.00	3.15	7.00	300.00	25.00	3.00	175.00	8.00	0.25
48.90	57.00	43.00	20.00	40.00	10.00	8.00	90.00	1.70	7.00	300.00	25.00	3.00	175.00	8.00	0.25
8.50	100.00	0.00	100.00	100.00	0.00	0.00	0.00	0.00	11.40	500.00	30.00	3.00	300.00	11.00	0.25
5.80	100.00	0.00	100.00	100.00	0.00	0.00	0.00	0.00	13.30	500.00	30.00	20.00	350.00	11.00	0.25
13.60	76.55	23.45	53.00	19.00	11.00	5.00	90.00	4.90	5.00	300.00	35.00	5.00	325.00	12.00	0.35
13.60	83.00	17.00	53.00	19.00	11.00	5.00	90.00	5.20	3.30	300.00	35.00	5.00	325.00	12.00	0.35
14.40	50.96	49.04	40.00	80.00	47.00	2.00	90.00	15.00	11.50	300.00	40.00	8.00	250.00	12.00	0.35
8.90	69.85	30.15	40.00	50.00	21.40	2.70	90.00	1.90	5.00	350.00	50.00	10.00	300.00	15.00	1.00
8.60	66.30	33.70	40.00	50.00	21.40	2.70	90.00	0.80	1.00	300.00	40.00	8.00	250.00	12.00	0.35
13.80	78.99	21.01	53.00	19.00	7.00	4.00	90.00	1.70	2.00	300.00	40.00	8.00	300.00	12.00	0.35
11.60	50.76	49.24	12.00	33.00	10.00	5.00	90.00	3.10	2.00	150.00	12.00	5.00	135.00	6.00	0.12
12.60	43.00	57.00	12.00	33.00	10.00	5.00	90.00	3.03	1.87	150.00	12.00	5.00	135.00	6.00	0.12
12.00	56.40	43.60	12.00	33.00	10.00	5.00	100.00	3.10	2.00	250.00	20.00	8.00	160.00	8.00	0.15
3.10	56.11	43.89	25.00	45.00	30.00	5.00	13.60	0.80	3.00	300.00	40.00	8.00	300.00	12.00	0.35
14.20	77.00	23.00	30.00	25.00	4.00	2.00	90.00	2.34	2.01	300.00	40.00	8.00	300.00	12.00	0.35
18.40	77.15	22.85	40.00	75.00	4.00	3.00	90.00	3.20	2.40	300.00	70.00	12.00	250.00	5.50	0.10
14.20	73.95	26.05	30.00	73.00	4.00	2.00	90.00	2.00	2.00	300.00	40.00	8.00	300.00	12.00	0.35
11.30	73.95	26.05	30.00	73.00	4.00	2.00	90.00	1.90	2.00	300.00	40.00	8.00	300.00	12.00	0.35
0.00	0.00	100.00	0.00	0.00	0.00	0.00	0.00	0.00	100.00	0.00	0.00	0.00	0.00	0.00	0.00
0.00	0.00	100.00	0.00	0.00	0.00	0.00	0.00	0.00	100.00	0.00	0.00	0.00	0.00	0.00	0.00
0.00	0.00	100.00	0.00	0.00	0.00	0.00	0.00	0.00	100.00	0.00	0.00	0.00	0.00	0.00	0.00
0.00	0.00	100.00	0.00	0.00	0.00	0.00	0.00	0.00	100.00	0.00	0.00	0.00	0.00	0.00	0.00
0.00	0.00	100.00	0.00	0.00	0.00	0.00	0.00	0.00	100.00	0.00	0.00	0.00	0.00	0.00	0.00
0.00	0.00	100.00	0.00	0.00	0.00	0.00	0.00	0.00	100.00	0.00	0.00	0.00	0.00	0.00	0.00
0.00	0.00	100.00	0.00	0.00	0.00	0.00	0.00	0.00	100.00	0.00	0.00	0.00	0.00	0.00	0.00
0.00	0.00	100.00	0.00	0.00	0.00	0.00	0.00	0.00	100.00	0.00	0.00	0.00	0.00	0.00	0.00
0.00	0.00	100.00	0.00	0.00	0.00	0.00	0.00	0.00	100.00	0.00	0.00	0.00	0.00	0.00	0.00
0.00	0.00	100.00	0.00	0.00	0.00	0.00	0.00	0.00	100.00	0.00	0.00	0.00	0.00	0.00	0.00
0.00	0.00	100.00	0.00	0.00	0.00	0.00	0.00	0.00	100.00	0.00	0.00	0.00	0.00	0.00	0.00

APPENDIX TABLE 1A   Feed Library (*Continued*)

Feed No.	Common Name	Int. Ref. No.	Cost $/ton AF	Conc. %DM	Forage %DM	DM %AF	NDF %DM	Lignin %NDF	eNDF %NDF	TDN %DM	ME Mcal/kg	NE$_{ma}$ Mcal/kg	NE$_{ga}$ Mcal/kg
448	Blank		0.00	100.00	0.00	100.00	0.00	0.00	0.00	0.00	0.00	0.00	0.00
449	Blank		0.00	100.00	0.00	100.00	0.00	0.00	0.00	0.00	0.00	0.00	0.00
450	Blank		0.00	100.00	0.00	100.00	0.00	0.00	0.00	0.00	0.00	0.00	0.00
501	Brewers Grain 21% Dry Matter	5-02-142	0.00	100.00	0.00	21.00	42.00	9.52	18.00	70.00	2.53	1.63	1.03
502	Brewers Grain Dehydrated	5-02-141	0.00	100.00	0.00	92.00	48.70	13.04	18.00	66.00	2.39	1.51	0.91
503	Canola Meal	5-03-871	0.00	100.00	0.00	92.00	27.20	12.76	23.00	69.00	2.49	1.60	1.00
504	Coconut Meal		0.00	100.00	0.00	92.00	56.00	17.86	23.00	64.00	2.31	1.44	0.86
505	Corn Gluten Feed	5-28-243	0.00	100.00	0.00	90.00	36.20	2.22	36.00	80.00	2.89	1.94	1.30
506	Corn Gluten Meal	5-02-900	0.00	100.00	0.00	91.00	37.00	2.70	36.00	84.00	3.04	2.06	1.40
507	Corn Gluten Meal 60%CP	5-28-242	0.00	100.00	0.00	91.00	8.90	7.14	36.00	89.00	3.22	2.21	1.52
508	Distillers Gr. + solubles	5-02-843	0.00	100.00	0.00	25.00	46.00	9.09	4.00	88.00	3.18	2.18	1.50
509	Distillers G. Dehy—Light	5-28-236	0.00	100.00	0.00	91.00	46.00	10.00	4.00	88.00	3.18	2.18	1.50
510	Distillers Gr. Dehy—Inter.	5-28-236	0.00	100.00	0.00	91.00	46.00	10.00	4.00	88.00	3.18	2.18	1.50
511	Distillers Gr. Dehy—Dark	5-28-236	0.00	100.00	0.00	91.00	46.00	10.00	4.00	88.00	3.18	2.18	1.50
512	Distillers Gr. Dehy—Very Dark	5-28-236	0.00	100.00	0.00	91.00	46.00	10.00	4.00	88.00	3.18	2.18	1.50
513	Distillers Gr. solubles dehy	5-28-844	0.00	100.00	0.00	91.00	23.00	4.35	4.00	88.00	3.18	2.18	1.50
514	Distillers Gr. Wet		0.00	100.00	0.00	25.00	40.00	10.00	4.00	90.00	3.25	2.24	1.55
515	Lupins		0.00	100.00	0.00	90.00	33.00	10.00	0.00	78.00	2.82	1.88	1.24
516	Peanut Meal	5-03-650	0.00	100.00	0.00	92.40	14.00	10.00	36.00	77.00	2.78	1.85	1.22
517	Soybean Meal—44	5-20-637	0.00	100.00	0.00	89.00	14.90	2.14	23.00	84.00	3.04	2.06	1.40
518	Soybean Meal—49	5-04-612	0.00	100.00	0.00	90.00	7.79	2.50	23.00	87.00	3.15	2.15	1.47
519	Soybean Whole	5-04-610	0.00	100.00	0.00	90.00	14.90	1.54	100.00	94.00	3.40	2.35	1.65
520	Soybean Whole Roasted		0.00	100.00	0.00	90.00	13.40	10.00	100.00	94.00	3.40	2.35	1.65
521	Sunflower Seed meal	5-04-739	0.00	100.00	0.00	90.00	40.00	30.00	23.00	65.00	2.35	1.47	0.88
522	Urea		0.00	100.00	0.00	99.00	0.00	0.00	0.00	0.00	0.00	0.00	0.00
523	Blank		0.00	100.00	0.00	100.00	0.00	0.00	0.00	0.00	0.00	0.00	0.00
524	Blank		0.00	100.00	0.00	100.00	0.00	0.00	0.00	0.00	0.00	0.00	0.00
525	Blank		0.00	100.00	0.00	100.00	0.00	0.00	0.00	0.00	0.00	0.00	0.00
526	Blank		0.00	100.00	0.00	100.00	0.00	0.00	0.00	0.00	0.00	0.00	0.00
527	Blank		0.00	100.00	0.00	100.00	0.00	0.00	0.00	0.00	0.00	0.00	0.00
528	Blank		0.00	100.00	0.00	100.00	0.00	0.00	0.00	0.00	0.00	0.00	0.00
529	Blank		0.00	100.00	0.00	100.00	0.00	0.00	0.00	0.00	0.00	0.00	0.00
530	Blank		0.00	100.00	0.00	100.00	0.00	0.00	0.00	0.00	0.00	0.00	0.00
531	Blank		0.00	100.00	0.00	100.00	0.00	0.00	0.00	0.00	0.00	0.00	0.00
532	Blank		0.00	100.00	0.00	100.00	0.00	0.00	0.00	0.00	0.00	0.00	0.00
533	Blank		0.00	100.00	0.00	100.00	0.00	0.00	0.00	0.00	0.00	0.00	0.00
534	Blank		0.00	100.00	0.00	100.00	0.00	0.00	0.00	0.00	0.00	0.00	0.00
535	Blank		0.00	100.00	0.00	100.00	0.00	0.00	0.00	0.00	0.00	0.00	0.00
536	Blank		0.00	100.00	0.00	100.00	0.00	0.00	0.00	0.00	0.00	0.00	0.00
537	Blank		0.00	100.00	0.00	100.00	0.00	0.00	0.00	0.00	0.00	0.00	0.00
538	Blank		0.00	100.00	0.00	100.00	0.00	0.00	0.00	0.00	0.00	0.00	0.00
539	Blank		0.00	100.00	0.00	100.00	0.00	0.00	0.00	0.00	0.00	0.00	0.00
540	Blank		0.00	100.00	0.00	100.00	0.00	0.00	0.00	0.00	0.00	0.00	0.00
541	Blank		0.00	100.00	0.00	100.00	0.00	0.00	0.00	0.00	0.00	0.00	0.00
542	Blank		0.00	100.00	0.00	100.00	0.00	0.00	0.00	0.00	0.00	0.00	0.00
543	Blank		0.00	100.00	0.00	100.00	0.00	0.00	0.00	0.00	0.00	0.00	0.00
544	Blank		0.00	100.00	0.00	100.00	0.00	0.00	0.00	0.00	0.00	0.00	0.00
545	Blank		0.00	100.00	0.00	100.00	0.00	0.00	0.00	0.00	0.00	0.00	0.00
546	Blank		0.00	100.00	0.00	100.00	0.00	0.00	0.00	0.00	0.00	0.00	0.00
547	Blank		0.00	100.00	0.00	100.00	0.00	0.00	0.00	0.00	0.00	0.00	0.00
548	Blank		0.00	100.00	0.00	100.00	0.00	0.00	0.00	0.00	0.00	0.00	0.00
549	Blank		0.00	100.00	0.00	100.00	0.00	0.00	0.00	0.00	0.00	0.00	0.00
550	Blank		0.00	100.00	0.00	100.00	0.00	0.00	0.00	0.00	0.00	0.00	0.00
601	Apple Pomace	4-00-424	0.00	0.00	100.00	22.00	41.00	2.00	34.00	68.90	2.49	1.60	1.00
602	Bakery Waste	4-00-466	0.00	100.00	0.00	92.00	18.00	5.56	0.00	89.00	3.22	2.21	1.52
603	Beet Pulp + Steffen's filt	4-00-675	0.00	100.00	0.00	91.00	42.00	4.76	33.00	66.00	2.39	1.51	0.91
604	Beet Pulp Dehydrated	4-00-669	0.00	100.00	0.00	91.00	44.60	3.70	33.00	74.00	2.68	1.76	1.14
605	Citrus Pulp Dehydrated	4-01-237	0.00	100.00	0.00	91.00	23.00	13.04	33.00	82.00	2.96	2.00	1.35
606	Grape Pomace	1-02-208	0.00	100.00	0.00	90.00	55.00	41.70	34.00	33.00	1.19	0.34	-0.18
607	Soybean Hulls	1-04-560	0.00	100.00	0.00	91.00	66.30	2.99	2.00	80.00	2.89	1.94	1.30
608	Blank		0.00	100.00	0.00	100.00	0.00	0.00	0.00	0.00	0.00	0.00	0.00
609	Blank		0.00	100.00	0.00	100.00	0.00	0.00	0.00	0.00	0.00	0.00	0.00
610	Blank		0.00	100.00	0.00	100.00	0.00	0.00	0.00	0.00	0.00	0.00	0.00
611	Blank		0.00	100.00	0.00	100.00	0.00	0.00	0.00	0.00	0.00	0.00	0.00
612	Blank		0.00	100.00	0.00	100.00	0.00	0.00	0.00	0.00	0.00	0.00	0.00
613	Blank		0.00	100.00	0.00	100.00	0.00	0.00	0.00	0.00	0.00	0.00	0.00

CP %DM	DIP %CP	UIP %CP	SolP %CP	NPN %SolP	NDFIP %CP	ADFIP %CP	Starch %NSC	Fat %DM	Ash %DM	A kd[a] %/hr	B1 kd[a] %/hr	B2 kd[a] %/hr	B1 kd[b] %/hr	B2 kd[b] %/hr	B3 kd[b] %/hr
0.00	0.00	100.00	0.00	0.00	0.00	0.00	0.00	0.00	100.00	0.00	0.00	0.00	0.00	0.00	0.00
0.00	0.00	100.00	0.00	0.00	0.00	0.00	0.00	0.00	100.00	0.00	0.00	0.00	0.00	0.00	0.00
0.00	0.00	100.00	0.00	0.00	0.00	0.00	0.00	0.00	100.00	0.00	0.00	0.00	0.00	0.00	0.00
26.00	40.86	59.14	8.00	50.00	38.00	10.00	100.00	6.50	10.00	300.00	38.00	6.00	150.00	8.00	0.50
29.20	34.12	65.88	4.00	75.00	40.00	12.00	100.00	10.80	4.00	300.00	38.00	6.00	150.00	6.00	0.50
40.90	67.85	32.15	32.40	65.00	10.64	6.38	90.00	3.47	7.10	300.00	40.00	6.00	230.00	12.00	0.20
21.50	61.64	38.36	14.00	75.00	10.00	3.00	90.00	7.40	7.00	300.00	40.00	6.00	230.00	12.00	
23.80	75.00	25.00	49.00	100.00	8.00	2.00	100.00	3.91	6.90	300.00	50.00	5.00	150.00	4.00	0.08
46.80	38.08	61.92	4.00	75.00	11.00	2.00	100.00	2.40	3.00	300.00	50.00	5.00	150.00	4.00	0.08
66.30	41.00	59.00	4.00	75.00	11.00	2.00	100.00	2.56	2.86	300.00	50.00	5.00	150.00	4.00	0.08
29.50	27.19	72.81	19.00	89.00	62.00	21.00	100.00	10.30	5.00	300.00	17.00	7.00	150.00	4.00	0.10
30.40	27.97	72.03	6.00	67.00	40.00	13.00	100.00	9.80	4.00	300.00	17.00	7.00	150.00	4.00	0.10
30.40	26.35	73.65	6.00	67.00	44.00	18.00	100.00	10.70	4.60	300.00	17.00	7.00	150.00	4.00	0.10
30.40	25.92	74.08	6.00	67.00	45.00	21.00	100.00	9.80	4.00	300.00	17.00	7.00	150.00	4.00	0.10
30.40	24.90	75.10	6.00	67.00	47.00	36.00	100.00	9.80	4.00	300.00	17.00	7.00	150.00	4.00	0.10
29.70	45.09	54.91	44.00	100.00	55.00	13.00	100.00	9.20	8.00	300.00	17.00	7.00	150.00	4.00	0.10
26.00	33.40	66.60	25.00	68.00	55.00	12.00	100.00	9.90	4.00	300.00	17.00	7.00	150.00	4.00	0.10
34.20	69.20	30.80	26.00	23.00	4.00	3.00	90.00	5.50	5.10	300.00	30.00	5.00	200.00	10.00	0.20
52.90	80.00	20.00	33.00	27.00	10.00	1.00	90.00	2.30	6.30	300.00	45.00	6.00	230.00	13.00	0.20
49.90	65.00	35.00	20.00	55.00	5.00	2.00	90.00	1.60	7.20	300.00	45.00	6.00	230.00	11.00	0.20
54.00	65.00	35.00	20.00	55.00	5.00	2.00	90.00	1.10	6.70	300.00	45.00	6.00	230.00	11.00	0.20
40.34	75.00	25.00	44.00	22.73	4.00	3.00	90.00	18.20	4.56	300.00	30.00	5.00	200.00	10.00	0.20
42.80	38.31	61.69	5.70	100.00	23.60	7.29	90.00	18.80	5.80	300.00	35.00	5.00	150.00	5.00	0.18
25.90	80.00	20.00	30.00					18.20	4.56	300.00	30.00	5.00	200.00	10.00	0.20
291.00	100.00	0.00	100.00	100.00	0.00	0.00	0.00	0.00	0.00	0.00	0.00	0.00	400.00	0.00	0.00
0.00	0.00	100.00	0.00	0.00	0.00	0.00	0.00	0.00	100.00	0.00	0.00	0.00	0.00	0.00	0.00
0.00	0.00	100.00	0.00	0.00	0.00	0.00	0.00	0.00	100.00	0.00	0.00	0.00	0.00	0.00	0.00
0.00	0.00	100.00	0.00	0.00	0.00	0.00	0.00	0.00	100.00	0.00	0.00	0.00	0.00	0.00	0.00
0.00	0.00	100.00	0.00	0.00	0.00	0.00	0.00	0.00	100.00	0.00	0.00	0.00	0.00	0.00	0.00
0.00	0.00	100.00	0.00	0.00	0.00	0.00	0.00	0.00	100.00	0.00	0.00	0.00	0.00	0.00	0.00
0.00	0.00	100.00	0.00	0.00	0.00	0.00	0.00	0.00	100.00	0.00	0.00	0.00	0.00	0.00	0.00
0.00	0.00	100.00	0.00	0.00	0.00	0.00	0.00	0.00	100.00	0.00	0.00	0.00	0.00	0.00	0.00
0.00	0.00	100.00	0.00	0.00	0.00	0.00	0.00	0.00	100.00	0.00	0.00	0.00	0.00	0.00	0.00
0.00	0.00	100.00	0.00	0.00	0.00	0.00	0.00	0.00	100.00	0.00	0.00	0.00	0.00	0.00	0.00
0.00	0.00	100.00	0.00	0.00	0.00	0.00	0.00	0.00	100.00	0.00	0.00	0.00	0.00	0.00	0.00
0.00	0.00	100.00	0.00	0.00	0.00	0.00	0.00	0.00	100.00	0.00	0.00	0.00	0.00	0.00	0.00
0.00	0.00	100.00	0.00	0.00	0.00	0.00	0.00	0.00	100.00	0.00	0.00	0.00	0.00	0.00	0.00
0.00	0.00	100.00	0.00	0.00	0.00	0.00	0.00	0.00	100.00	0.00	0.00	0.00	0.00	0.00	0.00
0.00	0.00	100.00	0.00	0.00	0.00	0.00	0.00	0.00	100.00	0.00	0.00	0.00	0.00	0.00	0.00
0.00	0.00	100.00	0.00	0.00	0.00	0.00	0.00	0.00	100.00	0.00	0.00	0.00	0.00	0.00	0.00
0.00	0.00	100.00	0.00	0.00	0.00	0.00	0.00	0.00	100.00	0.00	0.00	0.00	0.00	0.00	0.00
0.00	0.00	100.00	0.00	0.00	0.00	0.00	0.00	0.00	100.00	0.00	0.00	0.00	0.00	0.00	0.00
0.00	0.00	100.00	0.00	0.00	0.00	0.00	0.00	0.00	100.00	0.00	0.00	0.00	0.00	0.00	0.00
0.00	0.00	100.00	0.00	0.00	0.00	0.00	0.00	0.00	100.00	0.00	0.00	0.00	0.00	0.00	0.00
0.00	0.00	100.00	0.00	0.00	0.00	0.00	0.00	0.00	100.00	0.00	0.00	0.00	0.00	0.00	0.00
0.00	0.00	100.00	0.00	0.00	0.00	0.00	0.00	0.00	100.00	0.00	0.00	0.00	0.00	8.00	0.00
0.00	0.00	100.00	0.00	0.00	0.00	0.00	0.00	0.00	100.00	0.00	0.00	0.00	0.00	0.00	0.00
5.40	48.21	51.79	11.00	100.00	30.00	31.50	100.00	4.70	5.00	350.00	50.00	9.00	350.00	11.00	0.25
9.00	75.60	24.40	40.00	75.00	6.00	3.00	90.00	12.70	5.00	300.00	45.00	12.00	350.00	15.00	0.40
10.00	42.46	57.54	26.50	96.00	53.00	11.00	90.00	0.40	6.00	300.00	40.00	8.00	300.00	12.00	0.35
9.80	42.63	57.37	27.00	96.00	53.00	11.00	90.00	0.60	5.30	300.00	40.00	8.00	300.00	12.00	0.35
6.70	41.60	58.40	27.00	96.00	53.00	11.00	90.00	3.70	6.60	350.00	50.00	9.00	350.00	11.00	0.25
0.00	100.00	0.00	0.00	0.00	0.00	0.00	100.00	7.90	10.00	350.00	50.00	9.00	350.00	11.00	0.25
12.20	58.00	42.00	18.00	72.00	20.00	14.00	90.00	2.10	4.90	350.00	40.00	8.00	150.00	12.00	0.15
0.00	0.00	100.00	0.00	0.00	0.00	0.00	0.00	0.00	100.00	0.00	0.00	0.00	0.00	0.00	0.00
0.00	0.00	100.00	0.00	0.00	0.00	0.00	0.00	0.00	100.00	0.00	0.00	0.00	0.00	0.00	0.00
0.00	0.00	100.00	0.00	0.00	0.00	0.00	0.00	0.00	100.00	0.00	0.00	0.00	0.00	0.00	0.00
0.00	0.00	100.00	0.00	0.00	0.00	0.00	0.00	0.00	100.00	0.00	0.00	0.00	0.00	0.00	0.00
0.00	0.00	100.00	0.00	0.00	0.00	0.00	0.00	0.00	100.00	0.00	0.00	0.00	0.00	0.00	0.00
0.00	0.00	100.00	0.00	0.00	0.00	0.00	0.00	0.00	100.00	0.00	0.00	0.00	0.00	0.00	0.00

APPENDIX TABLE 1A   Feed Library (*Continued*)

Feed No.	Common Name	Int. Ref. No.	Cost $/ton AF	Conc. %DM	Forage %DM	DM %AF	NDF %DM	Lignin %NDF	eNDF %NDF	TDN %DM	ME Mcal/kg	NE$_{ma}$ Mcal/kg	NE$_{ga}$ Mcal/kg
614	Blank		0.00	100.00	0.00	100.00	0.00	0.00	0.00	0.00	0.00	0.00	0.00
615	Blank		0.00	100.00	0.00	100.00	0.00	0.00	0.00	0.00	0.00	0.00	0.00
616	Blank		0.00	100.00	0.00	100.00	0.00	0.00	0.00	0.00	0.00	0.00	0.00
617	Blank		0.00	100.00	0.00	100.00	0.00	0.00	0.00	0.00	0.00	0.00	0.00
618	Blank		0.00	100.00	0.00	100.00	0.00	0.00	0.00	0.00	0.00	0.00	0.00
619	Blank		0.00	100.00	0.00	100.00	0.00	0.00	0.00	0.00	0.00	0.00	0.00
620	Blank		0.00	100.00	0.00	100.00	0.00	0.00	0.00	0.00	0.00	0.00	0.00
701	Bloodmeal	5-00-380	0.00	100.00	0.00	90.00	0.92	0.00	0.00	66.00	2.39	1.51	0.91
702	Feather Meal	5-03-795	0.00	100.00	0.00	90.00	39.00	0.00	23.00	68.00	2.46	1.57	0.97
703	Fishmeal	5-02-009	0.00	100.00	0.00	90.00	2.00	0.00	10.00	73.00	2.64	1.73	1.11
704	Meat Meal	5-00-385	0.00	100.00	0.00	95.00	28.21	0.00	0.00	71.00	2.57	1.66	1.05
705	Tallow	4-00-376	0.00	100.00	0.00	99.00	0.00	0.00	0.00	177.00	6.40	4.75	3.51
706	Whey Acid	4-08-134	0.00	100.00	0.00	7.00	0.00	0.00	0.00	78.00	2.82	1.88	1.24
707	Whey Delact.	4-01-186	0.00	100.00	0.00	93.00	0.00	0.00	0.00	71.00	2.57	1.66	1.05
708	Blank		0.00	100.00	0.00	100.00	0.00	0.00	0.00	0.00	0.00	0.00	0.00
709	Blank		0.00	100.00	0.00	100.00	0.00	0.00	0.00	0.00	0.00	0.00	0.00
710	Blank		0.00	100.00	0.00	100.00	0.00	0.00	0.00	0.00	0.00	0.00	0.00
711	Blank		0.00	100.00	0.00	100.00	0.00	0.00	0.00	0.00	0.00	0.00	0.00
712	Blank		0.00	100.00	0.00	100.00	0.00	0.00	0.00	0.00	0.00	0.00	0.00
713	Blank		0.00	100.00	0.00	100.00	0.00	0.00	0.00	0.00	0.00	0.00	0.00
714	Blank		0.00	100.00	0.00	100.00	0.00	0.00	0.00	0.00	0.00	0.00	0.00
715	Blank		0.00	100.00	0.00	100.00	0.00	0.00	0.00	0.00	0.00	0.00	0.00
716	Blank		0.00	100.00	0.00	100.00	0.00	0.00	0.00	0.00	0.00	0.00	0.00
717	Blank		0.00	100.00	0.00	100.00	0.00	0.00	0.00	0.00	0.00	0.00	0.00
718	Blank		0.00	100.00	0.00	100.00	0.00	0.00	0.00	0.00	0.00	0.00	0.00
719	Blank		0.00	100.00	0.00	100.00	0.00	0.00	0.00	0.00	0.00	0.00	0.00
720	Blank		0.00	100.00	0.00	100.00	0.00	0.00	0.00	0.00	0.00	0.00	0.00
801	Ammonium Phos (Mono)	6-09-338	0.00	100.00	0.00	97.00	0.00	0.00	0.00	0.00	0.00	0.00	0.00
802	Ammonium Phos (Dibasic)	6-00-370	0.00	100.00	0.00	97.00	0.00	0.00	0.00	0.00	0.00	0.00	0.00
803	Ammonium Sulfate	6-09-339	0.00	100.00	0.00	100.00	0.00	0.00	0.00	0.00	0.00	0.00	0.00
804	Bone Meal	6-00-400	0.00	100.00	0.00	97.00	0.00	0.00	0.00	0.00	0.00	0.00	0.00
805	Calcium Carbonate	6-01-069	0.00	100.00	0.00	100.00	0.00	0.00	0.00	0.00	0.00	0.00	0.00
806	Calcium Sulfate	6-01-089	0.00	100.00	0.00	97.00	0.00	0.00	0.00	0.00	0.00	0.00	0.00
807	Cobalt Carbonate	6-01-566	0.00	100.00	0.00	99.00	0.00	0.00	0.00	0.00	0.00	0.00	0.00
808	Copper Sulfate	6-01-720	0.00	100.00	0.00	100.00	0.00	0.00	0.00	0.00	0.00	0.00	0.00
809	Dicalcium Phosphate	6-01-080	0.00	100.00	0.00	97.00	0.00	0.00	0.00	0.00	0.00	0.00	0.00
810	EDTA	6-01-842	0.00	100.00	0.00	98.00	0.00	0.00	0.00	0.00	0.00	0.00	0.00
811	Iron Sulfate	6-20-734	0.00	100.00	0.00	98.00	0.00	0.00	0.00	0.00	0.00	0.00	0.00
812	Limestone	6-02-632	0.00	100.00	0.00	100.00	0.00	0.00	0.00	0.00	0.00	0.00	0.00
813	Limestone Magnesium	6-02-633	0.00	100.00	0.00	99.00	0.00	0.00	0.00	0.00	0.00	0.00	0.00
814	Magnesium Carbonate	6-02-754	0.00	100.00	0.00	98.00	0.00	0.00	0.00	0.00	0.00	0.00	0.00
815	Magnesium Oxide	6-02-756	0.00	100.00	0.00	98.00	0.00	0.00	0.00	0.00	0.00	0.00	0.00
816	Manganese Oxide	6-03-056	0.00	100.00	0.00	99.00	0.00	0.00	0.00	0.00	0.00	0.00	0.00
817	Manganese Carbonate	6-03-036	0.00	100.00	0.00	97.00	0.00	0.00	0.00	0.00	0.00	0.00	0.00
818	Mono-Sodium Phosphate	6-04-288	0.00	100.00	0.00	97.00	0.00	0.00	0.00	0.00	0.00	0.00	0.00
819	Oystershell Ground	6-03-481	0.00	100.00	0.00	99.00	0.00	0.00	0.00	0.00	0.00	0.00	0.00
820	Phosphate Deflourinated	6-01-780	0.00	100.00	0.00	100.00	0.00	0.00	0.00	0.00	0.00	0.00	0.00
821	Phosphate Rock	6-03-945	0.00	100.00	0.00	100.00	0.00	0.00	0.00	0.00	0.00	0.00	0.00
822	Phosphate Rock—Low Fl	6-03-946	0.00	100.00	0.00	100.00	0.00	0.00	0.00	0.00	0.00	0.00	0.00
823	Phosphate Rock—Soft	6-03-947	0.00	100.00	0.00	100.00	0.00	0.00	0.00	0.00	0.00	0.00	0.00
824	Phosphate Mono-Mono	6-04-288	0.00	100.00	0.00	97.00	0.00	0.00	0.00	0.00	0.00	0.00	0.00
825	Phosphoric Acid	6-03-707	0.00	100.00	0.00	75.00	0.00	0.00	0.00	0.00	0.00	0.00	0.00
826	Potassium Bicarbonate	6-29-493	0.00	100.00	0.00	99.00	0.00	0.00	0.00	0.00	0.00	0.00	0.00
827	Potassium Iodide	6-03-759	0.00	100.00	0.00	100.00	0.00	0.00	0.00	0.00	0.00	0.00	0.00
828	Potassium Sulfate	6-06-098	0.00	100.00	0.00	98.00	0.00	0.00	0.00	0.00	0.00	0.00	0.00
829	Salt	6-04-152	0.00	100.00	0.00	100.00	0.00	0.00	0.00	0.00	0.00	0.00	0.00
830	Sodium Bicarbonate	6-04-272	0.00	100.00	0.00	100.00	0.00	0.00	0.00	0.00	0.00	0.00	0.00
831	Sodium Selenite	6-26-013	0.00	100.00	0.00	98.00	0.00	0.00	0.00	0.00	0.00	0.00	0.00
832	Sodium Sulfate	6-04-292	0.00	100.00	0.00	97.00	0.00	0.00	0.00	73.00	2.64	0.00	0.00
833	Zinc Oxide	6-05-553	0.00	100.00	0.00	100.00	0.00	0.00	0.00	0.00	0.00	0.00	0.00
834	Zinc Sulfate	6-05-555	0.00	100.00	0.00	99.00	0.00	0.00	0.00	0.00	0.00	0.00	0.00
835	Potassium Chloride	6-03-755	0.00	100.00	0.00	100.00	0.00	0.00	0.00	0.00	0.00	0.00	0.00
836	Calcium Phosphate (Mono)	6-01-082	0.00	100.00	0.00	97.00	0.00	0.00	0.00	0.00	0.00	0.00	0.00
837	Sodium TriPoly Phosphate	6-08-076	0.00	100.00	0.00	96.00	0.00	0.00	0.00	0.00	0.00	0.00	0.00
838	Blank		0.00	100.00	0.00	100.00	0.00	0.00	0.00	0.00	0.00	0.00	0.00
839	Blank		0.00	100.00	0.00	100.00	0.00	0.00	0.00	0.00	0.00	0.00	0.00
840	Blank		0.00	100.00	0.00	100.00	0.00	0.00	0.00	0.00	0.00	0.00	0.00

CP %DM	DIP %CP	UIP %CP	SolP %CP	NPN %SolP	NDFIP %CP	ADFIP %CP	Starch %NSC	Fat %DM	Ash %DM	A kd[a] %/hr	B1 kd[a] %/hr	B2 kd[a] %/hr	B1 kd[b] %/hr	B2 kd[b] %/hr	B3 kd[b] %/hr
0.00	0.00	100.00	0.00	0.00	0.00	0.00	0.00	0.00	100.00	0.00	0.00	0.00	0.00	0.00	0.00
0.00	0.00	100.00	0.00	0.00	0.00	0.00	0.00	0.00	100.00	0.00	0.00	0.00	0.00	0.00	0.00
0.00	0.00	100.00	0.00	0.00	0.00	0.00	0.00	0.00	100.00	0.00	0.00	0.00	0.00	0.00	0.00
0.00	0.00	100.00	0.00	0.00	0.00	0.00	0.00	0.00	100.00	0.00	0.00	0.00	0.00	0.00	0.00
0.00	0.00	100.00	0.00	0.00	0.00	0.00	0.00	0.00	100.00	0.00	0.00	0.00	0.00	0.00	0.00
0.00	0.00	100.00	0.00	0.00	0.00	0.00	0.00	0.00	100.00	0.00	0.00	0.00	0.00	0.00	0.00
0.00	0.00	100.00	0.00	0.00	0.00	0.00	0.00	0.00	100.00	0.00	0.00	0.00	0.00	0.00	0.00
93.79	25.00	75.00	5.00	0.00	1.00	1.00	0.00	1.70	2.62	0.00	0.00	0.00	75.00	3.00	0.09
85.80	30.00	70.00	9.00	89.00	50.00	32.00	90.00	7.21	3.50	0.00	0.00	0.00	125.00	3.00	0.09
67.90	40.00	60.00	21.00	0.00	1.00	1.00	90.00	10.70	20.60	0.00	0.00	0.00	150.00	1.00	0.80
58.20	45.00	55.00	13.36	26.50	56.41	3.16	0.00	11.00	21.30	0.00	0.00	0.00	150.00	5.00	0.12
0.00	100.00	0.00	0.00	0.00	0.00	0.00	0.00	99.20	0.00	0.00	0.00	0.00	0.00	0.00	0.00
14.20	100.00	0.00	100.00	100.00	0.00	0.00	0.00	0.70	10.00	350.00	0.00	0.00	350.00	0.00	0.00
17.90	100.00	0.00	100.00	100.00	0.00	0.00	0.00	1.10	17.00	350.00	0.00	0.00	350.00	0.00	0.00
0.00	0.00	100.00	0.00	0.00	0.00	0.00	0.00	0.00	100.00	0.00	0.00	0.00	0.00	0.00	0.00
0.00	0.00	100.00	0.00	0.00	0.00	0.00	0.00	0.00	100.00	0.00	0.00	0.00	0.00	0.00	0.00
0.00	0.00	100.00	0.00	0.00	0.00	0.00	0.00	0.00	100.00	0.00	0.00	0.00	0.00	0.00	0.00
0.00	0.00	100.00	0.00	0.00	0.00	0.00	0.00	0.00	100.00	0.00	0.00	0.00	0.00	0.00	0.00
0.00	0.00	100.00	0.00	0.00	0.00	0.00	0.00	0.00	100.00	0.00	0.00	0.00	0.00	0.00	0.00
0.00	0.00	100.00	0.00	0.00	0.00	0.00	0.00	0.00	100.00	0.00	0.00	0.00	0.00	0.00	0.00
0.00	0.00	100.00	0.00	0.00	0.00	0.00	0.00	0.00	100.00	0.00	0.00	0.00	0.00	0.00	0.00
0.00	0.00	100.00	0.00	0.00	0.00	0.00	0.00	0.00	100.00	0.00	0.00	0.00	0.00	0.00	0.00
0.00	0.00	100.00	0.00	0.00	0.00	0.00	0.00	0.00	100.00	0.00	0.00	0.00	0.00	0.00	0.00
0.00	0.00	100.00	0.00	0.00	0.00	0.00	0.00	0.00	100.00	0.00	0.00	0.00	0.00	0.00	0.00
70.90	100.00	0.00	100.00	100.00	0.00	0.00	0.00	0.00	100.00	0.00	0.00	0.00	0.00	0.00	0.00
115.90	100.00	0.00	100.00	100.00	0.00	0.00	0.00	0.00	100.00	0.00	0.00	0.00	0.00	0.00	0.00
134.10	100.00	0.00	100.00	100.00	0.00	0.00	0.00	0.00	100.00	0.00	0.00	0.00	0.00	0.00	0.00
13.20	40.00	60.00	40.00	0.00	0.00	0.00	0.00	0.00	100.00	0.00	0.00	0.00	500.00	0.00	0.00
0.00	0.00	100.00	0.00	0.00	0.00	0.00	0.00	0.00	100.00	0.00	0.00	0.00	0.00	0.00	0.00
0.00	0.00	100.00	0.00	0.00	0.00	0.00	0.00	0.00	100.00	0.00	0.00	0.00	0.00	0.00	0.00
0.00	0.00	100.00	0.00	0.00	0.00	0.00	0.00	0.00	100.00	0.00	0.00	0.00	0.00	0.00	0.00
0.00	0.00	100.00	0.00	0.00	0.00	0.00	0.00	0.00	100.00	0.00	0.00	0.00	0.00	0.00	0.00
0.00	0.00	100.00	0.00	0.00	0.00	0.00	0.00	0.00	100.00	0.00	0.00	0.00	0.00	0.00	0.00
0.00	0.00	100.00	0.00	0.00	0.00	0.00	0.00	0.00	100.00	0.00	0.00	0.00	0.00	0.00	0.00
0.00	0.00	100.00	0.00	0.00	0.00	0.00	0.00	0.00	100.00	0.00	0.00	0.00	0.00	0.00	0.00
0.00	0.00	100.00	0.00	0.00	0.00	0.00	0.00	0.00	100.00	0.00	0.00	0.00	0.00	0.00	0.00
0.00	0.00	100.00	0.00	0.00	0.00	0.00	0.00	0.00	100.00	0.00	0.00	0.00	0.00	0.00	0.00
0.00	0.00	100.00	0.00	0.00	0.00	0.00	0.00	0.00	100.00	0.00	0.00	0.00	0.00	0.00	0.00
0.00	0.00	100.00	0.00	0.00	0.00	0.00	0.00	0.00	100.00	0.00	0.00	0.00	0.00	0.00	0.00
0.00	0.00	100.00	0.00	0.00	0.00	0.00	0.00	0.00	100.00	0.00	0.00	0.00	0.00	0.00	0.00
0.00	0.00	100.00	0.00	0.00	0.00	0.00	0.00	0.00	100.00	0.00	0.00	0.00	0.00	0.00	0.00
0.00	0.00	100.00	0.00	0.00	0.00	0.00	0.00	0.00	100.00	0.00	0.00	0.00	0.00	0.00	0.00
0.00	0.00	100.00	0.00	0.00	0.00	0.00	0.00	0.00	100.00	0.00	0.00	0.00	0.00	0.00	0.00
0.00	0.00	100.00	0.00	0.00	0.00	0.00	0.00	0.00	100.00	0.00	0.00	0.00	0.00	0.00	0.00
0.00	0.00	100.00	0.00	0.00	0.00	0.00	0.00	0.00	100.00	0.00	0.00	0.00	0.00	0.00	0.00
0.00	0.00	100.00	9.00	0.00	0.00	0.00	0.00	0.00	100.00	0.00	0.00	0.00	0.00	0.00	0.00
0.00	0.00	100.00	0.00	0.00	0.00	0.00	0.00	0.00	100.00	0.00	0.00	0.00	0.00	0.00	0.00
0.00	0.00	100.00	0.00	0.00	0.00	0.00	0.00	0.00	100.00	0.00	0.00	0.00	0.00	0.00	0.00
0.00	0.00	100.00	0.00	0.00	0.00	0.00	0.00	0.00	100.00	0.00	0.00	0.00	0.00	0.00	0.00
0.00	0.00	100.00	0.00	0.00	0.00	0.00	0.00	0.00	100.00	0.00	0.00	0.00	0.00	0.00	0.00
0.00	0.00	100.00	0.00	0.00	0.00	0.00	0.00	0.00	100.00	0.00	0.00	0.00	0.00	0.00	0.00
0.00	0.00	100.00	0.00	0.00	0.00	0.00	0.00	0.00	100.00	0.00	0.00	0.00	0.00	0.00	0.00
0.00	0.00	100.00	0.00	0.00	0.00	0.00	0.00	0.00	100.00	0.00	0.00	0.00	0.00	0.00	0.00
0.00	0.00	100.00	0.00	0.00	0.00	0.00	0.00	0.00	100.00	0.00	0.00	0.00	0.00	0.00	0.00
0.00	0.00	100.00	0.00	0.00	0.00	0.00	0.00	0.00	100.00	0.00	0.00	0.00	0.00	0.00	0.00
0.00	0.00	100.00	0.00	0.00	0.00	0.00	0.00	0.00	100.00	0.00	0.00	0.00	0.00	0.00	0.00
0.00	0.00	100.00	0.00	0.00	0.00	0.00	0.00	0.00	100.00	0.00	0.00	0.00	0.00	0.00	0.00
0.00	0.00	100.00	0.00	0.00	50.00	32.00	90.00	7.21	100.00	0.00	0.00	0.00	0.00	3.00	0.09
0.00	0.00	100.00	21.00	0.00	1.00	1.00	90.00	10.70	100.00	0.00	0.00	0.00	0.00	1.00	0.80
0.00	0.00	100.00	0.00	0.00	0.00	0.00	0.00	0.00	100.00	0.00	0.00	0.00	0.00	0.00	0.00
0.00	0.00	100.00	0.00	0.00	0.00	0.00	0.00	0.00	100.00	0.00	0.00	0.00	0.00	0.00	0.00
0.00	0.00	100.00	0.00	0.00	0.00	0.00	0.00	0.00	100.00	0.00	0.00	0.00	0.00	0.00	0.00
0.00	0.00	100.00	0.00	0.00	0.00	0.00	0.00	0.00	100.00	0.00	0.00	0.00	0.00	0.00	0.00
0.00	0.00	100.00	0.00	0.00	0.00	0.00	0.00	0.00	100.00	0.00	0.00	0.00	0.00	0.00	0.00
0.00	0.00	100.00	0.00	0.00	0.00	0.00	0.00	0.00	100.00	0.00	0.00	0.00	0.00	0.00	0.00
0.00	0.00	100.00	0.00	0.00	0.00	0.00	0.00	0.00	100.00	0.00	0.00	0.00	0.00	0.00	0.00

APPENDIX TABLE 1A   Feed Library (*Continued*)

Feed No.	Common Name	Int. Ref. No.	Cost $/ton	Conc. AF %DM	Forage %DM	DM %AF	NDF %DM	Lignin %NDF	eNDF %NDF	TDN %DM	ME Mcal/kg	NE$_{ma}$ Mcal/kg	NE$_{ga}$ Mcal/kg
841	Blank		0.00	100.00	0.00	100.00	0.00	0.00	0.00	0.00	0.00	0.00	0.00
842	Blank		0.00	100.00	0.00	100.00	0.00	0.00	0.00	0.00	0.00	0.00	0.00
843	Blank		0.00	100.00	0.00	100.00	0.00	0.00	0.00	0.00	0.00	0.00	0.00
844	Blank		0.00	100.00	0.00	100.00	0.00	0.00	0.00	0.00	0.00	0.00	0.00
845	Blank		0.00	100.00	0.00	100.00	0.00	0.00	0.00	0.00	0.00	0.00	0.00
901	Blank		0.00	100.00	0.00	100.00	0.00	0.00	0.00	0.00	0.00	0.00	0.00
902	Blank		0.00	100.00	0.00	100.00	0.00	0.00	0.00	0.00	0.00	0.00	0.00
903	Blank		0.00	100.00	0.00	100.00	0.00	0.00	0.00	0.00	0.00	0.00	0.00
904	Blank		0.00	100.00	0.00	100.00	0.00	0.00	0.00	0.00	0.00	0.00	0.00
905	Blank		0.00	100.00	0.00	100.00	0.00	0.00	0.00	0.00	0.00	0.00	0.00
906	Blank		0.00	100.00	0.00	100.00	0.00	0.00	0.00	0.00	0.00	0.00	0.00
907	Blank		0.00	100.00	0.00	100.00	0.00	0.00	0.00	0.00	0.00	0.00	0.00
908	Blank		0.00	100.00	0.00	100.00	0.00	0.00	0.00	0.00	0.00	0.00	0.00
909	Blank		0.00	100.00	0.00	100.00	0.00	0.00	0.00	0.00	0.00	0.00	0.00
910	Blank		0.00	100.00	0.00	100.00	0.00	0.00	0.00	0.00	0.00	0.00	0.00
999	Minerals	X-XX-XXX	0.00	100.00	0.00	99.00	0.00	0.00	0.00	0.00	0.00	0.00	0.00

CP %DM	DIP %CP	UIP %CP	SolP %CP	NPN %SolP	NDFIP %CP	ADFIP %CP	Starch %NSC	Fat %DM	Ash %DM	A kd[a] %/hr	B1 kd[a] %/hr	B2 kd[a] %/hr	B1 kd[b] %/hr	B2 kd[b] %/hr	B3 kd[b] %/hr
0.00	0.00	100.00	0.00	0.00	0.00	0.00	0.00	0.00	100.00	0.00	0.00	0.00	0.00	0.00	0.00
0.00	0.00	100.00	0.00	0.00	0.00	0.00	0.00	0.00	100.00	0.00	0.00	0.00	0.00	0.00	0.00
0.00	0.00	100.00	0.00	0.00	0.00	0.00	0.00	0.00	100.00	0.00	0.00	0.00	0.00	0.00	0.00
0.00	0.00	100.00	0.00	0.00	0.00	0.00	0.00	0.00	100.00	0.00	0.00	0.00	0.00	0.00	0.00
0.00	0.00	100.00	0.00	0.00	0.00	0.00	0.00	0.00	100.00	0.00	0.00	0.00	0.00	0.00	0.00
0.00	0.00	100.00	0.00	0.00	0.00	0.00	0.00	0.00	100.00	0.00	0.00	0.00	0.00	0.00	0.00
0.00	0.00	100.00	0.00	0.00	0.00	0.00	0.00	0.00	100.00	0.00	0.00	0.00	0.00	0.00	0.00
0.00	0.00	100.00	0.00	0.00	0.00	0.00	0.00	0.00	100.00	0.00	0.00	0.00	0.00	0.00	0.00
0.00	0.00	100.00	0.00	0.00	0.00	0.00	0.00	0.00	100.00	0.00	0.00	0.00	0.00	0.00	0.00
0.00	0.00	100.00	0.00	0.00	0.00	0.00	0.00	0.00	100.00	0.00	0.00	0.00	0.00	0.00	0.00
0.00	0.00	100.00	0.00	0.00	0.00	0.00	0.00	0.00	100.00	0.00	0.00	0.00	0.00	0.00	0.00
0.00	0.00	100.00	0.00	0.00	0.00	0.00	0.00	0.00	100.00	0.00	0.00	0.00	0.00	0.00	0.00
0.00	0.00	100.00	0.00	0.00	0.00	0.00	0.00	0.00	100.00	0.00	0.00	0.00	0.00	0.00	0.00
0.00	0.00	100.00	0.00	0.00	0.00	0.00	0.00	0.00	100.00	0.00	0.00	0.00	0.00	0.00	0.00
0.00	0.00	100.00	0.00	0.00	0.00	0.00	0.00	0.00	100.00	0.00	0.00	0.00	0.00	0.00	0.00
0.00	0.00	100.00	0.00	0.00	0.00	0.00	0.00	0.00	100.00	0.00	0.00	0.00	0.00	0.00	0.00

APPENDIX TABLE 1B   Feed Library—Amino Acids, Minerals, and Vitamins

Feed No.	Common Name	Int. Ref. No.	MET %UIP	LYS %UIP	ARG %UIP	THR %UIP	LEU %UIP	ILE %UIP	VAL %UIP	HIS %UIP	PHE %UIP	TRP %UIP
101	Bahiagrass 30% Dry Matter	2-00-464	0.67	2.83	2.83	2.83	5.49	2.83	3.83	1.00	3.50	4.50
102	Bahiagrass Hay	1-00-462	0.67	2.83	2.83	2.83	5.49	2.83	3.83	1.00	3.50	4.50
103	Bermudagrass Late Vegetative	1-09-210	0.67	2.83	2.83	2.83	5.49	2.83	3.83	1.00	3.50	4.50
104	Brome Hay Pre-bloom	1-00-887	0.67	2.83	2.83	2.83	5.49	2.83	3.83	1.00	3.50	4.50
105	Brome Hay Mid Bloom	1-05-633	0.67	2.83	2.83	2.83	5.49	2.83	3.83	1.00	3.50	4.50
106	Brome Hay Late bloom	1-00-888	0.67	2.83	2.83	2.83	5.49	2.83	3.83	1.00	3.50	4.50
107	Brome Hay Mature	1-00-944	0.67	2.83	2.83	2.83	5.49	2.83	3.83	1.00	3.50	4.50
108	Fescue Meadow Hay	1-01-912	0.67	2.83	2.83	2.83	5.49	2.83	3.83	1.00	3.50	4.50
109	Fescue, Alta Hay	1-05-684	0.67	2.83	2.83	2.83	5.49	2.83	3.83	1.00	3.50	4.50
110	Fescue, K31 Hay	1-09-187	0.67	2.83	2.83	2.83	5.49	2.83	3.83	1.00	3.50	4.50
111	Fescue K31 Hay Full bloom	1-09-188	0.67	2.83	2.83	2.83	5.49	2.83	3.83	1.00	3.50	4.50
112	Fescue, K31 Mature	1-09-189	0.67	2.83	2.83	2.83	5.49	2.83	3.83	1.00	3.50	4.50
113	Napiergrass Fresh 30 day DM	2-03-158	0.67	2.83	2.83	2.83	5.49	2.83	3.83	1.00	3.50	4.50
114	Napiergrass Fresh 60 day DM	2-03-162	0.67	2.83	2.83	2.83	5.49	2.83	3.83	1.00	3.50	4.50
115	Orchardgrass Hay, Early bloom	1-03-425	0.67	2.83	2.83	2.83	5.49	2.83	3.83	1.00	3.50	4.50
116	Orchardgrass Hay, Late bloom	1-03-428	0.67	2.83	2.83	2.83	5.49	2.83	3.83	1.00	3.50	4.50
117	Pangolagrass Fresh	2-03-493	0.67	2.83	2.83	2.83	5.49	2.83	3.83	1.00	3.50	4.50
118	Red Top Fresh	2-03-897	0.67	2.83	2.83	2.83	5.49	2.83	3.83	1.00	3.50	4.50
119	Reed Canarygrass Hay	1-00-104	0.67	2.83	2.83	2.83	5.49	2.83	3.83	1.00	3.50	4.50
120	Ryegrass Hay	1-04-077	0.67	2.83	2.83	2.83	5.49	2.83	3.83	1.00	3.50	4.50
121	Sorghum Sudan Hay	1-04-480	0.67	2.83	2.83	2.83	5.49	2.83	3.83	1.00	3.50	4.50
122	Sorghum-Sudan Pasture	2-04-484	0.67	2.83	2.83	2.83	5.49	2.83	3.83	1.00	3.50	4.50
123	Sorghum-Sudan Silage	3-04-499	0.67	2.83	2.83	2.83	5.49	2.83	3.83	1.00	3.50	4.50
124	Timothy Hay Late Vegetative	1-04-881	0.67	2.83	2.83	2.83	5.49	2.83	3.83	1.00	3.50	4.50
125	Timothy Hay Early bloom	1-04-882	0.67	2.83	2.83	2.83	5.49	2.83	3.83	1.00	3.50	4.50
126	Timothy Hay Mid bloom	1-04-883	0.67	2.83	2.83	2.83	5.49	2.83	3.83	1.00	3.50	4.50
127	Timothy Hay Full bloom	1-04-884	0.67	2.83	2.83	2.83	5.49	2.83	3.83	1.00	3.50	4.50
128	Timothy Hay Seed stage	1-04-888	0.67	2.83	2.83	2.83	5.49	2.83	3.83	1.00	3.50	4.50
129	Wheatgrass crest., hay	1-05-351	0.67	2.83	2.83	2.83	5.49	2.83	3.83	1.00	3.50	4.50
130	Blank		0.00	0.00	0.00	0.00	0.00	0.00	0.00	0.00	0.00	0.00
131	Blank		0.00	0.00	0.00	0.00	0.00	0.00	0.00	0.00	0.00	0.00
132	Blank		0.00	0.00	0.00	0.00	0.00	0.00	0.00	0.00	0.00	0.00
133	Blank		0.00	0.00	0.00	0.00	0.00	0.00	0.00	0.00	0.00	0.00
134	Blank		0.00	0.00	0.00	0.00	0.00	0.00	0.00	0.00	0.00	0.00
135	Grass Pasture Spring	2-00-956	0.67	2.83	2.83	2.83	5.49	2.83	3.83	1.00	3.50	4.50
136	Grass Pasture Summer		0.67	2.83	2.83	2.83	5.49	2.83	3.83	1.00	3.50	4.50
137	Grass Pasture Fall		0.67	2.83	2.83	2.83	5.49	2.83	3.83	1.00	3.50	4.50
138	Mix Pasture Spring		0.70	4.43	4.61	3.92	7.38	4.42	5.49	1.81	4.91	3.17
139	Mix Pasture Summer		0.70	4.43	4.61	3.92	7.38	4.42	5.49	1.81	4.91	3.17
140	Range June Diet		0.67	2.83	2.83	2.83	5.49	2.83	3.83	1.00	3.50	4.50
141	Range July Diet		0.67	2.83	2.83	2.83	5.49	2.83	3.83	1.00	3.50	4.50
142	Range Aug. Diet		0.67	2.83	2.83	2.83	5.49	2.83	3.83	1.00	3.50	4.50
143	Range Sep. Diet		0.67	2.83	2.83	2.83	5.49	2.83	3.83	1.00	3.50	4.50
144	Range Winter		0.67	2.83	2.83	2.83	5.49	2.83	3.83	1.00	3.50	4.50
145	Meadow Spring		0.67	2.83	2.83	2.83	5.49	2.83	3.83	1.00	3.50	4.50
146	Meadow Fall		0.67	2.83	2.83	2.83	5.49	2.83	3.83	1.00	3.50	4.50
147	Meadow Hay		0.67	2.83	2.83	2.83	5.49	2.83	3.83	1.00	3.50	4.50
148	Prairie Hay	1-03-191	0.67	2.83	2.83	2.83	5.49	2.83	3.83	1.00	3.50	4.50
149	Blank		0.00	0.00	0.00	0.00	0.00	0.00	0.00	0.00	0.00	0.00
150	Blank		0.00	0.00	0.00	0.00	0.00	0.00	0.00	0.00	0.00	0.00
201	Alfalfa Hay Early Vegetative	1-00-54-S	0.73	6.02	6.39	5.00	9.26	6.01	7.14	2.62	6.32	1.84
202	Alfalfa Hay Early Vegetative	1-00-N	0.73	6.02	6.39	5.00	9.26	6.01	7.14	2.62	6.32	1.84
203	Alfalfa Hay Late Vegetative	1-00-059-S	0.73	6.02	6.39	5.00	9.26	6.01	7.14	2.62	6.32	1.84
204	Alfalfa Hay Late Vegetative	1-00-N	0.73	6.02	6.39	5.00	9.26	6.01	7.14	2.62	6.32	1.84
205	Alfalfa Hay Early Bloom	1-00-059-S	0.73	6.02	6.39	5.00	9.26	6.01	7.14	2.62	6.32	1.84
206	Alfalfa Hay Early Bloom	1-00-N	0.73	6.02	6.39	5.00	9.26	6.01	7.14	2.62	6.32	1.84
207	Alfalfa Hay Mid Bloom	1-00-063-S	0.73	6.02	6.39	5.00	9.26	6.01	7.14	2.62	6.32	1.84
208	Alfalfa Hay Mid Bloom	1-00-N	0.73	6.02	6.39	5.00	9.26	6.01	7.14	2.62	6.32	1.84
209	Alfalfa Hay Full bloom	1-00-068-S	0.73	6.02	6.39	5.00	9.26	6.01	7.14	2.62	6.32	1.84
210	Alfalfa Hay Full Bloom	1-00-N	0.73	6.02	6.39	5.00	9.26	6.01	7.14	2.62	6.32	1.84
211	Alfalfa Hay Late Bloom	1-00-070-S	0.73	6.02	6.39	5.00	9.26	6.01	7.14	2.62	6.32	1.84
212	Alfalfa Hay Late Bloom	1-00-N	0.73	6.02	6.39	5.00	9.26	6.01	7.14	2.62	6.32	1.84
213	Alfalfa Hay Mature	1-00-71-S	0.73	6.02	6.39	5.00	9.26	6.01	7.14	2.62	6.32	1.84
214	Alfalfa Hay Seeded		0.73	6.02	6.39	5.00	9.26	6.01	7.14	2.62	6.32	1.84
215	Alfalfa Hay Weathered		0.73	6.02	6.39	5.00	9.26	6.01	7.14	2.62	6.32	1.84
216	Alfalfa Meal dehydrated 15%CP	1-00-022	0.73	6.02	6.39	5.00	9.26	6.01	7.14	2.62	6.32	1.84
217	Alfalfa Silage Early Bloom	3-00-216	1.22	3.21	2.44	3.30	6.40	3.13	0.00	0.63	4.18	1.84
218	Alfalfa Silage Mid Bloom	3-00-217	1.22	3.21	2.44	3.30	6.40	3.13	0.00	0.63	4.18	1.84
219	Alfalfa Silage Full Bloom	3-00-218	1.22	3.21	2.44	3.30	6.40	3.13	0.00	0.63	4.18	1.84

NOTE: See the glossary for definitions of acronyms and Chapter 10 for a discussion of tabular energy and protein values, feed carbohydrate and protein fractions, and recommended analytical procedures.

Calcium %DM	Phosphorus %DM	Magnesium %DM	Chlorine %DM	Potassium %DM	Sodium %DM	Sulfur %DM	Cobalt mg/kg	Copper mg/kg	Iodine mg/kg	Iron mg/kg	Manganese mg/kg	Selenium mg/kg	Zinc mg/kg	Vit A 1000 IU/kg	Vit D 1000 U/kg	Vit E IU/kg
0.46	0.22	0.25	0.00	1.45	0.00	0.00	0.00	0.00	0.00	0.00	0.00	0.00	0.00	304.20	0.00	0.00
0.50	0.22	0.19	0.00	0.00	0.00	0.00	0.00	0.00	0.00	60.00	0.00	0.00	0.00	0.00	0.00	0.00
0.26	0.18	0.13	0.00	1.30	0.08	0.21	0.12	9.00	0.00	290.00	0.00	0.00	20.00	136.20	0.00	0.00
0.32	0.37	0.09	0.00	2.32	0.02	0.20	0.00	0.00	0.00	0.00	0.00	0.00	0.00	108.40	1.00	0.00
0.29	0.28	0.10	0.00	1.99	0.01	0.00	0.58	25.00	0.00	91.00	40.00	0.00	30.00	26.00	0.00	0.00
0.00	0.00	0.00	0.00	0.00	0.00	0.00	0.00	0.00	0.00	0.00	0.00	0.00	0.00	0.00	0.00	0.00
0.26	0.22	0.12	0.00	1.85	0.01	0.00	0.00	10.40	0.00	80.00	73.00	0.00	24.00	15.00	1.00	0.00
0.37	0.29	0.50	0.00	1.84	0.00	0.00	0.14	0.00	0.00	0.00	24.50	0.00	0.00	120.90	0.00	135.60
0.39	0.24	0.23	0.06	2.38	0.00	0.00	0.00	0.00	0.00	0.00	0.00	0.00	0.00	34.60	0.00	0.00
0.51	0.37	0.27	0.00	2.30	0.00	0.18	0.00	0.00	0.00	0.00	0.00	0.00	22.00	0.00	0.00	0.00
0.43	0.32	0.17	0.00	2.30	0.00	0.26	38.00	28.00	0.00	0.00	103.00	0.00	0.00	0.00	0.00	0.00
0.41	0.30	0.16	0.00	1.96	0.02	0.00	0.00	22.00	0.00	132.00	97.00	0.00	35.00	0.00	0.00	0.00
0.60	0.41	0.26	0.00	1.31	0.01	0.10	0.00	0.00	0.00	0.00	0.00	0.00	0.00	0.00	0.00	0.00
0.60	0.41	0.26	0.00	1.31	0.01	0.10	0.00	0.00	0.00	0.00	0.00	0.00	0.00	0.00	0.00	0.00
0.27	0.34	0.11	0.41	2.91	0.01	0.26	0.43	19.00	0.00	93.00	157.00	0.00	40.00	15.00	0.00	0.00
0.26	0.30	0.11	0.00	2.67	0.01	0.00	0.30	20.00	20.00	84.00	167.00	0.03	38.00	8.00	0.00	0.00
0.38	0.22	0.18	0.00	1.43	0.00	0.00	0.00	0.00	0.00	0.00	0.00	0.00	0.00	0.00	0.00	0.00
0.62	0.37	0.25	0.00	2.35	0.05	0.16	0.00	0.00	0.00	200.00	0.00	0.00	0.00	254.40	0.00	0.00
0.36	0.24	0.22	0.00	2.91	0.02	0.00	0.00	11.90	0.00	150.00	92.40	0.00	0.00	31.60	0.00	0.00
0.00	0.00	0.00	0.00	0.00	0.00	0.00	0.00	0.00	0.00	0.00	0.00	0.00	0.00	199.90	0.00	0.00
0.51	0.31	0.37	0.00	2.08	0.02	0.06	0.13	31.40	0.00	170.00	76.30	0.00	38.00	0.00	0.00	0.00
0.49	0.44	0.35	0.00	2.14	0.00	0.11	0.13	35.90	0.00	210.00	81.40	0.00	0.00	304.60	0.00	0.00
0.50	0.21	0.42	0.00	2.61	0.02	0.06	0.27	36.60	0.00	120.00	98.80	0.00	0.00	175.40	0.00	0.00
0.45	0.40	0.11	0.00	3.05	0.07	0.13	0.00	25.80	0.00	240.00	89.00	0.00	67.00	208.40	0.00	0.00
0.51	0.29	0.13	0.00	2.41	0.01	0.13	0.00	11.00	0.00	203.00	103.00	0.00	62.00	87.50	0.00	13.00
0.48	0.23	0.13	0.00	1.82	0.01	0.13	0.00	16.00	0.00	150.00	56.10	0.00	43.00	88.90	2.00	0.00
0.43	0.20	0.09	0.62	1.99	0.07	0.14	0.00	29.00	0.00	140.00	93.00	0.00	54.00	0.00	0.00	0.00
0.00	0.00	0.00	0.00	0.00	0.00	0.00	0.00	0.00	0.00	0.00	0.00	0.00	0.00	0.00	0.00	0.00
0.26	0.15	0.00	0.00	0.00	0.00	0.00	0.24	0.00	0.00	0.00	0.00	0.00	0.00	37.20	0.00	0.00
0.00	0.00	0.00	0.00	0.00	0.00	0.00	0.00	0.00	0.00	0.00	0.00	0.00	0.00	0.00	0.00	0.00
0.00	0.00	0.00	0.00	0.00	0.00	0.00	0.00	0.00	0.00	0.00	0.00	0.00	0.00	0.00	0.00	0.00
0.00	0.00	0.00	0.00	0.00	0.00	0.00	0.00	0.00	0.00	0.00	0.00	0.00	0.00	0.00	0.00	0.00
0.00	0.00	0.00	0.00	0.00	0.00	0.00	0.00	0.00	0.00	0.00	0.00	0.00	0.00	0.00	0.00	0.00
0.00	0.00	0.00	0.00	0.00	0.00	0.00	0.00	0.00	0.00	0.00	0.00	0.00	0.00	0.00	0.00	0.00
0.55	0.45	0.32	0.00	3.16	0.00	0.20	0.00	0.00	0.00	0.00	0.00	0.00	21.00	184.00	0.00	0.00
0.00	0.00	0.00	0.00	0.00	0.00	0.00	0.00	0.00	0.00	0.00	0.00	0.00	0.00	0.00	0.00	0.00
0.00	0.00	0.00	0.00	0.00	0.00	0.00	0.00	0.00	0.00	0.00	0.00	0.00	0.00	0.00	0.00	0.00
0.00	0.00	0.00	0.00	0.00	0.00	0.00	0.00	0.00	0.00	0.00	0.00	0.00	0.00	0.00	0.00	0.00
0.00	0.00	0.00	0.00	0.00	0.00	0.00	0.00	0.00	0.00	0.00	0.00	0.00	0.00	0.00	0.00	0.00
0.26	0.15	0.00	0.00	0.00	0.00	0.00	0.24	0.00	0.00	0.00	0.00	0.00	0.00	37.20	0.00	0.00
0.26	0.15	0.00	0.00	0.00	0.00	0.00	0.24	0.00	0.00	0.00	0.00	0.00	0.00	37.20	0.00	0.00
0.26	0.15	0.00	0.00	0.00	0.00	0.00	0.24	0.00	0.00	0.00	0.00	0.00	0.00	37.20	0.00	0.00
0.26	0.15	0.00	0.00	0.00	0.00	0.00	0.24	0.00	0.00	0.00	0.00	0.00	0.00	37.20	0.00	0.00
0.26	0.15	0.00	0.00	0.00	0.00	0.00	0.24	0.00	0.00	0.00	0.00	0.00	0.00	37.20	0.00	0.00
0.26	0.15	0.00	0.00	0.00	0.00	0.00	0.24	0.00	0.00	0.00	0.00	0.00	0.00	37.20	0.00	0.00
0.26	0.15	0.00	0.00	0.00	0.00	0.00	0.24	0.00	0.00	0.00	0.00	0.00	0.00	37.20	0.00	0.00
0.35	0.14	0.26	0.00	1.00	0.00	0.00	0.00	0.00	0.00	88.00	0.00	0.00	34.00	37.20	0.00	0.00
0.00	0.00	0.00	0.00	0.00	0.00	0.00	0.00	0.00	0.00	0.00	0.00	0.00	0.00	0.00	0.00	0.00
0.00	0.00	0.00	0.00	0.00	0.00	0.00	0.00	0.00	0.00	0.00	0.00	0.00	0.00	0.00	0.00	0.00
1.50	0.33	0.21	0.34	2.51	0.12	0.54	0.29	11.40	0.00	240.00	47.10	0.55	37.40	80.00	0.00	0.00
1.50	0.33	0.21	0.34	2.51	0.12	0.54	0.29	11.40	0.00	240.00	47.10	0.55	37.40	80.00	0.00	0.00
1.50	0.33	0.21	0.34	2.51	0.12	0.54	0.29	11.40	0.00	240.00	47.10	0.55	37.40	81.00	0.00	0.00
1.50	0.33	0.21	0.34	2.51	0.12	0.54	0.29	11.40	0.00	240.00	47.10	0.55	37.40	81.00	0.00	0.00
1.41	0.22	0.34	0.34	2.51	0.12	0.30	0.29	12.70	0.17	240.00	36.00	0.55	30.00	56.00	2.00	26.00
1.63	0.22	0.21	0.34	2.51	0.12	0.54	0.29	11.40	0.00	240.00	47.10	0.55	37.40	56.00	2.00	26.00
1.37	0.22	0.35	0.38	1.56	0.12	0.28	0.39	17.70	0.16	225.00	28.00	0.55	30.90	46.00	2.00	11.00
1.39	0.24	0.35	0.00	1.56	0.12	0.28	0.39	17.10	0.00	225.00	60.50	0.55	30.90	46.00	2.00	11.00
1.19	0.24	0.27	0.00	1.56	0.07	0.27	0.23	9.90	0.13	155.00	42.30	0.00	26.10	26.00	2.00	11.00
1.19	0.24	0.27	0.00	1.56	0.07	0.30	0.23	9.90	0.00	160.00	42.30	0.55	26.10	26.00	2.00	11.00
1.19	0.24	0.27	0.00	1.56	0.07	0.30	0.23	9.90	0.13	160.00	42.30	0.55	26.10	19.30	1.00	0.00
1.19	0.24	0.27	0.00	1.56	0.07	0.30	0.23	9.90	0.13	160.00	42.30	0.55	26.10	19.30	1.00	0.00
1.18	0.21	0.22	0.00	2.07	0.08	0.25	0.41	13.70	0.00	170.00	38.50	0.55	22.10	19.30	1.00	0.00
1.18	0.21	0.22	0.00	2.07	0.08	0.25	0.41	13.70	0.00	170.00	38.50	0.55	22.10	19.30	1.00	0.00
2.29	0.23	0.27	0.00	2.42	0.06	0.00	0.00	2.80	0.00	290.00	24.80	0.55	26.60	0.00	1.00	0.00
1.38	0.25	0.29	0.00	2.46	0.08	0.21	0.19	10.50	0.13	309.00	31.00	0.31	21.00	33.00	0.00	91.00
1.32	0.31	0.26	0.00	2.85	0.02	0.28	0.65	12.10	0.16	252.00	32.40	0.18	19.50	155.00	0.00	0.00
1.74	0.27	0.33	0.41	2.35	0.16	0.31	0.00	11.10	0.00	280.00	49.70	0.00	40.70	155.00	0.00	0.00
1.74	0.27	0.33	0.41	2.35	0.16	0.31	0.00	11.10	0.00	280.00	49.70	0.00	40.70	155.00	0.00	0.00

APPENDIX TABLE 1B   Feed Library—Amino Acids, Minerals, and Vitamins

Feed No.	Common Name	Int. Ref. No.	MET %UIP	LYS %UIP	ARG %UIP	THR %UIP	LEU %UIP	ILE %UIP	VAL %UIP	HIS %UIP	PHE %UIP	TRP %UIP
220	Birdsfoot Trefoil, hay	1-05-044	0.73	6.02	6.39	5.00	9.26	6.01	7.14	2.62	6.32	1.84
221	Clover Ladino Hay	1-01-378	0.73	6.02	6.39	5.00	9.26	6.01	7.14	2.62	6.32	1.84
222	Clover Red Hay	1-01-415	0.73	6.02	6.39	5.00	9.26	6.01	7.14	2.62	6.32	1.84
223	Vetch Hay	1-05-106	0.73	6.02	6.39	5.00	9.26	6.01	7.14	2.62	6.32	1.84
224	Blank		0.00	0.00	0.00	0.00	0.00	0.00	0.00	0.00	0.00	0.00
225	Blank		0.00	0.00	0.00	0.00	0.00	0.00	0.00	0.00	0.00	0.00
226	Blank		0.00	0.00	0.00	0.00	0.00	0.00	0.00	0.00	0.00	0.00
227	Blank		0.00	0.00	0.00	0.00	0.00	0.00	0.00	0.00	0.00	0.00
228	Blank		0.00	0.00	0.00	0.00	0.00	0.00	0.00	0.00	0.00	0.00
229	Blank		0.00	0.00	0.00	0.00	0.00	0.00	0.00	0.00	0.00	0.00
230	Leg Pasture Spring		0.73	6.02	6.39	5.00	9.26	6.01	7.14	2.62	6.32	1.84
231	Leg Pasture Summer	2-00-181	0.73	6.02	6.39	5.00	9.26	6.01	7.14	2.62	6.32	1.84
232	Blank		0.00	0.00	0.00	0.00	0.00	0.00	0.00	0.00	0.00	0.00
233	Blank		0.00	0.00	0.00	0.00	0.00	0.00	0.00	0.00	0.00	0.00
234	Blank		0.00	0.00	0.00	0.00	0.00	0.00	0.00	0.00	0.00	0.00
235	Blank		0.00	0.00	0.00	0.00	0.00	0.00	0.00	0.00	0.00	0.00
236	Blank		0.00	0.00	0.00	0.00	0.00	0.00	0.00	0.00	0.00	0.00
237	Blank		0.00	0.00	0.00	0.00	0.00	0.00	0.00	0.00	0.00	0.00
238	Blank		0.00	0.00	0.00	0.00	0.00	0.00	0.00	0.00	0.00	0.00
239	Blank		0.00	0.00	0.00	0.00	0.00	0.00	0.00	0.00	0.00	0.00
240	Blank		0.00	0.00	0.00	0.00	0.00	0.00	0.00	0.00	0.00	0.00
241	Blank		0.00	0.00	0.00	0.00	0.00	0.00	0.00	0.00	0.00	0.00
242	Blank		0.00	0.00	0.00	0.00	0.00	0.00	0.00	0.00	0.00	0.00
243	Blank		0.00	0.00	0.00	0.00	0.00	0.00	0.00	0.00	0.00	0.00
244	Blank		0.00	0.00	0.00	0.00	0.00	0.00	0.00	0.00	0.00	0.00
245	Blank		0.00	0.00	0.00	0.00	0.00	0.00	0.00	0.00	0.00	0.00
246	Blank		0.00	0.00	0.00	0.00	0.00	0.00	0.00	0.00	0.00	0.00
247	Blank		0.00	0.00	0.00	0.00	0.00	0.00	0.00	0.00	0.00	0.00
248	Blank		0.00	0.00	0.00	0.00	0.00	0.00	0.00	0.00	0.00	0.00
249	Blank		0.00	0.00	0.00	0.00	0.00	0.00	0.00	0.00	0.00	0.00
250	Blank		0.00	0.00	0.00	0.00	0.00	0.00	0.00	0.00	0.00	0.00
301	Barley Silage		1.73	3.65	1.73	3.94	6.35	3.65	5.48	1.83	3.94	1.35
302	Barley Straw	1-00-498	0.67	2.83	2.83	2.83	5.49	2.83	3.83	1.00	3.50	4.50
303	Corn Cobs Ground	1-28-234	0.76	1.14	1.90	3.42	14.40	3.42	4.56	2.66	4.90	0.38
304	Corn Silage 25% Grain	3-28-250-N	0.80	2.13	1.87	2.13	6.40	2.40	3.20	1.07	2.94	0.11
305	Corn Silage 25% Grain	3-28-250-S	0.80	2.13	1.87	2.13	6.40	2.40	3.20	1.07	2.94	0.11
306	Corn Silage 35% Grain	3-28-250	0.80	2.13	1.87	2.13	6.40	2.40	3.20	1.07	2.94	0.11
307	Corn Silage 40% Grain	3-28-250	0.80	2.13	1.87	2.13	6.40	2.40	3.20	1.07	2.94	0.11
308	Corn Silage 40% GR +NPN	3-28-250	0.80	2.13	1.87	2.13	6.40	2.40	3.20	1.07	2.94	0.11
309	Corn Silage 40% GR +NPN+Ca	3-28-250	0.80	2.13	1.87	2.13	6.40	2.40	3.20	1.07	2.94	0.11
310	Corn Silage 45% Grain	3-28-250	0.80	2.13	1.87	2.13	6.40	2.40	3.20	1.07	2.94	0.11
311	Corn Silage 45% GR +NPN		0.80	2.13	1.87	2.13	6.40	2.40	3.20	1.07	2.94	0.11
312	Corn Silage 45% GR +NPN+Ca		0.80	2.13	1.87	2.13	6.40	2.40	3.20	1.07	2.94	0.11
313	Corn Silage 50% Grain	3-28-250	0.80	2.13	1.87	2.13	6.40	2.40	3.20	1.07	2.94	0.11
314	CS50% + NPN+CA		0.80	2.13	1.87	2.13	6.40	2.40	3.20	1.07	2.94	0.11
315	Corn Silage Immature (no Ears)	3-28-252	0.80	2.13	1.87	2.13	6.40	2.40	3.20	1.07	2.94	0.11
316	Corn Silage Stalklage	3-28-251	0.80	2.13	1.87	2.13	6.40	2.40	3.20	1.07	2.94	0.11
317	Corn Stalks Grazing		0.80	2.13	1.87	2.13	6.40	2.40	3.20	1.07	2.94	0.11
318	Oat Silage Dough	3-03-296	2.12	2.02	4.38	2.16	7.70	3.84	0.00	1.80	5.86	1.28
319	Oat Straw	1-03-283	0.67	2.83	2.83	2.83	5.49	2.83	3.83	1.00	3.50	4.50
320	Oat Hay	1-03-280	0.67	2.83	2.83	2.83	5.49	2.83	3.83	1.00	3.50	4.50
321	Sorghum, Silage	3-04-323	0.75	3.61	7.07	2.26	4.29	3.01	2.78	1.35	2.78	0.75
322	Wheat Silage dough	3-05-184	0.98	3.00	4.33	2.82	13.64	3.98	4.50	2.23	4.84	1.06
323	Wheat Straw	1-05-175	0.67	2.83	2.83	2.83	5.49	2.83	3.83	1.00	3.50	4.50
324	Blank		0.00	0.00	0.00	0.00	0.00	0.00	0.00	0.00	0.00	0.00
325	Blank		0.00	0.00	0.00	0.00	0.00	0.00	0.00	0.00	0.00	0.00
326	Blank		0.00	0.00	0.00	0.00	0.00	0.00	0.00	0.00	0.00	0.00
327	Blank		0.00	0.00	0.00	0.00	0.00	0.00	0.00	0.00	0.00	0.00
328	Blank		0.00	0.00	0.00	0.00	0.00	0.00	0.00	0.00	0.00	0.00
329	Blank		0.00	0.00	0.00	0.00	0.00	0.00	0.00	0.00	0.00	0.00
330	Blank		0.00	0.00	0.00	0.00	0.00	0.00	0.00	0.00	0.00	0.00
331	Blank		0.00	0.00	0.00	0.00	0.00	0.00	0.00	0.00	0.00	0.00
332	Blank		0.00	0.00	0.00	0.00	0.00	0.00	0.00	0.00	0.00	0.00
333	Blank		0.00	0.00	0.00	0.00	0.00	0.00	0.00	0.00	0.00	0.00
334	Blank		0.00	0.00	0.00	0.00	0.00	0.00	0.00	0.00	0.00	0.00
335	Blank		0.00	0.00	0.00	0.00	0.00	0.00	0.00	0.00	0.00	0.00
336	Blank		0.00	0.00	0.00	0.00	0.00	0.00	0.00	0.00	0.00	0.00
337	Blank		0.00	0.00	0.00	0.00	0.00	0.00	0.00	0.00	0.00	0.00
338	Blank		0.00	0.00	0.00	0.00	0.00	0.00	0.00	0.00	0.00	0.00
339	Blank		0.00	0.00	0.00	0.00	0.00	0.00	0.00	0.00	0.00	0.00

Calcium %DM	Phosphorus %DM	Magnesium %DM	Chlorine %DM	Potassium %DM	Sodium %DM	Sulfur %DM	Cobalt mg/kg	Copper mg/kg	Iodine mg/kg	Iron mg/kg	Manganese mg/kg	Selenium mg/kg	Zinc mg/kg	Vit A 1000 IU/kg	Vit D 1000 U/kg	Vit E IU/kg
1.70	0.23	0.51	0.00	1.92	0.07	0.25	0.11	9.26	0.00	227.00	29.00	0.00	77.00	75.00	1.50	0.00
1.45	0.33	0.47	0.30	2.44	0.13	0.21	0.16	9.41	0.30	470.00	123.00	0.00	17.00	33.00	0.00	0.00
1.38	0.24	0.38	0.32	1.81	0.18	0.16	0.16	11.00	0.25	238.00	108.00	0.00	17.00	8.00	1.90	0.00
1.36	0.34	0.27	0.00	2.12	0.52	0.15	0.34	9.90	0.49	490.00	60.80	0.00	0.00	0.00	0.00	0.00
0.00	0.00	0.00	0.00	0.00	0.00	0.00	0.00	0.00	0.00	0.00	0.00	0.00	0.00	0.00	0.00	0.00
0.00	0.00	0.00	0.00	0.00	0.00	0.00	0.00	0.00	0.00	0.00	0.00	0.00	0.00	0.00	0.00	0.00
0.00	0.00	0.00	0.00	0.00	0.00	0.00	0.00	0.00	0.00	0.00	0.00	0.00	0.00	0.00	0.00	0.00
0.00	0.00	0.00	0.00	0.00	0.00	0.00	0.00	0.00	0.00	0.00	0.00	0.00	0.00	0.00	0.00	0.00
0.00	0.00	0.00	0.00	0.00	0.00	0.00	0.00	0.00	0.00	0.00	0.00	0.00	0.00	0.00	0.00	0.00
0.00	0.00	0.00	0.00	0.00	0.00	0.00	0.00	0.00	0.00	0.00	0.00	0.00	0.00	0.00	0.00	0.00
1.71	0.30	0.36	0.00	2.27	0.21	0.36	0.17	10.70	0.00	111.00	41.00	0.55	30.00	253.00	2.00	26.00
1.71	0.30	0.36	0.00	2.27	0.21	0.36	0.17	10.70	0.00	111.00	41.00	0.55	30.00	253.00	2.00	26.00
0.00	0.00	0.00	0.00	0.00	0.00	0.00	0.00	0.00	0.00	0.00	0.00	0.00	0.00	0.00	0.00	0.00
0.00	0.00	0.00	0.00	0.00	0.00	0.00	0.00	0.00	0.00	0.00	0.00	0.00	0.00	0.00	0.00	0.00
0.00	0.00	0.00	0.00	0.00	0.00	0.00	0.00	0.00	0.00	0.00	0.00	0.00	0.00	0.00	0.00	0.00
0.00	0.00	0.00	0.00	0.00	0.00	0.00	0.00	0.00	0.00	0.00	0.00	0.00	0.00	0.00	0.00	0.00
0.00	0.00	0.00	0.00	0.00	0.00	0.00	0.00	0.00	0.00	0.00	0.00	0.00	0.00	0.00	0.00	0.00
0.00	0.00	0.00	0.00	0.00	0.00	0.00	0.00	0.00	0.00	0.00	0.00	0.00	0.00	0.00	0.00	0.00
0.00	0.00	0.00	0.00	0.00	0.00	0.00	0.00	0.00	0.00	0.00	0.00	0.00	0.00	0.00	0.00	0.00
0.00	0.00	0.00	0.00	0.00	0.00	0.00	0.00	0.00	0.00	0.00	0.00	0.00	0.00	0.00	0.00	0.00
0.00	0.00	0.00	0.00	0.00	0.00	0.00	0.00	0.00	0.00	0.00	0.00	0.00	0.00	0.00	0.00	0.00
0.00	0.00	0.00	0.00	0.00	0.00	0.00	0.00	0.00	0.00	0.00	0.00	0.00	0.00	0.00	0.00	0.00
0.00	0.00	0.00	0.00	0.00	0.00	0.00	0.00	0.00	0.00	0.00	0.00	0.00	0.00	0.00	0.00	0.00
0.00	0.00	0.00	0.00	0.00	0.00	0.00	0.00	0.00	0.00	0.00	0.00	0.00	0.00	0.00	0.00	0.00
0.00	0.00	0.00	0.00	0.00	0.00	0.00	0.00	0.00	0.00	0.00	0.00	0.00	0.00	0.00	0.00	0.00
0.00	0.00	0.00	0.00	0.00	0.00	0.00	0.00	0.00	0.00	0.00	0.00	0.00	0.00	0.00	0.00	0.00
0.00	0.00	0.00	0.00	0.00	0.00	0.00	0.00	0.00	0.00	0.00	0.00	0.00	0.00	0.00	0.00	0.00
0.00	0.00	0.00	0.00	0.00	0.00	0.00	0.00	0.00	0.00	0.00	0.00	0.00	0.00	0.00	0.00	0.00
0.52	0.29	0.19	0.00	2.57	0.12	0.24	0.72	7.70	0.00	375.00	44.80	0.15	24.50	0.00	0.00	0.00
0.30	0.07	0.23	0.67	2.37	0.14	0.17	0.07	5.40	0.00	200.00	16.00	0.00	7.00	1.00	0.70	0.00
0.12	0.04	0.07	0.00	0.89	0.08	0.47	0.13	7.00	0.00	230.00	6.00	0.08	5.00	1.20	0.00	0.00
0.31	0.27	0.22	0.18	1.22	0.03	0.12	0.10	9.20	0.00	180.00	41.10	0.00	21.20	58.10	0.00	0.00
0.31	0.27	0.22	0.18	1.22	0.03	0.12	0.10	9.20	0.00	180.00	41.10	0.00	21.20	58.10	0.00	0.00
0.31	0.27	0.22	0.18	1.22	0.03	0.12	0.10	9.20	0.00	180.00	41.10	0.00	21.20	58.10	0.00	0.00
0.31	0.27	0.22	0.18	1.22	0.03	0.12	0.10	9.20	0.00	180.00	41.10	0.00	21.20	58.10	0.00	0.00
0.31	0.27	0.22	0.18	1.22	0.03	0.12	0.10	9.20	0.00	180.00	41.10	0.00	21.20	58.10	0.00	0.00
0.25	0.22	0.18	0.18	1.14	0.01	0.12	0.10	4.18	0.00	131.00	23.50	0.00	17.70	18.00	0.10	0.00
0.31	0.27	0.22	0.18	1.22	0.03	0.12	0.10	9.20	0.00	180.00	41.10	0.00	21.20	58.10	0.00	0.00
0.31	0.27	0.22	0.18	1.22	0.03	0.12	0.10	9.20	0.00	180.00	41.10	0.00	21.20	58.10	0.00	0.00
0.31	0.27	0.22	0.18	1.22	0.03	0.12	0.10	9.20	0.00	180.00	41.10	0.00	21.20	58.10	0.00	0.00
0.52	0.31	0.31	0.00	1.64	0.00	0.00	0.00	0.00	0.00	490.00	0.00	0.00	184.50	0.00	0.00	0.00
0.00	0.00	0.00	0.00	0.00	0.00	0.00	0.00	0.00	0.00	0.00	0.00	0.00	0.00	0.00	0.00	0.00
0.62	0.09	0.00	0.00	1.63	0.00	0.00	0.00	0.00	0.00	0.00	0.00	0.00	0.00	0.00	0.00	0.00
0.58	0.31	0.21	0.00	2.88	0.09	0.24	0.00	8.00	0.00	367.00	66.30	0.07	29.80	0.00	0.00	0.00
0.23	0.06	0.17	0.78	2.53	0.42	0.22	0.00	10.30	0.00	164.00	31.00	0.00	6.00	6.30	1.00	0.00
0.32	0.25	0.29	0.52	1.49	0.18	0.23	0.07	4.80	0.00	406.00	99.00	0.00	45.00	49.60	2.00	0.00
0.49	0.22	0.28	0.13	1.72	0.01	0.12	0.30	9.20	0.00	383.00	68.50	0.03	32.00	0.00	0.00	0.00
0.44	0.29	0.17	0.00	2.24	0.04	0.21	0.00	9.00	0.00	386.00	79.50	0.00	28.00	59.00	0.00	0.00
0.17	0.05	0.12	0.32	1.41	0.14	0.19	0.05	3.60	0.00	157.00	41.00	0.00	6.00	1.00	0.07	0.00
0.00	0.00	0.00	0.00	0.00	0.00	0.00	0.00	0.00	0.00	0.00	0.00	0.00	0.00	0.00	0.00	0.00
0.00	0.00	0.00	0.00	0.00	0.00	0.00	0.00	0.00	0.00	0.00	0.00	0.00	0.00	0.00	0.00	0.00
0.00	0.00	0.00	0.00	0.00	0.00	0.00	0.00	0.00	0.00	0.00	0.00	0.00	0.00	0.00	0.00	0.00
0.00	0.00	0.00	0.00	0.00	0.00	0.00	0.00	0.00	0.00	0.00	0.00	0.00	0.00	0.00	0.00	0.00
0.00	0.00	0.00	0.00	0.00	0.00	0.00	0.00	0.00	0.00	0.00	0.00	0.00	0.00	0.00	0.00	0.00
0.00	0.00	0.00	0.00	0.00	0.00	0.00	0.00	0.00	0.00	0.00	0.00	0.00	0.00	0.00	0.00	0.00
0.00	0.00	0.00	0.00	0.00	0.00	0.00	0.00	0.00	0.00	0.00	0.00	0.00	0.00	0.00	0.00	0.00
0.00	0.00	0.00	0.00	0.00	0.00	0.00	0.00	0.00	0.00	0.00	0.00	0.00	0.00	0.00	0.00	0.00
0.00	0.00	0.00	0.00	0.00	0.00	0.00	0.00	0.00	0.00	0.00	0.00	0.00	0.00	0.00	0.00	0.00
0.00	0.00	0.00	0.00	0.00	0.00	0.00	0.00	0.00	0.00	0.00	0.00	0.00	0.00	0.00	0.00	0.00
0.00	0.00	0.00	0.00	0.00	0.00	0.00	0.00	0.00	0.00	0.00	0.00	0.00	0.00	0.00	0.00	0.00

APPENDIX TABLE 1B   Feed Library—Amino Acids, Minerals, and Vitamins

Feed No.	Common Name	Int. Ref. No.	MET %UIP	LYS %UIP	ARG %UIP	THR %UIP	LEU %UIP	ILE %UIP	VAL %UIP	HIS %UIP	PHE %UIP	TRP %UIP
340	Blank		0.00	0.00	0.00	0.00	0.00	0.00	0.00	0.00	0.00	0.00
341	Blank		0.00	0.00	0.00	0.00	0.00	0.00	0.00	0.00	0.00	0.00
342	Blank		0.00	0.00	0.00	0.00	0.00	0.00	0.00	0.00	0.00	0.00
343	Blank		0.00	0.00	0.00	0.00	0.00	0.00	0.00	0.00	0.00	0.00
344	Blank		0.00	0.00	0.00	0.00	0.00	0.00	0.00	0.00	0.00	0.00
345	Blank		0.00	0.00	0.00	0.00	0.00	0.00	0.00	0.00	0.00	0.00
346	Blank		0.00	0.00	0.00	0.00	0.00	0.00	0.00	0.00	0.00	0.00
347	Blank		0.00	0.00	0.00	0.00	0.00	0.00	0.00	0.00	0.00	0.00
348	Blank		0.00	0.00	0.00	0.00	0.00	0.00	0.00	0.00	0.00	0.00
349	Blank		0.00	0.00	0.00	0.00	0.00	0.00	0.00	0.00	0.00	0.00
350	Blank		0.00	0.00	0.00	0.00	0.00	0.00	0.00	0.00	0.00	0.00
401	Barley Malt Sprouts w/hulls	4-00-545	1.17	3.50	11.70	2.83	6.67	3.67	4.50	2.17	5.17	1.00
402	Barley Grain Heavy	4-00-549	0.81	3.07	4.83	3.15	6.83	3.92	4.88	2.29	5.60	1.26
403	Barley Grain Light		0.81	3.07	4.83	3.15	6.83	3.92	4.88	2.29	5.60	1.26
404	Corn Hominy	4-02-887	1.11	3.20	5.42	3.67	10.83	3.91	5.19	2.87	4.88	0.11
405	Corn Grain Cracked	4-20-698	1.12	1.65	1.82	2.80	10.73	2.69	3.75	2.06	3.65	0.37
406	Corn Dry Ear 45 lb/bu		1.12	1.65	1.82	2.80	10.73	2.69	3.75	2.06	3.65	0.37
407	Corn Dry Ear 56 lb/bu	04-28-238	1.12	1.65	1.82	2.80	10.73	2.69	3.75	2.06	3.65	0.37
408	Corn Dry Grain 45 lb/bu		1.12	1.65	1.82	2.80	10.73	2.69	3.75	2.06	3.65	0.37
409	Corn Ground Grain 56 lb/bu	04-02-931	1.12	1.65	1.82	2.80	10.73	2.69	3.75	2.06	3.65	0.37
410	Corn Dry Grain 56 lb/bu	04-02-931	1.12	1.65	1.82	2.80	10.73	2.69	3.75	2.06	3.65	0.37
411	Corn Grain Flaked	4-20-224	1.12	1.65	1.82	2.80	10.73	2.69	3.75	2.06	3.65	0.37
412	Corn HM Ear 56 lb/bu		0.99	2.47	4.11	3.33	12.10	3.85	4.78	2.70	4.99	0.37
413	Corn HM Grain 45 lb/bu		0.99	2.47	4.11	3.33	12.10	3.85	4.78	2.70	4.99	0.37
414	Corn HM Grain 56 lb/bu	04-20-771	0.99	2.47	4.11	3.33	12.10	3.85	4.78	2.70	4.99	0.37
415	Cottonseed Black Whole	5-01-614	0.63	3.85	10.40	3.45	6.33	3.77	5.27	3.14	5.85	1.74
416	Cottonseed High Lint	5-01-614	0.63	3.85	10.40	3.45	6.33	3.77	5.27	3.14	5.85	1.74
417	Cottonseed Meal—mech	5-01-617	0.63	3.85	10.40	3.45	6.33	3.77	5.27	3.14	5.85	1.74
418	Cottonseed Meal—Sol-41%CP	5-07-873	0.63	3.85	10.40	3.45	6.33	3.77	5.27	3.14	5.85	1.74
419	Cottonseed Meal—Sol-43%CP	5-01-630	0.63	3.85	10.40	3.45	6.33	3.77	5.27	3.14	5.85	1.74
420	Molasses Beet	4-00-668	0.00	0.00	0.00	0.00	0.00	0.00	0.00	0.00	0.00	0.00
421	Molasses Cane	4-04-696	0.00	0.00	0.00	0.00	0.00	0.00	0.00	0.00	0.00	0.00
422	Oats 32 lb/bu	4-03-318	2.12	2.02	4.38	2.16	7.70	3.84	0.00	1.80	5.86	1.28
423	Oats 38 lb/bu	4-03-309	2.12	2.02	4.38	2.16	7.70	3.84	0.00	1.80	5.86	1.28
424	Rice Bran	4-03-928	1.88	4.31	7.01	3.50	6.64	3.28	4.96	2.55	4.38	0.88
425	Rice Grain Ground	4-03-938	2.20	4.30	7.40	3.60	7.40	3.70	5.40	2.60	4.80	1.00
426	Rice Grain Polished	4-03-932	2.20	4.30	7.40	3.60	7.40	3.70	5.40	2.60	4.80	1.00
427	Rye Grain	4-04-047	1.38	3.47	4.42	2.97	5.80	3.84	4.64	2.10	4.64	0.94
428	Sorghum, Dry grain	4-04-383	0.75	3.61	7.07	2.26	4.29	3.01	2.78	1.35	2.78	0.75
429	Sorghum, Rolled grain	4-04-383	0.75	3.61	7.07	2.26	4.29	3.01	2.78	1.35	2.78	0.75
430	Sorghum, Steam flaked		0.75	3.61	7.07	2.26	4.29	3.01	2.78	1.35	2.78	0.75
431	Tapioca		1.33	3.33	4.67	2.67	4.67	3.00	3.67	3.00	1.00	0.67
432	Wheat Ground	4-05-211	0.98	3.00	4.33	2.82	13.64	3.98	4.50	2.23	4.84	1.06
433	Wheat Middlings	4-05-205	1.02	3.77	6.96	3.67	7.37	4.09	5.79	2.41	4.74	1.20
434	Wheat Grain Hard red spring	4-05-268	0.98	3.00	4.33	2.82	13.64	3.98	4.50	2.23	4.84	1.06
435	Wheat Grain Soft white	4-05-337	0.98	3.00	4.33	2.82	13.64	3.98	4.50	2.23	4.84	1.06
436	Blank		0.00	0.00	0.00	0.00	0.00	0.00	0.00	0.00	0.00	0.00
437	Blank		0.00	0.00	0.00	0.00	0.00	0.00	0.00	0.00	0.00	0.00
438	Blank		0.00	0.00	0.00	0.00	0.00	0.00	0.00	0.00	0.00	0.00
439	Blank		0.00	0.00	0.00	0.00	0.00	0.00	0.00	0.00	0.00	0.00
440	Blank		0.00	0.00	0.00	0.00	0.00	0.00	0.00	0.00	0.00	0.00
441	Blank		0.00	0.00	0.00	0.00	0.00	0.00	0.00	0.00	0.00	0.00
442	Blank		0.00	0.00	0.00	0.00	0.00	0.00	0.00	0.00	0.00	0.00
443	Blank		0.00	0.00	0.00	0.00	0.00	0.00	0.00	0.00	0.00	0.00
444	Blank		0.00	0.00	0.00	0.00	0.00	0.00	0.00	0.00	0.00	0.00
445	Blank		0.00	0.00	0.00	0.00	0.00	0.00	0.00	0.00	0.00	0.00
446	Blank		0.00	0.00	0.00	0.00	0.00	0.00	0.00	0.00	0.00	0.00
447	Blank		0.00	0.00	0.00	0.00	0.00	0.00	0.00	0.00	0.00	0.00
448	Blank		0.00	0.00	0.00	0.00	0.00	0.00	0.00	0.00	0.00	0.00
449	Blank		0.00	0.00	0.00	0.00	0.00	0.00	0.00	0.00	0.00	0.00
450	Blank		0.00	0.00	0.00	0.00	0.00	0.00	0.00	0.00	0.00	0.00
501	Brewers Grain 21% Dry Matter	5-02-142	1.70	3.23	4.69	3.43	9.18	5.71	5.95	1.90	5.31	1.36
502	Brewers Grain Dehydrated	5-02-141	1.26	2.15	2.61	2.76	8.46	3.53	3.78	1.47	4.80	1.12
503	Canola Meal	5-03-871	1.40	6.67	6.78	4.85	7.99	4.94	6.44	4.04	4.68	1.22
504	Coconut Meal		0.20	1.45	2.60	10.26	2.29	6.15	3.28	4.74	1.88	4.58
505	Corn Gluten Feed	5-28-243	1.68	1.50	6.97	1.71	7.04	0.89	5.32	2.18	1.68	0.66
506	Corn Gluten Meal	5-02-900	2.09	1.24	3.17	2.93	16.22	4.34	5.04	2.45	6.48	0.37
507	Corn Gluten Meal 60%CP	5-28-242	2.09	1.24	3.17	2.93	16.22	4.34	5.04	2.45	6.48	0.37
508	Distillers Gr. + solubles	5-02-843	1.20	2.06	4.15	3.12	9.07	2.78	5.24	1.82	4.20	1.64

Calcium %DM	Phosphorus %DM	Magnesium %DM	Chlorine %DM	Potassium %DM	Sodium %DM	Sulfur %DM	Cobalt mg/kg	Copper mg/kg	Iodine mg/kg	Iron mg/kg	Manganese mg/kg	Selenium mg/kg	Zinc mg/kg	Vit A 1000 IU/kg	Vit D 1000 U/kg	Vit E IU/kg
0.00	0.00	0.00	0.00	0.00	0.00	0.00	0.00	0.00	0.00	0.00	0.00	0.00	0.00	0.00	0.00	0.00
0.00	0.00	0.00	0.00	0.00	0.00	0.00	0.00	0.00	0.00	0.00	0.00	0.00	0.00	0.00	0.00	0.00
0.00	0.00	0.00	0.00	0.00	0.00	0.00	0.00	0.00	0.00	0.00	0.00	0.00	0.00	0.00	0.00	0.00
0.00	0.00	0.00	0.00	0.00	0.00	0.00	0.00	0.00	0.00	0.00	0.00	0.00	0.00	0.00	0.00	0.00
0.00	0.00	0.00	0.00	0.00	0.00	0.00	0.00	0.00	0.00	0.00	0.00	0.00	0.00	0.00	0.00	0.00
0.00	0.00	0.00	0.00	0.00	0.00	0.00	0.00	0.00	0.00	0.00	0.00	0.00	0.00	0.00	0.00	0.00
0.00	0.00	0.00	0.00	0.00	0.00	0.00	0.00	0.00	0.00	0.00	0.00	0.00	0.00	0.00	0.00	0.00
0.00	0.00	0.00	0.00	0.00	0.00	0.00	0.00	0.00	0.00	0.00	0.00	0.00	0.00	0.00	0.00	0.00
0.00	0.00	0.00	0.00	0.00	0.00	0.00	0.00	0.00	0.00	0.00	0.00	0.00	0.00	0.00	0.00	0.00
0.00	0.00	0.00	0.00	0.00	0.00	0.00	0.00	0.00	0.00	0.00	0.00	0.00	0.00	0.00	0.00	0.00
0.00	0.00	0.00	0.00	0.00	0.00	0.00	0.00	0.00	0.00	0.00	0.00	0.00	0.00	0.00	0.00	0.00
0.19	0.68	0.18	0.39	0.27	0.95	0.85	0.00	6.30	0.00	200.00	31.70	0.45	60.70	0.00	0.00	4.00
0.05	0.35	0.12	0.13	0.57	0.01	0.15	0.35	5.30	0.05	59.50	18.30	0.18	13.00	3.80	0.00	26.20
0.06	0.39	0.15	0.13	0.52	0.03	0.17	0.19	8.60	0.05	90.00	18.10	0.18	44.40	3.80	0.00	26.20
0.05	0.57	0.26	0.06	0.65	0.09	0.03	0.06	15.10	0.00	80.00	16.10	0.00	0.00	0.00	0.00	0.00
0.03	0.32	0.12	0.05	0.44	0.01	0.11	0.31	2.51	0.03	54.50	7.90	0.14	24.20	1.00	0.00	25.00
0.07	0.27	0.14	0.05	0.53	0.02	0.16	0.31	8.00	0.03	54.50	14.00	0.14	14.00	0.00	0.00	0.00
0.07	0.27	0.14	0.05	0.53	0.02	0.16	0.19	8.00	0.03	91.00	23.00	0.07	14.00	0.00	0.00	0.00
0.04	0.30	0.15	0.06	0.32	0.01	0.12	0.43	2.50	0.00	30.00	5.80	0.00	0.00	0.00	0.00	0.00
0.03	0.31	0.11	0.06	0.33	0.01	0.14	0.43	4.80	0.00	30.00	6.40	0.00	0.00	0.00	0.00	0.00
0.03	0.31	0.11	0.06	0.33	0.01	0.14	0.43	4.80	0.00	30.00	6.40	0.00	0.00	0.00	0.00	0.00
0.03	0.31	0.11	0.06	0.33	0.01	0.14	0.43	4.80	0.00	30.00	6.40	0.00	0.00	0.00	0.00	0.00
0.07	0.27	0.14	0.05	0.53	0.02	0.16	0.31	8.00	0.03	910.00	14.00	0.09	14.00	0.00	0.00	0.00
0.04	0.30	0.15	0.06	0.32	0.01	0.12	0.43	2.50	0.00	30.00	5.80	0.00	0.00	0.00	0.00	0.00
0.03	0.31	0.11	0.06	0.33	0.01	0.14	0.43	4.80	0.00	30.00	6.40	0.00	0.00	0.00	0.00	0.00
0.16	0.62	0.35	0.00	1.22	0.03	0.26	0.00	7.90	0.00	160.00	12.20	0.00	37.70	0.00	0.00	0.00
0.17	0.62	0.38	0.00	1.24	0.01	0.27	0.00	7.90	0.00	107.00	131.00	0.00	37.70	0.00	0.00	0.00
0.16	0.76	0.35	0.00	1.22	0.03	0.26	0.00	53.90	0.00	160.00	12.20	0.00	0.00	0.00	0.00	0.00
0.20	1.16	0.65	0.00	1.65	0.07	0.42	0.53	16.50	0.00	162.00	26.90	0.98	74.00	0.00	0.00	0.00
0.16	0.76	0.35	0.00	1.22	0.03	0.26	0.00	53.90	0.00	160.00	12.20	0.00	0.00	0.00	0.00	0.00
0.15	0.03	0.29	1.64	6.06	1.48	0.60	0.47	21.60	0.00	87.00	6.00	0.00	18.00	0.00	0.00	0.00
1.00	0.10	0.42	3.04	4.01	0.22	0.47	1.59	65.70	2.10	263.00	59.00	0.00	21.00	0.00	0.00	7.00
0.07	0.30	0.16	0.10	0.45	0.06	0.23	0.06	6.70	0.13	80.00	40.10	0.24	39.20	0.20	0.00	15.00
0.01	0.41	0.16	0.10	0.51	0.02	0.21	0.06	8.60	0.13	94.10	40.30	0.24	40.80	0.20	0.00	15.00
0.10	1.73	0.97	0.08	1.89	0.03	0.20	1.53	12.20	0.00	229.00	396.00	0.44	33.00	0.00	0.00	66.70
0.07	0.36	0.14	0.08	0.53	0.07	0.05	0.05	3.00	0.05	0.00	20.20	0.00	16.90	0.00	0.00	15.70
0.03	0.13	0.10	0.04	0.26	0.02	0.09	0.96	6.10	0.00	20.00	33.40	0.00	15.40	0.00	0.00	11.70
0.07	0.36	0.14	0.03	0.52	0.03	0.17	0.00	8.60	0.00	80.00	82.30	0.44	32.20	0.20	0.00	16.60
0.05	0.34	0.14	0.09	0.47	0.04	0.12	0.53	4.90	0.07	60.00	17.90	0.23	19.10	0.05	0.00	12.00
0.04	0.34	0.17	0.09	0.44	0.01	0.14	0.53	4.70	0.07	80.80	15.40	0.46	16.00	0.05	0.00	12.00
0.05	0.34	0.14	0.09	0.35	0.04	0.12	0.53	4.90	0.07	60.00	17.90	0.23	19.10	0.05	0.00	12.00
0.00	0.00	0.00	0.00	0.00	0.00	0.00	0.00	0.00	0.00	0.00	0.00	0.00	0.00	0.00	0.00	0.00
0.05	0.44	0.13	0.09	0.40	0.01	0.14	0.50	6.50	0.10	45.10	36.60	0.05	38.10	0.00	0.00	17.00
0.15	1.00	0.38	0.04	1.10	0.01	0.19	0.11	11.00	0.12	110.00	128.30	0.83	109.10	5.80	0.00	26.90
0.05	0.42	0.16	0.09	0.41	0.02	0.17	0.14	6.80	0.00	70.00	42.20	0.30	43.30	0.00	0.00	14.40
0.07	0.33	0.11	0.09	0.43	0.02	0.13	0.15	7.80	0.00	40.00	40.00	0.05	30.00	0.00	0.00	34.20
0.00	0.00	0.00	0.00	0.00	0.00	0.00	0.00	0.00	0.00	0.00	0.00	0.00	0.00	0.00	0.00	0.00
0.00	0.00	0.00	0.00	0.00	0.00	0.00	0.00	0.00	0.00	0.00	0.00	0.00	0.00	0.00	0.00	0.00
0.00	0.00	0.00	0.00	0.00	0.00	0.00	0.00	0.00	0.00	0.00	0.00	0.00	0.00	0.00	0.00	0.00
0.00	0.00	0.00	0.00	0.00	0.00	0.00	0.00	0.00	0.00	0.00	0.00	0.00	0.00	0.00	0.00	0.00
0.00	0.00	0.00	0.00	0.00	0.00	0.00	0.00	0.00	0.00	0.00	0.00	0.00	0.00	0.00	0.00	0.00
0.00	0.00	0.00	0.00	0.00	0.00	0.00	0.00	0.00	0.00	0.00	0.00	0.00	0.00	0.00	0.00	0.00
0.00	0.00	0.00	0.00	0.00	0.00	0.00	0.00	0.00	0.00	0.00	0.00	0.00	0.00	0.00	0.00	0.00
0.00	0.00	0.00	0.00	0.00	0.00	0.00	0.00	0.00	0.00	0.00	0.00	0.00	0.00	0.00	0.00	0.00
0.00	0.00	0.00	0.00	0.00	0.00	0.00	0.00	0.00	0.00	0.00	0.00	0.00	0.00	0.00	0.00	0.00
0.00	0.00	0.00	0.00	0.00	0.00	0.00	0.00	0.00	0.00	0.00	0.00	0.00	0.00	0.00	0.00	0.00
0.00	0.00	0.00	0.00	0.00	0.00	0.00	0.00	0.00	0.00	0.00	0.00	0.00	0.00	0.00	0.00	0.00
0.00	0.00	0.00	0.00	0.00	0.00	0.00	0.00	0.00	0.00	0.00	0.00	0.00	0.00	0.00	0.00	0.00
0.00	0.00	0.00	0.00	0.00	0.00	0.00	0.00	0.00	0.00	0.00	0.00	0.00	0.00	0.00	0.00	0.00
0.00	0.00	0.00	0.00	0.00	0.00	0.00	0.00	0.00	0.00	0.00	0.00	0.00	0.00	0.00	0.00	0.00
0.29	0.70	0.27	0.13	0.58	0.15	0.34	0.10	11.30	0.07	270.00	40.90	0.00	106.00	0.80	0.00	29.00
0.29	0.70	0.27	0.17	0.58	0.15	0.40	0.08	11.30	0.07	221.00	44.00	0.76	82.00	0.80	0.00	29.00
0.70	1.20	0.57	0.00	1.37	0.03	1.17	0.00	7.95	0.00	211.00	55.80	0.00	71.50	0.00	0.00	0.00
0.63	0.21	0.65	0.33	0.00	1.80	0.04	0.37	0.14	18.20	0.00	750.00	76.60	0.00	53.00	0.00	0.00
0.07	0.95	0.40	0.25	1.40	0.26	0.47	0.10	6.98	0.07	226.00	22.10	0.30	73.30	1.00	0.00	94.00
0.16	0.51	0.06	0.07	0.03	0.10	0.22	0.09	30.30	0.00	430.00	8.50	1.11	190.20	29.80	0.00	32.00
0.07	0.61	0.15	0.07	0.48	0.06	0.90	0.09	4.76	0.00	159.00	20.60	0.00	61.40	14.00	0.00	26.00
0.32	0.83	0.33	0.28	1.07	0.24	0.40	0.18	10.56	0.09	560.00	27.60	0.40	67.80	1.20	0.00	49.40

APPENDIX TABLE 1B   Feed Library—Amino Acids, Minerals, and Vitamins

Feed No.	Common Name	Int. Ref. No.	MET %UIP	LYS %UIP	ARG %UIP	THR %UIP	LEU %UIP	ILE %UIP	VAL %UIP	HIS %UIP	PHE %UIP	TRP %UIP
509	Distillers Gr. Dehy—Light	5-28-236	1.20	2.06	4.15	3.12	9.07	2.78	5.24	1.82	4.20	1.64
510	Distillers Gr. Dehy—Inter.	5-28-236	1.20	2.06	4.15	3.12	9.07	2.78	5.24	1.82	4.20	1.64
511	Distillers Gr. Dehy—Dark	5-28-236	1.20	2.06	4.15	3.12	9.07	2.78	5.24	1.82	4.20	1.64
512	Distillers Gr. Dehy—Very Dark	5-28-236	1.20	2.06	4.15	3.12	9.07	2.78	5.24	1.82	4.20	1.64
513	Distillers Gr. solubles dehy	5-28-844	1.20	2.06	4.15	3.12	9.07	2.78	5.24	1.82	4.20	1.64
514	Distillers Gr. Wet		1.20	2.06	4.15	3.12	9.07	2.78	5.24	1.82	4.20	1.64
515	Lupins		1.01	7.06	9.37	3.53	6.93	4.08	4.62	2.48	4.62	0.76
516	Peanut Meal	5-03-650	1.10	3.14	10.88	2.53	6.06	3.16	3.82	2.18	4.92	0.98
517	Soybean Meal—44	5-20-637	1.01	5.36	6.55	3.52	7.23	4.65	5.09	2.82	4.94	1.64
518	Soybean Meal—49	5-04-612	0.83	6.08	7.69	3.03	6.13	4.25	3.79	2.27	3.88	1.64
519	Soybean Whole	5-04-610	1.01	5.36	6.55	3.52	7.23	4.65	5.09	2.82	4.94	1.64
520	Soybean Whole Roasted		1.02	5.77	6.42	3.56	7.15	4.61	4.91	2.96	4.81	1.64
521	Sunflower Seed meal	5-04-739	2.15	4.29	9.87	4.51	6.86	4.29	6.87	2.36	4.94	1.93
522	Urea		0.00	0.00	0.00	0.00	0.00	0.00	0.00	0.00	0.00	0.00
523	Blank		0.00	0.00	0.00	0.00	0.00	0.00	0.00	0.00	0.00	0.00
524	Blank		0.00	0.00	0.00	0.00	0.00	0.00	0.00	0.00	0.00	0.00
525	Blank		0.00	0.00	0.00	0.00	0.00	0.00	0.00	0.00	0.00	0.00
526	Blank		0.00	0.00	0.00	0.00	0.00	0.00	0.00	0.00	0.00	0.00
527	Blank		0.00	0.00	0.00	0.00	0.00	0.00	0.00	0.00	0.00	0.00
528	Blank		0.00	0.00	0.00	0.00	0.00	0.00	0.00	0.00	0.00	0.00
529	Blank		0.00	0.00	0.00	0.00	0.00	0.00	0.00	0.00	0.00	0.00
530	Blank		0.00	0.00	0.00	0.00	0.00	0.00	0.00	0.00	0.00	0.00
531	Blank		0.00	0.00	0.00	0.00	0.00	0.00	0.00	0.00	0.00	0.00
532	Blank		0.00	0.00	0.00	0.00	0.00	0.00	0.00	0.00	0.00	0.00
533	Blank		0.00	0.00	0.00	0.00	0.00	0.00	0.00	0.00	0.00	0.00
534	Blank		0.00	0.00	0.00	0.00	0.00	0.00	0.00	0.00	0.00	0.00
535	Blank		0.00	0.00	0.00	0.00	0.00	0.00	0.00	0.00	0.00	0.00
536	Blank		0.00	0.00	0.00	0.00	0.00	0.00	0.00	0.00	0.00	0.00
537	Blank		0.00	0.00	0.00	0.00	0.00	0.00	0.00	0.00	0.00	0.00
538	Blank		0.00	0.00	0.00	0.00	0.00	0.00	0.00	0.00	0.00	0.00
539	Blank		0.00	0.00	0.00	0.00	0.00	0.00	0.00	0.00	0.00	0.00
540	Blank		0.00	0.00	0.00	0.00	0.00	0.00	0.00	0.00	0.00	0.00
541	Blank		0.00	0.00	0.00	0.00	0.00	0.00	0.00	0.00	0.00	0.00
542	Blank		0.00	0.00	0.00	0.00	0.00	0.00	0.00	0.00	0.00	0.00
543	Blank		0.00	0.00	0.00	0.00	0.00	0.00	0.00	0.00	0.00	0.00
544	Blank		0.00	0.00	0.00	0.00	0.00	0.00	0.00	0.00	0.00	0.00
545	Blank		0.00	0.00	0.00	0.00	0.00	0.00	0.00	0.00	0.00	0.00
546	Blank		0.00	0.00	0.00	0.00	0.00	0.00	0.00	0.00	0.00	0.00
547	Blank		0.00	0.00	0.00	0.00	0.00	0.00	0.00	0.00	0.00	0.00
548	Blank		0.00	0.00	0.00	0.00	0.00	0.00	0.00	0.00	0.00	0.00
549	Blank		0.00	0.00	0.00	0.00	0.00	0.00	0.00	0.00	0.00	0.00
550	Blank		0.00	0.00	0.00	0.00	0.00	0.00	0.00	0.00	0.00	0.00
601	Apple Pomace	4-00-424	0.67	2.83	2.83	2.83	5.49	2.83	3.83	1.00	3.50	4.50
602	Bakery Waste	4-00-466	1.77	3.17	4.77	4.95	7.47	4.57	4.30	1.30	4.10	1.00
603	Beet Pulp +Steffen's filt	4-00-675	0.65	3.00	4.43	3.17	4.61	2.69	4.50	1.87	2.80	1.10
604	Beet Pulp Dehydrated	4-00-669	0.65	3.00	4.43	3.17	4.61	2.69	4.50	1.87	2.80	1.10
605	Citrus Pulp Dehydrated	4-01-237	0.65	3.00	4.43	3.17	4.61	2.69	4.50	1.87	2.80	1.10
606	Grape Pomace	1-02-208	0.00	0.00	0.00	0.00	0.00	0.00	0.00	0.00	0.00	0.00
607	Soybean Hulls	1-04-560	0.47	4.54	4.72	2.74	4.86	2.46	3.30	1.84	2.99	0.67
608	Blank		0.00	0.00	0.00	0.00	0.00	0.00	0.00	0.00	0.00	0.00
609	Blank		0.00	0.00	0.00	0.00	0.00	0.00	0.00	0.00	0.00	0.00
610	Blank		0.00	0.00	0.00	0.00	0.00	0.00	0.00	0.00	0.00	0.00
611	Blank		0.00	0.00	0.00	0.00	0.00	0.00	0.00	0.00	0.00	0.00
612	Blank		0.00	0.00	0.00	0.00	0.00	0.00	0.00	0.00	0.00	0.00
613	Blank		0.00	0.00	0.00	0.00	0.00	0.00	0.00	0.00	0.00	0.00
614	Blank		0.00	0.00	0.00	0.00	0.00	0.00	0.00	0.00	0.00	0.00
615	Blank		0.00	0.00	0.00	0.00	0.00	0.00	0.00	0.00	0.00	0.00
616	Blank		0.00	0.00	0.00	0.00	0.00	0.00	0.00	0.00	0.00	0.00
617	Blank		0.00	0.00	0.00	0.00	0.00	0.00	0.00	0.00	0.00	0.00
618	Blank		0.00	0.00	0.00	0.00	0.00	0.00	0.00	0.00	0.00	0.00
619	Blank		0.00	0.00	0.00	0.00	0.00	0.00	0.00	0.00	0.00	0.00
620	Blank		0.00	0.00	0.00	0.00	0.00	0.00	0.00	0.00	0.00	0.00
701	Bloodmeal	5-00-380	1.07	9.34	5.01	4.73	13.40	0.88	9.08	6.45	7.86	1.88
702	Feather Meal	5-03-795	0.49	2.57	7.42	4.17	8.31	4.60	7.95	0.94	5.21	0.80
703	Fishmeal	5-02-009	2.84	7.13	7.19	4.17	7.01	4.53	4.81	2.30	4.33	1.52
704	Meat Meal	5-00-385	1.34	5.06	6.36	3.37	6.36	2.98	4.57	1.86	3.51	0.52
705	Tallow	4-00-376	0.00	0.00	0.00	0.00	0.00	0.00	0.00	0.00	0.00	0.00
706	Whey Acid	4-08-134	0.00	0.00	0.00	0.00	0.00	0.00	0.00	0.00	0.00	0.00
707	Whey Delact.	4-01-186	0.00	0.00	0.00	0.00	0.00	0.00	0.00	0.00	0.00	0.00
708	Blank		0.00	0.00	0.00	0.00	0.00	0.00	0.00	0.00	0.00	0.00

Calcium %DM	Phosphorus %DM	Magnesium %DM	Chlorine %DM	Potassium %DM	Sodium %DM	Sulfur %DM	Cobalt mg/kg	Copper mg/kg	Iodine mg/kg	Iron mg/kg	Manganese mg/kg	Selenium mg/kg	Zinc mg/kg	Vit A 1000 IU/kg	Vit D 1000 U/kg	Vit E IU/kg
0.32	1.40	0.65	0.28	1.83	0.24	0.40	0.18	83.90	0.09	560.00	77.60	0.40	94.80	1.20	0.00	49.40
0.26	0.83	0.33	0.28	1.08	0.30	0.44	0.18	10.60	0.09	358.00	27.60	0.40	67.80	1.00	0.60	43.00
0.32	1.40	0.65	0.28	1.83	0.24	0.40	0.18	83.90	0.09	560.00	77.60	0.40	94.80	1.20	0.00	49.40
0.32	1.40	0.65	0.28	1.83	0.24	0.40	0.18	83.90	0.09	560.00	77.60	0.40	94.80	1.20	0.00	49.40
0.32	1.40	0.65	0.28	1.83	0.24	0.40	0.18	83.90	0.09	560.00	77.60	0.40	94.80	1.20	0.00	49.40
0.32	1.40	0.65	0.28	1.83	0.24	0.40	0.18	83.90	0.09	560.00	77.60	0.40	94.80	1.20	0.00	49.40
0.26	0.44	0.00	0.00	0.91	0.00	0.00	0.00	0.00	0.00	0.00	0.00	0.00	0.00	0.00	0.00	0.00
0.32	0.66	0.17	0.00	1.28	0.03	0.33	0.00	16.00	0.07	155.00	29.00	0.00	36.00	0.00	0.00	0.00
0.40	0.71	0.31	0.00	2.22	0.04	0.46	0.12	22.40	0.00	185.00	35.00	0.51	57.00	0.00	0.00	0.00
0.29	0.71	0.33	0.08	2.36	0.01	0.48	0.12	22.50	0.12	145.00	41.00	0.22	63.00	0.00	0.00	0.00
0.27	0.65	0.27	0.03	2.01	0.04	0.35	0.00	14.58	0.00	182.00	34.50	0.12	59.00	1.60	0.00	36.60
0.27	0.65	0.29	0.03	1.80	0.00	0.24	0.00	19.80	0.00	100.00	39.60	0.12	61.80	1.60	0.00	36.60
0.45	1.02	0.70	0.11	1.27	0.03	0.33	0.00	4.00	0.00	33.00	20.00	2.30	105.00	0.00	0.00	0.00
0.00	0.00	0.00	0.00	0.00	0.00	0.00	0.00	0.00	0.00	0.00	0.00	0.00	0.00	0.00	0.00	0.00
0.00	0.00	0.00	0.00	0.00	0.00	0.00	0.00	0.00	0.00	0.00	0.00	0.00	0.00	0.00	0.00	0.00
0.00	0.00	0.00	0.00	0.00	0.00	0.00	0.00	0.00	0.00	0.00	0.00	0.00	0.00	0.00	0.00	0.00
0.00	0.00	0.00	0.00	0.00	0.00	0.00	0.00	0.00	0.00	0.00	0.00	0.00	0.00	0.00	0.00	0.00
0.00	0.00	0.00	0.00	0.00	0.00	0.00	0.00	0.00	0.00	0.00	0.00	0.00	0.00	0.00	0.00	0.00
0.00	0.00	0.00	0.00	0.00	0.00	0.00	0.00	0.00	0.00	0.00	0.00	0.00	0.00	0.00	0.00	0.00
0.00	0.00	0.00	0.00	0.00	0.00	0.00	0.00	0.00	0.00	0.00	0.00	0.00	0.00	0.00	0.00	0.00
0.00	0.00	0.00	0.00	0.00	0.00	0.00	0.00	0.00	0.00	0.00	0.00	0.00	0.00	0.00	0.00	0.00
0.00	0.00	0.00	0.00	0.00	0.00	0.00	0.00	0.00	0.00	0.00	0.00	0.00	0.00	0.00	0.00	0.00
0.00	0.00	0.00	0.00	0.00	0.00	0.00	0.00	0.00	0.00	0.00	0.00	0.00	0.00	0.00	0.00	0.00
0.00	0.00	0.00	0.00	0.00	0.00	0.00	0.00	0.00	0.00	0.00	0.00	0.00	0.00	0.00	0.00	0.00
0.00	0.00	0.00	0.00	0.00	0.00	0.00	0.00	0.00	0.00	0.00	0.00	0.00	0.00	0.00	0.00	0.00
0.00	0.00	0.00	0.00	0.00	0.00	0.00	0.00	0.00	0.00	0.00	0.00	0.00	0.00	0.00	0.00	0.00
0.00	0.00	0.00	0.00	0.00	0.00	0.00	0.00	0.00	0.00	0.00	0.00	0.00	0.00	0.00	0.00	0.00
0.00	0.00	0.00	0.00	0.00	0.00	0.00	0.00	0.00	0.00	0.00	0.00	0.00	0.00	0.00	0.00	0.00
0.00	0.00	0.00	0.00	0.00	0.00	0.00	0.00	0.00	0.00	0.00	0.00	0.00	0.00	0.00	0.00	0.00
0.00	0.00	0.00	0.00	0.00	0.00	0.00	0.00	0.00	0.00	0.00	0.00	0.00	0.00	0.00	0.00	0.00
0.00	0.00	0.00	0.00	0.00	0.00	0.00	0.00	0.00	0.00	0.00	0.00	0.00	0.00	0.00	0.00	0.00
0.00	0.00	0.00	0.00	0.00	0.00	0.00	0.00	0.00	0.00	0.00	0.00	0.00	0.00	0.00	0.00	0.00
0.00	0.00	0.00	0.00	0.00	0.00	0.00	0.00	0.00	0.00	0.00	0.00	0.00	0.00	0.00	0.00	0.00
0.00	0.00	0.00	0.00	0.00	0.00	0.00	0.00	0.00	0.00	0.00	0.00	0.00	0.00	0.00	0.00	0.00
0.00	0.00	0.00	0.00	0.00	0.00	0.00	0.00	0.00	0.00	0.00	0.00	0.00	0.00	0.00	0.00	0.00
0.00	0.00	0.00	0.00	0.00	0.00	0.00	0.00	0.00	0.00	0.00	0.00	0.00	0.00	0.00	0.00	0.00
0.00	0.00	0.00	0.00	0.00	0.00	0.00	0.00	0.00	0.00	0.00	0.00	0.00	0.00	0.00	0.00	0.00
0.00	0.00	0.00	0.00	0.00	0.00	0.00	0.00	0.00	0.00	0.00	0.00	0.00	0.00	0.00	0.00	0.00
0.00	0.00	0.00	0.00	0.00	0.00	0.00	0.00	0.00	0.00	0.00	0.00	0.00	0.00	0.00	0.00	0.00
0.00	0.00	0.00	0.00	0.00	0.00	0.00	0.00	0.00	0.00	0.00	0.00	0.00	0.00	0.00	0.00	0.00
0.23	0.11	0.00	0.00	0.53	0.00	0.11	0.00	0.00	0.00	0.00	0.00	0.00	0.00	0.00	0.00	0.00
0.15	0.24	0.18	1.61	0.43	1.12	0.02	1.34	12.10	0.00	180.00	71.20	0.00	19.50	7.70	0.00	44.90
0.70	0.10	0.28	0.04	0.20	0.21	0.20	0.08	13.70	0.00	300.00	37.70	0.00	0.08	0.40	1.00	0.00
0.68	0.10	0.28	0.04	0.22	0.20	0.22	0.08	13.80	0.00	293.00	37.70	0.12	1.00	0.40	0.60	0.00
1.88	0.13	0.17	0.00	0.77	0.08	0.08	0.19	6.15	0.00	360.00	7.00	0.00	15.00	0.00	0.00	0.00
0.58	0.17	0.10	0.01	0.91	0.09	0.00	0.00	0.00	0.00	0.00	40.70	0.00	24.20	0.00	0.00	0.00
0.53	0.18	0.22	0.00	1.29	0.03	0.11	0.12	17.80	0.00	409.00	10.00	0.14	48.00	0.00	0.00	3.70
0.00	0.00	0.00	0.00	0.00	0.00	0.00	0.00	0.00	0.00	0.00	0.00	0.00	0.00	0.00	0.00	0.00
0.00	0.00	0.00	0.00	0.00	0.00	0.00	0.00	0.00	0.00	0.00	0.00	0.00	0.00	0.00	0.00	0.00
0.00	0.00	0.00	0.00	0.00	0.00	0.00	0.00	0.00	0.00	0.00	0.00	0.00	0.00	0.00	0.00	0.00
0.00	0.00	0.00	0.00	0.00	0.00	0.00	0.00	0.00	0.00	0.00	0.00	0.00	0.00	0.00	0.00	0.00
0.00	0.00	0.00	0.00	0.00	0.00	0.00	0.00	0.00	0.00	0.00	0.00	0.00	0.00	0.00	0.00	0.00
0.00	0.00	0.00	0.00	0.00	0.00	0.00	0.00	0.00	0.00	0.00	0.00	0.00	0.00	0.00	0.00	0.00
0.00	0.00	0.00	0.00	0.00	0.00	0.00	0.00	0.00	0.00	0.00	0.00	0.00	0.00	0.00	0.00	0.00
0.00	0.00	0.00	0.00	0.00	0.00	0.00	0.00	0.00	0.00	0.00	0.00	0.00	0.00	0.00	0.00	0.00
0.00	0.00	0.00	0.00	0.00	0.00	0.00	0.00	0.00	0.00	0.00	0.00	0.00	0.00	0.00	0.00	0.00
0.00	0.00	0.00	0.00	0.00	0.00	0.00	0.00	0.00	0.00	0.00	0.00	0.00	0.00	0.00	0.00	0.00
0.00	0.00	0.00	0.00	0.00	0.00	0.00	0.00	0.00	0.00	0.00	0.00	0.00	0.00	0.00	0.00	0.00
0.00	0.00	0.00	0.00	0.00	0.00	0.00	0.00	0.00	0.00	0.00	0.00	0.00	0.00	0.00	0.00	0.00
0.00	0.00	0.00	0.00	0.00	0.00	0.00	0.00	0.00	0.00	0.00	0.00	0.00	0.00	0.00	0.00	0.00
0.40	0.32	0.04	0.33	0.31	0.40	0.80	0.10	13.90	0.00	2281.00	11.70	0.80	33.00	0.00	0.00	0.00
1.19	0.68	0.06	0.30	0.20	0.24	1.85	0.13	14.20	0.05	702.00	12.00	0.98	105.00	0.00	0.00	0.00
5.46	3.14	0.16	1.37	0.77	0.44	0.58	0.12	11.30	1.19	594.00	40.00	2.34	157.00	0.00	0.00	13.00
9.13	4.34	0.27	0.00	0.49	0.80	0.51	0.00	21.40	0.00	758.00	174.00	0.00	265.00	0.00	0.00	1.00
0.57	0.06	0.06	0.00	0.32	0.01	0.00	0.57	15.00	0.68	482.00	47.00	0.00	42.00	0.00	0.00	0.00
0.81	0.71	0.00	0.00	2.75	0.00	0.00	0.00	0.00	0.00	290.00	3.20	0.00	0.00	0.00	0.00	0.00
1.60	1.18	0.23	1.10	3.16	1.54	1.15	0.00	7.50	10.55	270.00	8.60	0.06	8.40	0.00	0.00	0.00
0.00	0.00	0.00	0.00	0.00	0.00	0.00	0.00	0.00	0.00	0.00	0.00	0.00	0.00	0.00	0.00	0.00

APPENDIX TABLE 1B   Feed Library—Amino Acids, Minerals, and Vitamins

Feed No.	Common Name	Int. Ref. No.	MET %UIP	LYS %UIP	ARG %UIP	THR %UIP	LEU %UIP	ILE %UIP	VAL %UIP	HIS %UIP	PHE %UIP	TRP %UIP
709	Blank		0.00	0.00	0.00	0.00	0.00	0.00	0.00	0.00	0.00	0.00
710	Blank		0.00	0.00	0.00	0.00	0.00	0.00	0.00	0.00	0.00	0.00
711	Blank		0.00	0.00	0.00	0.00	0.00	0.00	0.00	0.00	0.00	0.00
712	Blank		0.00	0.00	0.00	0.00	0.00	0.00	0.00	0.00	0.00	0.00
713	Blank		0.00	0.00	0.00	0.00	0.00	0.00	0.00	0.00	0.00	0.00
714	Blank		0.00	0.00	0.00	0.00	0.00	0.00	0.00	0.00	0.00	0.00
715	Blank		0.00	0.00	0.00	0.00	0.00	0.00	0.00	0.00	0.00	0.00
716	Blank		0.00	0.00	0.00	0.00	0.00	0.00	0.00	0.00	0.00	0.00
717	Blank		0.00	0.00	0.00	0.00	0.00	0.00	0.00	0.00	0.00	0.00
718	Blank		0.00	0.00	0.00	0.00	0.00	0.00	0.00	0.00	0.00	0.00
719	Blank		0.00	0.00	0.00	0.00	0.00	0.00	0.00	0.00	0.00	0.00
720	Blank		0.00	0.00	0.00	0.00	0.00	0.00	0.00	0.00	0.00	0.00
801	Ammonium Phos (Mono)	6-09-338	0.00	0.00	0.00	0.00	0.00	0.00	0.00	0.00	0.00	0.00
802	Ammonium Phos (Dibasic)	6-00-370	0.00	0.00	0.00	0.00	0.00	0.00	0.00	0.00	0.00	0.00
803	Ammonium Sulfate	6-09-339	0.00	0.00	0.00	0.00	0.00	0.00	0.00	0.00	0.00	0.00
804	Bone Meal	6-00-400	0.00	0.00	0.00	0.00	0.00	0.00	0.00	0.00	0.00	0.00
805	Calcium Carbonate	6-01-069	0.00	0.00	0.00	0.00	0.00	0.00	0.00	0.00	0.00	0.00
806	Calcium Sulfate	6-01-089	0.00	0.00	0.00	0.00	0.00	0.00	0.00	0.00	0.00	0.00
807	Cobalt Carbonate	6-01-566	0.00	0.00	0.00	0.00	0.00	0.00	0.00	0.00	0.00	0.00
808	Copper Sulfate	6-01-720	0.00	0.00	0.00	0.00	0.00	0.00	0.00	0.00	0.00	0.00
809	Dicalcium Phosphate	6-01-080	0.00	0.00	0.00	0.00	0.00	0.00	0.00	0.00	0.00	0.00
810	EDTA	6-01-842	0.00	0.00	0.00	0.00	0.00	0.00	0.00	0.00	0.00	0.00
811	Iron Sulfate	6-20-734	0.00	0.00	0.00	0.00	0.00	0.00	0.00	0.00	0.00	0.00
812	Limestone	6-02-632	0.00	0.00	0.00	0.00	0.00	0.00	0.00	0.00	0.00	0.00
813	Limestone Magnesium	6-02-633	0.00	0.00	0.00	0.00	0.00	0.00	0.00	0.00	0.00	0.00
814	Magnesium Carbonate	6-02-754	0.00	0.00	0.00	0.00	0.00	0.00	0.00	0.00	0.00	0.00
815	Magnesium Oxide	6-02-756	0.00	0.00	0.00	0.00	0.00	0.00	0.00	0.00	0.00	0.00
816	Manganese Oxide	6-03-056	0.00	0.00	0.00	0.00	0.00	0.00	0.00	0.00	0.00	0.00
817	Manganese Carbonate	6-03-036	0.00	0.00	0.00	0.00	0.00	0.00	0.00	0.00	0.00	0.00
818	Mono-Sodium Phosphate	6-04-288	0.00	0.00	0.00	0.00	0.00	0.00	0.00	0.00	0.00	0.00
819	Oystershell Ground	6-03-481	0.00	0.00	0.00	0.00	0.00	0.00	0.00	0.00	0.00	0.00
820	Phosphate Deflournated	6-01-780	0.00	0.00	0.00	0.00	0.00	0.00	0.00	0.00	0.00	0.00
821	Phosphate Rock	6-03-945	0.00	0.00	0.00	0.00	0.00	0.00	0.00	0.00	0.00	0.00
822	Phosphate Rock—Low Fl	6-03-946	0.00	0.00	0.00	0.00	0.00	0.00	0.00	0.00	0.00	0.00
823	Phosphate Rock—Soft	6-03-947	0.00	0.00	0.00	0.00	0.00	0.00	0.00	0.00	0.00	0.00
824	Phosphate Mono-Mono	6-04-288	0.00	0.00	0.00	0.00	0.00	0.00	0.00	0.00	0.00	0.00
825	Phosphoric Acid	6-03-707	0.00	0.00	0.00	0.00	0.00	0.00	0.00	0.00	0.00	0.00
826	Potassium Bicarbonate	6-29-493	0.00	0.00	0.00	0.00	0.00	0.00	0.00	0.00	0.00	0.00
827	Potassium Iodide	6-03-759	0.00	0.00	0.00	0.00	0.00	0.00	0.00	0.00	0.00	0.00
828	Potassium Sulfate	6-06-098	0.00	0.00	0.00	0.00	0.00	0.00	0.00	0.00	0.00	0.00
829	Salt	6-04-152	0.00	0.00	0.00	0.00	0.00	0.00	0.00	0.00	0.00	0.00
830	Sodium Bicarbonate	6-04-272	0.00	0.00	0.00	0.00	0.00	0.00	0.00	0.00	0.00	0.00
831	Sodium Selenite	6-26-013	0.00	0.00	0.00	0.00	0.00	0.00	0.00	0.00	0.00	0.00
832	Sodium Sulfate	6-04-292	0.00	0.00	0.00	0.00	0.00	0.00	0.00	0.00	0.00	0.00
833	Zinc Oxide	6-05-553	0.00	0.00	0.00	0.00	0.00	0.00	0.00	0.00	0.00	0.00
834	Zinc Sulfate	6-05-555	0.00	0.00	0.00	0.00	0.00	0.00	0.00	0.00	0.00	0.00
835	Potassium Chloride	6-03-755	0.00	0.00	0.00	0.00	0.00	0.00	0.00	0.00	0.00	0.00
836	Calcium Phosphate (Mono)	6-01-082	0.00	0.00	0.00	0.00	0.00	0.00	0.00	0.00	0.00	0.00
837	Sodium TriPoly Phosphate	6-08-076	0.00	0.00	0.00	0.00	0.00	0.00	0.00	0.00	0.00	0.00
838	Blank		0.00	0.00	0.00	0.00	0.00	0.00	0.00	0.00	0.00	0.00
839	Blank		0.00	0.00	0.00	0.00	0.00	0.00	0.00	0.00	0.00	0.00
840	Blank		0.00	0.00	0.00	0.00	0.00	0.00	0.00	0.00	0.00	0.00
841	Blank		0.00	0.00	0.00	0.00	0.00	0.00	0.00	0.00	0.00	0.00
842	Blank		0.00	0.00	0.00	0.00	0.00	0.00	0.00	0.00	0.00	0.00
843	Blank		0.00	0.00	0.00	0.00	0.00	0.00	0.00	0.00	0.00	0.00
844	Blank		0.00	0.00	0.00	0.00	0.00	0.00	0.00	0.00	0.00	0.00
845	Blank		0.00	0.00	0.00	0.00	0.00	0.00	0.00	0.00	0.00	0.00
901	Blank		0.00	0.00	0.00	0.00	0.00	0.00	0.00	0.00	0.00	0.00
902	Blank		0.00	0.00	0.00	0.00	0.00	0.00	0.00	0.00	0.00	0.00
903	Blank		0.00	0.00	0.00	0.00	0.00	0.00	0.00	0.00	0.00	0.00
904	Blank		0.00	0.00	0.00	0.00	0.00	0.00	0.00	0.00	0.00	0.00
905	Blank		0.00	0.00	0.00	0.00	0.00	0.00	0.00	0.00	0.00	0.00
906	Blank		0.00	0.00	0.00	0.00	0.00	0.00	0.00	0.00	0.00	0.00
907	Blank		0.00	0.00	0.00	0.00	0.00	0.00	0.00	0.00	0.00	0.00
908	Blank		0.00	0.00	0.00	0.00	0.00	0.00	0.00	0.00	0.00	0.00
909	Blank		0.00	0.00	0.00	0.00	0.00	0.00	0.00	0.00	0.00	0.00
910	Blank		0.00	0.00	0.00	0.00	0.00	0.00	0.00	0.00	0.00	0.00
999	Minerals	X-XX-XXX	0.00	0.00	0.00	0.00	0.00	0.00	0.00	0.00	0.00	0.00

Calcium %DM	Phosphorus %DM	Magnesium %DM	Chlorine %DM	Potassium %DM	Sodium %DM	Sulfur %DM	Cobalt mg/kg	Copper mg/kg	Iodine mg/kg	Iron mg/kg	Manganese mg/kg	Selenium mg/kg	Zinc mg/kg	Vit A 1000 IU/kg	Vit D 1000 U/kg	Vit E IU/kg
0.00	0.00	0.00	0.00	0.00	0.00	0.00	0.00	0.00	0.00	0.00	0.00	0.00	0.00	0.00	0.00	0.00
0.00	0.00	0.00	0.00	0.00	0.00	0.00	0.00	0.00	0.00	0.00	0.00	0.00	0.00	0.00	0.00	0.00
0.00	0.00	0.00	0.00	0.00	0.00	0.00	0.00	0.00	0.00	0.00	0.00	0.00	0.00	0.00	0.00	0.00
0.00	0.00	0.00	0.00	0.00	0.00	0.00	0.00	0.00	0.00	0.00	0.00	0.00	0.00	0.00	0.00	0.00
0.00	0.00	0.00	0.00	0.00	0.00	0.00	0.00	0.00	0.00	0.00	0.00	0.00	0.00	0.00	0.00	0.00
0.00	0.00	0.00	0.00	0.00	0.00	0.00	0.00	0.00	0.00	0.00	0.00	0.00	0.00	0.00	0.00	0.00
0.00	0.00	0.00	0.00	0.00	0.00	0.00	0.00	0.00	0.00	0.00	0.00	0.00	0.00	0.00	0.00	0.00
0.00	0.00	0.00	0.00	0.00	0.00	0.00	0.00	0.00	0.00	0.00	0.00	0.00	0.00	0.00	0.00	0.00
0.00	0.00	0.00	0.00	0.00	0.00	0.00	0.00	0.00	0.00	0.00	0.00	0.00	0.00	0.00	0.00	0.00
0.00	0.00	0.00	0.00	0.00	0.00	0.00	0.00	0.00	0.00	0.00	0.00	0.00	0.00	0.00	0.00	0.00
0.00	0.00	0.00	0.00	0.00	0.00	0.00	0.00	0.00	0.00	0.00	0.00	0.00	0.00	0.00	0.00	0.00
0.00	0.00	0.00	0.00	0.00	0.00	0.00	0.00	0.00	0.00	0.00	0.00	0.00	0.00	0.00	0.00	0.00
0.28	24.74	0.46	0.00	0.01	0.06	1.46	10.00	10.00	0.00	17400.00	400.00	0.00	100.00	0.00	0.00	0.00
0.52	20.60	0.46	0.00	0.01	0.05	2.16	10.00	10.00	0.00	12400.00	400.00	0.00	100.00	0.00	0.00	0.00
0.00	0.00	0.00	0.00	0.00	0.00	24.10	0.00	1.00	0.00	10.00	1.00	0.00	0.00	0.00	0.00	0.00
30.71	12.86	0.33	0.00	0.19	5.69	2.51	0.00	0.00	0.00	26700.00	0.00	0.00	100.00	0.00	0.00	0.00
39.39	0.04	0.05	0.00	0.06	0.06	0.00	0.00	0.00	0.00	300.00	300.00	0.00	0.00	0.00	0.00	0.00
23.28	0.00	0.00	0.00	0.00	0.00	18.62	0.00	0.00	0.00	0.00	0.00	0.00	0.00	0.00	0.00	0.00
0.00	0.00	0.00	0.00	0.00	0.00	0.20	460000.00	0.00	0.00	500.00	0.00	0.00	0.00	0.00	0.00	0.00
0.00	0.00	0.00	0.00	0.00	0.00	12.84	0.00	254500.00	0.00	0.00	0.00	0.00	0.00	0.00	0.00	0.00
22.00	19.30	0.59	0.00	0.07	0.05	1.14	10.00	10.00	0.00	14400.00	300.00	0.00	100.00	0.05	0.00	0.00
0.00	0.00	0.00	0.00	0.00	0.00	0.00	0.00	0.00	803400.00	0.00	0.00	0.00	0.00	0.00	0.00	0.00
0.00	0.00	0.00	0.00	0.00	0.00	12.35	0.00	0.00	0.00	218400.00	0.00	0.00	0.00	0.00	0.00	0.00
34.00	0.02	2.06	0.03	0.12	0.06	0.04	0.00	0.00	0.00	3500.00	0.00	0.00	0.00	0.00	0.00	0.00
22.30	0.04	9.99	0.12	0.36	0.04	0.00	0.00	0.00	0.00	770.00	0.00	0.00	0.00	0.00	0.00	0.00
0.02	0.00	30.81	0.00	0.00	0.00	0.00	0.00	0.00	0.00	220.00	0.00	0.00	0.00	0.00	0.00	0.00
3.07	0.00	56.20	0.00	0.00	0.00	0.00	0.00	0.00	0.00	100.00	0.00	0.00	0.00	0.00	0.00	0.00
0.00	0.00	0.00	0.00	0.00	0.00	0.00	0.00	0.00	0.00	774500.00	0.00	0.00	0.00	0.00	0.00	0.00
0.00	0.00	0.00	0.00	0.00	0.00	0.00	0.00	0.00	0.00	478000.00	0.00	0.00	0.00	0.00	0.00	0.00
0.00	22.50	0.00	0.00	0.00	16.68	0.00	0.00	0.00	0.00	0.00	0.00	0.00	0.00	0.00	0.00	0.00
38.00	0.07	0.30	0.01	0.10	0.21	0.00	0.00	0.00	0.00	2870.00	100.00	0.00	0.00	0.00	0.00	0.00
32.00	18.00	0.42	0.00	0.08	4.90	0.00	10.00	20.00	0.00	6700.00	200.00	0.00	60.00	0.00	0.00	0.00
35.00	13.00	0.41	0.00	0.06	0.03	0.00	10.00	10.00	0.00	16800.00	200.00	0.00	100.00	0.00	0.00	0.00
36.00	14.00	0.00	0.00	0.00	0.00	0.00	0.00	0.00	0.00	0.00	0.00	0.00	0.00	0.00	0.00	0.00
17.00	9.00	0.38	0.00	0.00	0.10	0.00	0.00	0.00	0.00	19000.00	1000.00	0.00	0.00	0.00	0.00	0.00
0.00	22.50	0.00	0.00	0.00	16.68	0.00	0.00	0.00	0.00	0.00	0.00	0.00	0.00	0.00	0.00	0.00
0.05	31.60	0.51	0.00	0.02	0.04	1.55	10.00	10.00	0.00	17500.00	500.00	0.00	130.00	0.00	0.00	0.00
0.00	0.00	0.00	0.00	39.05	0.00	0.00	0.00	0.00	0.00	0.00	0.00	0.00	0.00	0.00	0.00	0.00
0.00	0.00	0.00	0.00	21.00	0.00	0.00	0.00	0.00	681700.00	0.00	0.00	0.00	0.00	0.00	0.00	0.00
0.15	0.00	0.61	1.55	41.84	0.09	17.35	0.00	0.00	0.00	710.00	10.00	0.00	0.00	0.00	0.00	0.00
0.00	0.00	0.00	60.66	0.00	39.34	0.00	0.00	0.00	0.00	0.00	0.00	0.00	0.00	0.00	0.00	0.00
0.00	0.00	0.00	0.00	0.00	27.00	0.00	0.00	0.00	0.00	0.00	0.00	0.00	0.00	0.00	0.00	0.00
0.00	0.00	0.00	0.00	0.00	26.60	0.00	0.00	0.00	0.00	0.00	456000.00	0.00	0.00	0.00	0.00	0.00
0.00	0.00	0.00	0.00	0.00	14.27	9.95	0.00	0.00	0.00	0.00	0.00	0.00	0.00	0.00	0.00	0.00
0.00	0.00	0.00	0.00	0.00	0.00	0.00	0.00	0.00	0.00	0.00	0.00	780000.00	0.00	0.00	0.00	0.00
0.02	0.00	0.00	0.02	0.00	0.00	17.68	0.00	0.00	0.00	10.00	10.00	0.00	363600.00	0.00	0.00	0.00
0.05	0.00	0.34	47.30	50.00	1.00	0.45	0.00	0.00	0.00	600.00	0.00	0.00	0.00	0.00	0.00	0.00
16.40	21.60	0.61	0.00	0.08	0.06	1.22	10.00	10.00	0.00	15800.00	360.00	0.00	90.00	0.00	0.00	0.00
0.00	25.00	0.00	0.00	0.00	31.00	0.00	0.00	0.00	0.00	40.00	0.00	0.00	0.00	0.00	0.00	0.00
0.00	0.00	0.00	0.00	0.00	0.00	0.00	0.00	0.00	0.00	0.00	0.00	0.00	0.00	0.00	0.00	0.00
0.00	0.00	0.00	0.00	0.00	0.00	0.00	0.00	0.00	0.00	0.00	0.00	0.00	0.00	0.00	0.00	0.00
0.00	0.00	0.00	0.00	0.00	0.00	0.00	0.00	0.00	0.00	0.00	0.00	0.00	0.00	0.00	0.00	0.00
0.00	0.00	0.00	0.00	0.00	0.00	0.00	0.00	0.00	0.00	0.00	0.00	0.00	0.00	0.00	0.00	0.00
0.00	0.00	0.00	0.00	0.00	0.00	0.00	0.00	0.00	0.00	0.00	0.00	0.00	0.00	0.00	0.00	0.00
0.00	0.00	0.00	0.00	0.00	0.00	0.00	0.00	0.00	0.00	0.00	0.00	0.00	0.00	0.00	0.00	0.00
0.00	0.00	0.00	0.00	0.00	0.00	0.00	0.00	0.00	0.00	0.00	0.00	0.00	0.00	0.00	0.00	0.00
0.00	0.00	0.00	0.00	0.00	0.00	0.00	0.00	0.00	0.00	0.00	0.00	0.00	0.00	0.00	0.00	0.00
0.00	0.00	0.00	0.00	0.00	0.00	0.00	0.00	0.00	0.00	0.00	0.00	0.00	0.00	0.00	0.00	0.00
0.00	0.00	0.00	0.00	0.00	0.00	0.00	0.00	0.00	0.00	0.00	0.00	0.00	0.00	0.00	0.00	0.00
0.00	0.00	0.00	0.00	0.00	0.00	0.00	0.00	0.00	0.00	0.00	0.00	0.00	0.00	0.00	0.00	0.00
0.00	0.00	0.00	0.00	0.00	0.00	0.00	0.00	0.00	0.00	0.00	0.00	0.00	0.00	0.00	0.00	0.00
0.00	0.00	0.00	0.00	0.00	0.00	0.00	0.00	0.00	0.00	0.00	0.00	0.00	0.00	0.00	0.00	0.00
0.00	0.00	0.00	0.00	0.00	0.00	0.00	0.00	0.00	0.00	0.00	0.00	0.00	0.00	0.00	0.00	0.00
0.00	0.00	0.00	0.00	0.00	0.00	0.00	0.00	0.00	0.00	0.00	0.00	0.00	0.00	0.00	0.00	0.00
0.00	0.00	0.00	0.00	0.00	0.00	0.00	0.00	0.00	0.00	0.00	0.00	0.00	0.00	0.00	0.00	0.00
0.00	0.00	0.00	0.00	0.00	0.00	0.00	0.00	0.00	0.00	0.00	0.00	0.00	0.00	0.00	0.00	0.00
0.00	0.00	0.00	0.00	0.00	0.00	0.00	0.00	0.00	0.00	0.00	0.00	0.00	0.00	0.00	0.00	0.00

APPENDIX TABLE 2   Cow Condition Score

Condition Score	Body fat, %[a]	Appearance of cow[b]
1	3.8	Emaciated—Bone structure of shoulder, ribs, back, hooks and pins sharp to touch and easily visible. Little evidence of fat deposits or muscling.
2	7.5	Very thin—Little evidence of fat deposits but some muscling in hindquarters. The spinous processes feel sharp to the touch and are easily seen, with space between them.
3	11.3	Thin—Beginning of fat cover over the loin, back and foreribs. Backbone still highly visible. Processes of the spine can be identified individually by touch and may still be visible. Spaces between the processes are less pronounced.
4	15.1	Borderline—Foreribs not noticeable; 12th and 13th ribs still noticeable to the eye, particularly in cattle with a big spring of rib and ribs wide apart. The transverse spinous processes can be identified only by palpation (with slight pressure) to feel rounded rather than sharp. Full but straightness of muscling in the hindquarters.
5	18.9	Moderate—12th and 13th ribs not visible to the eye unless animal has been shrunk. The transverse spinous processes can only be felt with firm pressure to feel rounded—not noticeable to the eye. Spaces between processes not visible and only distinguishable with firm pressure. Areas on each side of the tail head are fairly well filled but not mounded.
6	22.6	Good—Ribs fully covered, not noticeable to the eye. Hindquarters plump and full. Noticeable sponginess to covering of foreribs and on each side of the tail head. Firm pressure now required to feel transverse process.
7	26.4	Very Good—Ends of the spinous processes can only be felt with very firm pressure. Spaces between processes can barely be distinguished at all. Abundant fat cover on either side of tail head with some patchiness evident.
8	30.2	Fat—Animal taking on a smooth, blocky appearance; bone structure disappearing from sight. Fat cover thick and spongy with patchiness likely.
9	33.9	Very fat—Bone structure not seen or easily felt. Tail head buried in fat. Animal's mobility may actually be impaired by excess amount of fat.

[a]Based on the model presented in this chapter.
[b]Adapted from Herd and Sprott, 1986.

APPENDIX TABLE 3   Condition Score Resulting from Various Rates of Gain[a]

	Condition Score				
	1	3	5	7	9
	Description of condition score				
	Very thin	Average		Very fat	
Mature or Finishing Weight, lb	------------ Previous daily gain, lb/day ------------				
880	0.66	0.97	1.30	1.60	1.90
1030	0.73	1.06	1.39	1.72	2.05
1180	0.79	1.15	1.50	1.85	2.20
1325	0.84	1.21	1.60	1.97	2.35
1470	0.88	1.30	1.70	2.09	2.50

[a]Fox et al., 1988.

APPENDIX TABLE 4   Breed Maintenance Requirement Multipliers, Birth Weights, and Peak Milk Production[a]

Breed	Code	NE_m (BE)	Birth wt. lb (CBW)	Peak milk yield lb/day (PKYD)
Angus	1	1.0	68.3	17.6
Braford	2	.95	79.4	15.4
Brahman	3	.9	68.3	17.6
Brangus	4	.95	72.8	17.6
Braunvieh	5	.95	86.0	26.5
Charolais	6	1.0	86.0	19.8
Chianina	7	1.1	90.4	13.2
Devon	8	1.0	70.5	17.6
Galloway	9	1.0	79.4	17.6
Gelbvieh	10	1.0	86.0	25.4
Hereford	11	1.0	79.4	15.4
Holstein	12	1.2	94.8	33.1
Jersey	13	1.2	68.3	26.5
Limousin	14	1.0	81.6	19.8
Longhorn	15	1.0	72.8	11.0
Maine Anjou	16	1.0	88.2	19.8
Nellore	17	.9	88.2	15.4
Piedmontese	18	1.0	83.8	15.4
Pinzgauer	19	1.0	72.8	24.3
Polled Here.	20	1.0	72.8	15.4
Red Poll	21	1.0	79.4	22.0
Sahiwal	22	.9	83.8	17.6
Salers	23	1.0	77.2	19.8
S.Gertudis	24	.9	72.8	17.6
Shorthorn	25	1.0	81.6	18.7
Simmental	26	1.2	86.0	26.5
South Devon	27	1.0	72.8	17.6
Tarentaise	28	1.0	72.8	19.8

[a]Variable names (BE, CBW, PKYD) are used in various equations to predict cow requirements.

APPENDIX TABLE 5   Additive Codes and Adjustment Factors

Code	Additive	DMI	NE_{ma}
1	No anabolic implant or ionophore	.94	1.00
2	Ionophore only	.94	1.12
3	Implant only	1.00	1.00
4	Ionophore + implant	1.00	1.12

APPENDIX TABLE 6   Digestion Rates (%/hr) of Grains[a]

Ingredient	Carbohydrate			Protein		
	A	B1	B2	B1	B2	B3
Corn						
Dry, whole shell corn						
Whole	75–150	5–10	3–5	120–150	3–5	.06–.07
Corn, cracked	100–200	10–20	5–7	140–160	4–6	.08–.10
Corn, meal	200–300	20–30	7–9	150–175	6–9	.09–.12
High moisture corn						
>35% moisture						
Whole	150–200	10–15	5–7	140–160	4–6	.09–.12
Coarsely rolled	200–300	15–20	6–8	200–250	9–10	.10–.20
Intermediate rolled	300–400	20–30	6–8	200–250	10–11	.15–.25
Finely rolled	300–400	30–40	8–10	200–250	11–12	.20–.30
30–35% moisture						
Whole	100–150	10–15	4–6	125–150	4–7	.08–.09
Coarsely rolled	150–250	15–20	6–8	125–150	8–9	.09–.15
Intermediate rolled	250–350	20–30	6–8	125–250	9–10	.10–.20
Finely rolled	250–350	30–40	8–10	125–250	10–11	.15–.25
25–30% moisture						
Whole	75–125	10–15	4–6	120–150	3–5	.07–.08
Coarsely rolled	125–175	15–20	6–8	120–150	6–7	.09–.10
Intermediate rolled	250–350	20–30	6–8	120–150	8–9	.10–.15
Finely rolled	250–350	30–40	8–10	120–150	9–10	.10–.20
<25% moisture						
Whole	75–125	10–15	3–5	120–150	3–5	.06–.07
Coarsely rolled	150–200	15–20	6–8	120–150	5–6	.07–.08
Intermediate rolled	200–300	20–30	6–8	120–150	7–8	.08–.10
Finely rolled	250–350	30–40	6–8	200–300	8–9	.09–.15
Steam-flaked corn	150–200	20–30	6–8	120–150	5–6	.07–.08
Sorghum						
Dry, rolled	100–200	5–15	4–5	120–150	6–8	.09–.15
Steam-flaked	200–300	15–20	6–8	150–170	8–10	.10–.20
Oats						
Ground	250–350	30–40	4–6	300–350	12–15	.20–.50
Barley						
Rolled, dry	250–350	20–30	4–6	250–350	12–15	.20–.50
Rolled, wet	250–350	30–40	4–6	250–350	13–16	.25–.55
Wheat						
Dry, rolled	250–350	35–45	8–10	250–350	12–15	.20–.50
Steam-flaked	250–350	40–50	10–14	300–400	14–16	.25–.55

[a]Sniffen et al., 1992

APPENDIX TABLE 7   Digestion Rates (%/hr) of Proteinaceous Feeds[a]

Ingredient	Carbohydrate			Protein		
	A	B1	B2	B1	B2	B3
Soybean						
Whole, raw	250–350	25–35	2–4	150–250	8–10	.10–.30
Whole, heated	250–350	35–45	4–6	100–200	5–6	.15–.20
Meal, solvent	250–350	40–50	4–8	200–260	9–12	.10–.30
Meal, expeller	250–350	35–45	4–8	150–250	6–8	.15–.20
Canola, solvent	250–350	40–50	4–8	200–260	11–13	.10–.30
Peanut, solvent	250–350	40–50	4–8	200–260	12–14	.10–.30
Cottonseed						
Solvent	250–350	30–40	4–8	120–200	8–10	.10–.20
Expeller	250–350	25–35	4–8	100–150	6–8	.10–.15
Whole, linted	250–350	20–30	3–5	150–200	10–12	.20–.30
Whole, delinted	250–350	20–30	1–2	100–200	8–10	.20–.30
Corn gluten meal	250–350	40–60	4–6	100–200	2–4	.05–.10
Corn gluten feed	250–350	40–60	6–8	100–200	2–4	.05–.10
Corn distillers w/sol	250–350	15–20	6–8	100–200	3–4	.05–.15
Wheat middlings	250–350	60–85	10–15	200–300	5–6	.08–.15
Animal meals						
Fishmeal	250–350	15–20	3–5	100–200	5–6	.08–.15
Meat and bonemeal	250–350	15–20	3–5	100–200	5–6	.08–.15
Bloodmeal	250–350	15–20	3–5	50–100	2–4	.05–.08
Feathermeal	250–350	15–20	3–5	100–150	3–4	.05–.10
Brewers grain	250–350	35–40	4–8	100–200	6–8	.10–.20
Alfalfa meal, dehy	250–350	35–40	8–10	100–200	7–9	.10–.20
Whey	—	250–350	—	—	300–400	—

[a]Sniffen et al., 1992

APPENDIX TABLE 8   Digestion Rates (%/hr) of Forages[a]

Ingredient	Carbohydrate			Protein		
	A	B1	B2	B1	B2	B3
Corn silage						
>40% DM						
Coarsely chopped	200–300	10–20	3–6	150–250	8–9	.08–.10
Finely chopped	250–350	20–30	4–8	250–350	10–12	.10–.20
30–40% DM						
Coarsely chopped	200–300	15–25	4–8	200–300	9–10	.10–.20
Finely chopped	250–350	25–30	8–10	250–350	10–11	.15–.25
<30% DM						
Coarsely chopped	200–300	25–35	4–8	250–350	10–11	.15–.25
Finely chopped	250–350	35–40	8–10	250–350	10–12	.20–.30
Legumes						
Hay	200–300	25–35	3–6	100–200	8–10	1.0–1.5
Silage						
Coarsely chopped	200–300	30–40	4–7	100–200	10–12	1.5–2.0
Finely chopped	250–350	35–45	5–9	100–200	12–14	1.5–2.0
Grasses						
Hay	200–300	25–35	2–4	120–150	10–12	.08–.10
Silage						
Coarsely chopped	200–300	35–40	3–5	200–250	12–14	1.0–1.2
Finely chopped	200–300	40–45	4–6	250–300	13–15	1.1–1.3

[a]Sniffen et al., 1992

APPENDIX TABLE 9   Effective NDF values for feeds[a]

Ingredient	eNDF[b] % of NDF
Light weight concentrates	
Dried brewers grains	18
Wheat middlings	2
Soybean mill feed	33
Citrus pulp	33
Beet pulp	33
Wheat bran	33
Whole cottonseed	100
Whole soybeans	100
Dehy alfalfa	6
Corn cobs, ground	56
Intermediate weight concentrates	
Barley, ground	34
Wheat, ground	34
Oats, ground	34
Fish meal	9
Hominy feed	9
Distillers, w/sol	4
Corn and cobmeal	56
Blood meal	9
Heavy weight concentrates	
Whole dry corn	100
Corn meal	48
Cracked corn	60
High moisture corn	
Whole	100
Coarsely rolled	70
Intermediately rolled	60
Finely rolled	48
Soybean meal	23
Cottonseed meal	36
Corn gluten meal	36
Corn gluten feed	36
Peanut meal	36
Meat and bonemeal	8
Legumes	
High quality, 18–21% CP	
Long	92
20% > 1″ length	82
1/4″ length	67
Average quality, <18% CP	
Long	92
20% > 1″ length	82
1/4″ length	67
Grasses	
Long	98
20% > 1″ length	88
1/4″ length	73
Corn Silage	
Mature, >50% Grain	
Normal chop	71
Fine chop	61
Intermediate, 30–50% Grain	
Normal chop	81
Fine chop	71
Immature, <30% Grain	
Normal chop	81
Fine chop	71

[a]Sniffen et al., 1992
[b]Equals the proportion of the NDF that is effective in stimulating rumination, and is defined as the percent remaining on a 1.18 mm screen after dry sieving.

APPENDIX TABLE 10   Post-ruminal Starch Digestibilities (%)[a]

Feed	% Entering Intestines
Corn	
Whole corn	50–60
Dry, rolled	65–75
Cracked	70–80
Corn meal	80–90
High moisture, whole	80–90
High moisture, ground	85–95
Steam-flaked	92–97
Sorghum	
Dry, rolled	60–70
Dry, ground	70–80
Steam-flaked	90–95

[a]Sniffen et al., 1992

APPENDIX TABLE 11   Predicted Biological Values of Feeds with Different Digestion and Passage Rates[a]

Item	Unit	Corn Sil 40% Grain	Brome Hay Midbloom	Alfalfa Hay Midbloom	Corn Dry Grain 56	Corn HM Grain 56	Soybean Meal-49	Soybean Whole	Soybean Whl Roast
					@ Passage Rate of 2%/h				
DIP	% CP	79	63	71	64	77	84	87	58
UIP	% CP	21	37	29	36	23	16	13	42
TDN	% DM	70	60	60	85	86	86	86	84
NE_g	Mcal/kg	1.13	0.79	0.78	1.59	1.63	1.62	1.64	1.57
MTP[b]	g/kg	62	48	51	71	79	73	60	48
					@ Passage Rate of 4%/h				
DIP	% CP	75	58	63	52	72	75	81	46
UIP	% CP	25	42	37	48	28	25	19	54
TDN	% DM	65	53	57	82	85	85	85	84
NE_g	Mcal/kg	0.96	0.52	0.68	1.52	1.59	1.60	1.61	1.57
MTP	g/kg	55	36	46	61	74	66	55	43
					@ Passage Rate of 6%/h				
DIP	% CP	72	54	58	45	68	68	76	38
UIP	% CP	28	46	42	55	32	32	24	62
TDN	% DM	62	49	56	80	83	84	84	84
NE_g	Mcal/kg	0.86	0.36	0.61	1.46	1.56	1.58	1.58	1.57
MTP	g/kg	50	33	43	54	70	61	51	39
					@ Passage Rate of 8%/h				
DIP	% CP	69	51	54	39	65	63	72	33
UIP	% CP	31	49	46	61	35	37	28	67
@ pH = 6.5									
TDN	% DM	60	47	54	79	83	84	84	84
NE_g	Mcal/kg	0.79	0.26	0.57	1.42	1.53	1.57	1.57	1.56
MTP	g/kg	46	30	41	48	66	57	48	36
@ pH = 5.7[c]									
TDN	% DM	52	36	49	78	82	83	82	84
NE_g	Mcal/kg	0.49	−0.25	0.36	1.39	1.50	1.55	1.51	1.58
MTP	g/kg	21	10	20	27	38	33	27	21

[a]All values are predicted by the Level 2 model.
[b]MTP is microbial true protein yield,
[c]Microbial yield is reduced by 40% at pH 5.7.

APPENDIX TABLE 12   Predicting Peak Milk in Beef Cows[a]

Mature Weight (lb)	Peak Milk lb/day				
	6	12	18	24	30
	Avg. expected 7 month male calf weight (lb)				
880	398	444	477	—	—
950	416	460	493	—	—
1030	431	475	510	546	574
1100	449	491	526	561	590
1170	464	506	541	576	607
1250	477	521	557	590	623
1320	491	537	572	605	638
1400	504	550	587	620	656
1470	517	565	601	634	671

[a]Fox et al., 1988.

APPENDIX TABLE 13 Energy Reserves for Cows with Different Body Sizes and Condition Scores

| Body CS | Mature weight (lb) at body condition score 5 | | | | | | | | |
| | 800 | 900 | 1,000 | 1,100 | 1,200 | 1,300 | 1,400 | 1,500 | 1,600 |
	Mcal NE required or provided for each CS[a,b]								
2	101	114	126	139	151	164	177	189	202
3	114	129	143	157	172	186	200	214	229
4	131	147	163	180	196	212	229	245	261
5	151	170	188	207	226	245	264	283	301
6	176	198	220	242	264	286	308	330	351
7	208	234	260	285	311	337	363	389	415
8	249	280	311	342	373	405	436	467	498
9	304	342	380	418	456	494	532	570	608

[a]Represents the energy mobilized in moving to the next lower score, or required to move from the next lower score to this one. Each kg of SBW change contains 5.82 Mcal, and SBW at CS 1 through 9 are 76.5, 81.3, 86.7, 92.9, 100, 108.3, 118.1, 129.9, and 144.3% of CS 5 weight, respectively.

APPENDIX TABLE 14 Maintenance Requirement Multipliers for Representative Environmental Conditions[a,b]

| | Hair coat code[c] at 30° F | | Hair coat code[c] at 10° F | | Hair coat code[c] at −10° F | |
	1	3	1	3	1	3
Hide code[d]	Beef cow wintering ration (.60 Mcal NE_m/lb DM)					
			Wind @ 1.0 mph			
1	1.19	1.19	1.29	1.68	1.58	2.07
2	1.19	1.19	1.29	1.55	1.41	1.92
3	1.19	1.19	1.29	1.45	1.39	1.79
			Wind @ 10 mph			
1	1.22	1.48	1.60	1.94	1.98	2.39
2	1.19	1.41	1.47	1.84	1.82	2.27
3	1.19	1.34	1.36	1.75	1.69	2.17
	Typical calf wintering ration (.35 Mcal NE_g/lb DM)					
			Wind @ 1.0 mph			
1	1.19	1.47	1.50	1.93	1.87	2.39
2	1.19	1.37	1.36	1.80	1.69	2.23
3	1.19	1.28	1.29	1.69	1.55	2.09
			Wind @ 10 mph			
1	1.41	1.69	1.85	2.20	2.29	2.72
2	1.30	1.61	1.71	2.10	2.12	2.59
3	1.21	1.54	1.60	2.01	1.98	2.48
	Typical finishing ration (.62 Mcal NE_g/lb DM)					
			Wind @ 1.0 mph			
1	1.19	1.19	1.33	1.76	1.69	2.21
2	1.19	1.19	1.29	1.63	1.51	2.05
3	1.19	1.19	1.29	1.51	1.39	1.92
			Wind @ 10 mph			
1	1.24	1.52	1.67	2.03	2.11	2.54
2	1.19	1.44	1.54	1.93	1.95	2.42
3	1.19	1.36	1.42	1.83	1.81	2.31

[a]This table was developed from the Level 2 model on the computer disk, assuming a winter hair depth of 0.5 inches.
[b]Values given are NE_m required for conditions described, divided by no stress maintenance requirement (77 kcal/BW_{kg}$^{.75}$).
[c]1 is dry and clean and 3 is wet and matted.
[d]1 is thin (typical of Holstein and Zebu types), 2 is average, 3 is thick (hide thickness similar to Hereford types).

APPENDIX TABLE 15   Diet Nutrient Densities for Growing and Finishing Cattle

1000 @ finishing (28% body fat—for feedlot steers and heifers) or maturity (replacement heifers).

Body Weight (lb)	TDN % DM	NE$_m$ Mcal/lb	NE$_g$ Mcal/lb	DMI lb/day	ADG lb/day	CP % DM	Ca % DM	P % DM
550	50	0.45	0.20	15.2	0.64	7.1%	0.21%	0.13%
	60	0.61	0.35	16.1	1.77	9.8%	0.36%	0.19%
	70	0.76	0.48	15.7	2.68	12.4%	0.49%	0.24%
	80	0.90	0.61	14.8	3.34	14.9%	0.61%	0.29%
	90	1.04	0.72	13.7	3.75	17.3%	0.73%	0.34%
600	50	0.45	0.20	16.2	0.64	7.0%	0.21%	0.13%
	60	0.61	0.35	17.2	1.77	9.5%	0.34%	0.18%
	70	0.76	0.48	16.8	2.68	11.9%	0.45%	0.23%
	80	0.90	0.61	15.8	3.34	14.3%	0.56%	0.27%
	90	1.04	0.72	14.6	3.75	16.5%	0.66%	0.32%
650	50	0.45	0.20	17.3	0.64	6.9%	0.20%	0.12%
	60	0.61	0.35	18.2	1.77	9.2%	0.32%	0.17%
	70	0.76	0.48	17.8	2.68	11.5%	0.42%	0.21%
	80	0.90	0.61	16.8	3.34	13.7%	0.52%	0.26%
	90	1.04	0.72	15.5	3.75	15.9%	0.61%	0.30%
700	50	0.45	0.20	18.2	0.64	6.8%	0.19%	0.12%
	60	0.61	0.35	19.3	1.77	8.8%	0.30%	0.16%
	70	0.76	0.48	18.8	2.68	10.9%	0.39%	0.20%
	80	0.90	0.61	17.8	3.34	13.0%	0.48%	0.24%
	90	1.04	0.72	16.4	3.75	15.0%	0.56%	0.28%
750	50	0.45	0.20	19.2	0.64	6.7%	0.19%	0.12%
	60	0.61	0.35	20.3	1.77	8.5%	0.28%	0.16%
	70	0.76	0.48	19.8	2.68	10.3%	0.37%	0.19%
	80	0.90	0.61	18.7	3.34	12.2%	0.45%	0.23%
	90	1.04	0.72	17.3	3.75	14.0%	0.52%	0.26%
800	50	0.45	0.20	20.2	0.64	6.5%	0.19%	0.12%
	60	0.61	0.35	21.3	1.77	8.1%	0.27%	0.15%
	70	0.76	0.48	20.8	2.68	9.8%	0.34%	0.18%
	80	0.90	0.61	19.6	3.34	11.5%	0.42%	0.22%
	90	1.04	0.72	18.1	3.75	13.2%	0.48%	0.25%

APPENDIX TABLE 16   Diet Nutrient Densities for Growing and Finishing Cattle

1,100 @ finishing (28% body fat—for feedlot steers and heifers) or maturity (replacement heifers).

Body Weight (lb)	TDN % DM	NE$_m$ Mcal/lb	NE$_g$ Mcal/lb	DMI lb/day	ADG lb/day	CP % DM	Ca % DM	P % DM
605	50	0.45	0.20	16.3	0.68	7.2%	0.22%	0.13%
	60	0.61	0.35	17.3	1.88	10.0%	0.36%	0.19%
	70	0.76	0.48	16.9	2.86	12.7%	0.49%	0.24%
	80	0.90	0.61	15.9	3.56	15.3%	0.61%	0.29%
	90	1.04	0.72	14.7	4.00	17.8%	0.72%	0.34%
660	50	0.45	0.20	17.5	0.68	7.1%	0.21%	0.13%
	60	0.61	0.35	18.4	1.88	9.7%	0.34%	0.18%
	70	0.76	0.48	18.0	2.86	12.3%	0.45%	0.23%
	80	0.90	0.61	17.0	3.56	14.7%	0.56%	0.27%
	90	1.04	0.72	15.7	4.00	17.1%	0.66%	0.32%
715	50	0.45	0.20	18.5	0.68	6.9%	0.20%	0.13%
	60	0.61	0.35	19.6	1.88	9.2%	0.32%	0.17%
	70	0.76	0.48	19.1	2.86	11.5%	0.42%	0.21%
	80	0.90	0.61	18.1	3.56	13.7%	0.52%	0.26%
	90	1.04	0.72	16.7	4.00	15.9%	0.61%	0.30%
770	50	0.45	0.20	19.6	0.68	6.8%	0.20%	0.12%
	60	0.61	0.35	20.7	1.88	8.8%	0.30%	0.16%
	70	0.76	0.48	20.2	2.86	10.9%	0.39%	0.20%
	80	0.90	0.61	19.1	3.56	12.9%	0.48%	0.24%
	90	1.04	0.72	17.6	4.00	14.8%	0.56%	0.28%
825	50	0.45	0.20	20.6	0.68	6.6%	0.19%	0.12%
	60	0.61	0.35	21.8	1.88	8.4%	0.28%	0.16%
	70	0.76	0.48	21.3	2.86	10.3%	0.37%	0.19%
	80	0.90	0.61	20.1	3.56	12.1%	0.44%	0.23%
	90	1.04	0.72	18.6	4.00	13.9%	0.52%	0.26%
880	50	0.45	0.20	21.7	0.68	6.5%	0.19%	0.12%
	60	0.61	0.35	22.9	1.88	8.1%	0.27%	0.15%
	70	0.76	0.48	22.4	2.86	9.8%	0.34%	0.18%
	80	0.90	0.61	21.1	3.56	11.4%	0.42%	0.22%
	90	1.04	0.72	19.5	4.00	13.1%	0.48%	0.25%

APPENDIX TABLE 17   Diet Nutrient Densities for Growing and Finishing Cattle

Body Weight (lb)	TDN % DM	NE$_m$ Mcal/lb	NE$_g$ Mcal/lb	DMI lb/day	ADG lb/day	CP % DM	Ca % DM	P % DM
\multicolumn{9}{l}{1,200 @ finishing (28% body fat—for feedlot steers and heifers) or maturity (replacement heifers).}
660	50	0.45	0.20	17.5	0.72	7.3%	0.22%	0.13%
	60	0.61	0.35	18.4	2.00	10.2%	0.36%	0.19%
	70	0.76	0.48	18.0	3.04	13.0%	0.49%	0.24%
	80	0.90	0.61	17.0	3.78	15.8%	0.61%	0.29%
	90	1.04	0.72	15.7	4.25	18.4%	0.72%	0.34%
720	50	0.45	0.20	18.6	0.72	7.1%	0.21%	0.13%
	60	0.61	0.35	19.7	2.00	9.7%	0.34%	0.18%
	70	0.76	0.48	19.2	3.04	12.2%	0.45%	0.23%
	80	0.90	0.61	18.2	3.78	14.6%	0.56%	0.27%
	90	1.04	0.72	16.8	4.25	17.0%	0.66%	0.32%
780	50	0.45	0.20	19.8	0.72	6.9%	0.20%	0.13%
	60	0.61	0.35	20.9	2.00	9.2%	0.32%	0.17%
	70	0.76	0.48	20.4	3.04	11.4%	0.42%	0.21%
	80	0.90	0.61	19.3	3.78	13.6%	0.52%	0.26%
	90	1.04	0.72	17.8	4.25	15.8%	0.61%	0.30%
840	50	0.45	0.20	20.9	0.72	6.8%	0.20%	0.13%
	60	0.61	0.35	22.1	2.00	8.8%	0.30%	0.16%
	70	0.76	0.48	21.6	3.04	10.8%	0.39%	0.20%
	80	0.90	0.61	20.4	3.78	12.8%	0.48%	0.24%
	90	1.04	0.72	18.8	4.25	14.7%	0.56%	0.28%
900	50	0.45	0.20	22.0	0.72	6.6%	0.19%	0.12%
	60	0.61	0.35	23.3	2.00	8.4%	0.28%	0.16%
	70	0.76	0.48	22.7	3.04	10.2%	0.37%	0.19%
	80	0.90	0.61	21.5	3.78	12.0%	0.44%	0.23%
	90	1.04	0.72	19.8	4.25	13.8%	0.52%	0.26%
960	50	0.45	0.20	23.1	0.72	6.5%	0.19%	0.12%
	60	0.61	0.35	24.4	2.00	8.1%	0.27%	0.15%
	70	0.76	0.48	23.9	3.04	9.7%	0.34%	0.19%
	80	0.90	0.61	22.5	3.78	11.3%	0.41%	0.22%
	90	1.04	0.72	20.8	4.25	13.0%	0.48%	0.25%

APPENDIX TABLE 18   Diet Nutrient Densities for Growing and Finishing Cattle

Body Weight (lb)	TDN % DM	NE$_m$ Mcal/lb	NE$_g$ Mcal/lb	DMI lb/day	ADG lb/day	CP % DM	Ca % DM	P % DM
1,300 @ finishing (28% body fat—for feedlot steers and heifers) or maturity (replacement heifers).								
715	50	0.45	0.20	18.5	0.76	7.3%	0.22%	0.13%
	60	0.61	0.35	19.6	2.11	10.2%	0.36%	0.19%
	70	0.76	0.48	19.1	3.21	13.0%	0.49%	0.24%
	80	0.90	0.61	18.1	3.99	15.7%	0.61%	0.29%
	90	1.04	0.72	16.7	4.48	18.3%	0.72%	0.34%
780	50	0.45	0.20	19.8	0.76	7.1%	0.21%	0.13%
	60	0.61	0.35	20.9	2.11	9.6%	0.34%	0.18%
	70	0.76	0.48	20.4	3.21	12.1%	0.45%	0.23%
	80	0.90	0.61	19.3	3.99	14.5%	0.56%	0.27%
	90	1.04	0.72	17.8	4.48	16.9%	0.66%	0.32%
845	50	0.45	0.20	21.0	0.76	6.9%	0.21%	0.13%
	60	0.61	0.35	22.2	2.11	9.1%	0.32%	0.17%
	70	0.76	0.48	21.7	3.21	11.4%	0.42%	0.22%
	80	0.90	0.61	20.5	3.99	13.6%	0.51%	0.26%
	90	1.04	0.72	18.9	4.48	15.7%	0.60%	0.30%
910	50	0.45	0.20	22.2	0.76	6.7%	0.20%	0.13%
	60	0.61	0.35	23.5	2.11	8.7%	0.30%	0.17%
	70	0.76	0.48	22.9	3.21	10.7%	0.39%	0.20%
	80	0.90	0.61	21.6	3.99	12.7%	0.48%	0.24%
	90	1.04	0.72	20.0	4.48	14.6%	0.56%	0.28%
975	50	0.45	0.20	23.4	0.76	6.6%	0.20%	0.13%
	60	0.61	0.35	24.7	2.11	8.3%	0.28%	0.16%
	70	0.76	0.48	24.1	3.21	10.2%	0.37%	0.19%
	80	0.90	0.61	22.8	3.99	11.9%	0.44%	0.23%
	90	1.04	0.72	21.0	4.48	13.7%	0.52%	0.26%
1,040	50	0.45	0.20	24.5	0.76	6.5%	0.19%	0.13%
	60	0.61	0.35	25.9	2.11	8.0%	0.27%	0.15%
	70	0.76	0.48	25.3	3.21	9.6%	0.34%	0.19%
	80	0.90	0.61	23.9	3.99	11.3%	0.41%	0.22%
	90	1.04	0.72	22.1	4.48	12.9%	0.48%	0.25%

APPENDIX TABLE 19 Diet Nutrient Densities for Growing and Finishing Cattle

1,400 @ finishing (28% body fat—for feedlot steers and heifers) or maturity (replacement heifers).

Body Weight (lb)	TDN % DM	NE$_m$ Mcal/lb	NE$_g$ Mcal/lb	DMI lb/day	ADG lb/day	CP % DM	Ca % DM	P % DM
770	50	0.45	0.20	19.6	0.80	7.3%	0.22%	0.13%
	60	0.61	0.35	20.7	2.22	10.1%	0.36%	0.19%
	70	0.76	0.48	20.2	3.38	12.9%	0.49%	0.24%
	80	0.90	0.61	19.1	4.20	15.6%	0.61%	0.29%
	90	1.04	0.72	17.6	4.72	18.1%	0.72%	0.34%
840	50	0.45	0.20	20.9	0.80	7.1%	0.21%	0.13%
	60	0.61	0.35	22.1	2.22	9.6%	0.34%	0.18%
	70	0.76	0.48	21.6	3.38	12.1%	0.45%	0.23%
	80	0.90	0.61	20.4	4.20	14.5%	0.56%	0.27%
	90	1.04	0.72	18.8	4.72	16.8%	0.65%	0.32%
910	50	0.45	0.20	22.2	0.80	6.9%	.21%	0.13%
	60	0.61	0.35	23.5	2.22	9.1%	0.32%	0.17%
	70	0.76	0.48	22.9	3.38	11.3%	0.42%	0.22%
	80	0.90	0.61	21.6	4.20	13.5%	0.51%	0.26%
	90	1.04	0.72	20.0	4.72	15.6%	0.60%	0.30%
980	50	0.45	0.20	23.5	0.80	6.7%	0.20%	0.13%
	60	0.61	0.35	24.8	2.22	8.7%	0.30%	0.17%
	70	0.76	0.48	24.2	3.38	10.7%	0.39%	0.20%
	80	0.90	0.61	22.9	4.20	12.6%	0.47%	0.24%
	90	1.04	0.72	21.1	4.72	14.5%	0.56%	0.28%
1,050	50	0.45	0.20	24.7	0.80	6.6%	0.20%	0.13%
	60	0.61	0.35	26.1	2.22	8.3%	0.28%	0.16%
	70	0.76	0.48	25.5	3.38	10.1%	0.37%	0.20%
	80	0.90	0.61	24.1	4.20	11.9%	0.44%	0.23%
	90	1.04	0.72	22.2	4.72	13.6%	0.51%	0.26%
1,120	50	0.45	0.20	25.9	0.80	6.5%	0.19%	0.13%
	60	0.61	0.35	27.4	2.22	8.0%	0.27%	0.16%
	70	0.76	0.48	26.8	3.38	9.6%	0.34%	0.19%
	80	0.90	0.61	25.3	4.20	11.2%	0.41%	0.22%
	90	1.04	0.72	23.3	4.72	12.8%	0.48%	0.25%

APPENDIX TABLE 20 Diet Nutrient Density Requirements of Pregnant Replacement Heifers

	Months Since Conception								
	1	2	3	4	5	6	7	8	9
1,000 lb Mature Weight									
TDN, % DM	50.1	50.2	50.4	50.7	51.3	52.3	54.0	56.8	61.3
ME, mcal/lb	0.46	0.46	0.46	0.46	0.47	0.49	0.52	0.56	0.63
NE_m, mcal/lb	0.21	0.21	0.21	0.21	0.22	0.24	0.26	0.30	0.37
DMI, lb	16.7	17.2	17.7	18.2	18.7	19.4	20.0	20.7	21.3
Target ADG	0.73	0.73	0.73	0.73	0.73	0.73	0.73	0.73	0.73
Shrunk Body Wt.	622	644	667	689	711	733	756	778	800
CP % DM	7.18	7.16	7.16	7.21	7.32	7.56	7.99	8.74	10.02
Ca % DM	0.22	0.22	0.22	0.21	0.21	0.20	0.32	0.31	0.31
P % DM	0.17	0.17	0.17	0.17	0.17	0.16	0.23	0.23	0.22
1,100 lb Mature Weight									
TDN, % DM	50.3	50.4	50.5	50.8	51.3	52.3	53.9	56.5	60.6
ME, mcal/lb	0.46	0.46	0.46	0.47	0.48	0.49	0.52	0.56	0.62
NE_m, mcal/lb	0.21	0.21	0.21	0.22	0.22	0.24	0.26	0.30	0.36
DMI, lb	18.0	18.5	19.0	19.5	20.1	20.8	21.5	22.3	22.9
Target ADG	0.80	0.80	0.80	0.80	0.80	0.80	0.80	0.80	0.80
Shrunk Body Wt.	684	709	733	758	782	807	831	856	880
CP % DM	7.20	7.17	7.17	7.21	7.32	7.54	7.93	8.63	9.80
Ca % DM	0.23	0.22	0.22	0.22	0.21	0.21	0.32	0.31	0.30
P % DM	0.18	0.17	0.17	0.17	0.17	0.17	0.23	0.22	0.22
1,200 lb Mature Weight									
TDN, % DM	50.5	50.5	50.7	50.9	51.4	52.3	53.8	56.2	59.9
ME, mcal/lb	0.46	0.46	0.46	0.47	0.48	0.49	0.51	0.55	0.61
NE_m, mcal/lb	0.21	0.21	0.21	0.22	0.23	0.24	0.26	0.30	0.35
DMI, lb	19.3	19.8	20.3	20.9	21.5	22.2	23.0	23.7	24.4
Target ADG	0.88	0.88	0.88	0.88	0.88	0.88	0.88	0.88	0.88
Shrunk Body Wt.	747	773	800	827	853	880	907	933	960
CP % DM	7.21	7.19	7.18	7.22	7.31	7.52	7.89	8.53	9.62
Ca % DM	0.23	0.23	0.22	0.22	0.22	0.21	0.31	0.31	0.30
P % DM	0.18	0.18	0.18	0.17	0.17	0.17	0.23	0.22	0.22
1,300 lb Mature Weight									
TDN,% DM	50.6	50.7	50.8	51.0	51.5	52.4	53.7	56.0	59.5
ME, mcal/lb	0.46	0.46	0.47	0.47	0.48	0.49	0.51	0.55	0.60
NE_m, mcal/lb	0.21	0.21	0.22	0.22	0.23	0.24	0.26	0.29	0.34
DMI, lb	20.5	21.0	21.6	22.2	22.9	23.6	24.4	25.2	25.9
Target ADG	0.95	0.95	0.95	0.95	0.95	0.95	0.95	0.95	0.95
Shrunk Body Wt.	809	838	867	896	924	953	982	1011	1040
CP % DM	7.23	7.20	7.20	7.22	7.31	7.50	7.85	8.45	9.46
Ca % DM	0.24	0.23	0.23	0.22	0.22	0.22	0.31	0.30	0.30
P % DM	0.18	0.18	0.18	0.18	0.18	0.17	0.23	0.22	0.22
1,400 lb Mature Weight									
TDN, % DM	50.7	50.8	50.9	51.2	51.6	52.4	53.7	55.8	59.0
ME, mcal/lb	0.47	0.47	0.47	0.47	0.48	0.49	0.51	0.55	0.60
NE_m, mcal/lb	0.22	0.22	0.22	0.22	0.23	0.24	0.26	0.29	0.34
DMI, lb	21.7	22.3	22.9	23.5	24.2	24.9	25.8	26.6	27.4
Target ADG	1.02	1.02	1.02	1.02	1.02	1.02	1.02	1.02	1.02
Shrunk Body Wt.	871	902	933	964	996	1027	1058	1089	1120
CP % DM	7.25	7.22	7.21	7.23	7.31	7.48	7.81	8.38	9.33
Ca % DM	0.24	0.24	0.23	0.23	0.22	0.22	0.31	0.30	0.30
P % DM	0.18	0.18	0.18	0.18	0.18	0.18	0.23	0.22	0.22

APPENDIX TABLE 21 Diet Nutrient Density Requirements of Beef Cows

	Months since Calving											
	1	2	3	4	5	6	7	8	9	10	11	12
1,000 lb Mature Weight, 10 lb Peak Milk												
TDN, % DM	55.8	56.6	54.3	53.4	52.5	51.8	44.9	45.7	47.0	49.1	52.0	55.7
ME, mcal/lb	0.93	0.95	0.91	0.89	0.88	0.86	0.75	0.76	0.79	0.82	0.87	0.93
NE_m, mcal/lb	0.55	0.56	0.52	0.51	0.49	0.48	0.37	0.38	0.40	0.44	0.49	0.54
DM, lb	21.6	22.1	23.0	22.5	22.1	21.7	21.1	21.0	20.9	20.8	21.0	21.4
Milk, lb/day	8.3	10.0	9.0	7.2	5.4	3.9	0.0	0.0	0.0	0.0	0.0	0.0
CP %DM	8.70	9.10	8.41	7.97	7.51	7.14	5.98	6.16	6.47	6.95	7.66	8.67
Ca % DM	0.24	0.25	0.23	0.22	0.20	0.19	0.15	0.15	0.15	0.24	0.24	0.24
P % DM	0.17	0.17	0.16	0.15	0.14	0.14	0.11	0.11	0.11	0.15	0.15	0.15
1,000 lb Mature Weight, 20 lb Peak Milk												
TDN, % DM	59.6	60.9	58.6	57.0	55.4	54.0	44.9	45.7	47.0	49.1	52.0	55.7
ME, mcal/lb	1.00	1.02	0.98	0.95	0.92	0.90	0.75	0.76	0.79	0.82	0.87	0.93
NE_m, mcal/lb	0.60	0.62	0.59	0.56	0.54	0.52	0.37	0.38	0.40	0.44	0.49	0.54
DM, lb	24.0	25.0	25.4	24.4	23.5	22.7	21.1	21.0	20.9	20.8	21.0	21.4
Milk, lb/day	16.7	20.0	18.0	14.4	10.8	7.8	0.0	0.0	0.0	0.0	0.0	0.0
CP %DM	10.54	11.18	10.38	9.65	8.86	8.17	5.98	6.16	6.47	6.95	7.66	8.67
Ca % DM	0.30	0.32	0.30	0.27	0.24	0.22	0.15	0.15	0.15	0.24	0.24	0.24
P % DM	0.20	0.21	0.19	0.18	0.17	0.15	0.11	0.11	0.11	0.15	0.15	0.15
1,000 lb Mature Weight, 30 lb Peak Milk												
TDN, % DM	62.8	64.5	62.1	60.1	57.9	55.9	44.9	45.7	47.0	49.1	52.0	55.7
ME, mcal/lb	1.05	1.08	1.04	1.00	0.97	0.93	0.75	0.76	0.79	0.82	0.87	0.93
NE_m, mcal/lb	0.65	0.68	0.64	0.61	0.58	0.55	0.37	0.38	0.40	0.44	0.49	0.54
DM, lb	26.4	27.8	27.8	26.4	24.9	23.7	21.1	21.0	20.9	20.8	21.0	21.4
Milk, lb/day	25.0	30.0	27.0	21.6	16.2	11.7	0.0	0.0	0.0	0.0	0.0	0.0
CP %DM	12.06	12.86	12.00	11.07	10.04	9.09	5.98	6.16	6.47	6.95	7.66	8.67
Ca % DM	0.35	0.38	0.35	0.32	0.28	0.25	0.15	0.15	0.15	0.24	0.24	0.24
P % DM	0.22	0.24	0.22	0.21	0.19	0.17	0.11	0.11	0.11	0.15	0.15	0.15

APPENDIX TABLE 22   Diet Nutrient Density Requirements of Beef Cows

	1	2	3	4	5	Months since Calving 6	7	8	9	10	11	12
1,200 lb Mature Weight, 10 lb Peak Milk												
TDN, % DM	55.3	56.0	53.7	52.9	52.1	51.5	44.9	45.8	47.1	49.3	52.3	56.2
ME, mcal/lb	0.92	0.94	0.90	0.88	0.87	0.86	0.75	0.76	0.79	0.82	0.87	0.94
NE$_m$, mcal/lb	0.54	0.55	0.51	0.50	0.49	0.48	0.37	0.38	0.41	0.44	0.49	0.55
DM, lb	24.4	24.9	26.0	25.6	25.1	24.8	24.2	24.1	24.0	23.9	24.1	24.6
Milk, lb/day	8.3	10.0	9.0	7.2	5.4	3.9	0.0	0.0	0.0	0.0	0.0	0.0
CP % DM	8.43	8.79	8.13	7.73	7.33	7.00	5.99	6.18	6.50	7.00	7.73	8.78
Ca % DM	0.24	0.25	0.23	0.21	0.20	0.19	0.15	0.15	0.15	0.26	0.25	0.25
P % DM	0.17	0.17	0.16	0.15	0.14	0.14	0.12	0.12	0.12	0.16	0.16	0.16
1,200 lb Mature Weight, 20 lb Peak Milk												
TDN, % DM	58.7	59.9	57.6	56.2	54.7	53.4	44.9	45.8	47.1	49.3	52.3	56.2
ME, mcal/lb	0.98	1.00	0.96	0.94	0.91	0.89	0.75	0.76	0.79	0.82	0.87	0.94
NE$_m$, mcal/lb	0.59	0.61	0.57	0.55	0.53	0.51	0.37	0.38	0.41	0.44	0.49	0.55
DM, lb	26.8	27.8	28.4	27.4	26.5	25.7	24.2	24.1	24.0	23.9	24.1	24.6
Milk, lb/day	16.7	20.0	18.0	14.4	10.8	7.8	0.0	0.0	0.0	0.0	0.0	0.0
CP % DM	10.10	10.69	9.92	9.25	8.54	7.92	5.99	6.18	6.50	7.00	7.73	8.78
Ca % DM	0.29	0.31	0.29	0.26	0.24	0.22	0.15	0.15	0.15	0.26	0.25	0.25
P % DM	0.19	0.21	0.19	0.18	0.17	0.15	0.12	0.12	0.12	0.16	0.16	0.16
1,200 lb Mature Weight, 30 lb Peak Milk												
TDN, % DM	61.6	63.2	60.8	59.0	57.0	55.2	44.9	45.8	47.1	49.3	52.3	56.2
ME, mcal/lb	1.03	1.06	1.02	0.99	0.95	0.92	0.75	0.76	0.79	0.82	0.87	0.94
NE$_m$, mcal/lb	0.64	0.66	0.62	0.59	0.56	0.54	0.37	0.38	0.41	0.44	0.49	0.55
DM, lb	29.2	30.6	30.8	29.4	27.9	26.7	24.2	24.1	24.0	23.9	24.1	24.6
Milk, lb/day	25.0	30.0	27.0	21.6	16.2	11.7	0.0	0.0	0.0	0.0	0.0	0.0
CP % DM	11.51	12.25	11.41	10.55	9.61	8.75	5.99	6.18	6.50	7.00	7.73	8.78
Ca % DM	0.34	0.36	0.34	0.31	0.27	0.25	0.15	0.15	0.15	0.26	0.25	0.25
P % DM	0.22	0.23	0.22	0.20	0.18	0.17	0.12	0.12	0.12	0.16	0.16	0.16

APPENDIX TABLE 23   Diet Nutrient Density Requirements of Beef Cows

						Months since Calving						
	1	2	3	4	5	6	7	8	9	10	11	12
1,400 lb Mature Weight, 10 lb Peak Milk												
TDN, % DM	54.9	55.5	53.3	52.5	51.8	51.2	45.0	45.8	47.3	49.5	52.6	56.6
ME, mcal/lb	0.92	0.93	0.89	0.88	0.86	0.86	0.75	0.77	0.79	0.83	0.88	0.95
NE$_m$, mcal/lb	0.53	0.54	0.51	0.49	0.48	0.47	0.37	0.39	0.41	0.44	0.49	0.56
DM, lb	27.1	27.6	28.9	28.5	28.0	27.7	27.2	27.0	26.9	26.8	27.0	27.6
Milk, lb/day	8.3	10.0	9.0	7.2	5.4	3.9	0.0	0.0	0.0	0.0	0.0	0.0
CP %DM	8.23	8.56	7.91	7.55	7.19	6.90	6.00	6.20	6.53	7.04	7.80	8.88
Ca % DM	0.23	0.25	0.23	0.21	0.20	0.19	0.16	0.16	0.16	0.27	0.26	0.26
P % DM	0.17	0.17	0.16	0.15	0.15	0.14	0.12	0.12	0.12	0.17	0.17	0.16
1,400 lb Mature Weight, 20 lb Peak Milk												
TDN, % DM	58.0	59.1	56.8	55.5	54.1	53.0	45.0	45.8	47.3	49.5	52.6	56.6
ME, mcal/lb	0.97	0.99	0.95	0.93	0.90	0.89	0.75	0.77	0.79	0.83	0.88	0.95
NE$_m$, mcal/lb	0.58	0.60	0.56	0.54	0.52	0.50	0.37	0.39	0.41	0.44	0.49	0.56
DM, lb	29.5	30.5	31.3	30.3	29.4	28.6	27.2	27.0	26.9	26.8	27.0	27.6
Milk, lb/day	16.7	20.0	18.0	14.4	10.8	7.8	0.0	0.0	0.0	0.0	0.0	0.0
CP %DM	9.76	10.31	9.56	8.94	8.29	7.73	6.00	6.20	6.53	7.04	7.80	8.88
Ca % DM	0.28	0.30	0.28	0.26	0.24	0.22	0.16	0.16	0.16	0.27	0.26	0.26
P % DM	0.19	0.20	0.19	0.18	0.17	0.16	0.12	0.12	0.12	0.17	0.17	0.16
1,400 lb Mature Weight, 30 lb Peak Milk												
TDN, % DM	60.7	62.2	59.8	58.1	56.2	54.7	45.0	45.8	47.3	49.5	52.6	56.6
ME, mcal/lb	1.01	1.04	1.00	0.97	0.94	0.91	0.75	0.77	0.79	0.83	0.88	0.95
NE$_m$, mcal/lb	0.62	0.64	0.61	0.58	0.55	0.53	0.37	0.39	0.41	0.44	0.49	0.56
DM, lb	31.9	33.3	33.7	32.3	30.8	29.6	27.2	27.0	26.9	26.8	27.0	27.6
Milk, lb/day	25.0	30.0	27.0	21.6	16.2	11.7	0.0	0.0	0.0	0.0	0.0	0.0
CP %DM	11.07	11.77	10.95	10.15	9.27	8.49	6.00	6.20	6.53	7.04	7.80	8.88
Ca % DM	0.33	0.35	0.32	0.30	0.27	0.24	0.16	0.16	0.16	0.27	0.26	0.26
P % DM	0.22	0.23	0.21	0.20	0.18	0.17	0.12	0.12	0.12	0.17	0.17	0.16

Glossary

a_1 thermal neutral maintenance requirement (Mcal/day per $BW^{0.75}$)

a_2 adjustment for previous temperature

AA proportion of empty body ash

AAA_{si} total amount of the ith absorbed amino acid supplied by dietary and bacterial sources, g/day

$AABCW_i$ i^{th} amino acid content of rumen bacteria cell wall protein, g/100 g (Table 10-7)

$AABNCW_i$ i^{th} amino acid content of rumen bacteria noncell wall protein, g/100 g (Table 10-7)

$AAINSP_{ij}$ i^{th} amino acid content of the insoluble protein for the jth feedstuff, g/100 g

$AALACT_i$ i^{th} amino acid content of milk true protein, g/100 g (Table 10-5)

AAN proportion of ash at the next body condition score

$AATISS_i$ amino acid composition of body tissue, g/100 g (Table 10-5)

ACADG after calving target ADG, kg/day

ADF acid detergent fiber

ADG average daily gain

ADG_{preg} average daily gain due to pregnancy

ADIP acid detergent insoluble protein

$ADIP_j$ (%CP) percentage of the crude protein of the jth feedstuff that is acid detergent insoluble protein

ADTV feed additive adjustment factor for DMI (Table 10-4)

AF proportion of empty body fat

AF1 proportion of empty body fat @ CS = 1

age age of cow, years

AP proportion of empty body protein

AP1 proportion of empty body protein @ CS = 1

APADG postpubertal target ADG, kg/day

ASH_j (%DM) percentage of ash of the jth feedstuff

AW proportion of empty body water

$BACT_j$ yield of bacteria from the jth feedstuff, g/day

$BACTN_j$ bacterial nitrogen, g/day

BCP bacterial (microbial) crude protein

BE breed effect on NE_m requirement (Table 10-1)

BFAF body fat adjustment factor (Table 10-4)

BI breed adjustment factor for DMI (Table 10-4)

BPADG prepubertal target ADG, kg/day

BTP bacterial true protein

BW body weight

$BW^{0.75}$ metabolic body weight

CA_j (%DM) percentage of dry matter of the jth feedstuff that is sugar

$CB1_j$ (%DM) percentage of dry matter of the jth feedstuff that is starch

$CB2_j$ percentage of dry matter of the jth feedstuff that is available fiber

CBW calf birth weight, kg

CC_j percentage of dry matter in the jth feedstuff that is unavailable fiber

CHO_j (%DM) percentage of carbohydrate of the jth feedstuff

CI calving interval, days

COMP effect of previous plane of nutrition on NE_m requirement

CP_j (%DM) percentage of crude protein of the jth feedstuff

CS body condition score

CW conceptus weight

D duration of lactation, weeks

DE digestible energy (gross energy of the food minus the energy lost in the feces)

$DIGBAA_i$ amount of the ith absorbed bacterial amino acid

$DIGBC_j$ digested bacterial carbohydrate produced from the jth feedstuff, g/day

$DIGBF_j$ digestible bacterial fat from the jth feedstuff, g/day

$DIGBNA_j$ digestible bacterial nucleic acids produced from the jth feedstuff, g/day

DIGBTP$_j$ digestible bacterial true protein produced from the jth feedstuff, g/day

DIGC$_j$ digestible carbohydrate from the jth feedstuff, g/day

DIGFAA$_i$ amount of the ith absorbed amino acid from dietary escaping rumen degradation, g/day

DIGFC$_j$ intestinally digested feed carbohydrate from the jth feedstuff, g/day

DIGFF$_j$ digestible feed fat from the jth feedstuff, g/day

DIGF$_j$ digestible fat from the jth feedstuff, g/day

DIGFP$_j$ digestible feed protein from the jth feedstuff, g/day

DIGPB1$_j$ digestible B1 protein from the jth feedstuff, g/day

DIGPB2$_j$ digestible B2 protein from the jth feedstuff, g/day

DIGPB3$_j$ digestible B3 protein from the jth feedstuff, g/day

DIGP$_j$ digestible protein from the jth feedstuff, g/day

DIP   degraded intake protein

DM   dry matter

DMI   dry matter intake

DOP   days on pasture

E   energy content of milk, Mcal (NE$_m$)/kg

e   base of natural logarithms

EAAG$_i$ efficiency of use of the ith amino acid for growth, g/g (Table 10-6)

EAAL$_i$ efficiency of use of the ith amino acid for milk protein formation, g/g (Table 10-6)

EAAP$_i$ efficiency of use of the ith amino acid for gestation, g/g (Table 10-6)

EAT   effective ambient temperature (°C)

EBG   empty body gain, kg

EBW$^{0.75}$ metabolic body weight based on empty body weight, kg

EBWN   EBW at the next body condition score

EI   external insulation value, °C/Mcal/m²/day

EN   nitrogen in excess of rumen bacterial nitrogen and tissue needs, g/day

EQEBW   equivalent empty body weight, kg

EQSBW   equivalent shrunk body weight, kg

ER   energy reserves, Mcal

FA   daily forage allowance, kg/day

FAT$_j$ fat composition of the jth feedstuff, g/day

FAT$_j$ (%DM)   percentage of fat of the jth feedstuff

FCBACT$_j$ yield of fiber carbohydrate bacteria from the jth feedstuff, g/day

FDM$_j$ amount of indigestible DM in feces from the jth feedstuff, g/day

FE   fecal energy

FEASH$_j$ amount of ash in feces from the jth feedstuff, g/day

FEBACT$_j$ amount of bacteria in feces from the jth feedstuff, g/day

FEBASH$_j$ amount of bacterial ash in feces from the jth feedstuff, g/day

FEBC$_j$ amount of bacterial carbohydrate in feces from the jth feedstuff, g/day

FEBCP$_j$ amount of fecal bacterial protein from the jth feedstuff, g/day

FEBCW$_j$ amount of fecal bacterial cell wall protein from the jth feedstuff, g/day

FEBF$_j$ amount of bacterial fat in feces from the jth feedstuff, g/day

FECB1$_j$ amount of feed starch in feces from the jth feedstuff, g/day

FECB2$_j$ amount of feed available fiber in feces from the jth feedstuff, g/day

FECC$_j$ amount of feed unavailable fiber in feces from the jth feedstuff, g/day

FECHO$_j$ amount of carbohydrate in feces from the jth feedstuff, g/day

FEENGA$_j$ amount of endogenous ash in feces from the jth feedstuff, g/day

FEENGF$_j$ amount of endogenous fat in feces from the jth feedstuff, g/day

FEENGP$_j$ amount of endogenous protein in feces from the jth feedstuff, g/day

FEFA$_j$ amount of undigested feed ash in feces from the jth feedstuff, g/day

FEFAT$_j$ amount of fat in feces from the jth feedstuff, g/day

FEFC$_j$ amount of feed carbohydrate in feces from the jth feedstuff, g/day

FEFP$_j$ amount of feed protein in feces from the jth feedstuff, g/day

FEPB3$_j$ amount of feed B3 protein fraction in feces from the jth feedstuff, g/day

FEPC$_j$ amount of feed C protein fraction in feces from the jth feedstuff, g/day

FEPROT$_j$ amount of fecal protein from the jth feedstuff, g/day

FFMtotal   feed for maintenance (adjusted for stress), kg DM/day

FHP   fasting heat production

FM   mobilizable fat, kg

FORAGE   forage concentration of the diet, %

FSBW   final shrunk body weight at maturity for breeding heifers or at the body fat end point selected for feedlot steers and heifers

GF   green forage availability (ton/ha)

GRAZE   forage availability factor if grazing, %

GU   grazing unit size, hectare

HAIR   effective hair depth, cm

HE   heat production, Mcal/day

H$_e$ minimal total evaporative heat loss

HIDE   hide adjustment factor for external insulation (1 = thin, 2 = average, 3 = thick)

H$_j$E heat of activity associated with obtaining food

IE intake energy

IFN international feed number

Ij intake of the jth feedstuff, g/day

I$_m$ intake for maintenance (no stress), kg DM/day

IMP$_j$ percentage improvement in bacterial yield, due to the ratio of peptides-to-peptides plus nonfiber CHO in jth feedstuff

I$_m$total intake for maintenance with stress, kg/DM day

IN total insulation (°C/Mcal/m²/day)

IPM initial pasture mass (kg DM/ha)

IVOMD dietary in vitro organic matter disappearance

Kd degradation rate of feedstuff component

Kd′ pH adjusted feed specific degradation rate of available fiber fraction (decimal form)

Kd$_{1j}$ rumen rate of digestion of the rapidly degraded protein fraction of the jth feedstuff, h^{-1}

Kd$_{2j}$ rumen rate of digestion of the intermediately degraded protein fraction of the jth feedstuff, h^{-1}

Kd$_{3j}$ rumen rate of digestion of the slowly degraded protein fraction of the jth feedstuff, h^{-1}

Kd$_{4j}$ rumen rate of sugar digestion of the jth feedstuff, h^{-1}

Kd$_{5j}$ rumen rate of starch digestion of the jth feedstuff, h^{-1}

Kd$_{6j}$ rumen rate of available fiber digestion of the jth feedstuff, h^{-1}

k$_m$ efficiency of utilization of ME for maintenance

KM$_1$ maintenance rate of the fiber carbohydrate bacteria, 0.05 g FC/g bacteria/h

KM$_2$ maintenance rate of the nonfiber carbohydrate bacteria, 0.15 g NFC/g bacteria/h

Kp$_j$ rate of passage from the rumen of the jth feedstuff, h^{-1}

L lactation effect on NE$_m$ requirement (1 if dry or 1.2 if lactating)

LCT animal's lower critical temperature, °C

LIGNIN$_j$ (%NDF) percentage of lignin of the jth feedstuff's NDF

LPAA$_i$ metabolizable requirement for lactation for the ith absorbed amino acid, g/day

MCP microbial crude protein, g/day

ME metabolizable energy

ME$_{aj}$ metabolizable energy available from the jth feedstuff, Mcal/day

MEC metabolizable energy concentration of the diet, Mcal/kg

MEC$_j$ metabolizable energy concentration of the jth feedstuff, Mcal, kg

MEcs metabolizable energy required due to cold stress, Mcal/day

MEI metabolizable energy intake, Mcal/day

ME$_m$ ME required for maintenance

MF milk fat composition, %

MM milk production, kg/day

MP$_{req}$ metabolizable protein requirement, g/day

MP metabolizable protein

MP$_a$ metabolizable protein available in the diet, g/day

MPAA$_i$ metabolizable requirement for gestation for the ith absorbed amino acid, g/day (Table 10-6)

MP$_g$ metabolizable protein requirement, g/day

MP$_{maint}$ metabolizable protein requirement for maintenance, g/day

MUD1 mud adjustment factor for DMI (Table 10-4)

MUD2 mud adjustment factor for external insulation

MW mature weight, kg

N number of animals

NDF neutral detergent fiber

NDF$_j$ (%DM) percentage of the jth feedstuff that is neutral detergent fiber

NDIP$_j$ (%DM) percentage of neutral detergent insoluble protein of the jth feedstuff

NE net energy

NEFG net energy required for gain

NE$_g$ net energy required for gain

NE$_{ga}$ net energy content of diet for gain, Mcal/kg

NEg$_{aj}$ net energy for gain content of the jth feedstuff, Mcal/kg

NE$_m$ net energy required for maintenance adjusted for acclimatization

NE$_{ma}$ net energy value of diet for maintenance, Mcal/kg

NE$_{mact}$ activity effect on NE$_m$ requirement

NE$_{maj}$ net energy for maintenance content of the jth feedstuff, Mcal/kg

NE$_{mcs}$ net energy required due to cold stress, Mcal/day

NE$_{mhs}$ net energy required due to heat stress, Mcal/day

NE$_m$total net energy for maintenance required adjusted for breed, lactation, sex, grazing, acclimatization, and stress effects

NE$_{preg}$ net energy retained as gravid uterus

NFCBACT$_j$ yield of nonstructural carbohydrate bacteria from the jth feedstuff, g/day

NFCBACTN$_j$ nonfiber carbohydrate bacterial nitrogen, g/day

NP net protein

NP$_g$ net protein requirement for growth, g/day

NPN nonprotein nitrogen

NPN$_j$ (%CP) percentage of crude protein of the jth feedstuff that is nonprotein nitrogen times 6.25

OMD organic matter digestibility

OMI organic matter intake

PA$_j$ (%DM) percentage of dry matter in the jth feedstuff that is nonprotein nitrogen

pAVAIL pasture mass available for grazing, T/ha

PB protein content of empty body gain, g/100 g

PB1$_j$ (%DM) percentage of dry matter in the jth feedstuff that is rapidly degraded protein

PB2$_j$ (%DM) percentage of dry matter in the jth feedstuff that is intermediately degraded protein

PB3$_j$ (%DM) percentage of dry matter in the jth feedstuff that is slowly degraded protein

PC$_j$ (%DM)   percentage of dry matter in the jth feedstuff that is bound protein

PEPUP$_j$ bacterial peptide from the jth feedstuff, g/day

PEPUPN$_j$ bacterial peptide nitrogen from the jth feedstuff, g/day

pH   ruminal pH

pI   pasture dry matter intake, kg/day

PKYD   peak milk yield, kg/day (Table 10-1)

PM   mobilizable protein, kg

RATIO$_j$ ratio of peptides-to-peptides plus NFC in the jth feedstuff

RD   proportion of component of a feedstuff degraded in the rumen

RDCA$_j$ amount of ruminally degraded sugar from the jth feedstuff, g/day

RDCB1$_j$ amount of ruminally degraded starch from the jth feedstuff, g/day

RDCB2$_j$ amount of ruminally degraded available fiber from the jth feedstuff, g/day

RDPA$_j$ amount of ruminally degraded NPN in the jth feedstuff, g/day

RDPB1$_j$ amount of ruminally degraded B1 true protein in the jth feedstuff, g/day

RDPB2$_j$ amount of ruminally degraded B2 true protein in the jth feedstuff, g/day

RDPB3$_j$ amount of ruminally degraded B3 true protein in the jth feedstuff, g/day

RDPEP$_j$ amount of ruminally degraded peptides from the jth feedstuff, g/day

RE   retained energy, Mcal/day

REAA$_i$ total amount of the ith amino acid appearing at the duodenum, g/day

REBAA$_i$ amount of the ith bacterial amino acid appearing at the duodenum, g/day

REBASH$_j$ amount of bacterial ash passed to the intestines by the jth feedstuff, g/day

REBCHO$_j$ amount of bacterial carbohydrate passed to the intestines by the jth feedstuff, g/day

REBCW$_j$ amount of bacterial cell wall protein passed to the intestines by the jth feedstuff, g/day

REBFAT$_j$ amount of bacterial fat passed to the intestines by the jth feedstuff, g/day

REBNA$_j$ amount of bacterial nucleic acids passed to the intestines by the jth feedstuff, g/day

REBTP$_j$ amount of bacterial true protein passed to the intestines by the jth feedstuff, g/day

RECA$_j$ amount of ruminally escaped sugar from the jth feedstuff, g/day

RECB1$_j$ amount of ruminally escaped starch from the jth feedstuff, g/day

RECB2$_j$ amount of ruminally escaped available fiber from the jth feedstuff, g/day

RECC$_j$ amount of ruminally escaped unavailable fiber from the jth feedstuff, g/day

REFAA$_i$ amount of the ith dietary amino acid appearing at the duodenum, g/day

REFAT$_j$ amount of ruminally escaped fat from the jth feedstuff, g/day

relY   relative yield adjustment

REPB1$_j$ amount of ruminally escaped B1 true protein in the jth feedstuff, g/day

REPB2$_j$ amount of ruminally escaped B2 true protein in the jth feedstuff, g/day

REPB3$_j$ amount of ruminally escaped B3 true protein in the jth feedstuff, g/day

REPC$_j$ amount of ruminally escaped bound C protein from the jth feedstuff, g/day

RESC   proportion of component of feedstuff escaping ruminal degradation

RPAA$_i$ metabolizable requirement for growth for the ith absorbed amino acid, g/day

RPN   net protein required for growth, g/day

SA   surface area, m^2

SBW   shrunk body weight, kg

SBW$^{0.75}$ metabolic body weight based on shrunk body weight, kg

SD   standard deviation

SEX   maintenance adjustment for bulls

SNF   milk solids not fat composition, %

SOLP$_j$ (%CP)   percentage of the crude protein of the jth feedstuff that is soluble protein

soluble nitrogen   NPN plus soluble true protein

SRW   standard reference weight for the expected final body fat

STARCH$_j$ (%NFC)   percentage of starch in the non-structural carbohydrate of the jth feedstuff

stdig   postruminal starch digestibility, g/g

SWG   shrunk weight gain, kg

t   day of pregnancy

TA   total kg ash at the current body condition score

T$_{age}$ heifer age, days

Tc   current temperature, °C

TCA   target calving age in days

TCW1   target first calf calving weight, kg

TCW2   target second calf calving weight, kg

TCW3   target third calf calving weight, kg

TCW4   target fourth calf calving weight, kg

TCWx   current target calving weight, kg

TCWxx   next target calving weight, kg

TDN   total digestible nutrient content of the diet, % or g/day

TDNAPP$_j$ apparent TDN from the jth feedstuff, g/day

TEMP1   temperature adjustment factor for DMI (Table 10-4)

TERRAIN   terrian factor (1 = level land, 2 = hilly land)

TF   total fat, kg

TF1   total body fat @ CS = 1, kg

TI   tissue (internal) insulation value, °C/Mcal/m^2/day

TotalProt   total protein yield for lactation, kg

TotalE   total energy yield for lactation, Mcal

TotalFat   total fat yield for lactation, kg

TotalY   total milk yield for lactation, kg

TP   total protein, kg

$\mathbf{T_p}$ previous ambient temperature, °C

TP1   total body protein @ CS = 1, kg

TPA   target puberty age, days

TPW   target puberty weight, kg

U   urea nitrogen recycled (percent of nitrogen intake)

UCT   upper critical temperature

UE   urinary energy

UIP   undegraded intake protein

VFA   volatile fatty acids

W   current week of lactation

WIND   wind speed, kph

X   diet crude protein, as a percentage of diet dry matter

Y   original yield for each feed

Y′   new yield for each feed

$\mathbf{Y_{1j}}$ yield efficiency of FC bacteria from the available fiber fraction of the jth feedstuff, g FC bacteria/g FC digested

$\mathbf{Y_{2j}}$ yield efficiency of NFC bacteria from the sugar fraction of the jth feedstuff, g NFC bacteria/g NFC digested

$\mathbf{Y_{3j}}$ yield efficiency of NFC bacteria from the starch fraction of the jth feedstuff, g NFC bacteria/g NFC digested

Ye   relationship of energy content of the gravid uterus

YEn   daily energy secreted in milk at current stage of lactation, Mcal (NE$_m$)/day

YFatn   daily milk fat yield at current stage of lactation, kg/day

$\mathbf{YG_1}$ theoretical maximum yield of the fiber carbohydrate bacteria, 0.4 g bacteria/g FC

$\mathbf{YG_2}$ theoretical maximum yield of the nonfiber carbohydrate bacteria, 0.4 g bacteria/g NFC

Yn   daily milk yield at current week of lactation, kg/day

YPN   net protein required for gestation, g/day

YProtn   daily milk protein yield at current stage of lactation, kg/day

Z   age in years

About the Authors

Jock Buchanan-Smith (*Chair*) is professor of animal science at the University of Guelph. He received his Ph.D. degree in animal science from Oklahoma State University. Research interests include utilization of silages by ruminants and digesta passage.

Larry L. Berger is a professor of ruminant nutrition at the University of Illinois. He received his Ph.D. degree in ruminant nutrition from the University of Nebraska. Research interests include low quality roughages, feedlot-feed additives, and bypass proteins.

Calvin Ferrell is research leader of the Nutrition Unit at the Roman L. Hruska U.S. Meat Animal Research Center at Clay Center, Nebraska. He received his Ph.D. degree in nutrition from the University of California at Davis. Research interests include energy and protein requirements of ruminant animals.

Danny G. Fox, since 1977, has been professor of animal science at Cornell University. He received his Ph.D. degree in ruminant nutrition from Ohio State University. Research interests include protein and energy requirements of feeding systems for various cattle types and development of computer models to predict nutrient requirements and performance under widely varying conditions.

Michael Galyean currently is professor of animal nutrition at Clayton Research Center at New Mexico State University, where he has taught since 1977. He received his Ph.D. degree in animal nutrition from Oklahoma State University. Research interests include manipulation of rumen fermentation, metabolic profile of stressed ruminants, nonprotein nitrogen utilization, grain processing for livestock, and rate of passage of nutrients through the gastrointestinal tract of ruminants.

David P. Hutcheson owns and operates Animal Agricultural Consulting, Inc., in Amarillo, Texas. He received his Ph.D. degree in animal husbandry from the University of Missouri at Columbia. Research interests include development of nutrient recommendations for transportation stress, respiratory diseased cattle, and the interaction of nutrition with immunology.

Terry J. Klopfenstein is professor of beef cattle nutrition research at University of Nebraska. He received his Ph.D. degree in nutrition from Ohio State University. Research interests include enhancing nutritional value of low quality forage, crop residues for beef cattle, bypass protein sources for growing beef cattle, and alternative feedstuffs.

Jerry Spears is professor of animal science at North Carolina State University, where he has concurrently served as nutrition coordinator since 1989. He received his Ph.D. degree in animal nutrition from the University of Illinois. Research interests include mineral bioavailability in ruminants, interactions between minerals and ionophores, and anion-cation balance in beef cattle.

Index

Other Titles
in the Series